Lecture Notes in Computer Science 2265
Edited by G. Goos, J. Hartmanis, and J. van Leeuwen

T0145083

Springer
Berlin
Heidelberg
New York
Barcelona
Hong Kong
London
Milan
Paris
Tokyo

Petra Mutzel Michael Jünger
Sebastian Leipert (Eds.)

Graph Drawing

9th International Symposium, GD 2001
Vienna, Austria, September 23-26, 2001
Revised Papers

Series Editors

Gerhard Goos, Karlsruhe University, Germany
Juris Hartmanis, Cornell University, NY, USA
Jan van Leeuwen, Utrecht University, The Netherlands

Volume Editors

Petra Mutzel
Technische Universität Wien
Abteilung für Algorithmen und Datenstrukturen
Institut für Computergraphik und Algorithmen
Favoritenstr. 9-11 E186, 1040 Wien Austria
E-mail: mutzel@ads.tuwien.ac.at

Michael Jünger
Universität zu Köln
Institut für Informatik
Pohligstr. 1, 50969 Köln, Germany
E-mail: mjuenger@informatik.uni-koeln.de

Sebastian Leipert
caesar - center of advanced european studies and research
Friedensplatz 16, 53111 Bonn, Germany
E-mail: leipert@caesar.de

Cataloging-in-Publication Data applied for

Die Deutsche Bibliothek - CIP-Einheitsaufnahme

Graph drawing : 9th international symposium ; revised papers / GD 2001, Vienna, Austria, September 23 - 26, 2001. Petra Mutzel ... (ed.). - Berlin ; Heidelberg ; New York ; Barcelona ; Hong Kong ; London ; Milan ; Paris ; Tokyo : Springer, 2002
(Lecture notes in computer science ; Vol. 2265)
ISBN 3-540-43309-0

CR Subject Classification (1998): G.2, I.3, F.2

ISSN 0302-9743
ISBN 3-540-43309-0 Springer-Verlag Berlin Heidelberg New York

This work is subject to copyright. All rights are reserved, whether the whole or part of the material is concerned, specifically the rights of translation, reprinting, re-use of illustrations, recitation, broadcasting, reproduction on microfilms or in any other way, and storage in data banks. Duplication of this publication or parts thereof is permitted only under the provisions of the German Copyright Law of September 9, 1965, in its current version, and permission for use must always be obtained from Springer-Verlag. Violations are liable for prosecution under the German Copyright Law.

Springer-Verlag Berlin Heidelberg New York
a member of BertelsmannSpringer Science+Business Media GmbH

http://www.springer.de

© Springer-Verlag Berlin Heidelberg 2002
Printed in Germany

Typesetting: Camera-ready by author, data conversion by PTP-Berlin, Stefan Sossna
Printed on acid-free paper SPIN 10846204 06/3142 5 4 3 2 1 0

Preface

With 133 registered participants from 27 countries, including 19 participants from industry, the International Symposium on Graph Drawing 2001 (GD 2001) that took place in Vienna, September 23–26, 2001, clearly demonstrated that the graph drawing community is still growing. The 31 contributed talks that had been selected out of 66 paper submissions by the program committee reflect the many facets of graph drawing and the high activity in our scientific discipline. In addition, we had the pleasure of enjoying invited presentations by Alexander Schrijver and Eduard Gröller that mark extreme points of the wide spectrum of graph drawing, the mathematical foundations and the computer graphics, respectively. We have compiled the written versions of these contributions in the same order as they were presented during the conference.

We have added a correct version of Joan P. Hutchinson's contribution to GD 2000 that had been misprinted in the GD 2000 proceedings.

GD 2001 hosted a software exhibition that gave participants and guests the opportunity for hands-on experience with state-of-the-art graph drawing tools. Out of the 26 submitted software tools, 24 were presented at the conference and received considerable attention by the participants. Each of them is represented here by a two-page-summary.

In a special session on graph exchange formats, organized by Giuseppe Liotta, the GXL and the GraphML projects were presented by Andreas Winter and Ulrik Brandes, respectively, and then a lively discussion followed. The written versions of the two reports are also included here.

The final contribution in this volume is a report on a traditional component of all graph drawing conferences that is a serious and a fun event at the same time: the Graph Drawing Contest 2001 organized by Franz Brandenburg.

We would like to thank all contributors for the pleasant cooperation.

December 2001

Petra Mutzel
Michael Jünger
Sebastian Leipert

Organization

Program Committee

Petra Mutzel	Vienna University of Technology - chair
Michael Jünger	University of Cologne - co-chair
Franz Aurenhammer	Graz University of Technology
Therese Biedl	University of Waterloo
Giuseppe Di Battista	University of Rome III
Franz Brandenburg	University of Passau
Yefim Dinitz	Ben Gurion University
Peter Eades	University of Sydney
Herbert Fleischner	Austrian Academy of Sciences
Hubert de Fraysseix	CNRS Paris
Mike Goodrich	Johns Hopkins University
Jan Kratochvíl	Charles University Prague
Giuseppe Liotta	University of Perugia
Brendan Madden	Tom Sawyer Software
Shin-ichi Nakano	Gunma University

Organization Committee

Petra Mutzel	Vienna University of Technology - chair
Michael Jünger	University of Cologne - co-chair
Leonid Dimitrov	Austrian Academy of Sciences
Barbara Hufnagel	Vienna University of Technology
Gunnar Klau	Vienna University of Technology
Sebastian Leipert	Research Center caesar, Bonn
René Weiskircher	Vienna University of Technology
Emanuel Wenger	Austrian Academy of Sciences

Software Exhibition Organizers

Martin Gruber	Vienna University of Technology
Thomas Lange	University of Cologne

Steering Committee

Franz J. Brandenburg	University of Passau
Giuseppe Di Battista	University of Rome
Peter Eades	University of Sydney
Michael Jünger	University of Cologne
Joe Marks	Mitsubishi Electrical Research Labs
Petra Mutzel	Technical University of Vienna
Takao Nishizeki	Tohoku University
Pierre Rosenstiehl	Ecole des Hautes Etudes en Sciences Sociales
Roberto Tamassia	Brown University
Ioannis G. Tollis	University of Texas, Dallas

External Referees

Oswin Aichholzer
Robert Babilon
Christian Bachmeier
Broňa Brejová
Christoph Buchheim
Erik Demaine
Emilio Di Giacomo
Walter Didimo
Michael Forster
Carsten Friedrich
Yashar Ganjali
Carsten Gutwenger
Patrick Healy
Seokhee Hong
Maolin Huang
John Johansen
Matya Katz
Gunnar Klau
Karsten Klein
Hannes Krasser
Sebastian Leipert
Xuemin Lin
Anna Lubiw
Alessandro Marcandalli
Hugo di Nascimento
Nikola Nikolov
Maurizio Patrignani
Merijam Percan
Andreas Pick
Maurizio Pizzonia
Aaron Quigley
Marcus Raitner
Franz Rendl
Joe Sawada
Falk Schreiber
Tomáš Vinař
Imrich Vrťo
Richard Webber
René Weiskircher
David Wood

Student Assistants

Christoph Dorn
Georg Kraml
Anna Potocka
Katarzyna Potocka
Barbara Reitgruber
Patrick Seidelmann
Barbara Schuhmacher

Sponsoring Institutions

We gratefully acknowledge the contributions of the following sponsors of the Graph Drawing Conference 2001:

bm:bwk

COMSORT

SIEMENS

Table of Contents

Hierarchical Drawing

Planarity

Crossing Theory

Compaction

Planar Graphs

Symmetries

Interactive Drawing

Representations

Aesthetics

2D- and 3D-Embeddings

Graph Exchange Formats

Graph Drawing Contest

A Fixed-Parameter Approach to Two-Layer Planarization*

V. Dujmović[1], M. Fellows[4], M. Hallett[1], M. Kitching[1], Giuseppe Liotta[2], C. McCartin[5], N. Nishimura[3], P. Ragde[3], F. Rosamond[4], M. Suderman[1], S. Whitesides[1], and David R. Wood[6]

[1] McGill University, Canada
[2] Università di Perugia, Italy
[3] University of Waterloo, Canada
[4] University of Victoria, Canada
[5] Victoria University of Wellington, New Zealand
[6] University of Sydney, Australia

Abstract. A bipartite graph is *biplanar* if the vertices can be placed on two parallel lines (*layers*) in the plane such that there are no edge crossings when edges are drawn straight. The 2-LAYER PLANARIZATION problem asks if k edges can be deleted from a given graph G so that the remaining graph is biplanar. This problem is NP-complete, as is the 1-LAYER PLANARIZATION problem in which the permutation of the vertices in one layer is fixed. We give the following fixed parameter tractability results: an $O(k \cdot 6^k + |G|)$ algorithm for 2-LAYER PLANARIZATION and an $O(3^k \cdot |G|)$ algorithm for 1-LAYER PLANARIZATION, thus achieving *linear* time for fixed k.

1 Introduction

In a *2-layer drawing* of a bipartite graph $G = (A, B; E)$, the vertices in A and B are positioned on two distinct parallel lines in the plane, and the edges are drawn straight. Such drawings have applications in visualization [1,10], DNA mapping [19], and VLSI layouts [11]; a recent survey [13] gives more details.

A *biplanar* graph is a bipartite graph that admits a 2-layer drawing with no edge crossings. Consider a 2-layer drawing of a bipartite graph produced by first drawing a maximum biplanar subgraph with no crossings and then drawing all the remaining edges. Such a drawing will almost certainly have more crossings than is necessary; however, there is some experimental evidence to suggest that

* Research initiated at the International Workshop on Fixed Parameter Tractability in Graph Drawing, Bellairs Research Institute of McGill University, Holetown, Barbados, Feb. 9-16, 2001, organized by S. Whitesides. Research of Canada-based authors supported by NSERC. Research of G. Liotta supported by CNR and MURST. Research of D. Wood supported by the ARC, and completed while visiting McGill University.

Contact author: S. Whitesides, School of Computer Science, McGill University, 3480 University St., Montréal, Canada, sue@cs.mcgill.ca

P. Mutzel, M. Jünger, and S. Leipert (Eds.): GD 2001, LNCS 2265, pp. 1–15, 2002.

© Springer-Verlag Berlin Heidelberg 2002

drawings in which all crossings occur in a few edges are more readable than drawings with fewer total crossings [12]. Maximizing the size of a biplanar subgraph is equivalent to minimizing the number of edges not in it. This leads naturally to the definition of the 2-LAYER PLANARIZATION problem: given a graph G (not necessarily bipartite), and integer k, can G be made biplanar by deleting at most k edges? This problem is the focus of this paper.

Two-layer drawings are of fundamental importance to the "Sugiyama" approach to multilayer drawing [16]. This method first assigns vertices to layers, then makes repeated sweeps up and down the layers to determine an ordering of the vertices in one layer given the ordering for the preceding layer. This involves solving the 1-LAYER PLANARIZATION problem: given a bipartite graph $G = (A, B; E)$, a permutation π of A, and an integer k, can at most k edges be deleted to permit G to be drawn without crossings with π as the ordering of vertices in A? In this paper, we present results on this problem as well.

Instead of deleting edges, one can seek to minimize the number of crossings in a 2-layer drawing (here the input graph must be bipartite). The corresponding problems are called 1- and 2-LAYER CROSSING MINIMIZATION. Both of these well studied problems are NP-complete [7,6]. The 2-LAYER PLANARIZATION problem is NP-complete [5,17] even for planar biconnected bipartite graphs with vertices in respective bipartitions having degree two and three [5]. The 1-LAYER PLANARIZATION problem is NP-complete even for graphs with only degree-1 vertices in the fixed layer and vertices of degree at most 2 in the other layer [5], i.e., for collections of 1- and 2-paths. With the order of the vertices in both layers fixed the problem can be solved in polynomial time [5,14].

Integer linear programming algorithms have been presented for 1- and 2-LAYER CROSSING MINIMIZATION [9,18]. Jünger and Mutzel [9] survey numerous heuristics proposed for both problems, and experimentally compare their performance with the optimal solutions. They report that the *iterated barycentre* method of Sugiyama *et al.* [16] performs best in practice. However, from a theoretical point of view the *median heuristic* of Eades and Wormald [6] is a better approach for 1-LAYER CROSSING MINIMIZATION. In particular, the median heuristic is a linear time 3-approximation algorithm, whereas the barycentre heuristic is a $\Theta(\sqrt{n})$-approximation algorithm [6]. Furthermore, for graphs with maximum degree three in the free layer, the median heuristic is a 2-approximation algorithm for this problem [1]. Recently, Shahrokhi *et al.* [15] devised a polynomial time algorithm that approximates 2-LAYER CROSSING MINIMIZATION within a factor of $O(\log n)$ for a wide class of n-vertex graphs.

Despite the practical significance of the problems, 1- and 2-LAYER PLANARIZATION have received less attention in the graph drawing literature than their crossing minimization counterparts. Integer linear programming algorithms have been presented [12,14]. For acyclic graphs G, Shahrokhi *et al.* [15] present an $O(n)$ time dynamic programming algorithm for 2-LAYER PLANARIZATION of weighted acyclic graphs, for which the objective is to minimize the total weight of deleted edges.

1.1 Fixed Parameter Tractability and Our Results

When the maximum number k of allowed edge deletions is small, an algorithm for 1- or 2-LAYER PLANARIZATION whose running time is exponential in k but polynomial in the size of the graph may be useful. The theory of parameterized complexity [2] addresses complexity issues of this nature, in which a problem is specified in terms of one or more parameters. Such a problem with input size n and parameter size k is *fixed parameter tractable*, or in the class *FPT*, if there is an algorithm to solve the problem in $f(k) \cdot n^\alpha$ time, for some function f and constant α. A problem in FPT is thus solvable in polynomial time for fixed k.

In a companion paper [3], we proved using bounded pathwidth techniques that the h-layer generalizations of the 2-LAYER CROSSING MINIMIZATION and 2-LAYER PLANARIZATION problems are in FPT, where h is also considered a parameter of the problem. The 1-layer versions of these problems can also be solved using this approach. Unfortunately, a pathwidth-based approach is not particularly practical, since the running time of the algorithms is $O(2^{32(h+2k)^3} n)$.

In this paper we apply other methods from the theory of fixed parameter tractability to obtain more practical algorithms for the 1- and 2-LAYER PLANARIZATION problems. In particular, using a "kernelization" approach we obtain an $O(\sqrt{k} \cdot 17^k + |G|)$ time algorithm for 2-LAYER PLANARIZATION in a graph G, which we improve to $O(k \cdot 6^k + |G|)$ using a "bounded search tree" approach combined with kernelization. Here $|G| = |V| + |E|$ for a graph $G = (V, E)$. For small values of k, the 2-LAYER PLANARIZATION problem is thus solvable optimally in a reasonable amount of time. We then refine this second algorithm to solve the 1-LAYER PLANARIZATION problem in $O(3^k \cdot |G|)$ time.

This paper is organized as follows. After definitions and preliminary results in Section 2, we present the "kernelization" approach for 2-LAYER PLANARIZATION in Section 3. Section 4 describes our "bounded search tree" algorithm for the same problem. In Section 5 we consider the 1-LAYER PLANARIZATION problem, and present a bounded search tree algorithm for its solution. We conclude in Section 6.

2 Preliminaries

In this paper each graph $G = (V, E)$ is simple and undirected. The subgraph of G induced by a subset E' of edges is denoted by $G[E']$. A vertex with degree one is a *leaf*. If vw is the edge incident to a leaf w, then we say w is a *leaf at* v and vw is a *leaf-edge at* v. The number of non-leaf edges at a vertex v is denoted by $\deg'_G(v)$, or $\deg'(v)$ if the graph is clear from the context. A bipartite graph is *biplanar* if it admits a biplanar drawing. A *claw* is a complete bipartite subgraph $K_{1,3}$. The *2-claw* is the graph consisting of one degree-3 vertex (the *centre*), which is adjacent to three degree-2 vertices, each of which is adjacent to the centre and one leaf. A graph is a *caterpillar* if deleting all leaves produces a (possibly empty) path. This path is the *spine* of the caterpillar. If v is an endpoint of the spine of a caterpillar then there is at least one leaf at v (otherwise v itself would be a leaf.) These graphs are illustrated in Fig. 1.

We define $V_3 = \{v \in V : \deg'(v) \geq 3\}$, and $V_3' = \{w \in V \setminus V_3 : \deg(w) \geq 2, \exists v \in V_3,\ s.t.\ vw \in E\}$. That is, V_3 is the set of vertices with at least three non-leaf neighbours, and V_3' is the set of non-leaf neighbours of vertices in V_3 that are not themselves in V_3. These sets will be important in the proofs of correctness of our algorithms. Observe that the centre of a 2-claw is in V_3. In Fig. 1 and subsequent illustrations, vertices in V_3 are black and vertices in V_3' are gray.

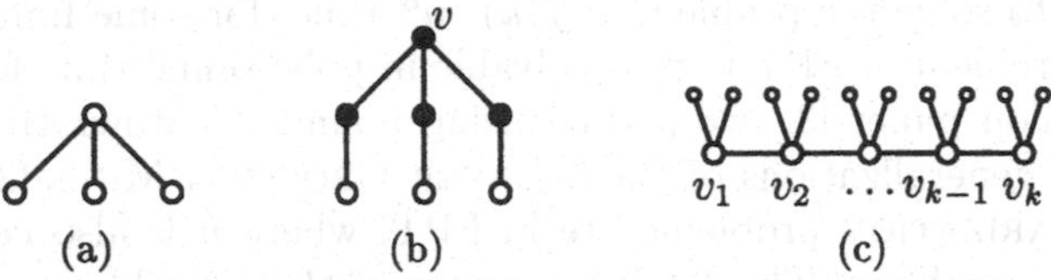

Fig. 1. (a) Claw, (b) 2-claw centred at v, (c) caterpillar.

A set of edges S of a (not necessarily bipartite) graph G is called a *biplanarizing set* if $G \setminus S$ is biplanar. The *bipartite planarization number* of a graph G, denoted by $\mathsf{bpr}(G)$, is the size of a minimum biplanarizing set for G. Thus the 2-LAYER PLANARIZATION problem is: given a graph G and integer k, is $\mathsf{bpr}(G) \leq k$? For a given bipartite graph $G = (A, B; E)$ and permutation π of A, the *1-layer bipartite planarization number* of G and π, denoted $\mathsf{bpr}(G, \pi)$, is the minimum number of edges in G whose deletion produces a graph that admits a biplanar drawing with π as the ordering of the vertices in A. The 1-LAYER PLANARIZATION problem asks if $\mathsf{bpr}(G, \pi) \leq k$.

Biplanar graphs are easily characterized, and there is a simple linear-time algorithm to recognize biplanar graphs, as the next lemma makes clear.

Lemma 1 ([4,8,17]). *Let G be a graph. The following are equivalent:*
(a) G is biplanar.
(b) G is a forest of caterpillars (see Fig. 2).
(c) G is acyclic and contains no 2-claw.
(d) The graph obtained from G by deleting all leaves is a forest and contains no vertex of degree three or greater.

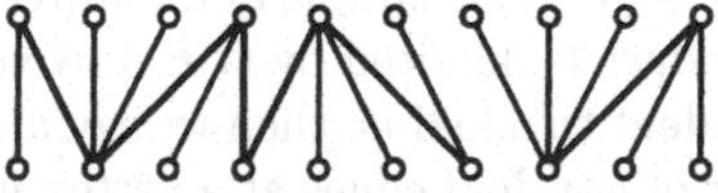

Fig. 2. A biplanar graph is a forest of caterpillars. Spine edges are dark

Since our algorithms seek to find biplanar subgraphs, some terminology concerning caterpillars will prove useful. A path $(v_1, v_2, \ldots, v_k)$ with $\deg_G(v_1) \geq 3$, $\deg_G(v_k) = 1$, and $\deg_G(v_i) = 2$, $1 < i < k$, is called a *pendant* path. A path $(v_1, v_2, \ldots, v_k)$ with $\deg_G(v_1) \geq 3$, $\deg_G(v_k) \geq 3$, and $\deg_G(v_i) = 2$, $1 < i < k$, is called an *internal* path. A *component* caterpillar of a graph is a connected component that is a caterpillar. A *pendant* caterpillar is a caterpillar subgraph C with spine $(v_2, v_3, \ldots, v_k)$ such that there is a leaf edge v_1v_2 of C with $\deg'_G(v_1) \geq 3$, $\deg'_G(v_i) = 2$ for all i, $1 < i < k$, and $\deg'_G(v_k) = 1$. An *internal* caterpillar is a caterpillar subgraph C with spine $(v_2, v_3, \ldots, v_{k-1})$ such that there are leaf-edges v_1v_2 and $v_{k-1}v_k$ of C with $\deg'_G(v_1) \geq 3$, $\deg'_G(v_k) \geq 3$, and $\deg'_G(v_i) = 2$ for all i, $1 < i < k$. A pendant (or internal) path (or caterpillar) is said to be *connected* at v_1 (and v_k), its *connection points.*

A graph consisting of a cycle and some number of leaves (possibly zero) is a *wreath.* A *component wreath* of a graph is a connected component that is a wreath. A *pendant wreath connected at* v_1 is a wreath subgraph with cycle $(v_1, v_2, \ldots, v_k)$ such that $\deg'_G(v_1) \geq 3$ and $\deg'_G(v_i) = 2$ for all i, $1 < i \leq k$. These graphs are illustrated in Fig. 3.

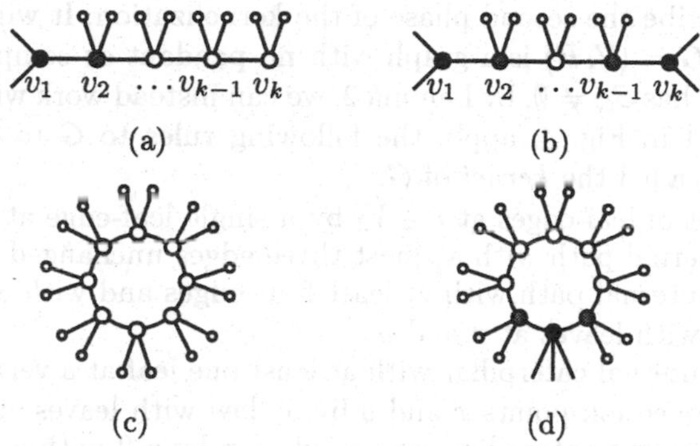

Fig. 3. (a) Pendant caterpillar, (b) internal caterpillar, (c) wreath, (d) pendant wreath.

3 Kernelization

A basic method for developing FPT algorithms is to reduce a parameterized problem instance I to an "equivalent" instance I_{kr} (the *kernel*), where the size of I_{kr} is bounded by some function of the parameter. Then the instance I_{kr} is solved using an exhaustive search method, and its solution determines a solution to the original instance I. Downey and Fellows [2, Chapter 3.2] survey this general approach, which is known as *kernelization.*

Here is an overview of our kernelization algorithm for the 2-LAYER PLANARIZATION problem, which operates in two phases. In the first phase we identify a set of edges S_G contained in the pendant and component wreaths of G that may as well be in a biplanarizing set. In the second phase, we collapse "long" internal and pendant caterpillars to "short" internal and pendant caterpillars. Intuitively, if a "short" internal or pendant caterpillar can be drawn without crossings then so can a "long" internal or pendant caterpillar. We obtain a graph G_{kr} such that $\mathsf{bpr}(G) = \mathsf{bpr}(G_{\mathsf{kr}}) + |S_G|$. We then prove that if G_{kr} satisfies a necessary condition for $\mathsf{bpr}(G_{\mathsf{kr}}) \leq k - |S_G|$ then $|G_{\mathsf{kr}}| \in O(k)$; the condition can be checked in $O(|G_{\mathsf{kr}}|)$ time. Exhaustive search determines if G_{kr} has a biplanarizing set with at most $k - |\mathcal{S}_G|$ edges; if so, this set plus $|\mathcal{S}_G|$ forms a biplanarizing set for G with at most k edges.

We now describe the first phase of the kernelization which constructs $\mathcal{S}_G$:

1. For each component wreath, add to $\mathcal{S}_G$ a cycle-edge from it.
2. For each pendant 3-cycle, add to $\mathcal{S}_G$ the edge not incident with its connection point.
3. For each pendant wreath that is not a 3-cycle, add to $\mathcal{S}_G$ a cycle-edge incident to its connection point.

The proof of the following lemma is relatively long and similar to that of the upcoming Lemma 3. Due to space limitations it is omitted here.

Lemma 2. *For every graph* G, $\mathsf{bpr}(G) \leq k$ *if and only if* $\mathsf{bpr}(G \setminus \mathcal{S}_G) \leq k - |\mathcal{S}_G|$. □

We now describe the second phase of the kernelization. It will be convenient to assume that $G = (V, E)$ is a graph with no pendant or component wreaths; i.e., $\mathcal{S}_G = \emptyset$. If G has $\mathcal{S}_G \neq \emptyset$, by Lemma 2, we can instead work with $G' = G \backslash \mathcal{S}_G$.

As illustrated in Fig. 4, apply the following rules to G to obtain a graph $G_{\mathsf{kr}} = (V_{\mathsf{kr}}, E_{\mathsf{kr}})$ called the *kernel* of G.

(a) Replace a set of leaf-edges at $v \in V_3$ by a single leaf-edge at v.
(b) Leave an internal path with at most three edges unchanged in G_{kr} .
(c) Replace an internal path with at least four edges and with endpoints x and y by a claw with leaves at x and y.
(d) Replace an internal caterpillar with at least one leaf at a vertex on the spine and with connection points x and y by a claw with leaves at x and y.
(e) Replace a pendant caterpillar connected at x by a 2-path connected at x.
(f) Delete a component caterpillar.

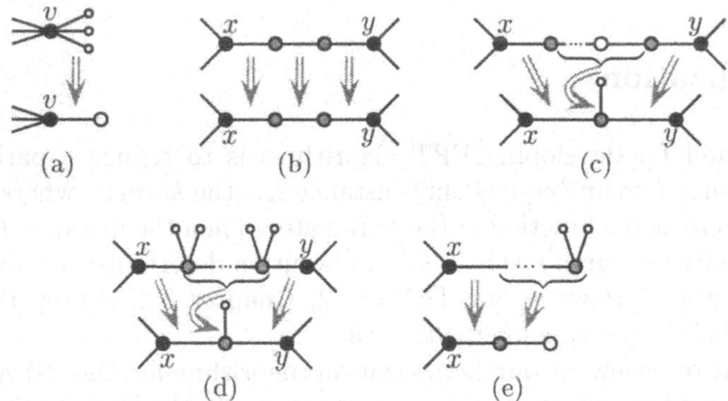

Fig. 4. Second phase of the kernelization.

Clearly G_{kr} can be constructed in $O(|G|)$ time, and every edge of G is mapped to an edge of G_{kr}. Thus, the construction of G_{kr} defines a function $f : E \to E_{\mathsf{kr}}$, indicated by arrows in Fig. 4. Observe that the non-leaf edges incident to vertices in V_3 are preserved under f. Let $E_{2,3}$ be the set of edges with both endpoints in $V_3 \cup V_3'$. Let $E_{2,3}^*$ be the set of edges in $E_{2,3}$ with at least one endpoint in V_3. Clearly f restricted to $E_{2,3}^*$ is 1-1.

Lemma 3. *For every graph G with $\mathcal{S}_G = \emptyset$, $\mathsf{bpr}(G) \leq k$ if and only if $\mathsf{bpr}(G_{\mathsf{kr}}) \leq k$.*

Proof. First we prove that if $\mathsf{bpr}(G) \leq k$ then $\mathsf{bpr}(G_{\mathsf{kr}}) \leq k$. Let T be a biplanarizing set for G with $|T| = k$. Let $T' = f(T)$. We now prove that T' is a biplanarizing set for G_{kr}. Suppose it is not. By Lemma 1(c), $G_{\mathsf{kr}} \setminus T'$ contains a subgraph C that is either a cycle or a 2-claw. If C is a cycle, then $G \setminus T$ contains a cycle, which contradicts the fact that T is a biplanarizing set for G. If C is a 2-claw, it must be centred at a vertex $v \in V_3$. Let w_1, w_2, w_3 be the neighbours of v in C with corresponding leaves x_1, x_2, x_3. Then for all i, $1 \leq i \leq 3$, $f^{-1}(vw_i) \in E^*_{2,3}$, and f maps at least one edge $ab \in G \setminus T$ to $w_i x_i$ such that ab and $f^{-1}(vw_i)$ are incident to a common vertex. Thus $f^{-1}(C)$ contains a 2-claw in $G \setminus T$, which is a contradiction. Thus, T' is a biplanarizing set for G_{kr}, and since $|T'| \leq |T| = k$, $\mathsf{bpr}(G_{\mathsf{kr}}) \leq k$.

We now prove that if $\mathsf{bpr}(G_{\mathsf{kr}}) \leq k$ then $\mathsf{bpr}(G) \leq k$. Let T be a minimal biplanarizing set for G_{kr} with $|T| \leq k$. T can be transformed into a biplanarizing set contained in $f(E_{2,3})$ by replacing edges not in $f(E_{2,3})$ as follows. Any edge vw in T but not in $f(E_{2,3})$ is a leaf edge of G_{kr} (say at v). Since T is minimal, $(G_{\mathsf{kr}} \setminus T) \cup \{vw\}$ is not biplanar; that is, $(G_{\mathsf{kr}} \setminus T) \cup \{vw\}$ contains a subgraph C that is either a cycle or 2-claw. Removing a leaf-edge from $G_{\mathsf{kr}} \setminus T$ cannot introduce a cycle; thus, C is a 2-claw and vw is a leaf-edge of C. Let vx be the other edge in C incident to v. Since x is the centre of a 2-claw, $\deg'_G(x) \geq 3$ and $vx \in f(E_{2,3})$. Every other edge vy incident to v must be in T, as otherwise $(C \setminus \{vw\}) \cup \{vy\}$ is a 2-claw in $G_{\mathsf{kr}} \setminus T$. Let $T' = (T \setminus \{vw\}) \cup \{vx\}$, as illustrated in Fig. 5. In $G_{\mathsf{kr}} \setminus T'$, vw is a connected component, and all other connected components are subgraphs of connected components of $G_{\mathsf{kr}} \setminus T$. Thus, T' is a biplanarizing set for G_{kr}. Replace T by T', and repeat this step until $T \subseteq f(E_{2,3})$.

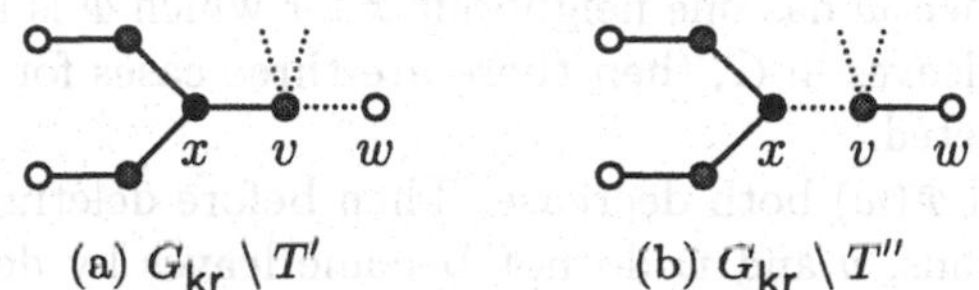

Fig. 5. Replacing the edge vw in a biplanarizing set.

Since $T \subseteq f(E_{2,3})$, it is easily seen that $|f^{-1}(T)| \leq |T| \leq k$. A 2-claw or cycle in $G \setminus f^{-1}(T)$ implies a 2-claw or cycle in $G_{\mathsf{kr}} \setminus T$. Thus $f^{-1}(T)$ is a biplanarizing set for G, and $\mathsf{bpr}(G) \leq k$. □

The proof of Lemma 3 describes how to obtain a biplanarizing set for G from a biplanarizing set for G_{kr} in $O(|G|)$ time. To enable us to prove that $|G_{\mathsf{kr}}| \in O(k)$ (assuming $\mathsf{bpr}(G_{\mathsf{kr}}) \leq k$ and $\mathcal{S}_G = \emptyset$), we introduce the following potential function, whose definition is suggested by Lemma 1(d). For a graph $G = (V, E)$, define

$$\Phi_G(v) = \max\{\deg'_G(v) - 2, 0\}, \text{ and } \Phi(G) = \sum_{v \in V} \Phi(v) .$$

Intuitively, $\Phi(v)$ approximates the number of edges in the distance-2 neighbourhood of v that must be deleted for G to become biplanar. Graphs G with $\Phi(G) = 0$ are of particular interest, as they provide another characterization of biplanar graphs.

Lemma 4. *For every graph G,*
(a) G is biplanar if and only if G is acyclic and $\Phi(G) = 0$, and
(b) if $\Phi(G) = 0$ then every component of G containing a cycle is a wreath.

Proof. $\Phi(G) = 0$ if and only if every vertex has non-leaf degree at most two. Thus, part (a) follows immediately from Lemma 1(d), and part (b) follows from the fact that a connected cyclic graph with maximum degree two is a cycle. □

For graphs G with $\Phi(G) = 0$, a minimum biplanarizing set for G consists of one cycle-edge from each component wreath. For graphs with $\Phi(G) > 0$ the observation in the following lemma will be useful.

Let the average non-leaf degree of vertices in V_3 be denoted by d.

Lemma 5. *Let G be a graph with $\Phi(G) > 0$ (that is, $V_3 \neq \emptyset$) then $|V_3| = \frac{\Phi(G)}{d-2}$.*

Proof. By definition, $d|V_3| = \sum_{v \in V_3} \deg'(v) = \sum_{v \in V_3}(\Phi_G(v)+2) = \Phi(G)+2|V_3|$. Thus, $(d-2)|V_3| = \Phi(G)$, and the result follows. □

We now prove that Φ provides a lower bound for $\mathsf{bpr}(G)$.

Lemma 6. *For every graph G, $\mathsf{bpr}(G) \geq \frac{1}{2}\Phi(G)$.*

Proof. The result follows from Lemma 4(a) if we prove that deleting one edge vw from G reduces $\Phi(G)$ by at most two. If at least one of v and w (say v) is a leaf, then $\Phi(v) = 0$ and $\Phi(w)$ does not change by deleting vw. If w becomes a leaf by deleting v, then w has one neighbour x for which Φ is reduced by one. If neither v nor w are leaves in G, then there are three cases for what can happen when edge vw is deleted.

Case 1. $\Phi(v)$ and $\Phi(w)$ both decrease: Then before deleting vw, $\deg'(v) \geq 3$ and $\deg'(w) \geq 3$. Thus, v and w do not become leaves by deleting vw, and Φ does not decrease for any other vertices.

Case 2. Exactly one of $\Phi(v)$ and $\Phi(w)$, say $\Phi(v)$, decreases: Then $\deg'(v) \geq 3$ and $\deg'(w) \leq 2$ before deleting vw. Thus, for at most one neighbour x ($\neq v$) of w is $\Phi(x)$ reduced, and $\Phi(x)$ is reduced by at most one. For no neighbour of v, except possibly x, is Φ reduced.

Case 3. Both $\Phi(v)$ and $\Phi(w)$ do not decrease: Thus, $\deg'(v) \leq 2$ and $\deg'(w) \leq 2$ before deleting vw. There is at most one neighbour of each of v and w for which Φ may decrease, and Φ may decrease by at most one for each neighbour (or by two if these neighbours coincide). □

Consider an instance (G, k) of the 2-Layer Planarization problem with $S_G = \emptyset$. If $\Phi(G_{\mathsf{kr}}) > 2k$ then we can immediately conclude from Lemma 6 that $\mathsf{bpr}(G_{\mathsf{kr}}) > k$ and hence $\mathsf{bpr}(G) > k$. On the other hand, if $\Phi(G_{\mathsf{kr}}) \leq 2k$ then, as we now prove, the size of the kernel is bounded by a function solely of k.

Lemma 7. *For every graph G and integer k, if $\mathcal{S}_G = \emptyset$ and $\Phi(G_{\mathsf{kr}}) \leq 2k$ then the kernel has size $|G_{\mathsf{kr}}| \in O(k)$.*

Proof. By counting the edges in G_{kr} with respect to vertices in V_3 we have

$$|E_{\mathsf{kr}}| \leq \sum_{v \in V_3} (2 \deg'_{G_{\mathsf{kr}}}(v) + 1) = \sum_{v \in V_3} (2\Phi_{G_{\mathsf{kr}}}(v) + 5) = 2\Phi(G_{\mathsf{kr}}) + 5|V_3| .$$

If $V_3 = \emptyset$ then $|E_{\mathsf{kr}}| \leq 2\Phi(G_{\mathsf{kr}})$. Otherwise, by Lemma 5 and since $d \geq 3$, $|V_3| \leq \Phi(G_{\mathsf{kr}})$ and $|E_{\mathsf{kr}}| \leq 7\Phi(G_{\mathsf{kr}})$. Since $\Phi(G_{\mathsf{kr}}) \leq 2k$, $|E_{\mathsf{kr}}| \leq 14k$, and since G_{kr} has no isolated vertices, $|G_{\mathsf{kr}}| \in O(k)$. □

One solution to the 2-LAYER PLANARIZATION problem is to search through all subsets of size k in E_{kr}. We obtain an algorithm whose running time is $O(k \cdot \binom{14k}{k} + |G|)$. However, as asserted in the following lemma, one need only search through a subset $\mathcal{K}$ of the edges in G_{kr}. Let $\mathcal{K} = f(E_{2,3})$ except in the case of an internal 3-path, in which case $\mathcal{K}$ contains only the middle edge in this 3-path. The set $\mathcal{K}$ is called the *sub-kernel* of G_{kr}.

Lemma 8. *Let (G, k) be an instance of* 2-LAYER PLANARIZATION *such that $\mathcal{S}_G = \emptyset$ and $0 < \Phi(G_{\mathsf{kr}}) \leq 2k$. If $\mathsf{bpr}(G_{\mathsf{kr}}) \leq k$ then there exists a biplanarizing set for G_{kr} with at most k edges contained in $\mathcal{K}$ and $|\mathcal{K}| \leq 2k(\frac{d}{d-2})$.* □

Due to space limitations, we have omitted the proof of this lemma, which is similar to that of Lemma 3.

Since $d \geq 3$, the size of the sub-kernel $|\mathcal{K}|$ is at most $6k$. There are a number of useful observations (omitted here) that may further reduce the size of the sub-kernel and improve the running time in practice. However, there is a pathological family of graphs for which our bound for the size of the sub-kernel is tight even with such improvements. Consider the graph $G_{p,q}$ $(p, q \in \mathbb{N})$ consisting of an *inner* cycle $(v_1, \ldots, v_{2p})$ and and *outer* cycle $(w_1, \ldots, w_{2p})$ with v_{2i} connected by q 2-paths to w_{2i} for all i, $1 \leq i \leq p$, as illustrated in Fig. 6(a). All vertices in V_3 have non-leaf degree $d = q + 2$. $G_{p,q}$ has $(d+2)p$ vertices and $2dp$ edges. The sub-kernel of $G_{p,q}$ is the whole graph. The largest biplanar subgraph in $G_{p,q}$ is its spanning caterpillar (see Fig. 6(b)), which has $p(d+2) - 1$ edges. Thus $\mathsf{bpr}(G_{p,q}) = 2dp - (p(d+2) - 1) = p(d-2) + 1$. The ratio of the number of edges in the sub-kernel of $G_{p,q}$ to $\mathsf{bpr}(G_{p,q})$ is $\frac{2dp}{p(d-2)+1} \rightarrow \frac{2d}{d-2}$ as $p \rightarrow \infty$. Thus the analysis of the size of the sub-kernel in Lemma 8 is tight for all d.

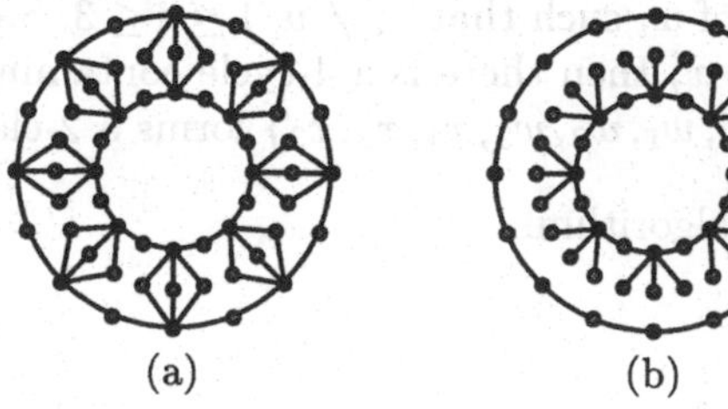

Fig. 6. The graph $G_{8,3}$ and a spanning caterpillar of $G_{8,3}$.

We now present our algorithm for the 2-LAYER PLANARIZATION problem based solely on kernelization.

Algorithm Kernelization (*input*: graph $G = (V, E)$; *parameter*: integer k)

determine $\mathcal{S}_G$ and let $k' = k - |\mathcal{S}_G|$;
determine the kernel $G_{\mathsf{kr}} = (V_{\mathsf{kr}}, E_{\mathsf{kr}})$ of $G \setminus \mathcal{S}_G$;
if $\Phi(G_{\mathsf{kr}}) > 2k'$ **then** return NO.
determine the sub-kernel $\mathcal{K} \subseteq E_{\mathsf{kr}}$ of G_{kr};
return YES iff $\exists T \subseteq \mathcal{K}$ s.t. $|T| \le k'$, $G_{\mathsf{kr}} \setminus T$ is acyclic, and $\Phi(G_{\mathsf{kr}} \setminus T) = 0$;

Combining the previous lemmas and using Sterling's Formula gives:

Theorem 1. *Given a graph $G = (V, E)$ and integer k, the algorithm* Kernelization (G, k) *determines if* $\mathsf{bpr}(G) \le k$. *If $\Phi(G) = 0$ then the running time is $O(|G|)$; otherwise it is $O(\sqrt{k} \cdot (\frac{2\mathrm{e}\,d}{d-2})^k + |G|)$ time, where d is the average non-leaf degree of vertices in V_3, and $\mathbf{e}$ is the base of the natural logarithm.* □

Clearly, if Kernelization(G, k) returns YES then a biplanarizing set for G with at most k edges can easily be computed. Since $d \ge 3$, in the worst-case the running time of algorithm Kernelization is $O(\sqrt{k} \cdot (6\mathrm{e})^k + |G|) \in O(\sqrt{k} \cdot 17^k + |G|)$.

4 Bounded Search Trees

A second approach to producing FPT algorithms is called the *method of bounded search trees* [2, Chapter 3.1]. Here one uses exhaustive search in a tree whose size is bounded by a function of the parameter. In this section we present an algorithm for the 2-LAYER PLANARIZATION problem based on a bounded search tree approach. At each node of the search tree a 2-claw or small cycle C is identified. At least one of the edges in C is in every biplanarizing set. Our algorithm then recursively solves $|C|$ subproblems such that one of the edges in C is deleted from the graph in each subproblem. The following lemma provides a sufficient condition for the existence of such a set C.

Lemma 9. *If there exists a vertex v in a graph G_0 such that $\deg'_{G_0}(v) \ge 3$ then G_0 contains a 2-claw or a 3- or 4-cycle containing v.*

Proof. Let w_1, w_2, w_3 be three distinct non-leaf neighbours of v. If some w_i is adjacent to w_j then there is a 3-cycle containing v. Suppose no w_i is adjacent to a w_j. Let x_i be a neighbour of w_i such that $x_i \ne v$, $1 \le i \le 3$. Such an x_i exists since w_i is not a leaf. If $x_i = x_j$ then there is a 4-cycle containing v. Otherwise the x_i's are distinct. Then $\{v, w_1, w_2, w_3, x_1, x_2, x_3\}$ forms a 2-claw. □

This yields the following algorithm.

Algorithm Bounded Search Tree (*input*: graph $G_0 = (V_0, E_0)$; *parameter*: integer k)

1 **if** $k = 0$ **then** return YES iff G_0 is acyclic and $\Phi(G_0) = 0$;
2 **else if** $\Phi(G_0) = 0$ **then** return YES iff $k \geq \#$ component wreaths of G_0;
3 **else** ($\exists v \in V_0$ such that $\deg'_{G_0}(v) \geq 3$)
  a) find a 2-claw, 3-cycle or 4-cycle C in G_0 containing v as described in Lemma 9;
  b) **for each** edge $xy \in C$ **do**
    if Bounded Search Tree($G_0 \setminus \{xy\}$, $k-1$) **then** return YES;
  c) return NO;

We could solve 2-Layer Planarization by running Bounded Search Tree (G, k). However, we apply Bounded Search Tree to the kernel of G so that the running time at each node of the search tree is $O(k)$ rather than $O(|G|)$.

Theorem 2. *Given a graph G and integer k, let G_{kr} be the kernel of $G \setminus \mathcal{S}_G$. The algorithm* Bounded Search Tree $(G_{\mathsf{kr}}, k - |\mathcal{S}_G|)$ *determines if* $\mathsf{bpr}(G) \leq k$ *in* $O(k \cdot 6^k + |G|)$ *time.*

Proof. We prove the correctness of the algorithm by induction on k with the following inductive hypothesis: "Bounded Search Tree (G_0, k) returns YES if and only if $\mathsf{bpr}(G_0) \leq k$". The basis of the induction with $k = 0$ (Step 1) follows immediately from Lemma 4(a). Assume the inductive hypothesis holds for $k-1$. If $\Phi(G_0) = 0$ (as in Step 2) then by Lemma 4, every connected component is a caterpillar or a wreath. Caterpillars and wreaths have bipartite planarization numbers of 0 and 1, respectively. Thus $\mathsf{bpr}(G_0)$ is the number of component wreaths of G_0, and hence Step 2 of the algorithm is valid.

Now assume $k > 0$ and $\Phi(G_0) > 0$; that is, there exists a vertex $v \in V$ such that $\deg'_{G_0}(v) \geq 3$. By Lemma 9, G_0 contains a 2-claw or a 3- or 4-cycle C. Every biplanarizing set for G_0 must contain an edge in C. Thus $\mathsf{bpr}(G_0) \leq k$ if and only if there exists an edge $xy \in C$ such that $\mathsf{bpr}(G_0 \setminus \{xy\}) \leq k-1$. By induction, Bounded Search Tree $(G_0 \setminus \{xy\}, k-1)$ determines if $\mathsf{bpr}(G_0 \setminus \{xy\}) \leq k-1$. Therefore the algorithm determines if $\mathsf{bpr}(G_0) \leq k$, and the inductive hypothesis holds. In particular, Bounded Search Tree $(G_{\mathsf{kr}}, k - |\mathcal{S}_G|)$ determines if $\mathsf{bpr}(G_{\mathsf{kr}}) \leq k - |\mathcal{S}_G|$, which holds if and only if $\mathsf{bpr}(G) \leq k$ by Lemmas 2 and 3.

In each recursive call k is reduced by one. Thus the height of the search tree is at most k. At each node of the search tree, there are $|C|$ branches. Since $|C| \leq 6$, the search tree has at most 6^k nodes.

A non-recursive implementation maintains the current copy of G_0 under the operations of the deletion and insertion of edges. Moving from one copy to the next requires $O(|G_0|)$ time. At any given node of the search tree, the algorithm takes $O(|G_0|)$ time. Each G_0 is a subgraph of G_{kr}; thus, by Lemma 7, the time taken at each node of the search tree is $O(k)$. Therefore the running time of the algorithm is $O(k \cdot 6^k + |G|)$. □

The exponential component of the time bound for Kernelization is $(\frac{2e\,d}{d-2})^k$, and for Bounded Search Tree is 6^k. In the worst case with $d = 3$ the Kernelization bound is approximately 17^k, which is considerably more than 6^k. However, for $d \geq 22$, $\frac{2e\,d}{d-2} < 6$, and the Kernelization algorithm provides a better upper bound on the running time than the Bounded Search Tree algorithm.

5 1-Layer Planarization

We now consider the 1-LAYER PLANARIZATION problem: given a bipartite graph $G = (A, B; E)$ and permutation π of A, is $\text{bpr}(G, \pi) \leq k$? It is not clear how one would apply a kernelization strategy to this problem. In this section, we apply a bounded search tree approach to the (unkernelized) graph G. The following result characterizes biplanar graphs with a fixed permutation of one bipartition.

Lemma 10. *There is a biplanar drawing of a bipartite graph $G = (A, B; E)$ with a permutation π of A iff G is acyclic and the following condition holds.*

For every path (x, v, y) of G with $x, y \in A$, if $u \in A$ is between x and y in π, then the only edge incident to u (if any) is uv. $(\star)$

Proof. The necessity of condition $(\star)$ is easily verified by observing that if $(\star)$ does not hold for some path (x, v, y) and edge uw $(w \neq v)$, then regardless of the relative positions of w and v in the permutation of B, uw must cross xv or yv, as illustrated in Fig. 7(a). This observation was also made by Mutzel and Weiskircher [14].

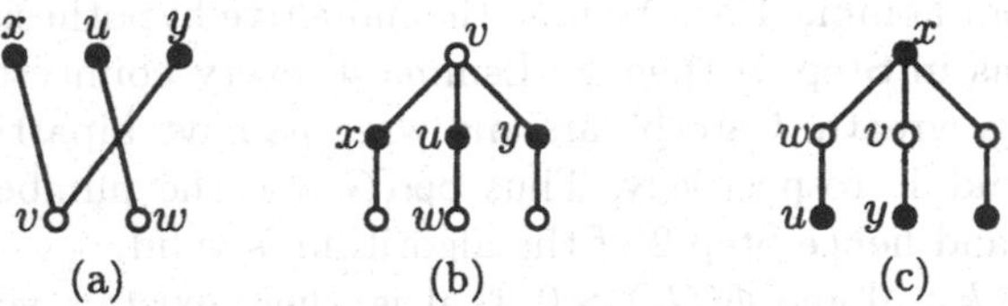

Fig. 7. Forbidden sub-structures for 1-layer biplanarity; vertices in A are gray.

Suppose condition $(\star)$ holds. Suppose for the sake of contradiction that G is not a forest of caterpillars. Since G is acyclic, by Lemma 1, G contains a 2-claw C. We consider two cases; in the first, C contains three vertices in A, as illustrated in Fig. 7(b). Let these vertices be x, u, y in the order they appear in π, and let v be the centre of C. Then condition $(\star)$ does not hold for the path (x, v, y) and the leaf-edge in C incident to u. Now consider the case in which C contains four vertices in A, as illustrated in Fig. 7(c). Let x be the centre of C. Then $x \in A$. Let u and y be two other vertices in $C \cap A$ both to the right or both to the left of x in π. Such a pair of vertices exists by symmetry. Without loss of generality, u is between x and y. Let v be the vertex in C adjacent to x and y, and let w be the vertex in C adjacent to x and u. Then condition $(\star)$ does not hold for the path (x, v, y) and edge uw. Thus G is a forest of caterpillars.

To construct a 2-layer drawing of G, all that remains is to describe the permutation of B. Let $(a_1, \ldots, a_n)$ be the ordering of A defined by π. For each vertex $v \in B$, define $L(v) = \min\{i : a_i v \in E\}$; that is, $L(v)$ is the leftmost neighbour of v in the fixed permutation of A. We say a vertex v with $L(v) = i$ *belongs* to a_i, as does the edge $a_i v$. Order the vertices $v \in B$ by the value of $L(v)$, breaking ties as follows. For each i, $1 \leq i \leq n$, there are at most two non-leaf vertices belonging to a_i, as otherwise there would be a 2-claw centred at a_i. Suppose there are exactly two non-leaf vertices v and w belonging to a_i. Then we can label v and w such that in π, the neighbours of v (not counting a_i) are all to the left of a_i, and the neighbours of w (not counting a_i) are all to the right of a_i, as otherwise ($\star$) is not satisfied. Let $(v, x_1, \ldots, x_p, w)$ be the order of the vertices belonging to a_i, where $\{x_1, \ldots, x_p\}$ are the leaves belonging to a_i, as illustrated in Fig. 8(a). This defines a 2-layer drawing of G.

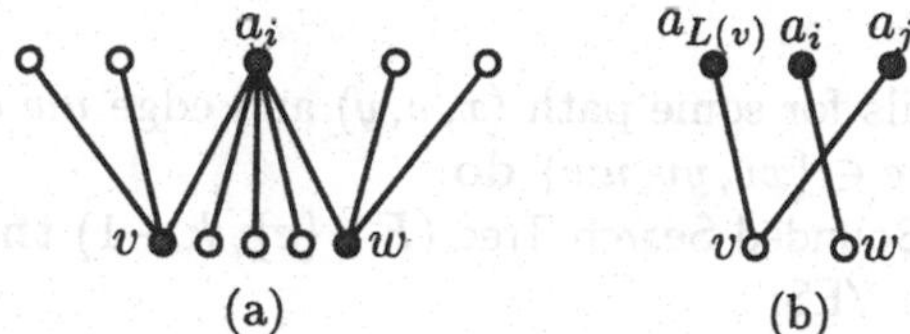

Fig. 8. Construction of the permutation of B.

Suppose there is a crossing between some edges $a_i w$ and $a_j v$ with $a_i, a_j \in A$ ($i < j$) and $v, w \in B$. Then v is to the left of w in the permutation of B, and thus $L(v) \leq i$. Since the edges belonging to a_i do not cross, $L(v) < i$. This implies that condition ($\star$) is not satisfied for the path $(a_{L(v)}, v, a_j)$ and the edge $a_i w$, as illustrated in Fig. 8(b). Thus there is no crossing in the drawing of G. □

Lemma 11. *If $G = (A, B; E)$ is a bipartite graph and π is a permutation of A which satisfies condition ($\star$), then all cycles of G are 4-cycles and every pair of non-edge-disjoint cycles shares exactly two edges. Moreover, the degree of all vertices in B which appear in a cycle is exactly two.* □

It is not difficult to show that this lemma follows from Lemma 10. We omit details due to space limitations.

Let $G = (A, B; E)$ be a bipartite graph with a fixed permutation of A which satisfies condition ($\star$). A complete bipartite subgraph $H = K_{2,p}$ of G with $|H \cap A| = 2$, $|H \cap B| = p$, and $\deg_G(v) = 2$ for every $v \in H \cap B$, is called a *p-diamond*, see Fig. 9. It follows from Lemma 11 that every cycle of G is in some diamond. We can now directly determine the 1-layer bipartite planarization numbers of graphs and permutations satisfying condition ($\star$).

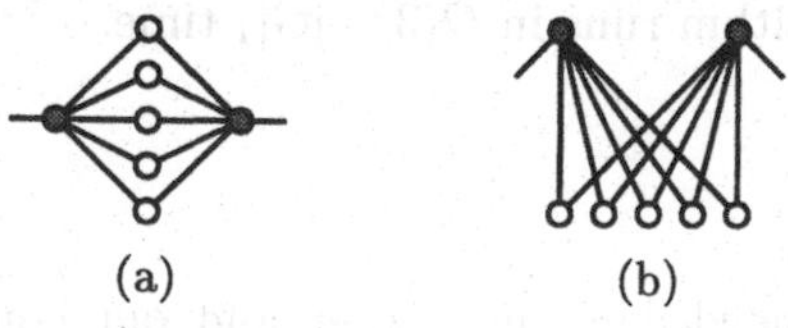

Fig. 9. (a) 5-diamond, (b) 2-layer drawing of a 5-diamond.

Lemma 12. *If $G = (A, B; E)$ is a bipartite graph and π is permutation of A satisfying condition $(\star)$ then* $\mathsf{bpr}(G, \pi) = \sum_{p\text{-diamonds of } G} (p-1)$.

Proof. For each p-diamond H of G, delete $p-1$ of the edges incident to one of the vertices in $H \cap A$. The resulting graph is acyclic and satisfies condition $(\star)$, and thus, by Lemma 10, has a biplanar drawing using π. To remove all cycles from G requires the deletion of at least $p-1$ edges from each p-diamond. □

The following algorithm solves the 1-LAYER PLANARIZATION problem:

Algorithm 1-Layer Bounded Search Tree (*input*: set $F \subseteq E$; *parameter*: integer k)

1 **if** $k > 0$ and $(\star)$ fails for some path (x, v, y) and edge uw of $G[F]$ **then**
  a) **for each** edge $e \in \{xv, yv, uw\}$ **do**
    if 1-Layer Bounded Search Tree $(F \setminus \{e\}, k-1)$ **then**
      return YES;
  b) return NO;
2 return YES iff $k \geq \sum_{p\text{-diamonds of } G[F]} (p-1)$.

Theorem 3. *Given a bipartite graph $G = (A, B; E)$, fixed permutation π of A, and integer k, algorithm* 1-Layer Bounded Search Tree (E, k) *determines if* $\mathsf{bpr}(G, \pi) \leq k$ *in $O(3^k \cdot |G|)$ time.*

Proof. The correctness of the algorithm follows from Lemmas 10 and 12. We now describe how to check if condition $(\star)$ holds in $O(|A|)$ time. We can assume that the adjacency lists of vertices in B are ordered by π. For each vertex $v \in B$, by simultaneously traversing the adjacency list of v and a list of vertices in A ordered by π, we can identify whenever the condition $(\star)$ does not hold for some 2-path centred at v. Repeating this step for each vertex $v \in B$ or until an appropriate 2-path is found, we can check if $(\star)$ is satisfied. This algorithm runs in $O(|A|)$ time since if the section of the ordered list of A traversed with respect to some vertex $v \in B$ overlaps the corresponding section for some other vertex $w \in B$, then condition $(\star)$ is not satisfied, and the algorithm will immediately terminate. To count the number and size of the diamonds in $G[F]$ takes $O(|G|)$ time. Thus, the algorithm takes $O(|G|)$ time at each node of the search tree. Since each node of the search tree has three children, and the height of the tree is at most k, the algorithm runs in $O(3^k \cdot |G|)$ time. □

6 Conclusion

To the best of our knowledge, this paper and our companion paper [3] give the first applications of fixed-parameter tractability methods to graph drawing problems. Here, we study two problems, 1- and 2-LAYER PLANARIZATION, which are of fundamental interest in layered graph drawing. Our methods yield linear-time algorithms to determine if $\mathsf{bpr}(G) \leq k$ and $\mathsf{bpr}(G, \pi) \leq k$, for fixed k. When k is allowed to vary, the running time of the algorithms is, not surprisingly, exponential in k, but the base of the exponent is small. We believe that further study will demonstrate practical as well as theoretical benefit from this approach, and that these methods will prove more widely applicable.

References

1. G. Di Battista, P. Eades, R. Tamassia, and I. G. Tollis. *Graph Drawing: Algorithms for the Visualization of Graphs.* Prentice-Hall, 1999.
2. R. G. Downey and M. R. Fellows. *Parametrized complexity.* Springer, 1999.
3. V. Dujmović, M. Fellows, M. Hallett, M. Kitching, G. Liotta, C. McCartin, N. Nishimura, P. Ragde, F. Rosemand, M. Suderman, S. Whitesides, and D. R. Wood. On the parameterized complexity of layered graph drawing. In *Proc. 9th European Symp. on Algorithms (ESA '01)*, to appear.
4. P. Eades, B. D. McKay, and N. C. Wormald. On an edge crossing problem. In *Proc. 9th Australian Computer Science Conference*, pages 327–334. Australian National University, 1986.
5. P. Eades and S. Whitesides. Drawing graphs in two layers. *Theoret. Comput. Sci.*, 131(2):361–374, 1994.
6. P. Eades and N. C. Wormald. Edge crossings in drawings of bipartite graphs. *Algorithmica*, 11(4):379–403, 1994.
7. M. R. Garey and D. S. Johnson. Crossing number is NP-complete. *SIAM J. Algebraic Discrete Methods*, 4(3):312–316, 1983.
8. F. Harary and A. Schwenk. A new crossing number for bipartite graphs. *Utilitas Math.*, 1:203–209, 1972.
9. M. Jünger and P. Mutzel. 2-layer straightline crossing minimization: performance of exact and heuristic algorithms. *J. Graph Algorithms Appl.*, 1(1):1–25, 1997.
10. M. Kaufmann and D. Wagner, editors. *Drawing Graphs: Methods and Models*, volume 2025 of *Lecture Notes in Comput. Sci.* Springer, 2001.
11. T. Lengauer. *Combinatorial Algorithms for Integrated Circuit Layout.* Wiley, 1990.
12. P. Mutzel. An alternative method to crossing minimization on hierarchical graphs. *SIAM J. Optimization*, 11(4):1065–1080, 2001.
13. P. Mutzel. Optimization in leveled graphs. In P. M. Pardalos and C. A. Floudas, editors, *Encyclopedia of Optimization.* Kluwer, to appear.
14. P. Mutzel and R. Weiskircher. Two-layer planarization in graph drawing. In K. Y. Chwa and O. H. Ibarra, editors, *Proc. 9th International Symp. on Algorithms and Computation (ISAAC'98)*, volume 1533 of *Lecture Notes in Comput. Sci.*, pages 69–78. Springer, 1998.
15. F. Shahrokhi, O. Sýkora, L. A. Székely, and I. Vrťo. On bipartite drawings and the linear arrangement problem. *SIAM J. Comput.*, 30(6):1773–1789, 2001.
16. K. Sugiyama, S. Tagawa, and M. Toda. Methods for visual understanding of hierarchical system structures. *Trans. Systems Man Cybernet.*, 11(2):109–125, 1981.
17. N. Tomii, Y. Kambayashi, and S. Yajima. On planarization algorithms of 2-level graphs. *Papers of tech. group on elect. comp., IECEJ*, EC77-38:1–12, 1977.
18. V. Valls, R. Marti, and P. Lino. A branch and bound algorithm for minimizing the number of crossing arcs in bipartite graphs. *J. Operat. Res.*, 90:303–319, 1996.
19. M. S. Waterman and J. R. Griggs. Interval graphs and maps of DNA. *Bull. Math. Biol.*, 48(2):189–195, 1986.

How to Layer a Directed Acyclic Graph

Patrick Healy and Nikola S. Nikolov

CSIS Department, University of Limerick, Limerick, Republic of Ireland,
fax +353-61-202734
{patrick.healy,nikola.nikolov}@ul.ie

Abstract. We consider the problem of partitioning a directed acyclic graph into layers such that all edges point unidirectionally. We perform an experimental analysis of some of the existing layering algorithms and then propose a new algorithm that is more realistic in the sense that it is possible to incorporate specific information about node and edge widths into the algorithm. The goal is to minimize the total sum of edge spans subject to dimension constraints on the drawing. We also present some preliminary results from experiments we have conducted using our layering algorithm on over 5900 example directed acyclic graphs.

1 Introduction

The layering problem for directed acyclic graphs (DAGs) arises as one of the steps of the classical Sugiyama algorithm for drawing directed graphs [6]. If the nodes of a DAG are not pre-assigned to specific layers then it is necessary to separate them into such layers in order to draw the DAG in Sugiyama fashion. We call an algorithm that finds a layering of a DAG a *layering algorithm*. Normally a layering algorithm must find a layering of a DAG subject to certain aesthetic criteria important to the final drawing. While these may be subjective, some are generally agreed upon [4]: the drawing should be compact; large edge spans should be avoided; and, the edges should be as straight as possible. Compactness can be achieved by specifying bounds W and H on the width and the height of the layering respectively. Short edge spans are desirable aesthetically because they increase the readability of the drawing but also because the forced introduction of dummy nodes, when an edge spans multiple layers, degrades further stages of drawing algorithms. The span of edge (u, v) with $u \in V_i$ and $v \in V_j$ is $i - j$. Further, the dummy nodes may also lead to additional bends on edges since edge bends mainly occur at dummy nodes.

At present there are three widely-used layering algorithms which find layerings of a DAG subject to *some* of the above aesthetic criteria. They all have polynomial running time: the *Longest Path* algorithm finds a layering with minimal height [4]; the *Coffman-Graham* algorithm finds a layering of width at most W and height $h \leq (2 - 2/W)h_{min}$, where h_{min} is the minimum height of a layering of width W [3]; and the ILP algorithm of *Gansner et al.* finds a layering with minimum number of dummy nodes [5]. An upper bound on the width of the layering can be specified only in the Coffman-Graham algorithm. In the classical

P. Mutzel, M. Jünger, and S. Leipert (Eds.): GD 2001, LNCS 2265, pp. 16–30, 2002.
© Springer-Verlag Berlin Heidelberg 2002

version of the Coffman-Graham algorithm the width of a layer is considered to be the number of real nodes the layer contains while neglecting the introduced dummy nodes. The algorithm can easily be modified to take into account the widths of the real nodes, but the width of the final drawing may still be much greater than expected, because of the contribution of the dummy nodes to it. The algorithm of Gansner et al. usually results in compact layerings, but the dimensions of the drawing are not controlled and they may be undesirable.

After introducing some basic definitions related to the layering problem we compare the performance of the existing layering algorithms on over 5900 example DAGs in Section 3. Then in Section 4 and Section 5 we introduce a new approach to the layering problem based on Integer Linear Programming (ILP) and identify a set of valid inequalities (some of which define facets) of the associated layering polytope. This approach allow us to construct a new ILP layering algorithm which we study and compare to Gansner et al.'s layering algorithm in Section 6. In Section 7 we discuss the results of this work and suggest further research directions.

2 Basic Definitions

Definition 1. *Given a DAG $G = (V, E)$, where each node $v \in V$ has a positive width w_v, a layering of G is a partition of its node set V into disjoint subsets $V_1, V_2, \ldots, V_h$, such that if $(u, v) \in E$ where $u \in V_i$ and $v \in V_j$ then $i > j$. A DAG with a layering is called a layered digraph.*

Definition 2. *The height of a layered digraph is the number of layers, h.*

Definition 3. *The width of layer V_k is traditionally defined as $w(V_k) = \sum_{v \in V_k} w_v$ and the width of a layered digraph (layering) is $w = \max_{1 \leq k \leq h} w(V_k)$.*

Layered digraphs are conventionally drawn so that all nodes in layer V_k lie on the horizontal line $y = k$ and all edges point downwards. In the process of drawing a layered digraph when an edge spans multiple layers, it is common to introduce dummy nodes with in- and out-degree 1 in the intermediate layers.

We denote by $d^-(v)$ and $d^+(v)$ the in- and out-degree of node $v \in V$, respectively. For $G = (V, E)$ with unitary edge lengths, define $l_p(v)$ to be the length of the longest path from any node u to v where $d^-(u) = 0$. Similarly, define $l_s(v)$ to be the longest path from v to any node u where $d^+(u) = 0$. That is, the values $l_p(v)$ and $l_s(v)$ refer, respectively, to the length of the longest path from any predecessor to v and to the length of the longest path from v to any successor. Let the node set V of G be constrained to be partitioned into at most H layers. Then for each node $v \in V$ there is a set of consecutive layers where the node can be potentially placed in if all the edges are required to point downwards. The following three definitions describe this set.

Definition 4. *The roof of node v is the number of the highest layer node v can be placed in. We denote the roof of v by $\rho(v)$, i.e. $\rho(v) = H - l_p(v)$.*

Definition 5. *The floor of v is the lowest level node v can be placed in. We denote the floor of v by $\varphi(v)$, i.e. $\varphi(v) = l_s(v) + 1$.*

Definition 6. *The layer span of node v is $L(v) = \{k \in \mathbb{N} : \varphi(v) \leq k \leq \rho(v)\}$. That is, $L(v)$ refers to the set of layers in which node v can be potentially placed if all the edges are required to point downwards.*

By definition the roof and the layer span of node v depend on the upper bound on the number of layers H. We do not include H in the notations of the roof and the layer span for simplicity of notation. Normally, it is clear from the context what is the value of H.

3 Performance of Existing Layering Algorithms

In this section we look at the behavior of the algorithms described earlier on a variety of inputs. As motivation for what is to follow we look at the output of the three algorithms on a specific graph, firstly. Then we consider their aggregate performance over a range of measures.

3.1 Three Layerings of a Graph

Figure 1 illustrates how the same DAG, Grafo1012.22 from the graph database introduced by Di Battista et al. [1], is layered by the three algorithms described earlier.

The layering in Figure 1(a) is the output of the Longest Path algorithm and it clearly shows its weakness: the width of the bottom layers is much larger than the width of the top layers. The layering in Figure 1(b) is the output of the Coffman-Graham algorithm with an input parameter $W = 5$ (i.e. maximum 5 nodes in a layer). As can be seen, the final width of the drawing can exceed the input parameter. The third layering in Figure 1(c) is the output of the Gansner et al.'s ILP layering algorithm. In this case the drawing has the minimum number of dummy nodes, but this algorithm does not put any bounds on the dimensions of the layering, which may result in a final drawing which does not fit the drawing area. For instance, if we need the same DAG Grafo1012.22 drawn on less than 9 layers (9 is the number of layers in Figure 1(c)) we may prefer to have the DAG layered as it is in Figure 1(d) which shows a more compact drawing of Grafo1012.22 having three dummy nodes more than the minimum.

3.2 Aggregate Performance of Layering Algorithms

Each of the three algorithms described above has some positive attribute: the Longest Path algorithm finds a layering of minimum height; Gansner et al.'s algorithm minimizes the number of dummy nodes; and, the Coffman-Graham algorithm permits one to specify a bound on the width of the drawing. We investigate the three algorithms' performance now on a large sample of graphs according to some accepted aesthetics. These aesthetics are 1) the area of the layering and 2) the number of dummy nodes that the algorithms introduce. We ran all three algorithms on 5911 connected DAGs from the graph database of Di Battista et al. [1]. The DAGs have node counts ranging from 10 to 100 and the average number of nodes is 48.34. The node labels in all DAGs are numbers, which makes all the nodes equally wide and allow us to set the node width equal to 1. We separated the graphs into "buckets" according to their node count, putting a graph of n nodes into bucket $\lfloor n/2 \rfloor$. (A similar separation according to edge count was rejected because of the number of buckets with just a single graph.)

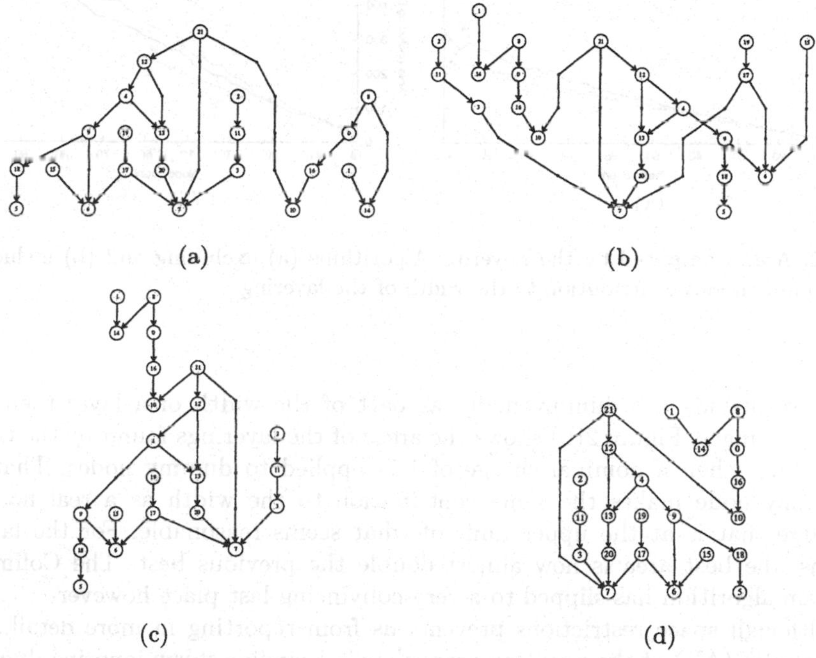

Fig. 1. Grafo1012.22 layered with (a) the Longest Path algorithm (b) the Coffman-Graham algorithm and (c) Gansner et al.'s algorithm. The drawing in (d) is a compact representation of Grafo1012.22 with three dummy nodes more than the minimum number of dummy nodes.

In the following, we refer to the three algorithms as Longest Path, Coffman-Graham and Gansner. For Coffman-Graham, the width bound for a graph was specified to be equal to Gansner's width for the same graph.

Figure 2(a) illustrates the three algorithms' performance according to the area aesthetic. For this figure, area was calculated to be the product of the number of layers and the width of the graph in terms of real nodes. Although Longest

Path's height is optimal, its width is so poor that it results in an increasingly poor performance. On the other hand, Gansner maintains a very close watch on Coffman-Graham in spite of there being no explicit mechanism to control its dimensions. This may be due to the following reason: the Coffman-Graham algorithm requires a bounding width, W, as input and the bound that we provided in all cases was the width resulting from the Gansner layering.

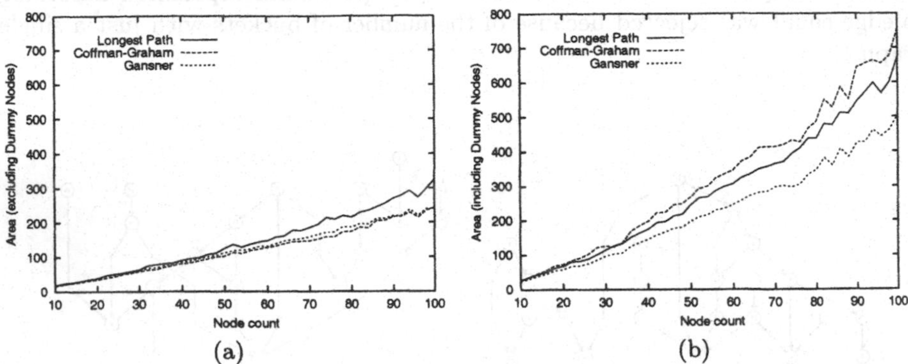

Fig. 2. Area Computed by the Layering Algorithms (a) excluding and (b) including the dummy nodes contribution to the width of the layering.

If we include the dummy nodes as part of the width of a layer then the picture changes. Figure 2(b) shows the areas of the layerings found by the three algorithms when a nominal charge of 1 is applied to dummy nodes. That is, a dummy node makes the same contribution to the width as a real node – a charge that is at the upper limit of what seems reasonable. For the larger graphs, the best area is now almost double the previous best. The Coffman-Graham algorithm has slipped to a very convincing last place however.

Although space restrictions prevents us from reporting in more detail, the aspect ratio (AR) of the resulting graphs[1] tells a similar story: ignoring dummy nodes, Gansner and Coffman-Graham behave similarly and better than Longest Path in that they have AR closer to unitary; when dummy nodes are factored into the calculations, Gansner is considerably better than the other two, which behave broadly similarly. Since Longest Path always has minimum height, its AR will be biased towards larger values. (An interesting observation about AR is that in all three algorithms, it peaks in the data at about $|V| = 75$ and then declines.)

To investigate further the apparent impact of dummy nodes on aesthetics such as area and aspect ratio, we computed both the average and maximum ratios of dummy nodes to real nodes in each of the graphs. We define these two parameters as follows.

[1] Aspect ratio is computed as the ratio of width to height.

Definition 7. *Let $G = (V, E)$ be a layered DAG. Maximum Layer Bloat (MLB) of a layering of G is the maximum ratio of the number of dummy nodes in a layer to the number of real nodes in the same layer over the layers of a graph.*

Definition 8. *Let $G = (V, E)$ be a layered DAG. Average Layer Bloat (ALB) of a layering of G is the average of the ratios of the number of dummy nodes in a layer to the number of real nodes in the same layer over the layers of a graph.*

Table 1 summarises the maximum and average bloat values for the 5911 DAGs. A more detailed breakdown is shown in Figure 3 where it can be seen that Gansner is a clear winner: the *maximum* bloat of this algorithm behaves similar to the average of the other two algorithms. The aspect ratio and area aesthetics which are determined by the widest layer will be poorest with either Coffman-Graham or Longest Path. However, even for Gansner the number of dummy nodes is becoming significant for larger node counts.

Table 1. Average values of MLB and ALB for 5911 sample DAGs.

Layering	MLB	ALB
Longest Path	6.41	2.34
Coffman-Graham	5.76	2.01
Gansner	2.38	0.81

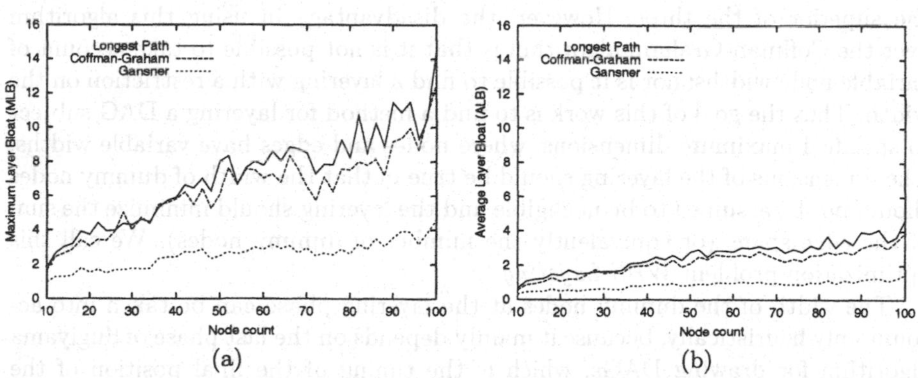

Fig. 3. (a) Maximum and (b) Average Layer Bloats Computed by the Layering Algorithms.

The number of dummy nodes introduced by a layering algorithm impacts on later stages of hierarchical graph drawing methods so we investigate this parameter. Figure 4(a) displays the number of dummy nodes introduced by the three algorithms normalized by the number of nodes in the original graph. Gansner computes the optimal number of dummy nodes and, although growing at a super-linear rate, it easily outperforms the other two. In this context again, their performance is quite similar.

Finally, we compare the running times of the three layering algorithms in Figure 4(b). As can be seen, Longest Path has the best running time, followed by Gansner, and Coffman-Graham is the slowest, but still very fast. (We attribute the stratified nature of the plots to clock resolution.)

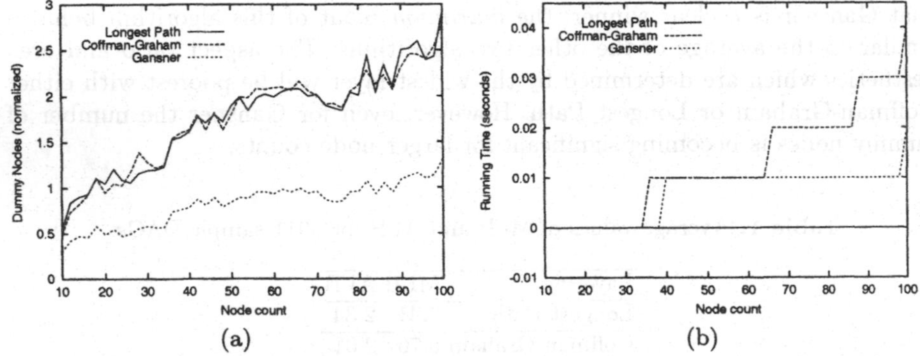

Fig. 4. (a) Normalized Number of Dummy Nodes introduced by the Layering Algorithms (b) Running times.

On the basis of these experiments we conclude that Gansner's algorithm is the superior of the three. However, the disadvantage in using this algorithm over the Coffman-Graham algorithm is that it is not possible to take account of variable node widths; nor is it possible to find a layering with a restriction on the width. Thus the goal of this work is to find a method for layering a DAG subject to specified maximum dimensions, where nodes and edges have variable widths. The dimensions of the layering should be true in that the width of dummy nodes should not be assumed to be negligible and the layering should minimize the sum of the edge spans (or equivalently the number of dummy nodes). We call this optimization problem *WHS-Layering*.

The width of the dummy nodes at the layering phase can be taken into account only heuristically, because it mainly depends on the last phase of Sugiyama algorithm for drawing DAGs, which is the tuning of the final position of the nodes. We see two different heuristic approaches. First, we may decide to assign a unit width to all dummy nodes and use this unit width to express the width of the other nodes of the DAG. In such case all the dummy nodes have width 1 at the layering phase. Alternatively, each edge may have different width of the dummy nodes which has to be placed on it. The experimental results we present in this paper are based on the assumption that all the dummy nodes of a DAG have unitary width.

4 An ILP Approach to *WHS-Layering*

Taking into account the dimensions constraints (width and height) alone makes *WHS-Layering* NP-hard since the Precedence Constrained Multiprocessor Scheduling problem can be reduced straightforwardly to it [3]. We have taken a combinatorial optimisation approach to *WHS-Layering* aiming to construct an exact ILP algorithm which finds a layering with minimum number of dummy nodes for specified upper bounds on the height and the width of the layering.

Let $G = (V, E)$ be a DAG the nodes of which have to be partitioned into at most $H > 0$ layers $V_1, V_2, \ldots, V_H$ of width at most $W > 0$. We construct a *layering DAG* or *LDAG* $\mathcal{L}_G^H = (V_\mathcal{L}, E_\mathcal{L})$ as follows. For each $v \in V$ and each $k \in L(v)$ there is a node $\lambda_{vk} \in V_\mathcal{L}$. That is, node λ_{vk} corresponds to node $v \in V$ placed in layer V_k. The pair $(\lambda_{uk}, \lambda_{vl}) \in E_\mathcal{L}$ if and only if $(u, v) \in E$.

Property 1. Let $\mathcal{L}_G^H = (V_\mathcal{L}, E_\mathcal{L})$ be an LDAG of $G = (V, E)$, $H > 0$. If $(u, v) \in E$ then $\varphi(u) > \varphi(v)$ and $\rho(u) > \rho(v)$.

Property 2. Let $\mathcal{L}_G^H = (V_\mathcal{L}, E_\mathcal{L})$ be an LDAG of $G = (V, E)$, $V = n$, $E = m$, $H > 0$. Then $|V_\mathcal{L}| = O(n^2)$ and $|E_\mathcal{L}| = O(mn^2)$.

Definition 9. *The set $F \subseteq V_\mathcal{L}$ partially represents G if 1) $\lambda_{uk} \in F$, $\lambda_{vl} \in F$ and $k \neq l$ imply $u \neq v$ and 2) all the edges of $\mathcal{L}_G^H[F]$, i.e. the subgraph of $\mathcal{L}_G^H$ induced by F, point downwards. We call F a partial layering of G.*

Definition 10. *$F \subseteq V_\mathcal{L}$ represents G if F partially represents G and for each node $v \in V$ there is a node $\lambda_{vk} \in F$ for some k.*

Note that if F partially represents G then $\mathcal{L}_G^H[F]$ is a layered digraph, where each node $v \in V$ is represented by at most one node of $\mathcal{L}_G^H$. The pair $(V_\mathcal{L}, \mathcal{I})$, where $\mathcal{I}$ is the family of all the subsets of $V_\mathcal{L}$ which represent G and induce layered digraphs of width at most W, is an independence system. The problem of finding a layering of G on at most H layers of width not greater than W and minimum total sum of edge spans can be expressed as an optimization problem over the independence system $(V_\mathcal{L}, \mathcal{I})$ as follows: $\min\{C(F) : F \in \mathcal{I}\}$, where $C(F) = \sum_{\lambda \in F} c(\lambda)$ and $c(\lambda)$ is a weight associated with each node $\lambda \in V_\mathcal{L}$.

We need such a weight function C with co-domain $\mathbb{R}$, so that $C(F)$ reaches its minimum at a set F which induces a layered digraph with minimum total sum of edge spans. In order to do this consider the binary variables x_{vk} for each node $v \in V$ and each $k \in L(v)$. Let x_{vk} be 1 if node v has to be placed in layer V_k, and 0 otherwise. Then the expression

$$\sum_{(u,v)\in E} \Big(\sum_{k=\varphi(u)}^{\rho(u)} k x_{uk} - \sum_{k=\varphi(v)}^{\rho(v)} k x_{vk} \Big) = \sum_{v\in V} \sum_{k=\varphi(v)}^{\rho(v)} k\Big(d^+(v) - d^-(v)\Big) x_{vk}$$

represents the sum of edge spans of $\mathcal{L}_G^H[F]$. If we set $c(\lambda_{vk}) = k(d^+(v) - d^-(v))$ then the minimum of the weight function would correspond to a layering with minimum total sum of edge spans. But the minimum can potentially be reached at a partial layering which does not represent G. To ensure that the minimum will be reached at a set that represents G we set $c(\lambda_{vk}) = k(d^+(v) - d^-(v)) - M$, where M is an appropriately large positive number, for instance, $M = H \times |E|$. Then the optimization problem over $(V_\mathcal{L}, \mathcal{I})$ takes the form

$$\min\left\{ \sum_{\lambda_{vk} \in F} [k(d^+(v) - d^-(v)) - M] : F \in \mathcal{I} \right\} .$$

We call the polytope associated with the family of subsets $\mathcal{I}$, the *graph layering polytope* and we denote it by $\mathcal{GLP}(\mathcal{L}_G^H, W)$.

In the next section we summarize the properties of $\mathcal{GLP}(\mathcal{L}_G^H, W)$.

5 Properties of the Layering Polytope

So far we have identified a number of nontrivial generic types of valid inequalities for the layering polytope $\mathcal{GLP}(\mathcal{L}_G^H, W)$, the most interesting of which are listed below.

Assignment inequalities

$$\sum_{k=\varphi(v)}^{\rho(v)} x_{vk} \leq 1 \qquad (1)$$

for all $v \in V$. These inequalities express the fact that if F is a partial layering of the DAG G then each node of $\mathcal{L}_G^H[F]$ corresponds to at most one node of G.

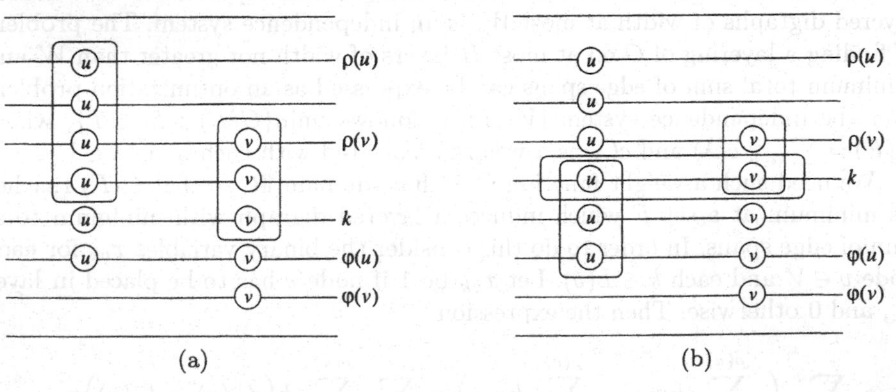

Fig. 5. Illustration of (a) weak RO and (b) strong RO inequalities.

Weak relative-ordering (RO) inequalities

$$\sum_{i=k}^{\rho(v)} x_{vi} - \sum_{i=k+1}^{\rho(u)} x_{ui} \leq 0 \tag{2}$$

with $\max(\varphi(v), \varphi(u) - 1) \leq k \leq \rho(v)$. These inequalities are valid when we have established that node u must be placed above node v. This might be true either because there is an edge (u, v) or because not placing u above v leads to a layering of width greater than W. In this case if $\sum_{i=k}^{\rho(v)} x_{vi} = 0$, then $-\sum_{i=k+1}^{\rho(u)} x_{ui} \leq 0$; and if $\sum_{i=k}^{\rho(v)} x_{vi} = 1$ then node v is placed in one of the layers from V_k to $V_{\rho(v)}$ and in that case, node u is placed above node v, i.e. in one of the layers from V_{k+1} to $V_{\rho(u)}$ and therefore $\sum_{i=k+1}^{\rho(u)} x_{ui} = 1$. (See Figure 5(a).)

Strong relative-ordering (RO) inequalities

$$\sum_{i=\varphi(u)}^{k} x_{ui} + \sum_{i=k}^{\rho(v)} x_{vi} \leq 1 \tag{3}$$

with $\varphi(u) \leq k \leq \rho(v)$. These inequalities are also valid in case we have established that node u must be placed above node v. (See their illustration in Figure 5(b).)

Path-augmented layer (PAL) inequalities

$$\sum_{i=1}^{r} x_{v_i k} + \sum_{i=1}^{m_p} x_{u_i^p k} + \sum_{i=1}^{m_s} x_{u_i^s k} \leq r - 1 \tag{4}$$

where

- nodes $v_1, v_2, \ldots, v_r$ are pairwise independent (without a direct path between any two of them) which cannot be placed together into the same layer without causing the width of the layering to exceed the upper bound W;
- $m_p \geq 0$ and nodes $u_1^p, \ldots, u_{m_p}^p$ form a directed path $(u_{m_p}^p, \ldots, u_1^p)$ where u_1^p is a common immediate predecessor to each of the nodes $v_1, \ldots, v_r$;
- $m_s \geq 0$ and nodes $u_1^s, \ldots, u_{m_s}^s$ form a directed path $(u_1^s, \ldots, u_{m_s}^s)$ where u_1^s is a common immediate successor to each of the nodes $v_1, \ldots, v_r$;
- $k \in \left(\bigcap_{i=1}^{r} L(v_i)\right) \bigcap \left(\bigcap_{i=1}^{m_p} L(u_i^p)\right) \bigcap \left(\bigcap_{i=1}^{m_s} L(u_i^s)\right)$.

Capacity inequalities

$$\sum_{v \in V(k)} w_v x_{vk} + \sum_{(u,v) \in E} w_e \Big(\sum_{l \in L(u),\, l > k} x_{ul} - \sum_{l \in L(v),\, l \geq k} x_{vl} \Big) \leq W \tag{5}$$

where $1 \leq k \leq H$ and w_e is the width of the dummy nodes placed in edge e. In this way we are able to localize dummy node widths for each edge of a DAG. Inequalities (5), called *capacity constraints*, restrict the width of each layer (including the dummy nodes) to be less than or equal to W: the first sum on the left hand side represents the contribution of the real nodes to layer V_k while the second sum is the contribution of the dummy nodes. The value of the expression inside the brackets in (5) is 1 if and only if the edge (u, v) spans layer V_k. Otherwise it is 0.

The two lemmas[2] below describe some trivial properties of the layering polytope $\mathcal{GLP}(\mathcal{L}_G^H, W)$.

Lemma 1. *The dimension of the layering polytope of* $G = (V, E)$ *is equal to* $\sum_{u \in V}(\rho(u) - \varphi(u) + 1)$, *so it is full dimensional. For each node* λ_{vk} *of of* $\mathcal{L}_G^H$ *the inequalities* $x_{vk} \geq 0$ *define facets of* $\mathcal{GLP}(\mathcal{L}_G^H, W)$.

Lemma 2. *The weak RO inequalities (2) are not facet defining for the layering polytope.*

The following two theorems describe facet-defining properties of the assignment inequalities and the strong RO inequalities.

Theorem 1. *The assignment inequalities (1) are facet defining for the layering polytope.*

Theorem 2. *Let* $G = (V, E)$ *be a DAG. The strong RO (relative-ordering) inequalities (3) are facet-defining for* $\mathcal{GLP}(\mathcal{L}_G^H, W)$ *in the following two cases.*

- $(u, v) \in E$ *is a non-transitive edge.*
- u *and* v *are two independent nodes (without a directed path between them) and* u *has to placed above* v *in order for the layering to have width at most* W.

Note that the strong RO inequalities for all the non-transitive edges are sufficient to ensure that all the edges of the layered digraph point downwards.

6 Experimental Results

We have constructed an ILP formulation, called `ULair`, which models the layering problem *WHS-Layering* employing the assignment inequalities (1), the strong RO inequalities (3) and the capacity inequalities (5). `ULair` is solved currently by running an ILP solver directly from CPLEX 7.0 on an Intel Pentium III/Red Hat Linux 6.2 platform. As we have shown in Section 3 Gansner outperforms the other present layering algorithms aesthetically. Thus we compared `ULair`'s performance to Gansner's. We ran Gansner and `ULair` with the same 5911 DAGs,

[2] Proofs of all lemmas and theorems in this section are available upon request.

described in Section 3.2, measuring a variety of parameters of the layering solutions: MLB, ALB, maximum edge density between two adjacent layers, aspect ratio of the width and the height of the layerings as well as their area.

`ULair` has two input parameters: an upper bound on the height of the layering, H, and an upper bound on the width of the layering, W. We conducted two groups of experiments over the 5911 example DAGs.

First group of experiments. Firstly we ran Gansner on all the 5911 DAGs. Suppose Gansner gives a solution with height H_G and width W_G and $W_G \geq H_G$ (alternatively $H_G > W_G$) for a DAG G. Then we took $W = W_G$ and $H = W/AR$ (alternatively $H = H_G$ and $W = H/AR$), where AR is the desired aspect ratio of the width and the height, as input parameters to `ULair`. We have tried two values of AR: the golden mean 1.618 and 1.0. In the case $AR = 1.618$ `ULair` reported solutions for 84.88% (5017 DAGs) of all the 5911 DAGs and in the case $AR = 1.0$ `ULair` reported solutions for all the DAGs except one. The values of all the parameters of the layerings we watched are slightly better than the parameters of the solutions given by Gansner, which shows that `ULair` gives the same quality of solutions when the dimensions are approximately the same as the dimensions of Gansner's solutions. In Figure 6 we present the maximum edge density between adjacent layers in Gansner's and `ULair` solutions. The edge density is normalized (i.e. divided by the total number of edges). In this group of experiments, the maximum edge density is the parameter on which Gansner and `ULair` differ most. The better values of edge density in `ULair`'s layerings are encouraging, because they suggest a more even distribution of the graph over the drawing area and perhaps to a fewer number of edge crossings at a later stage of the Sugiyama algorithm. This is a subject of further research.

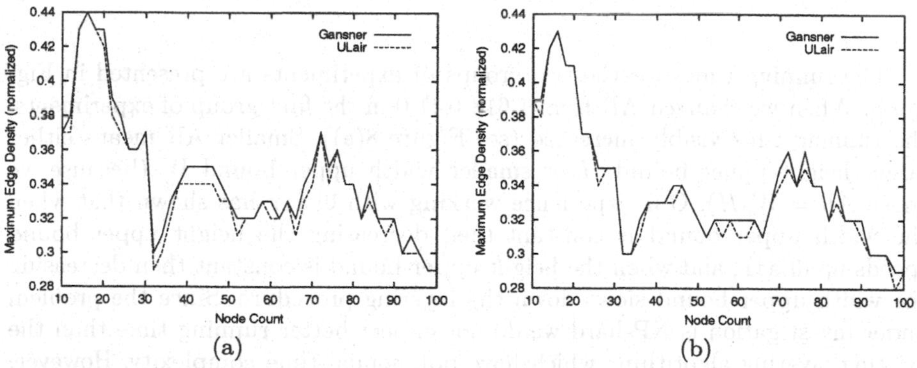

Fig. 6. First group of experiments: Maximum edge density when (a) $AR = 1.618$ and (b) $AR = 1.0$.

Second group of experiments. We then ran Gansner and `ULair` independently for the 5911 example DAGs. The input parameters of `ULair` were set according to a certain aspect ratio AR of the width and the height of a layering, assuming that the nodes are distributed evenly over the layers and the expected number of dummy nodes is equal to the number of real nodes. That is, H is the larger of $\sqrt{2n/AR}$ and the longest path in the DAG, and $W = H \times AR$. We performed these experiments for $AR = 1.618$ and `ULair` reported solutions for 70.39% (4161 DAGs) of all the 5911 DAGs. Here `ULair` performed better than in the previous group. Figure 7(a) compares the area and Figure 7(b) compares the edge density of the layerings of Gansner and `ULair`.

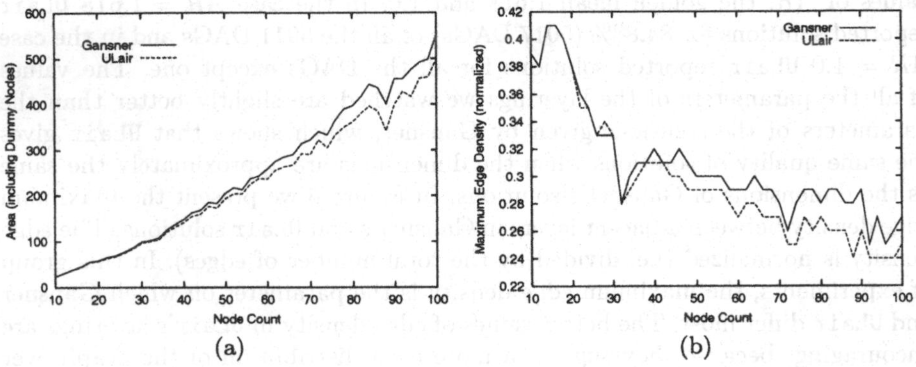

Fig. 7. Second group of experiments: (a) area and (b) maximum edge density of the layering solutions ($AR = 1.618$).

The running times for the two groups of experiments are presented in Figure 8. When we changed AR from 1.618 to 1.0 in the first group of experiments, the running time visibly increased (see Figure 8(a)). Smaller AR means either larger height upper bound H or smaller width upper bound W (because we chose $AR = W/H$). Our experience working with `ULair` has shown that when the width upper bound is constant then decreasing the height upper bound speeds up `ULair`, and when the height upper bound is constant then decreasing the width upper bound slows down the layering procedure. Since the problem under investigation is NP-hard we do not expect better running time than the existing layering algorithms which have polynomial time complexity. However, we believe that `ULair` – as it presently stands – is a better alternative for DAGs having up to 100 nodes (and possibly more), as well as in any case when the time for drawing is less important than the quality of the final picture.

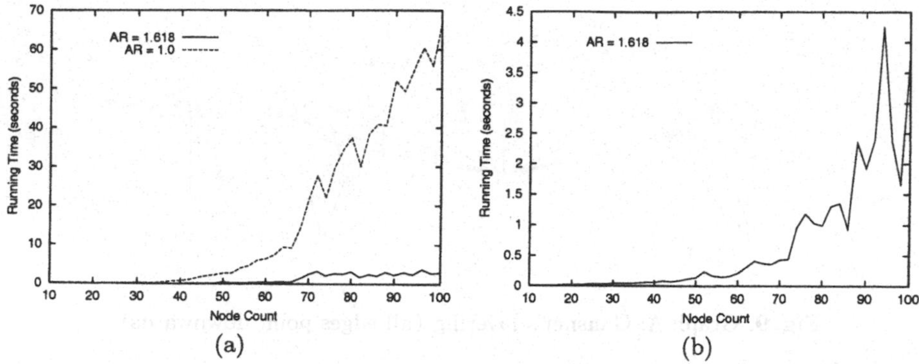

Fig. 8. Running time of ULair for (a) the first group and (b) the second group of experiments.

7 Conclusions

The experimental results show that ULair combines the positive attributes of the existing layering algorithms, namely, Longest Path, Coffman-Graham and Gansner. The width bound we specify when solving ULair represents the width of the final drawing much more accurately than the width bound used in the Coffman-Graham algorithm. The variety of parameters we can specify – width and height bounds, and the widths of nodes and edges (dummy nodes) – makes possible a family of layering solutions, permitting the user to choose the best one of them. This also makes ULair a good tool for studying from experimental point of view the relationship between various aesthetic parameters like the dimensions of a layering, the number of dummy nodes and the number of edge crossings in the final drawing.

As a final example we show two layerings of one of the small connected components of Graph A from the Graph Drawing Contest 2000 [2]. Graph A represents the structure of a software system from programmer's point view and contains large node labels. The layering in Figure 9 is a result of Gansner and the layering in Figure 10 is one of the alternative solutions given by a ULair when we draw the DAG putting different upper bounds on its width.

Our further work will be a more detailed study of the structure of the layering polytope $\mathcal{GLP}(\mathcal{L}_G^H, W)$ in terms of valid inequalities and facets in order to develop a branch-and-cut algorithm for solving ULair, and comparing its performance to the present solution method. This, we believe, will make it possible to accurately layer graphs well in excess of the 100-node examples that we have been working with.

Acknowledgments. We would like to thank Petra Mutzel and Gunnar Klau for the graph database they kindly let us use for performing our experiments.

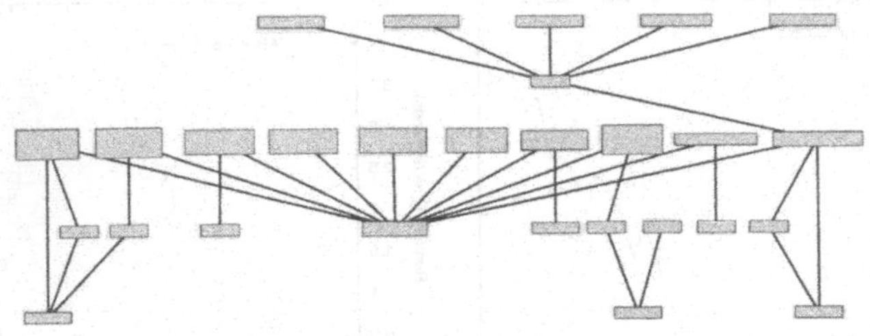

Fig. 9. Graph A: Gansner's layering (all edges point downwards).

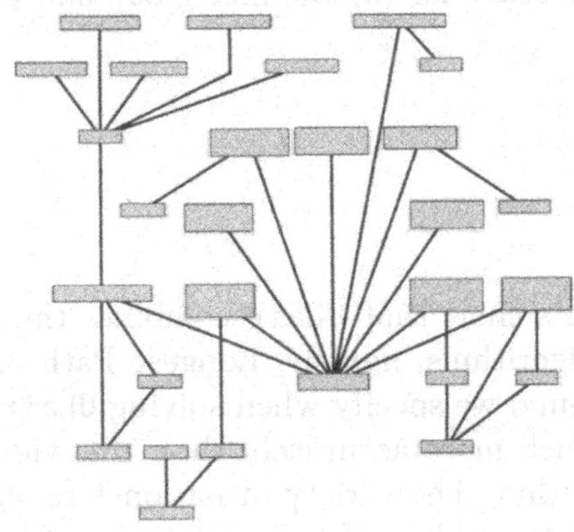

Fig. 10. Graph A: `ULair`'s layering (all edges point downwards).

References

1. G. Di Battista, A. Garg, G. Liotta, R. Tamassia, E. Tassinari, and F. Vargiu. An experimental comparison of four graph drawing algorithms. *Computational Geometry: Theory and Applications*, (7):303 – 316, 1997.
2. Fr. Brandenburg, Ul. Brandes, M. Himsolt, and M. Raitner. Graph-drawing contest report. In Joe Marks, editor, *Graph Drawing: Proceedings of 8th International Symposium, GD 2000*, pages 410 – 418. Springer-Verlag, 2000.
3. E. G. Coffman and R. L. Graham. Optimal scheduling for two processor systems. *Acta Informatica*, 1:200–213, 1972.
4. P. Eades and K. Sugiyama. How to draw a directed graph. *Journal of Information Processing*, 13(4):424–437, 1990.
5. E. Gansner, E. Koutsofios, S. North, and K. Vo. A technique for drawing directed graphs. *IEEE Transactions on Software Engineering*, 19(3):214–229, March 1993.
6. K. Sugiyama, S. Tagawa, and M. Toda. Methods for visual understanding of hierarchical system structures. *IEEE Transaction on Systems, Man, and Cybernetics*, SMC-11(2):109–125, February 1981.

Fast and Simple Horizontal Coordinate Assignment

Ulrik Brandes and Boris Köpf

Department of Computer & Information Science,
University of Konstanz, Box D 188, 78457 Konstanz, Germany
{Ulrik.Brandes|Boris.Koepf}@uni-konstanz.de

Abstract. We present a simple, linear-time algorithm to determine horizontal coordinates in layered layouts subject to a given ordering within each layer. The algorithm is easy to implement and compares well with existing approaches in terms of assignment quality.

1 Introduction

In layered graph layout, vertices are placed on parallel lines corresponding to an ordered partition into layers. W.l.o.g. these lines are horizontal, and we assume a polyline representation in which edges may bend where they intersect a layer. The standard approach for layered graph layout consists of three phases [18]: layer assignment (vertices are assigned to layers), crossing reduction (vertices and bend points are permuted) and coordinate assignment (coordinates are assigned to vertices and bend points).

The third phase is usually constrained to preserve the ordering determined in the second phase, and to introduce a minimum separation between layers, and between vertices and bend points within a layer. Criteria for readable layout include length and slope of edges, straightness of long edges, and balancing of edges incident to the same vertex. Note that, if an application does not prescribe vertical coordinates, it is easy to determine layer distances that bring about a minimum edge slope. We therefore confine ourselves to horizontal coordinate assignment.

Previous approaches for horizontal coordinate assignment either optimize a constrained objective function of coordinate differences, iteratively improve a candidate layout using one or more of various heuristics, or do both [18,10,6, 17,8,14,7,15,16,5]. A recently introduced method [3] successfully complements some of the above heuristics with new ideas to determine visually compelling assignments in time $\mathcal{O}(N \log^2 N)$, where N is the total number of vertices, bend points and edges. We present a much simpler algorithm that runs in time $\mathcal{O}(N)$ without compromising on layout quality.

In Sect. 2, we define some terminology to state the horizontal coordinate assignment problem formally. Several important ideas introduced in previous approaches are reviewed in Sect. 3, and our new method is described in Sect. 4.

P. Mutzel, M. Jünger, and S. Leipert (Eds.): GD 2001, LNCS 2265, pp. 31–44, 2002.
© Springer-Verlag Berlin Heidelberg 2002

2 Preliminaries

A *layering* $L = (L_1, \ldots, L_h)$ of a graph $G = (V, E)$ is an ordered partition of V into non-empty *layers* L_i such that adjacent vertices are in different layers. Let $L(v) = i$ if $v \in L_i$. Oblivious whether G is directed or undirected, an edge incident to $u, v \in V$ is denoted by (u, v) if $L(u) < L(v)$. An edge (u, v) is *short* if $L(v) - L(u) = 1$, otherwise it is *long* and *spans* layers $L_{L(u)+1}, \ldots, L_{L(v)-1}$. Let $N_v^- = \{u : (u,v) \in E\}$ ($N_v^+ = \{w : (v,w) \in E\}$) denote the *upper* (*lower*) *neighbors* and $d_v^- = |N_v^-|$ ($d_v^+ = |N_v^+|$) the *upper* (*lower*) *degree* of $v \in V$.

A layering is *proper*, if there are no long edges. Any layering can be turned into a proper one by subdividing long edges (u, v) with *dummy vertices* $b_i \in L_i$, $i = L(u) + 1, \ldots, L(v) - 1$, that represent potential edge bends.

A *layered graph* $G = (V \cup B, E; L)$ is a graph G together with a proper layering L, where the vertices are either *original* vertices in V or dummy vertices (having upper and lower degree one) in B. The edges of a layered graph are often called (edge) *segments*, and segments between two dummy vertices are called *inner* segments. A maximal path in G whose internal vertices are all dummy vertices is also called a long edge. Let $N = |V \cup B| + |E|$ denote the size of a layered graph.

An *ordering* of a layered graph is a partial order $\prec$ of $V \cup B$ such that either $u \prec v$ or $v \prec u$ if and only if $L(u) = L(v)$. We sometimes denote the vertices by $v_j^{(i)}$, where $L_i = \{v_1^{(i)}, \ldots, v_{|L_i|}^{(i)}\}$ with $v_1^{(i)} \prec \ldots \prec v_{|L_i|}^{(i)}$. The *position* $\mathrm{pos}[v_j^{(i)}] = j$ and the *predecessor* of $v_j^{(i)}$ with $j > 1$ is $\mathrm{pred}[v_j^{(i)}] = v_{j-1}^{(i)}$. An edge segment (u, v) is said to *cross* an edge segment (u', v'), if $u, u' \in L_i$, $v, v' \in L_{i+1}$, and either $u \prec u'$ and $v' \prec v$, or $u' \prec u$ and $v \prec v'$.

Given a layered graph together with an ordering, the *horizontal coordinate assignment problem* is to assign coordinates to the vertices such that the ordering and a *minimum separation* $\delta > 0$ are respected, i.e.

Horizontal Coordinate Assignment Problem: For a layered graph $G = (V \cup B, E; L)$ with ordering $\prec$, find real values $x(v)$, $v \in V \cup B$, such that $x(u) + \delta \leq x(v)$ if $u \prec v$ (*minimum separation constraint*).

A horizontal coordinate assignment should additionally satisfy two main criteria which appear to govern the readability of a layered drawing with given ordering:

- edges should have small length,
- vertex positions should be balanced between upper and lower neighbors,
- and long edges should be as straight as possible.

3 Previous Work

Apparently, an early algorithm to determine horizontal coordinates is used in a system for control flow diagrams, but we were unable to find a detailed description [11]. In the following, we summarize more recent approaches, since they nicely illustrate the rationale behind our approach. Horizontal coordinate assignment is also discussed in [4, Section 9.3] and [1].

Optimization approaches. In their seminal paper [18], Sugiyama et al. present a quadratic program for the horizontal coordinate assignment problem. The objective function is a weighted sum of terms

$$\sum_{(u,v)\in E} (x(u) - x(v))^2 \tag{1}$$

and

$$\sum_{v\in V}\left(x(v) - \sum_{u\in N_v^-}\frac{x(u)}{d_v^-}\right)^2 + \sum_{v\in V}\left(x(v) - \sum_{w\in N_v^+}\frac{x(w)}{d_v^+}\right)^2. \tag{2}$$

The first term penalizes large edge lengths, while the second serves to balance the influence of upper and lower neighbors. The objective function is subject to the minimum separation constraint and, to enforce vertical inner segments, $x(u) = x(v)$ if $(u,v) \in E$ for $u,v \in B$. Note that the quadratic program is infeasible if $\prec$ implies a crossing between inner segments of long edges.

A related approach is introduced in [5]. A necessary condition for an optimal solution of (1) is that all partial derivatives are zero. Equivalently, every vertex is placed at the mean coordinate of its neighbors. The system of linear equations thus obtained is modified so that upper and lower neighbors contribute equally,

$$x(v) = \frac{\sum_{u\in N_v^-}\frac{x(u)}{d_v^-} + \sum_{w\in N_v^+}\frac{x(w)}{d_v^+}}{2}. \tag{3}$$

If the coordinates of vertices in the top and bottom layer are fixed, say equidistantly, to exclude the trivial assignment of all-equal coordinates, the system has a unique solution that can be approximated quickly using Gauß-Seidel iteration. Though a given ordering may not be preserved, the resulting layouts have interesting properties with respect to planarity and symmetry [5].

Another quadratic program is discussed in [4, p. 293f]. For every directed path $v_1, \ldots, v_k$ where $v_1, v_k \in V$ and $v_2, \ldots, v_{k-1} \in B$, terms

$$\left(x(v_i) - x(v_1) - \frac{i-1}{k-1}(x(v_k) - x(v_1))\right)^2,$$

$i = 2, \ldots, k-1$, are introduced to make long edges straight, and only the minimum separation constraints are enforced.

Some popular implementations [9,13] use a piecewise linear objective function introduced in [8] reading

$$\sum_{e=(u,v)\in E} \omega(e) \cdot |x(u) - x(v)|, \tag{4}$$

subject to the minimum separation constraint. The weight $\omega(e)$ reflects the importance of drawing edge e vertically, and weights of 1, 2, and 8 are used for edges incident to 0, 1, and 2 dummy vertices, respectively. The corresponding integer optimization problem is solved to optimality using a clever transformation and the network simplex method. This objective function is the main justification for our heuristic.

Iterative heuristics. After an initial, say leftmost, placement respecting the ordering and minimum separation constraint, several heuristics can be applied to improve the assignment with respect to the criteria stated in Sect. 2.

The method proposed in [18] sweeps up and down the layering. While sweeping down, vertices in each layer are considered in order of non-increasing upper degree. Each vertex is shifted toward the average coordinate of its upper neighbors, but without violating the minimum separation constraint. To increase the available space, lower priority vertices may be shifted together with the current one. The reverse sweep is carried out symmetrically. A post-processing to straighten long edges without moving original vertices is applied in [7].

In addition to the average coordinates of upper or lower neighbors, the average of all neighbors is considered in [14,16]. Instead of a priority order based on degree, violation of the minimum separation constraint is avoided by grouping vertices and averaging over their independent movements. In [15], this kind of grouping is extended to paths of dummy vertices, so as to constrain inner segments of long edges to be vertical.

Several heuristic improvements are proposed in [8]. They are applied iteratively and the best assignment with respect to (4) is kept after each iteration. One of these heuristics straightens long edges in apparently the same way as [7]. Another one is similar to the method of [18], but, since the objective function consists of absolute instead of squared differences, uses the median instead of the average of the neighbor's coordinates.

A fast non-iterative heuristic is presented in [3] and implemented in the AGD library [13]. Similar to [15], the dummy vertices of each long edge are grouped, and leftmost and rightmost top-to-bottom placements of all vertices subject to this grouping are determined. Dummy vertices are fixed at the mean of their two positions thus obtained. In a rather involved second phase, original vertices are placed so as to minimize the length of some short edges as measured by (4) without changing positions of dummy vertices.

Since local improvements, such as straightening edges after y-coordinates have been determined [2], can be applied to layouts obtained from any method, all examples in this paper have been prepared without such postprocessing to better facilitate comparison.

4 The Algorithm

We present a heuristic approach for the horizontal coordinate assignment problem that guarantees vertical inner segments, and yields small edge lengths and a fair balance with respect to upper and lower neighbors.

We essentially follow the inherent objective of (4), i.e. we define the length of an edge segment (u, v) by $|x(u) - x(v)|$. Recall that $\sum_{i=1}^{k} |x - x_i|$ is minimized if x is the median of the x_i. Therefore, we align each vertex vertically with its median neighbor wherever possible. To achieve a balance between upper and lower neighbors similar to the one that motivated (2) and (3), their medians are considered separately and the results are combined.

The algorithm consists of three basic steps. The first two steps are carried out four times. In the first step, referred to as vertical alignment, we try to align each vertex with either its median upper or its median lower neighbor, and we resolve alignment conflicts (of type 0) either in a leftmost or a rightmost fashion. We thus obtain one vertical alignment for each combination of upward and downward alignment with leftmost and rightmost conflict resolution. In the second step, called horizontal compaction, aligned vertices are constrained to obtain the same horizontal coordinate, and all vertices are placed as close as possible to the next vertex in the preferred horizontal direction of the alignment. Finally, the four assignments thus obtained are combined to balance their biases.

Details on vertical alignment are given in the following two sections, though only for the case of upward alignment to the left. The other three cases are symmetric. Balanced combination of assignments is described in Sect. 4.3. thus obtained.

4.1 Vertical Alignment

We want to align each vertex with a median upper neighbor. Two alignments are conflicting if their corresponding edge segments cross or share a vertex. We classify conflicts according to the number of inner segments involved.

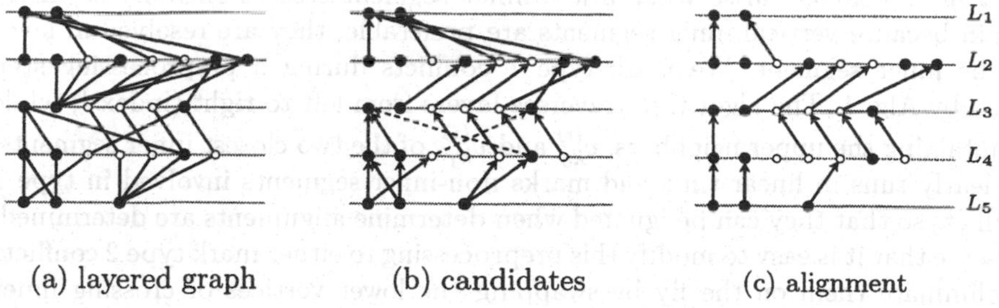

Fig. 1. Leftmost alignment with median upper neighbors (dummy vertices are outlined, non-inner segments involved in type 1 conflicts are dashed)

Type 2 conflicts correspond to a pair of crossing inner segments and prevent at least one of them from being vertical. One or both of the involved segments can therefore be marked as non-vertical and ignored when alignments are determined. Since vertical inner segments appear to improve readability dramatically, however, we assume that type 2 conflicts have been avoided in the crossing reduction phase (as, e.g., in [15]). Alternatively, one can eliminate type 2 conflicts in a preprocessing step prior to the horizontal coordinate assignment, e.g. by swapping the two lower vertices involved until the crossing is no longer between two inner segments [3,2]. Note that this changes the ordering, and potentially the number of crossings. If the ordering is more important than vertical inner segments, the original ordering can be restored in the final layout. Finally, type 2 conflicts can also be treated as described below for type 0 conflicts.

Alg. 1: Preprocessing (mark type 1 conflicts)

```
for i ← 2,...,h − 2 do
    k_0 ← 0;  l ← 1;
    for l_1 ← 1,...,|L_{i+1}| do
        if l_1 = |L_{i+1}| or v_{l_1}^{(i+1)} incident to inner segment between L_{i+1} and L_i
        then
            k_1 ← |L_i|;
            if v_{l_1}^{(i+1)} incident to inner segment between L_{i+1} and L_i then
                k_1 ← pos[upper neighbor of v_{l_1}^{(i+1)}];
            while l ≤ l_1 do
                foreach upper neighbor v_k^{(i)} of v_l^{(i+1)} do
                    if k < k_0 or k > k_1 then mark segment (v_k^{(i)}, v_l^{(i+1)});
                l ← l + 1;
            k_0 ← k_1;
```

Type 1 conflicts arise when a non-inner segment crosses an inner segment. Again because vertical inner segments are preferable, they are resolved in favor of the inner segment. We mark type 1 conflicts during a preprocessing step given by Alg. 1. The algorithm traverses layers from left to right (index l) while maintaining the upper neighbors, $v_{k_0}^{(i)}$ and $v_{k_1}^{(i)}$, of the two closest inner segments. It clearly runs in linear time and marks non-inner segments involved in type 1 conflicts so that they can be ignored when determine alignments are determined. Observe that it is easy to modify this preprocessing to either mark type 2 conflicts or eliminate them on the fly by swapping the lower vertices of crossing inner segments.

Finally, a *type 0* conflict corresponds to a pair of non-inner segments that either cross or share a vertex. We say that a segment (u, v) is *left* of a segment (u', v'), if either $v \prec v'$, or $v = v'$ and $u \prec u'$. Type 0 conflicts are resolved greedily in a leftmost fashion, i.e. in every layer we process the vertices from left to right and for each vertex we consider its median upper neighbor (its left and right median upper neighbor, in this order, if there are two). The pair is aligned, if no conflicting alignment is left of this one. The resulting bias is mediated by the fact that the symmetric bias is applied in one of the other three assignments.

By executing Alg. 2 we obtain a leftmost alignment with upper neighbors. A maximal set of vertically aligned vertices is called a *block*, and we define the *root* of a block to be its topmost vertex. Observe that blocks are represented

Alg. 2: Vertical alignment

```
initialize root[v] ← v, v ∈ V ∪ B;
initialize align[v] ← v, v ∈ V ∪ B;
for i ← 1,...,h do
    r ← 0;
    for k ← 1,...,|L_i| do
        if v_k^(i) has upper neighbors u_1 ≺ ... ≺ u_d with d > 0 then
            for m ← ⌊(d+1)/2⌋, ⌈(d+1)/2⌉ do
                if align[v_k^(i)] = v_k^(i) then
                    if (u_m, v_k^(i)) not marked and r < pos[u_m] then
                        align[u_m] ← v_k^(i);
                        root[v_k^(i)] ← root[u_m];
                        align[v_k^(i)] ← root[v_k^(i)];
                        r = pos[u_m];
```

by cyclically linked lists, where each vertex has a reference to its lower aligned neighbor, and the lowest vertex refers back to the topmost. Moreover, each vertex has an additional reference to the root of its block. These data structures are sufficient for the actual placement described in the next section.

4.2 Horizontal Compaction

In the second step of our algorithm, a horizontal coordinate assignment is determined subject to a vertical alignment, i.e. all vertices of a block are assigned the coordinate of the root.

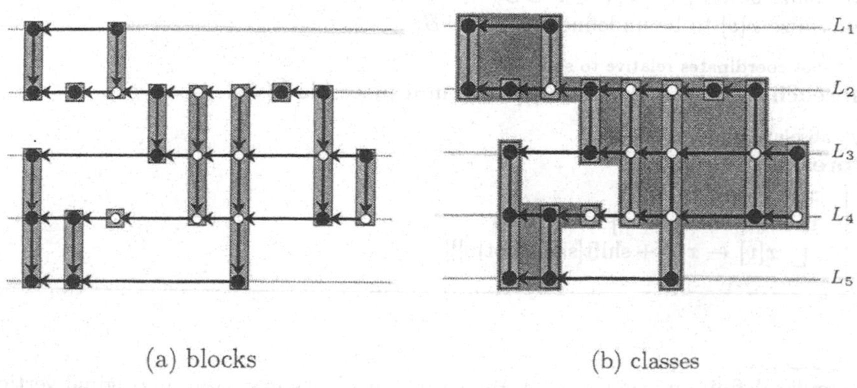

Fig. 2. Blocks and classes with respect to the alignment of Fig. 1(c)

Consider the *block graph* obtained by introducing directed edges between each vertex and its predecessor (if any) and contracting blocks into single vertices. See Fig. 2(a) and note that the root of a block that is a sink in this acyclic graph is always a leftmost vertex in its layer, so that there is at most one vertex of this kind in each layer.

We partition the block graph into *classes*. The class of a block is defined by that reachable sink which has the topmost root.[1] Within each class, we apply a longest path layering, i.e. the relative coordinate of a block with respect to the defining sink is recursively determined to be the maximum coordinate of the preceding blocks in the same class, plus minimum separation.

Alg. 3: Horizontal compaction

```
function place_block(v)
begin
    if x[v] undefined then
        x[v] ← 0;   w ← v;
        repeat
            if pos[w] > 1 then
                u ← root[pred[w]];
                place_block(u);
                if sink[v] = v then sink[v] = sink[u];
                if sink[v] ≠ sink[u] then
                    shift[sink[u]] ← min{shift[sink[u]], x[v] − x[u] − δ};
                else
                    x[v] ← max{x[v], x[u] + δ};
            w ← align[w];
        until w = v;
end

initialize sink[v] ← v, v ∈ V ∪ B;
initialize shift[v] ← ∞, v ∈ V ∪ B;
initialize x[v] to be undefined, v ∈ V ∪ B;

// root coordinates relative to sink
foreach v ∈ V ∪ B do if root[v] = v then place_block(v);

// absolute coordinates
foreach v ∈ V ∪ B do
    x[v] ← x[root[v]];
    if shift[sink[root[v]]] < ∞ then
        x[v] ← x[v] + shift[sink[root[v]]]
```

[1] A similar definition is given in [3]. However, our blocks may contain original vertices and non-inner segments, and therefore give rise to bigger classes. Note that, within a class, coordinates are easy to determine.

For each class, from top to bottom, we then compute the absolute coordinates of its members by placing the class with minimum separation from previously placed classes.

The entire compaction step is implemented in Alg. 3. The first iteration invokes a recursive version of a technique known as longest path layering to determine the relative coordinates of all roots with respect to the sink of their corresponding classes. Another variant of this algorithm is used to determine visibility representations of planar layered graphs [12]. At the same time, we determine for each sink the minimum distance of a vertex in its class from its neighboring vertex in a class with a higher sink. The second iteration distributes this information from the roots to all vertices to obtain the absolute coordinates.

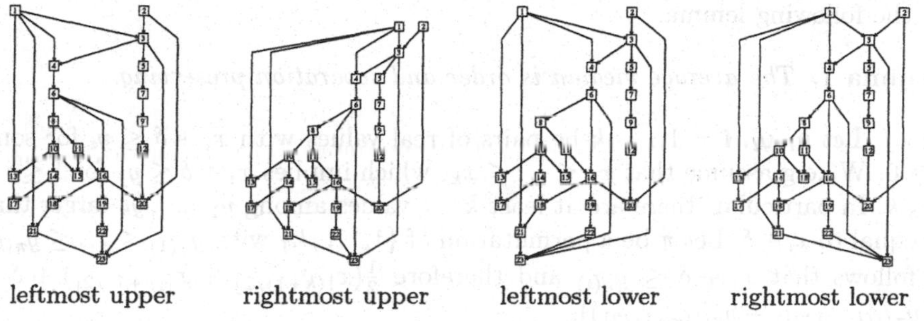

Fig. 3. Biased assignments resulting from leftmost/rightmost alignments with median upper/lower neighbors for the running example of [2,3]

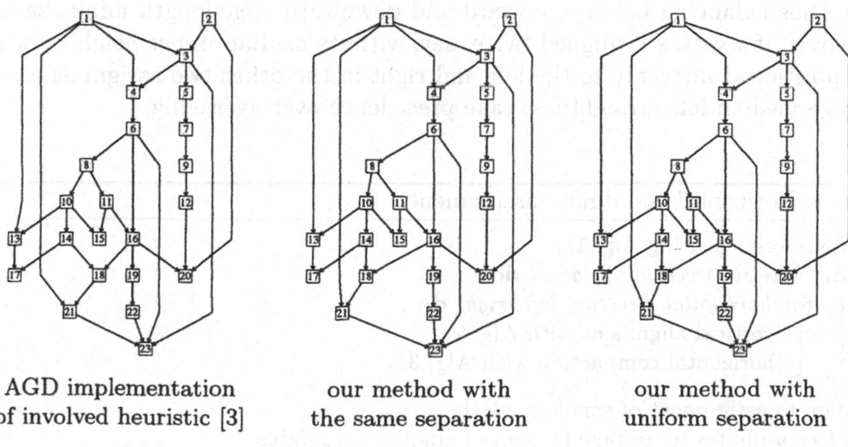

Fig. 4. Final assignments compared

4.3 Balancing

Although the coordinate assignments computed in the first two steps result in vertical inner segments and in general display short edge lengths, they are governed by their specific choices of a vertical alignment direction and horizontal preference. We even out their directional tendencies by combining them into a balanced horizontal coordinate assignment.

First, we align the layouts to the one with smallest width by shifting the two assignments for the leftmost (rightmost) alignments so that their minimum (maximum) coordinate agrees with the minimum coordinate in the smallest-width assignment. Out of the four resulting candidate coordinates we fix, for each vertex separately, the final coordinate to be the *average median*, which for k values $x_1 \leq \ldots \leq x_k$ equals $(x_{\lfloor (k+1)/2 \rfloor} + x_{\lceil (k+1)/2 \rceil})/2$. This choice is justified by the following lemma.

Lemma 1. *The average median is order and separation preserving.*

Proof. Let x_i, y_i, $i = 1, \ldots, k$ be pairs of real values with $x_i + \delta \leq y_i$ for some $\delta \geq 0$. W.l.o.g assume that $x_1 \leq \ldots \leq x_k$, which implies $x_i + \delta \leq y_j$ for $1 \leq i \leq j \leq k$. In particular, there are at least $k - i$ values among $y_1, \ldots, y_k$ larger than or equal to $x_i + \delta$. Let π be a permutation of $\{1, \ldots, k\}$ with $y_{\pi(1)} \leq \ldots \leq y_{\pi(k)}$. It follows that $x_i + \delta \leq y_{\pi(i)}$ and therefore $\frac{1}{2}(x_{\lfloor (k+1)/2 \rfloor} + x_{\lceil (k+1)/2 \rceil}) + \delta \leq \frac{1}{2}(y_{\pi(\lfloor (k+1)/2 \rfloor)} + y_{\pi(\lceil (k+1)/2 \rceil)})$. □

We have chosen the average median over the mean because it appears to suit better the way that biased assignments are determined. Extreme coordinates specific for a particular combination of vertical and horizontal directions of preference are dropped, and the two closer ones are averaged. In ideal situations, vertices end up at the average coordinate of their median upper and lower neighbors, thus balancing between upward and downward edge length minimization. Moreover, if a vertex is aligned twice, say, with its median upper neighbor while it is positioned unevenly to the left and right in the other two assignments, the average median lets straightness take precedence over averaging.

Alg. 4: Horizontal coordinate assignment

```
preprocessing using Alg. 1;
for vertical direction up, down do
    for horizontal direction left, right do
        vertical alignment with Alg. 2;
        horizontal compaction with Alg. 3;
align to assignment of smallest width;
set coordinates to average median of aligned candidates;
```

Our method for horizontal coordinate assignment is summarized in Alg. 4.

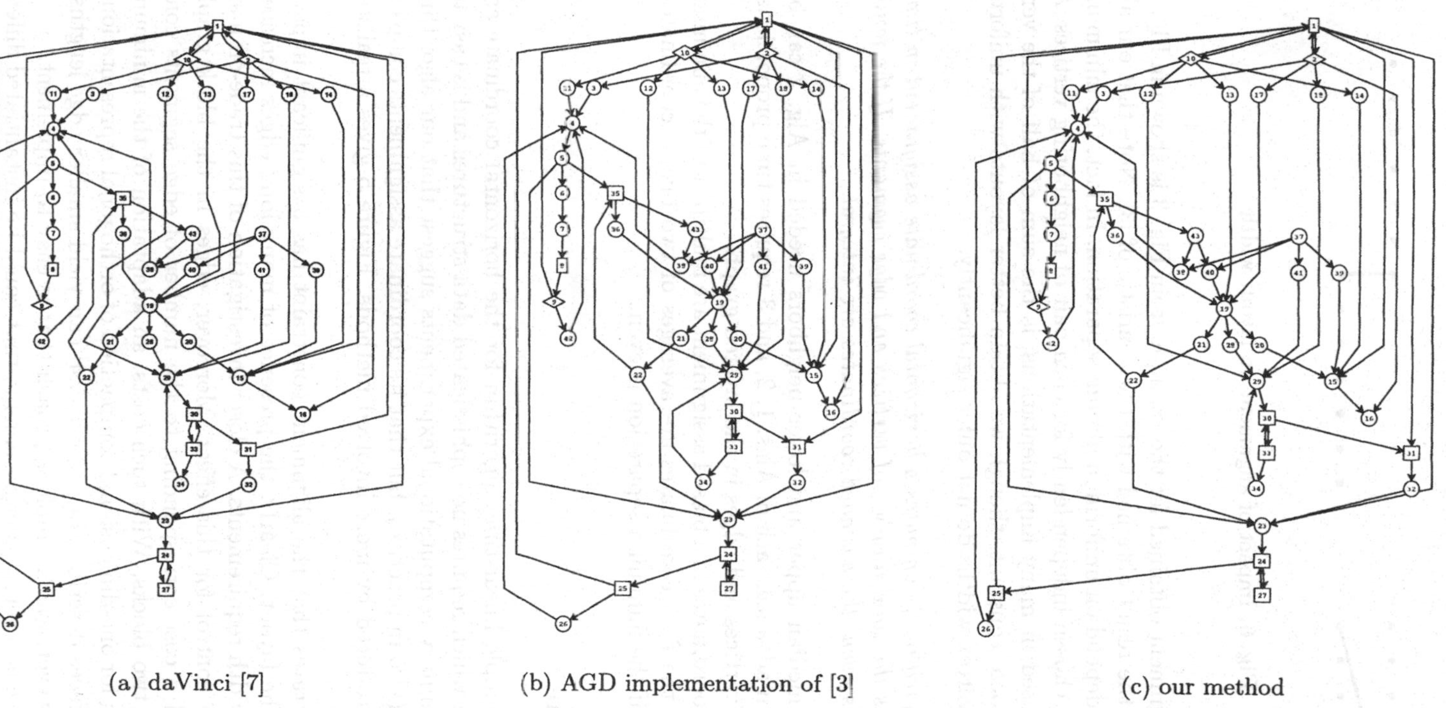

(a) daVinci [7]

(b) AGD implementation of [3]

(c) our method

Fig. 5. Example graph used in several publications on layered layout [7, p. 5]. The layout of the daVinci system, where horizontal coordinates are determined using the iterative heuristic of [18] with additional edge straightening, is used as a reference to fix layers, ordering, and y-coordinates. The other two layouts are computed with the minimum separation values of [3] and without postprocessing

Fig. 6. Impact of alignment on layout width

The final assignment obtained for the example from Fig. 3 is shown in Fig. 4 and compared to the result obtained with the heuristic of [3]. Note that our algorithm does not depend on uniform minimum separation. In fact, the minimum separation can be chosen independently for each pair of neighboring vertices. A common choice, used in many implementations, is the sum of half of the vertex widths plus some constant, though we obtain better layouts with uniform separation when vertex widths do not differ significantly.

Theorem 1. *Algorithm 4 computes a horizontal coordinate assignment in time* $\mathcal{O}(N)$, *where* N *is the total number of vertices and edge segments. If the minimum separation is even, the assigned coordinates are integral.*

Proof. Since the median upper and lower neighbors needed in Alg. 2 can be determined once in advance, each of Algs. 1, 2, and 3 requires time proportional to the number of vertices and edges in the layered graph.

Note that all coordinates in a biased assignment are multiples of the minimum separation. Since the final coordinates are averages of two biased coordinates, they are integral if the minimum separation is even. □

5 Discussion

We presented a simple linear-time algorithm for the horizontal coordinate assignment problem which requires no sophisticated data structures and is easy to implement. Preliminary computational experiments suggest that our algorithm is not only faster (also in practice), but that its coordinate assignments compare well with those produced by more involved methods. Figure 5 gives a realistic example.

Figure 6 illustrates that the alignment constraint has a significant impact on the width of the layout. Clearly, the presence of many long edges increases chances of large width requirements. A closer investigation of this trade-off may suggest means to control for this effect. Moreover, edges in the block graph could be assigned a cost corresponding to the number of edge segments connecting the same two blocks. With such costs, an adaptation of the minimum cost flow approach for one-dimensional compaction of orthogonal representations with rectangular faces described in [4, Sect. 5.4] can yield smaller edge lengths. Our aim here, however, was a simple approach that is easy to implement.

Our algorithm has several generic elements that could be instantiated differently. While it has already been mentioned that crossing inner segments can be dealt with in many ways, one could alter also the alignment (e.g., to only consider one of two medians, or to break conflicts in favor of high degree vertices), the method of compaction (e.g., using adaptive schemes of separation), and the final combination (e.g., aligning the central axes and fixing the average).

References

1. Oliver Bastert and Christian Matuszewski. Layered drawings of digraphs. In Michael Kaufmann and Dorothea Wagner, editors, *Drawing Graphs: Methods and Models*, volume 2025 of *Lecture Notes in Computer Science*, pages 104–139. Springer, 2001.
2. Christoph Buchheim, Michael Jünger, and Sebastian Leipert. A fast layout algorithm for k-level graphs. Technical Report 99-368, Department of Economics and Computer Science, University of Cologne, 1999.
3. Christoph Buchheim, Michael Jünger, and Sebastian Leipert. A fast layout algorithm for k-level graphs. In Joe Marks, editor, *Proceedings of the 8th International Symposium on Graph Drawing (GD 2000)*, volume 1984 of *Lecture Notes in Computer Science*, pages 229–240. Springer, 2001.
4. Giuseppe Di Battista, Peter Eades, Roberto Tamassia, and Ioannis G. Tollis. *Graph Drawing: Algorithms for the Visualization of Graphs*. Prentice Hall, 1999.
5. Peter Eades, Xuemin Lin, and Roberto Tamassia. An algorithm for drawing a hierarchical graph. *International Journal of Computational Geometry & Applications*, 6:145–156, 1996.
6. Peter Eades and Kozo Sugiyama. How to draw a directed graph. *Journal of Information Processing*, 13(4):424–437, 1990.
7. Michael Fröhlich and Mattias Werner. The graph visualization system daVinci — a user interface for applications. Technical Report 5/94, Department of Computer Science, University of Bremen, 1994.
8. Emden R. Gansner, Eleftherios Koutsofios, Stephen C. North, and Kiem-Phong Vo. A technique for drawing directed graphs. *IEEE Transactions on Software Engineering*, 19(3):214–230, 1993.
9. Emden R. Gansner and Stephen C. North. An open graph visualization system and its applications to software engineering. *Software—Practice and Experience*, 30(11):1203–1233, 2000.
10. Emden R. Gansner, Stephen C. North, and Kiem-Phong Vo. DAG – A program that draws directed graphs. *Software—Practice and Experience*, 17(1):1047–1062, 1988.
11. Lois M. Haibt. A program to draw multilevel flow charts. In *Proceedings of the Western Joint Computer Conference*, volume 15, pages 131–137, 1959.
12. Xuemin Lin and Peter Eades. Area minimization for grid visibility representation of hierarchically planar graphs. In Takao Asano, Hiroshi Imai, Der-Tsai Lee, Shin-ichi Nakano, and Takeshi Tokuyama, editors, *Proceedings of the 5th International Conference on Computing and Combinatorics (COCOON '99)*, volume 1627 of *Lecture Notes in Computer Science*, pages 92–102. Springer, 1999.
13. Petra Mutzel, Carsten Gutwenger, Ralf Brockenauer, Sergej Fialko, Gunnar W. Klau, Michael Krüger, Thomas Ziegler, Stefan Näher, David Alberts, Dirk Ambras, Gunter Koch, Michael Jünger, Christoph Buchheim, and Sebastian Leipert. A library of algorithms for graph drawing. In Sue H. Whitesides, editor, *Proceedings of the 6th International Symposium on Graph Drawing (GD '98)*, volume 1547 of *Lecture Notes in Computer Science*, pages 456–457. Springer, 1998.

14. Georg Sander. Graph layout through the VCG tool. In Roberto Tamassia and Ioannis G. Tollis, editors, *Proceedings of the DIMACS International Workshop on Graph Drawing (GD '94)*, volume 894 of *Lecture Notes in Computer Science*, pages 194–205. Springer, 1995.
15. Georg Sander. A fast heuristic for hierarchical Manhattan layout. In Franz J. Brandenburg, editor, *Proceedings of the 3rd International Symposium on Graph Drawing (GD '95)*, volume 1027 of *Lecture Notes in Computer Science*, pages 447–458. Springer, 1996.
16. Georg Sander. Graph layout for applications in compiler construction. *Theoretical Computer Science*, 217(2):175–214, 1999.
17. Kozo Sugiyama and Kazuo Misue. Visualization of structural information: Automatic drawing of compound digraphs. *IEEE Transactions on Systems, Man and Cybernetics*, 21(4):876–892, 1991.
18. Kozo Sugiyama, Shojiro Tagawa, and Mitsuhiko Toda. Methods for visual understanding of hierarchical system structures. *IEEE Transactions on Systems, Man and Cybernetics*, 11(2):109–125, February 1981.

Automated Visualization of Process Diagrams*

Janet M. Six and Ioannis G. Tollis

CAD & Visualization Lab
Department of Computer Science
The University of Texas at Dallas
P.O. Box 830688 EC31
Richardson, TX 75083
Fax: (972)883-2349
{janet,tollis}@utdallas.edu

Abstract. In this paper, we explore the problem of producing process diagrams and introduce a linear time technique for creating them. Each edge has at most 3 bends and portions of the edge routing have optimal height. While developing a solution, we explore the subproblems of determining the order of the layers in the diagram, assigning x and y coordinates to nodes, and routing the edges.

1 Introduction

Computer scientists have been using flowcharts for many years as aids for designing, debugging, and documenting algorithmic solutions and system architectures [5,12,15]. These diagrams show the flow of resources, tasks, and time through processes and systems. A *process graph* is a directed graph plus a partitioning of the nodes into groups. Like the structures depicted in flowcharts, the directed graph in a process graph also represents flow through a process or system. The node partition groups can represent any type of classification: e.g., program modules, system components, periods of time, geographic or geometric regions, or social groups. For example, the set of courses required for a degree, their prerequisites, and a grouping of the courses by level can be modeled with a process graph. A visualization of this structure could be quite useful for students as they form their degree plans. A *process diagram* is a drawing of a process graph. See Figure 1 for an example process diagram of the required courses and their prerequisites in the Undergraduate Computer Science Program at the University of Texas at Dallas. The groupings represent the Freshman, Sophomore, Junior, and Senior levels in the program. This process diagram was produced automatically by an implementation of our process diagramming technique.

A process diagram could also be used to model a project management scenario: (a) the nodes and edges could represent tasks and prerequisite constraints

* The research was supported in part by the Texas Advanced Research Program under grant number 009741-040 and a stipend from the Provost's Office at the University of Texas at Dallas.

P. Mutzel, M. Jünger, and S. Leipert (Eds.): GD 2001, LNCS 2265, pp. 45–59, 2002.
© Springer-Verlag Berlin Heidelberg 2002

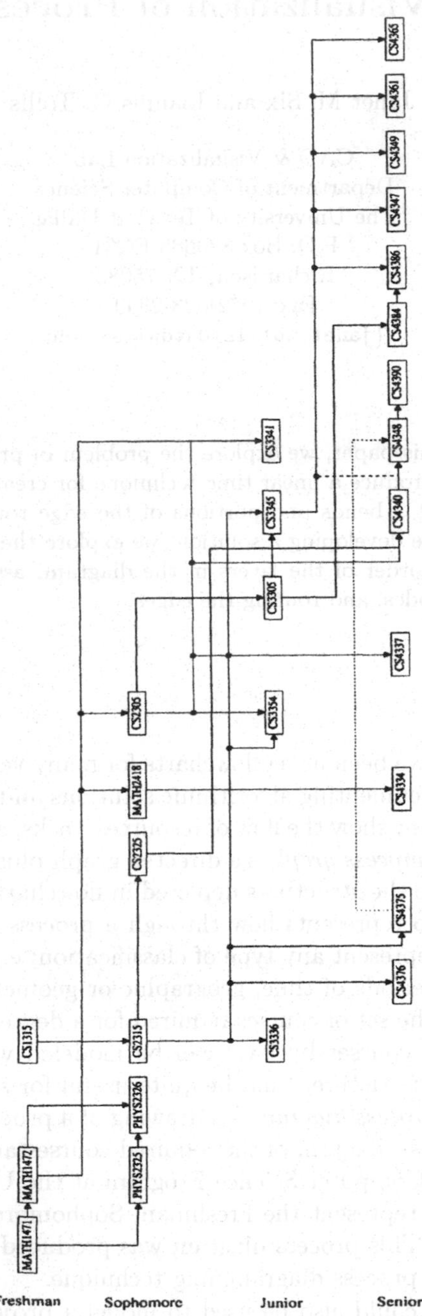

Fig. 1. Process Diagram of the University of Texas at Dallas Undergraduate Computer Science Program requirements where the layers represent the Freshman, Sophomore, Junior, and Senior levels.

and (b) the groups represent the technical and management teams working on the project. Alternatively, (a) the nodes could represent milestones and (b) the groups represent periods of time in which those milestones should be reached. A drawing of a project and its performance with respect to tasks or milestones would be helpful to a project manager.

Define a *layer* to be a horizontal line in the process diagram. Nodes which belong to the same node partition in a process graph are placed on the same layer in a process diagram. Define an *intralayer edge* to be incident to two nodes which are members of the same layer. Likewise, an *interlayer edge* is incident to two nodes which are not members of the same layer.

Since process diagramming is inspired by flowcharting, we present a technique which produces drawings that include the traditional characteristics of flowcharts [5,12,15]. One property of a good process diagram is that the flow through the elements of the graph should start in the top-left corner of the drawing and proceed to the bottom-right corner. Also, the edges should be drawn in an orthogonal manner to promote clarity. The number of bends, number of edge crossings, and edge length should be minimum. Also, nodes of a partition group must not appear on a layer with nodes of another group. The order of the layers can be specified by the user or the technique should order the layers such that the interlayer edges are directed downwards in the resulting process diagram.

Unfortunately, we cannot achieve all of the above properties and conventions simultaneously. For example, it is well known that the minimization of edge bends and crossings often contradict each other [2]. In addition, the problems of minimizing bends, crossings, and edge length are NP-Complete [2]. Therefore, we relax our requirements and strive to produce process diagrams with a low number of bends, crossings, and short edge length. Furthermore, process graphs often contain cycles which include interlayer edges and therefore it is not possible to draw all their edges in a downward direction. Since we cannot guarantee to produce a process diagram with all interlayer edges directed downward, we strive to find a process diagram with the minimum number of upward edges. We also provide a facility to the user to effectively order the layers.

Previously developed techniques do not meet the requirements of process diagramming. Hierarchical graph drawing techniques [2,20] are not suitable to produce process diagrams since the user has no control over the assignment of nodes to layers, no intralayer edges are allowed, the approach is not designed to show rightward flow, and the edges are not orthogonal. Sander presents an algorithm in [18] that draws edges in an orthogonal manner, but this technique does not solve the other problems of process diagramming. Orthogonal graph drawing techniques [2,16] cannot be used to produce process diagrams since they do not take edge directions into account and also that the user has no control over the assignment of nodes to layers.

Knuth presented a technique for automatically drawing flowcharts in [12]. In this algorithm, all nodes are placed in a single column and all edges are drawn to the right of this column. This technique does not allow for a user-based node partitioning. Also, this algorithm was designed to conserve memory by not requiring the entire data structure to be in memory at one time. Allowing a graph drawing algorithm to analyze an entire structure at once can lead to a more readable drawing.

Process diagrams can be created manually with a tool such as [17], but the effort required of the user can be extensive. It is of benefit to provide a fully automatic tool so that a user does not have to spend a significant amount of time and effort to produce the process diagram.

The GRADE tool [7,11] offers manual, semi-automatic, and automatic avenues for the creation of process diagrams. The tool has the feature of allocating nodes to "lanes" and placing nodes of a lane within a horizontal or vertical box. These lanes can be used to handle the node partitions in a process graph. However, the automatic layout technique described in [11] first uniformly places the nodes in the layout and then uses the *barycenter* method to finalize the node positions. The barycenter method for graph layout iteratively places each node in the midst of its neighbors. From the layout description given in [11], the worst case time requirement is unclear. Furthermore, for every $n > 1$, there exists a graph with n nodes such that the barycenter method requires exponential area [2]. Also, the edge routing is not described and therefore we do not know of an upper bound for the number of bends per edge.

Showbiz [22] is a process flow modeling tool. A facility is given to the user to visualize a flow with the additional quality that each node is placed in a lane that represents an attribute of that node. The x coordinates of the nodes are determined with the first phase of the Sugiyama hierarchical algorithm [20] and the y coordinate is given based on the location of the lane. Sometimes nodes can overlap and then the lane is widened so that those nodes appear one above the other. This layout technique visualizes rightward, but not downward flow. Also, some non-orthogonal edges are used in the edge routing. This technique is not sufficient to solve the process diagramming problem.

In this paper, we present a systematic $O(m)$ time approach for producing process diagrams, where m is the number of edges. These diagrams have at most 3 bends per edge and portions of the edge routing have optimal height.

2 A Linear Time Technique for Producing Process Diagrams and Determining the Order of the Layers

The problem of creating a process diagram can be divided into the following tasks:

1. Determine an order of the layers as they should appear from top to bottom.
2. Assign x and y coordinates to each node.
3. Route the edges.

We will now look at each task in detail and present algorithms to solve them. Then we will put these solutions together in Section 5 and present a linear time algorithm for producing process diagrams.

Let $G = (V, E, P)$ be a process graph, where V is the set of nodes, E the set of edges, and P the node partitioning. Our first task is to determine an ordering of the layers, corresponding to each partition, as they should appear from top to bottom in the process diagram Γ. We present a technique which finds an ordering with the following properties:

1. The layers which contain source nodes appear at the top of the drawing.
2. The layers which contain sink nodes, but no source nodes, appear at the bottom of the drawing. If there is a layer which contains both source and sink nodes, that layer is treated as one which contains source nodes and appears at the top of the drawing.
3. The layers are ordered such that most interlayer edges of the input graph are directed downwards in the resulting process diagram.

Given a process graph $G = (V, E, P)$, we find an ordering of the layers with the above properties by first creating a *layer graph*, $G' = (V', E')$. The nodes in V' represent the node partitions in P. The set of edges in E' is determined in the following manner:

For each pair of nodes in V', L_i and L_j

1. Examine the set E_{ij} of edges in E which have one incident node in L_i and L_j.
2. If neither L_i nor L_j contain a source or sink node then add a dominant type edge to E'. (If the majority of edges in E_{ij} are directed from L_i to L_j, we say that (L_i, L_j) is the *dominant type edge.* Otherwise, (L_j, L_i) is the dominant type edge.)
3. If L_i (or L_j) contains a source node then add (L_i, L_j) (or (L_j, L_i)) to E'.
4. If L_i (or L_j) contains a sink node then add (L_j, L_i) (or (L_i, L_j)) to E'.
5. If both L_i and L_j contain source (or sink) nodes then add a dominant type edge to E', as described in Step 2.

If a node represents a layer which contains both source and sink nodes, we treat it as a layer with source nodes. Not only does this process assure that no multiedges exist in G', it also prevents the inclusion of cycles of length two (*two-cycles*).

If G' is a directed acyclic graph, then we can apply a *topological sort* [1] to find a good layer order. Often times layer graphs contain cycles and therefore it is necessary to employ a more advanced method to obtain the order. We can transform a graph with cycles into an acyclic graph by reversing a set of edges. In order to produce a process diagram with the maximum number of interlayer edges directed downward, we want to reverse the minimum number of edges in the layer graph. This is equivalent to the NP-Complete *feedback arc set problem* [2,6,10] in which a minimum cardinality set of edges is removed in order to make the graph acyclic. In [3], a $O(m)$ time heuristic algorithm for reversing a small number of edges in order to make a graph acyclic is presented. The algorithm, *Greedy-Cycle-Removal*, orders the nodes of the input graph such that the number of edges going from a node later in the order to a node earlier in the order is small. These edges are called *backward* and form a set which can be reversed in order to produce an acyclic graph. The other edges are *forward*. The input graph to this algorithm is assumed to be connected, however it can easily be extended to handle disconnected graphs by performing Greedy-Cycle-Removal on each connected component successively.

There are properties of the resulting sequence S which are guaranteed if the input graph does not contain any two-cycles. Since layer graphs do not contain two-cycles, we can also guarantee the same performance.

Theorem 1. *[3] Suppose that G is a connected directed graph with n nodes and m edges, and no two-cycles. Then Greedy-Cycle-Removal computes a node ordering S of G, with at most $m/2 - n/6$ backward edges.*

This result is strengthened for graphs whose underlying undirected graph contains no nodes with degree greater than three. In those cases, Greedy-Cycle-Removal computes a node ordering of G with at most $m/3$ backward edges [2, 4]. In our case, we are applying Greedy-Cycle-Removal to G', which is typically a small-sized graph. In fact, in some cases the graph may be so small that an exponential search to find the best set of edges to reverse can be conducted in a reasonable amount of time.

These theorems show that Greedy-Cycle-Removal has a guaranteed performance much better than the worst case of another technique for making a graph acyclic which performs a DFS on a given directed graph and reverses all the back edges. In fact, in the worst case, $m - n - 1$ edges may be reversed. In addition to having a guaranteed performance, Greedy-Cycle-Removal can be implemented to run in linear time and space [3].

We now present an $O(m)$ time technique for determining the order of layers for a process diagram which has been inspired by the algorithm in [3]. In addition to finding an order of the layers, the nodes of the input graph are also given a y coordinate according to the order which is found.

Algorithm 1 *DetermineOrderOfLayers*
Input: A process graph, $G = (V, E, P)$.
Output: An ordering of the layers.

1. Create a layer graph G' from process graph G as described above.
2. Initialize S_l and S_r to be empty lists.
3. While G' is not empty do
   a) While G' contains a source u, remove u and append it to S_l.
   b) While G' contains a sink v, remove v and prepend it S_r.
   c) If G' is not empty then choose a node w, such that the value *outdegree*(w) − *indegree*(w) is maximum, remove w from G' and append it to S_l.
4. Concatenate S_l with S_r to form S.
5. Output S.

Applying Algorithm 1 to the process graph shown in Figure 1, the resulting layer graph would have 4 nodes (Freshman, Sophomore, Junior, and Senior) and 3 edges ((Freshman, Sophomore), (Sophomore, Junior), (Junior, Senior)). In Steps 4 and 5 of the algorithm, we would find S to be (Freshman, Sophomore, Junior, Senior).

3 Determining the x Coordinates of the Nodes

We divide the problem of assigning x coordinates to nodes into two subtasks: first for each layer, we create groups of nodes. Secondly, we order those groups in the layer such that most intralayer edges are rightward in the process diagram.

First, we discuss Algorithm *PartialOrdering*. To increase readability we wish to place nodes which are members of a long path consecutively on a layer. A node v is *reachable* from u if there exists some path from u to v. Nodes which are reachable from node u form the *reachability group* of u. We find the reachability groups in the subgraph $G_i = (V_i, E_i)$ induced by the nodes assigned to Layer L_i by employing a DFS. If G_i is not strongly connected, there will be multiple reachability groups. These groups can contain nodes of different strongly connected components, but not necessarily be the same as the connected components of the underlying undirected graph.

In order to find a good set of reachability groups, we will always start the DFS at a node of smallest indegree. If the DFS algorithm must choose a node to start an additional tree in the depth first spanning forest, it will also start at an unvisited node of lowest indegree. The result of these steps, in addition to the depth first spanning forest, is a set of groups each of which contains members of a tree from the forest. The nodes in each group are in the same order as visited in the DFS, thus providing a partial ordering. This process requires $O(m)$ time since it is a variant of DFS. Figure 2 shows the application of Algorithm PartialOrdering on the subgraph of the Sophomore layer in Figure 1. The left side of the figure shows the elements of the subgraph while the right side shows the grouping of the elements into reachability groups.

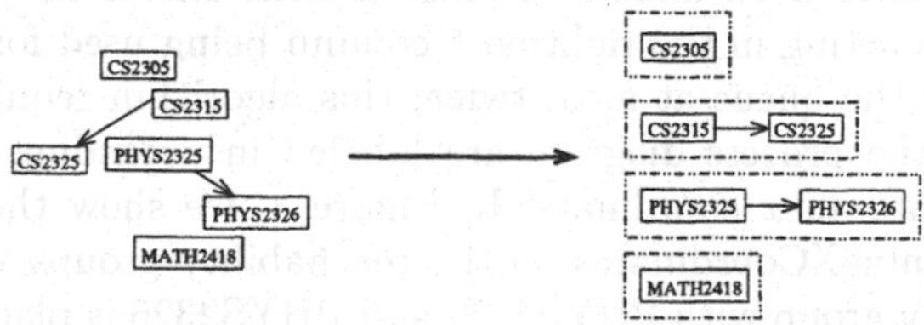

Fig. 2. The application of Algorithm PartialOrdering on the subgraph of the Sophomore layer in Figure 1.

The second subtask in determining the x coordinates is to order the reachability groups found with Algorithm PartialOrdering. A *weak dominance drawing* has the property that for any two vertices u and v, if there is a directed path from u to v then v is placed below and/or to the right of u in the drawing. This is a variant of the *dominance drawing* technique for planar *st*-graphs discussed in [2]. In a dominance drawing, a node v is given x and y coordinates greater than or equal to that of another node u if and only if there exists a directed path from u to v. Weak dominance drawing captures the down-and-to-the-right flow we want to see in process diagrams. For planar acyclic graphs, dominance drawing algorithms presented in [2] could be used to create a process diagram. However, the technique presented in this paper works on all process graphs.

We next discuss Algorithm *DetermineXCoordinates* which will order the groups. We place the reachability groups of all layers in the order they are found by Algorithm PartialOrdering into a queue Q. We set *currentSourceColumn* to be the leftmost column. We process the members of Q and place the groups which contain a source node. At each iteration, we dequeue a group α: if it contains a source node, then we place the first node of α in *currentSourceColumn* and the remaining nodes in the next $|\alpha| - 1$ columns. We insert columns as necessary to avoid nodes of the same layer being assigned to the same column. After the nodes of α are placed, we increment *currentSourceColumn* by $|\alpha|$. If α does not contain a source node, we enqueue it back into Q.

After all the groups with source nodes have been placed, we process the members of Q again and place the remaining groups. We dequeue a group α and determine if it contains a node which is incident to a node which has been placed. Due to the construction of Q and the expected down-and-to-the-right flow in the process graph, it is likely for a neighbor of a node in α to already be placed. If this is the case, we find the rightmost placed neighbor u of a node in α. If there is no obstacle in the column of u between the row of u and the row of α and a node has not already been placed in the column of u and the row of α, then place α starting in column of u. If there is an obstacle in the column of u between the row of u and the row of α and a node has not already been placed in the column of $u + 1$ and the row of α, then place α starting in column of $u + 1$. Columns are inserted as necessary to avoid the placement of groups being intermingled. Otherwise, if we have been unable to place α with one of the above methods, then we place α starting in the rightmost column being used for the layer of α. Since, we process the queue at most twice, this algorithm requires $O(m)$ time. The columns in the process diagram are labeled in ascending order from left to right and denote an x coordinate. In Figure 3, we show the application of Algorithm DetermineXCoordinates on the reachability groups shown in Figure 2. The reachability group with PHYS2325 and PHYS2326 is placed to the left of the reachability group with CS2315 and CS2325 since a relative of PHYS2325, MATH1471, was placed to the left of a relative of CS2315, CS1315, on a higher layer. MATH2418 and CS2305 are placed in columns to the right due to obstacles. The relatives of these 6 nodes which belong to lower layers will be placed below and to the right of these nodes on lower layers. This type of placement facilitates the down-and-to-the-right flow present in the process diagrams created by the approach in this paper.

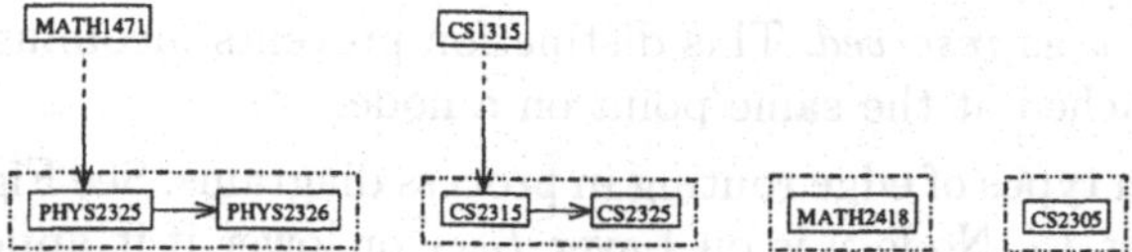

Fig. 3. The application of Algorithm DetermineXCoordinates on the reachability groups in Figure 2.

4 Routing the Edges and Determining Final Coordinates

The task of routing the edges in a process diagram is related to the VLSI design problem *Channel Width Minimization within the Jog-Free Manhattan Model* (CWM) [14]. In CWM, the terminals (or nodes) are given fixed placements on two horizontal lines and the nets (or edges) are routed in the area between the two lines. This area is called the *channel*. Horizontal wire segments occupy *tracks* and the number of tracks used is referred to as the *width* of the channel. The CWM is defined as follows: given two sets of terminals which have been placed on two different horizontal lines and a set of nets, find a routing such that each net has at most one horizontal segment and the channel is of minimum width [14]. The CWM is NP-Hard [13]. Define a *horizontal constraint* to exist between two nets if they would overlap if placed in the same track. Define a *vertical constraint* to exist between two nets if one terminal of each net resides in the same column. The general CWM is NP-Hard. However an optimal solution can be found in linear time for those instances of CWM which have no vertical constraints [14].

Theorem 2. *[14] Given an instance of CWM which has no vertical constraints, there does exist an algorithm which finds a channel of minimum width. Furthermore, this solution can be found in $O(m)$ time, where m is the number of nets which need to be routed.*

Hashimoto and Stevens present the *left-edge algorithm* in [9] as a solution to CWM. The main loop of the algorithm scans the channel from left to right and places each net in a track such that no two nets overlap. It packs the horizontal segments of nets as tightly as possible. We can apply the left-edge algorithm to our edge routing problem by treating the area between two layers as a channel and assigning a set of edges to each channel. In order to encourage the appearance of a down-and-to-the-right flow in process diagrams, we place the attachment points of incoming edges on the left and top sides of their target nodes. Likewise, outgoing edges emanate from the bottom and right sides of their source nodes. If we hold these properties strictly, we would add edge length or bends to process diagrams while routing edges from lower to higher layers. To avoid these problems, we route these edges emanating from the top of their source nodes. We also route outgoing edges from the top of the source node if they are directed rightward to a node on the same layer. If a node u has an outgoing edge to a node on a higher layer or directed rightward to a node on the same layer, we mark the top of u as *reserved.* This distinction prevents incoming and outgoing edges being attached at the same point on a node.

There are ten types of edge routing in process diagrams. See Figure 4. Assume u to be on Layer L_i. Node v is on Layer L_{i+1} or lower if it appear below u or on Layer L_{i-1} or higher if it appear above u. Routings (a) and (b) are used for downward and rightward edges for which the source and target nodes share a row or column. Routing types (c), (d), and (e) are for edges with their source and target on the same layer. Types (f), (g), (h), and (i) are for edges with their source on a higher layer than the target. Type (j) is for edges with their source on a lower level than the target. These edge routings have at most three bends per edge. As is shown in Figure 4, downward edges can be routed with routing types (a), (f), (g), (h), or (i). We can reduce the number of bends and possibly the area if we use routing type (a), but this routing is not always feasible. Therefore, we try to route a downward edge with first a type (a), then (f), (g), (h), and (i) type routings. Similarly, we try to route a rightward edge between nodes of the same layer with edge routing type (b) then (c) and (d). These sequences are used in order to reduce the number of bends, edge length, and area of process diagrams.

Let Channel $L_{i-1}L_i$ represent the area above Layer L_i. Likewise, let Channel L_iL_{i+1} represent the area below Layer L_i. We will assign horizontal segments of type (c), (d), (e), (h), (i), and (j) routings to some channel and then place them with the left-edge algorithm. In order to use the left-edge algorithm, we need to guarantee that no vertical constraints exist. In our edge routing problem, we have a vertical constraint if a Layer L_i node u in column x_u is the source of an edge routed with type (e), (h), or (i) and a Layer L_{i+1} node v is in column x_u and is the source of an edge routed with type (c), (d), or (j) or is the target of an edge routed with type (h) routing. We can avoid a type (e, h, i) / (h) vertical constraint, by routing the edge directed toward v with a type (i) routing. We can avoid type (e, h, i) / (c, d, j) vertical constraints by using an edge routing with more bends, however we can maintain our three bends per edge property by moving v to a new column x_v+1 and still using types (c), (d), or (e) routings. In order to avoid adding many columns, we add a new column after x_v at most one time. This new column can resolve multiple vertical constraints.

We will now examine each edge routing type in detail. If the source and target of a downward or rightward edge $e = (u, v)$ reside in the same row or column, we attempt to route e with a type (a) or (b) routing. Let x_u and y_u be the x and y coordinates of node u. Let x_u $(y_u)$to denote the column (row) of u. If e is downward, the top of v is not reserved, and there are no obstacles in the process diagram from (x_u, y_u) to (x_v, y_v), then we route e with a straight edge from (x_u, y_u) to (x_v, y_v). Likewise, if e is rightward and there are no obstacles in the process diagram from (x_u, y_u) to (x_v, y_v), then we route e with a straight edge from (x_u, y_u) to (x_v, y_v). If obstacles are encountered, we will route the edge with another routing type.

If we are unable to route a rightward edge $e = (u, v)$ with a type (b) routing, we will route the edge with a type (c) or (d) routing. If the top of v is not

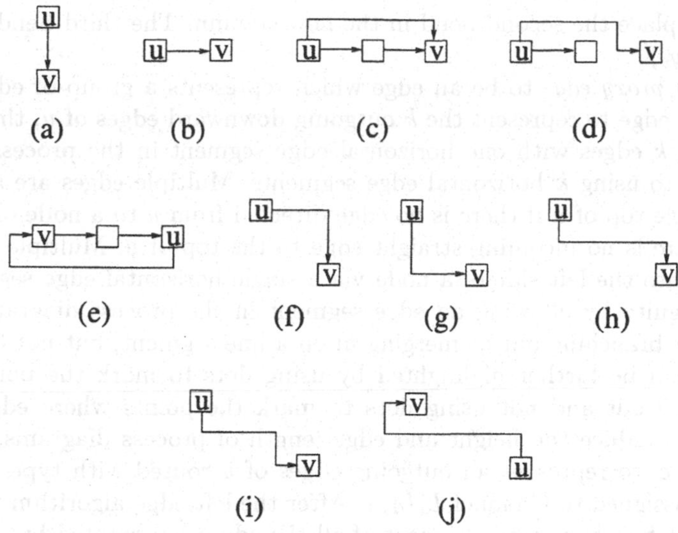

Fig. 4. Ten possible edge routings for the edge (u, v).

reserved, then we can route e with type (c), else we route with type (d). In a both types (c) and (d) routings, we place the first bend in the column of u. In a type (c) routing, we place the second bend in the column of v. We insert a new column, $x_v - 1$, for the placement of the second bend in a type (d) routing. The third bend will be placed at $(x_v - 1, y_v)$. Since we want edges routed with type (c) or (d) to be drawn above Layer L_i, we place the first horizontal segment into Channel $L_{i-1}L_i$. We will then use the left-edge algorithm to place this segment and thus determine the y coordinates of the first two bends.

Type (e) routing is for a leftward edge $e = (u, v)$ with source and target nodes in the same layer L_i and is very similar to type (d) routing. A new column is introduced at $x_v - 1$ and the first two bends placed in the columns x_u and $x_v - 1$. The third bend is placed at $(x_v - 1, y_v)$. Since we want these edges to appear below layer L_i, we assign edge e to Channel L_iL_{i+1} and use the left-edge algorithm to place the first horizontal segment of e.

Routing types (f), (g), (h), and (i) are used to route downward edges. Both types (f) and (g) attempt to route a downward edge $e = (u, v)$ with a one bend edge. Routing type (f) is used if the top of v is not reserved, u does not have an outgoing edge routed with type (b), and there are no obstacles in the process diagram from (x_u, y_u) to (x_v, y_u) and from (x_v, y_u) to (x_v, y_v). Alternatively, type (g) can be used if there are no obstacles in the process diagram from (x_u, y_u) to (x_u, y_v) and from (x_u, y_v) to (x_v, y_v). If neither type (f) nor (g) can be used, then we will route with type (h) or (i). The first bend in a (h) or (i) type routing will be placed in the column of u. If the top of v is not being used and a type (h) routing will not create a vertical constraint, then we will use a type (h) routing and place the second bend in the column of v. Else, we insert a new column

$x_v - 1$ and place the second bend in the new column. The third bend is placed at $(x_v - 1, y_v)$.

Define a *proxy edge* to be an edge which represents a group of edges. If we use a proxy edge to represent the k outgoing downward edges of u, then we can route those k edges with one horizontal edge segment in the process diagram as opposed to using k horizontal edge segments. Multiple edges are allowed to come into the top of u if there is no edge directed from u to a node on a higher layer or there is no incoming straight edge to the top of u. Multiple edges can also come into the left side of a node via a single horizontal edge segment. We avoid ambiguity by allowing an edge segment in the process diagram to have edges either branching out or merging in on a line segment, but not both. This difference can be further highlighted by using dots to mark the points where edges branch out and not using dots to mark the points where edges merge together. To reduce the height and edge length of process diagrams, we use a proxy edge e' to represent all outgoing edges of u routed with type (h) or (i). Edge e' is assigned to Channel L_iL_{i+1}. After the left-edge algorithm is applied, we place the first horizontal segment of all the edges represented by proxy edge e' in the track assigned to e'.

Upward edges are routed with type (j) from Figure 4. For an upward edge $e = (u, v)$, we place the first bend in the column of u and insert a new column, $x_v - 1$, for the second bend. The third bend is placed at $(x_v - 1, y_v)$. The first horizontal segment of e is assigned to Channel $L_{i-1}L_i$.

We now give an algorithm for the routing of edges in process diagrams.

Algorithm 2 *RouteEdges*
Input: A process graph, $G = (V, E, P)$ and
A set of $(x,\ y)$ coordinates for the nodes in V.
Output: A routing for all edges in E.

1. Resolve any vertical constraints by adding new columns as described in the previous text.
2. Attempt to route downward and rightward edges with source and target nodes on the same layer with type (a) and (b) routings.
3. Route rightward edges not routed in Step 2 with a type (c) or (d) routing.
4. Route leftward edges with source and target nodes on the same level with type (e) routings.
5. Route downward edges not routed with Step 2 with type (f), (g), (h), or (i) routings.
6. Route upward edges with type (j) routings.
7. For each channel, L_iL_{i+1},
   a) Set *currentRow* to be the row of Layer L_i.
   b) Set *currentColumn* to be the leftmost column.
   c) Set *maxColumn* to the rightmost column.
   d) While edges in L_iL_{i+1} not routed and *currentColumn* $\neq$ *maxColumn*
      i. Set *newRow* to be a newly inserted row below *currentRow*.
      ii. Set *currentRow* to *newRow*.

iii. Find the next column which contains an edge to be routed, place that edge in *currentRow*, and set *currentColumn* to the column after the column of the rightmost bend of the newly placed edge. If no edge can be routed on the current row, then set *currentColumn* to *maxColumn*.

Resolving the vertical constraints in Step 1 requires $O(n)$ time. Assigning a routing type to each edge in Steps 2 - 6 requires $O(m)$ time. The remaining steps are a variant of the left-edge algorithm [9]. From Theorem 2, we know that these steps require $O(m)$ time. The left-edge algorithm can be implemented to run in linear time if any available track can be used to route an edge as opposed to using the available track which is closest to the top of the channel [8].

Due to Theorem 2, we know that Algorithm 2 will find channels of minimum width given the assignment of edge segments to channels. In order to prove this approach finds an optimal solution [14], let m_j be a edge that is assigned to the track t_j which is furthest from the first line of nodes α. For each track, t_k, closer to α, there must be some edge, m_k, which has a horizontal constraint with m_j, otherwise m_j would be placed on track t_k due to the construction on the algorithm and the stipulation that there are no vertical constraints.

Define the *closed density* δ of an instance of CWM to be equal to $\max_{x}|E_x|$, where the columns of the channel are labeled in ascending order from left to right, x is a number, and E_x is the set of edges whose routings must pass the column labeled x under the constraints of the jog-free manhattan model. The maximum channel width found with the left-edge algorithm is the closed density δ. Moreover, we know that δ is the minimum channel width necessary to allow a legal routing. Therefore, the left-edge algorithm produces an optimal result, with respect to channel width, when given an instance of CWM without vertical constraints.

5 Creating a Process Diagram in Linear Time

We put the previous algorithms together in this section and present a $O(m)$ time technique for creating process diagrams.

Algorithm 3 *CreateProcessDiagram*
Input: A process graph $G = (V, E, P)$, and
An ordering of the layers (optional).
Output: A process diagram of G, Γ, with the conventions and properties discussed in the earlier text.

1. If the ordering of the layers is not given, then determine the ordering of the layers with Algorithm DetermineOrderOfLayers.
2. Determine the partial orderings of the nodes within each layer with Algorithm PartialOrdering.

3. Determine a x coordinate for each node with Algorithm DetermineXCoordinates.

4. Route the edges with Algorithm RouteEdges.

Algorithm 3 requires $O(m)$ time. Each edge has at most three bends in Γ and the height of the channels are minimum. This optimality does not guarantee that our drawing is of minimum height since it is possible that edges need to be assigned to other channels. The height of the drawing is bound by the number of partitions in P plus n (to represent the type (h) and (i) edge routings) plus the number of rightward, leftward, and upward edges. The width of the drawing is bound by the two times the number of nodes plus the number of edges. A sample process diagram as produced by an implementation of Algorithm 3 is shown in Figure 1.

We have implemented Algorithm 3 in C++ and include some preliminary results below. The first two process graphs are course flow charts for The University of Texas at Dallas. The third graph is a modified flowchart. The fourth and fifth process graphs are not acyclic. The criteria included for these process diagrams include the size of the process graph (number of nodes, edges, and partitions), total number of edge bends, maximum number of bends per edge, and number of edge crossings. The number of edges that are drawn with a down-and-to-the-right flow is also shown. This criterion shows how well our process diagrams capture the down-and-to-the-right flow in these process graphs. Finally, we show the number of rows saved with the use of proxy edges.

Symbol	Criterion Represented
β	The number of edge bends
η	The maximum number of bends per edge
χ	The number of edge crossings
ρ	The number of edges with down-and-to-the-right-flow
ϵ	The number of rows saved by using proxy edges

	$\lvert V\rvert$	$\lvert E\rvert$	$\lvert P\rvert$	β	η	χ	ρ	ϵ
Ex. 1	27	34	4	48	3	24	34	13
Ex. 2	35	42	4	59	3	25	41	9
Ex. 3	18	24	9	7	1	0	24	0
Ex. 4	18	25	8	24	3	8	19	0
Ex. 5	70	96	11	113	3	95	88	7

More details of this algorithm and extensions of this technique to handle nodes of special type are discussed in [19].

References

1. T. H. Cormen, C. E. Leiserson and R. L. Rivest, *Introduction to Algorithms*, McGraw-Hill, 1990.
2. G. Di Battista, P. Eades, R. Tamassia and I. G. Tollis, *Graph Drawing: Algorithms for the Visualization of Graphs*, Prentice-Hall, 1999.
3. P. Eades, X. Lin and W. F. Smyth, A Fast and Effective Heuristic for the Feedback Arc Set Problem, *Information Processing Letters*, 47, pp. 319-323, 1993.
4. P. Eades and X. Lin, A New Heuristic for the Feedback Arc Set Problem, *Australian Jrnl. of Combinatorics*, 12, pp. 15-26, 1995.
5. M. V. Farino, *Flowcharting*, Prentice-Hall, 1970.
6. M. Garey and D. Johnson, *Computers and Intractability: A Guide to the Theory of NP-Completeness*, Freeman, 1979.
7. GRADE User Manual, Infologistik GmbH, 2000. Also available at `http://www.gradetools.com`.
8. U. I. Gupta, D. T. Lee and J. Y.-T. Leung, An Optimal Solution for the Channel Assignment Problem, *IEEE Trans. Computers*, C-28(11), pp.807-810, 1979.
9. A. Hashimoto and J. Stevens, Wire Routing by Optimizing Channel Assignment with Large Apertures, *Proc. 8th ACM/IEEE Design Automation Workshop*, pp. 155-169, 1971.
10. G. Isaak, Tournaments as Feedback Arc Sets, *Electronic Jrnl. of Combinatorics*, 2(1), #R20, 1995. Also available at `http://www.combinatorics.org`.
11. P. Kikusts and P. Rucevskis, Layout Algorithms of Graph-Like Diagrams for GRADE Windows Graphic Editors, *Proc. of GD '95*, LNCS 1027, Springer-Verlag, pp. 361-364, 1996.
12. D. E. Knuth, Computer Drawn Flowcharts, *Comm. of the ACM*, 6(9), 1963.
13. A. S. La Paugh, *Algorithms for Integrated Circuit Layout: An Analytic Approach*, PhD Thesis, Massachusettes, 1980.
14. T. Lengauer, *Combinatorial Algorithms for Integrated Circuit Layout*, John Wiley and Sons, 1990.
15. J. Martin and C. McClure, *Diagramming Techniques for Analysts and Programmers*, Prentice-Hall, 1985.
16. A. Papakostas, *Information Visualization: Orthogonal Drawings of Graphs*, Ph.D. Thesis, The University of Texas at Dallas, 1996.
17. K. Ryall, J. Marks, and S. Shieber, An Interactive System for Drawing Graphs, *Proc. GD '96*, LNCS 1190, Springer-Verlag, pp. 387-94, 1997.
18. G. Sander, A Fast Heuristic for Hierarchical Manhattan Layout, *Proc. GD '95*, vol 1027 of *LNCS*, pp. 447-458, Springer-Verlag, 1996.
19. J. M. Six, *VisTool: A Tool For Visualizing Graphs*, Ph.D. Thesis, The University of Texas at Dallas, 2000.
20. K. Sugiyama, S. Tagawa and M. Toda, Methods for Visual Understanding of Hierarchical Systems, *IEEE Trans. Systems, Man and Cybernetics*, SMC-11, no. 2, pp. 109-125, 1981.
21. R. Tamassia, On Embedding a Graph in the Grid with the Minimum Number of Bends, *SIAM J. Comput.*, 16, pp. 421-444, 1987.
22. K. Wittenburg and L. Weitzman, Qualitative Visualization of Processes: Attributed Graph Layout and Focusing Techniques, *Proc. GD '96*, LNCS 1190, Springer-Verlag, pp. 401-8, 1997.

Planarization of Clustered Graphs
(Extended Abstract)*

Giuseppe Di Battista, Walter Didimo, and A. Marcandalli

Dipartimento di Informatica e Automazione, Università di Roma Tre, via della Vasca Navale 79, 00146 Roma, Italy. {gdb,didimo,marcanda}@dia.uniroma3.it

Abstract. We propose a planarization algorithm for clustered graphs and experimentally test its efficiency and effectiveness. Further, we integrate our planarization strategy into a complete topology-shape-metrics algorithm for drawing clustered graphs in the orthogonal drawing convention.

1 Introduction

Several application domains require to draw graphs in such a way that vertices are grouped together into *clusters*. For example, a large computer network is often partitioned into areas, and it is usual to represent the systems (routers, switches, etc.) belonging to the same area inside the same region (rectangle or convex polygon) of the drawing. Further, areas are recursively partitioned into sub-areas with a structure that can have many levels. Other examples come from the Computer Aided Software Engineering field, where in the diagrams representing the design process it is often required to highlight the "cohesion" among certain components.

Such application requirements motivated the study of algorithms for drawing graphs with recursive clustering structures over the vertices, such as *compound digraphs* and *clustered graphs*. Fig. 1 shows an example of clustered graph. Some of the papers on the subject are discussed below. In [23] Sugiyama and Misue proposed an algorithm for drawing compound digraphs within a drawing convention that is related to the one commonly adopted for hierarchical graphs [24]. Feng, Cohen, and Eades [14] studied the concept of planarity for clustered graphs (*c-planarity*) and provided the first c-planarity testing algorithm. The construction of orthogonal drawings of c-planar clustered graphs, based on visibility representations and bend-stretching transformations, is studied by Eades et al. [11]. Straight line and orthogonal drawing algorithms, three dimensional visualization techniques, and force directed methods are presented in [13,10,20,3], in [9], and in [17], respectively. The related problem of constructing balanced clusters on a given planar graph is studied by Duncan et al. [8]; the constructed clustered

* Research supported in part by the project "Algorithms for Large Data Sets: Science and Engineering" of the Italian Ministry of University and Scientific and Technological Research.

P. Mutzel, M. Jünger, and S. Leipert (Eds.): GD 2001, LNCS 2265, pp. 60–74, 2002.
© Springer-Verlag Berlin Heidelberg 2002

graph satisfies the conditions for c-planarity. However, as far as we know, a complete algorithm, based on the topology-shape-metrics approach [5], for drawing clustered graphs in the orthogonal drawing convention is currently not available.

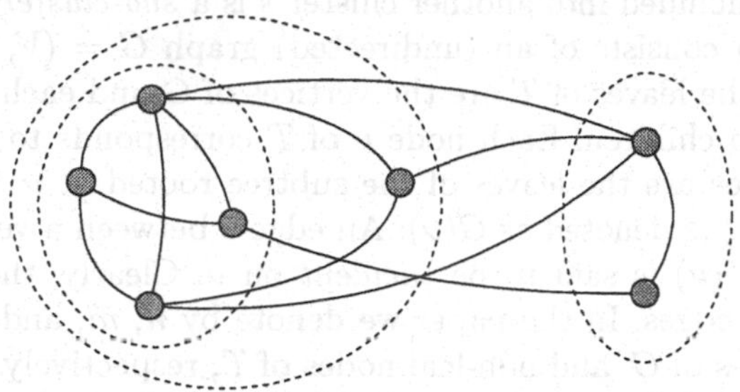

Fig. 1. A clustered graph: clusters are dashed.

The main results presented in this paper are the following:

- We propose a planarization algorithm for a clustered graph C with n vertices, m edges, and c clusters. It runs in $O(m\chi + m^2c + mnc)$ time, where χ is the number of crossings inserted in C by the algorithm. This is, as far as we know, the first complete planarization algorithm for clustered graphs.
- As a by-product of the planarization algorithm, we present an algorithm for constructing a "spanning tree" of C in $O(m)$ time. A definition of the term spanning tree for a clustered graph is given in Section 3.
- We present an implementation of the planarization algorithm with efficient data structures and show an experimental study that compares the effectiveness and the efficiency of the algorithm with those of simpler planarization techniques. The computational results put in evidence performance that are reasonable in many application domains.
- We describe the implementation of a complete drawing algorithm for clustered graphs, based on the topology-shape-metrics approach.

The rest of this paper is organized as follows. In Section 2 we introduce the basic terminology on clustered graphs. The planarization algorithm is presented in Sections 3 and 4. A simple planarization algorithm, whose behaviour is used as a reference point in our experiments, is described in Section 5. An experimental analysis of the effectiveness and of the efficiency of our planarization algorithm is presented in Section 6. The implementation of a complete drawing algorithm for clustered graphs is sketched in Section 7. Open problems are addressed in Section 8.

2 Preliminaries on Clustered Graphs and c-Planarity

We assume familiarity with connectivity and planarity of graphs [12,21,5]. Since we consider only planar graphs, we use the term *embedding* instead of *planar embedding*. Also, we assume that an embedding determines a choice for the external face.

We import several definitions from the papers on c-planarity by Cohen, Eades, and Feng [14,13]. Given a graph, we call *cluster* a subset of its vertices. A cluster that is included into another cluster s is a *sub-cluster* of s. A *clustered graph* $C = (G,T)$ consists of an (undirected) graph $G = (V,E)$ and a rooted tree T such that the leaves of T are the vertices of G and each non-leaf node of T has at least two children. Each node ν of T corresponds to the cluster $V(\nu)$ of G whose vertices are the leaves of the subtree rooted at ν. The subgraph of G induced by $V(\nu)$ is denoted as $G(\nu)$. An edge e between a vertex of $V(\nu)$ and a vertex of $V - V(\nu)$ is said to be *incident* on ν. Clearly, the root of T does not have incident edges. In the paper we denote by n, m, and c the number of vertices of G, edges of G, and non-leaf nodes of T, respectively. Note that, since each non-leaf node of T has at least two children, we have $c < n$.

Graph G and tree T are called the *underlying graph* and the *inclusion tree* of C, respectively. Observe that, given two nodes μ and ν of T such that μ is an ancestor of ν, $V(\nu)$ is a sub-cluster of $V(\mu)$.

Clustered graph C is *connected* if for each node ν of T we have that $G(\nu)$ is connected. In a connected clustered graph $m \geq n - 1$.

Suppose that $C_1 = (G_1, T_1)$ and $C_2 = (G_2, T_2)$ are two clustered graphs such that T_1 is a subtree of T_2 and for each node ν of T_1, $G_1(\nu)$ is a subgraph of $G_2(\nu)$; then C_1 is a *sub-clustered-graph* of C_2.

In a *drawing* of a clustered graph $C = (G,T)$ each vertex of G is a point and each edge is a simple curve between its end-vertices. For each node ν of T, $G(\nu)$ is drawn inside a simple closed region $R(\nu)$ such that: (i) for each node μ of T that is neither an ancestor nor a descendant of ν, $R(\mu)$ is completely contained in the exterior of $R(\nu)$; (ii) an edge e incident on ν crosses the boundary of $R(\nu)$ exactly once. We say that edge e and region R have an *edge-region crossing* if both endpoints of e are outside R and e crosses the boundary of R more than once; we assume that a drawing of a clustered graph does not have edge-region crossings.

A drawing of a clustered graph is *c-planar* if it does not have edge crossings. A clustered graph is *c-planar* if it has a c-planar drawing.

Theorem 1. [14] *A connected clustered graph $C = (G,T)$ is c-planar if and only if G is planar and there exists a planar drawing of G such that for each node ν of T all the vertices and edges of $G - G(\nu)$ are in the external face of $G(\nu)$.*

Theorem 2. [14] *Let C be a connected clustered graph whose underlying graph has n vertices. There exists an $O(nc)$ time algorithm for testing the c-planarity of C.*

Given a graph $G = (V,E)$ that is (in general) not planar a *planarization* of G is an embedded planar graph $G' = (V', E')$ such that:

- $V' = V \cup D$; the vertices of D represent crossings between edges, and are called *dummy vertices*;

- dummy vertices have degree equal to four;
- each edge (u, v) of E is associated with a path $u, d_1, \ldots, d_k, v$ $(k \geq 0)$ of G' such that $d_i \in D$ $(i = 1, \ldots, k)$; we call such a path an *edge path* of G'; and
- each dummy vertex is incident on two distinct edge paths and the edges of the same path are not consecutive in the embedding around the dummy vertex.

Given a clustered graph $C = (G, T)$ that is not c-planar, a *planarization* of C is a c-planar clustered graph $C' = (G', T')$ such that:

- G' is a planarization of G;
- T' is a tree obtained from T by adding one leaf for each dummy vertex of G';
- let d be a dummy vertex of G' and let u and v be the end-vertices of any edge path containing d. Vertex d is a child of a node of T' (i.e. it belongs to a cluster of C') that lies on the path of T' between u and v.

3 Computing a Maximal c-Planar Sub-clustered Graph

We now describe a planarization algorithm for clustered graphs, which we call `ClusteredGraphPlanarizer`. It follows the usual planarization strategy of the topology-shape-metrics approach. First, a maximal c-planar sub-clustered-graph of the given clustered graph is computed (Algorithm `MaximalcPlanar`). Second, the edges removed in the first step are reinserted (Algorithm `Reinsertion`) by suitably adding "dummy" vertices representing crossings. In this section we concentrate on Algorithm `MaximalcPlanar`, while Algorithm `Reinsertion` will be discussed in Section 4.

Although the problem of finding a maximal or a maximum planar subgraph of a given graph has been deeply studied (see, e.g., [7,19,18]), the problem of determining a maximal c-planar sub-clustered-graph of a clustered graph $C = (G, T)$ has not been investigated yet.

A possibility for solving the problem could be the one of inserting, starting from the empty graph, one-by-one the edges of G, repeating a c-planarity testing for each edge insertion and discarding the edges that cause crossings. This technique, although attractive for its simplicity, cannot be easily adopted. In fact, the only existing c-planarity testing algorithm [14] works only for connected clustered-graphs, while the intermediate clustered-graphs produced by the technique might be non-connected.

We adopt a different strategy. Instead of starting from the empty graph, we start from a connected c-planar sub-clustered-graph $C' = (G', T)$ that contains all the vertices of G. Such a sub-clustered-graph is computed so that its underlying graph contains a spanning tree of G. The computation of C' consists of two steps:

1. We compute a sub-clustered-graph C'' of C so that its underlying graph is a spanning tree of G. We call `SpanningTree` this step of the algorithm.

2. We compute C' from C'' by inserting edges that do not violate the c-planarity. We call `SimpleReinsertion` this step of the algorithm.

Intuitively, Algorithm `SpanningTree` constructs C'' by associating a spanning tree with each node of T (cluster of C) and by merging all such spanning trees. More formally, for each edge (u, v) of G we define the *allocation node* of (u, v) as $lca(u, v)$ in T, where $lca(u, v)$ denotes the *lowest common ancestor* of u and v. Note that $\nu = lca(u, v)$ is the deepest node of T such that cluster $V(\nu)$ contains both u and v. See Fig. 2 to have an example.

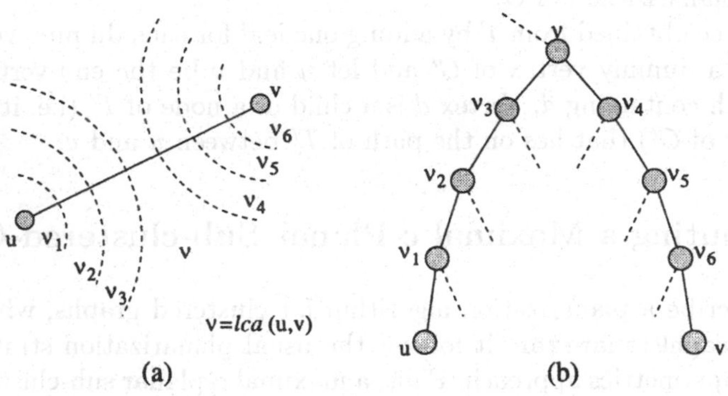

Fig. 2. (a) A fragment of a clustered graph. (b) A fragment of the inclusion tree, with vertices u and v and their lowest common ancestor (allocation node of (u, v)).

Let ν be a non-leaf node of T with children $\nu_1, \ldots, \nu_k$. Graph $F(\nu)$ is defined as follows (see for example Fig. 3):

- The vertices of $F(\nu)$ are $\nu_1, \ldots, \nu_k$.
- For each edge (u, v) such that $u \in V(\nu_i)$ and $v \in V(\nu_j)$ $(i \neq j)$ there is an edge (ν_i, ν_j) in $F(\nu)$. Edge (ν_i, ν_j) is the *representative* of (u, v) in $F(\nu)$.

Property 1. For each edge (u, v) of G there exists exactly one node ν of T such that $F(\nu)$ contains a representative of (u, v); ν is the allocation node of (u, v).

Property 2. The total number of vertices of graphs $F(\nu)$ is $n + c - 1$, and the total number of edges of graphs $F(\nu)$ is m.

We denote by $S(\nu)$ a spanning tree of $F(\nu)$. We construct a subgraph ST of G by selecting in G only the edges (u, v) such that, if ν is the allocation node of (u, v), then (u, v) has its representative in $S(\nu)$.

Property 3. Graph ST is a spanning tree of G.

We call $ST(\nu)$ the graph obtained from the intersection of ST and $G(\nu)$.

Property 4. Graph $ST(\nu)$ is a spanning tree of $G(\nu)$, for each ν in T.

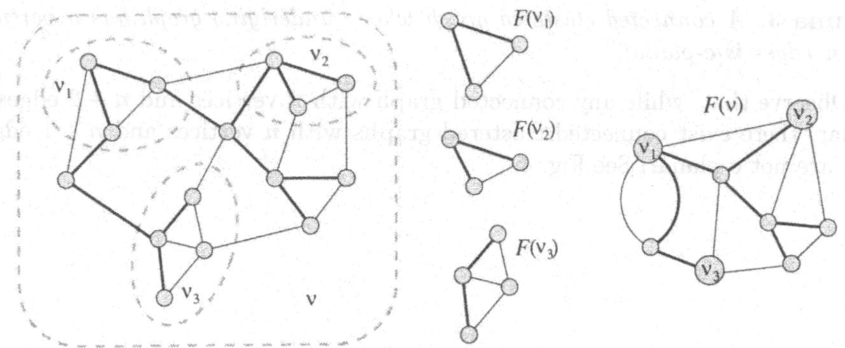

Fig. 3. A clustered graph and graphs $F(\nu)$, $F(\nu_1)$, $F(\nu_2)$, and $F(\nu_3)$. The edges in bold are those of the spanning trees.

Lemma 1. *A connected clustered graph whose underlying graph is a tree is c-planar.*

Proof. Follows from Theorem 1 and from the fact that, for any drawing, all the vertices of a tree stay on the same face.

Lemma 2. *Clustered graph $C'' = (ST, T)$ is a c-planar connected sub-clustered-graph of C.*

Proof. From Property 4 it follows that each $ST(\nu)$ is connected. Hence C'' is connected and its underlying graph is a tree. From Lemma 1 it follows that C'' is also c-planar.

Theorem 3. *Let $C = (G, T)$ be a connected clustered graph and let m be the number of edges of G. Algorithm* **SpanningTree** *computes a connected sub-clustered-graph of C whose underlying graph is a spanning tree of G in $O(m)$ time.*

Proof. For each node ν of T the construction of $F(\nu)$ is performed by visiting the edges of G and by assigning each edge to a specific $F(\nu)$. This requires the computation of the allocation node $\nu = lca(u, v)$ of each edge (u, v) and the computation of the children of ν in the paths from ν to u and from ν to v. A trivial implementation of this step would require $O(mc)$ time, where c is the number of non-leaf nodes of T. However, we can do it in $O(m)$ time by using a variation of the Schieber and Vishkin data structure [22]. The original data structure allows us, after a linear time preprocessing, to perform lowest common ancestor queries on a tree in constant time. Unfortunately, it does not give primitives to determine the required children of the lowest common ancestor. It is possible to suitably enrich the information associated with the "inlabel paths" of the data structure to solve the problem.

The computation of the spanning trees can be done in linear time.

Algorithm `SimpleReinsertion` constructs C' by reinserting into $C'' = (ST, T)$ some edges that do not cause crossings. This is done with the purpose of reducing the number of times the c-planarity testing is executed. The reinsertion strategy is based on the following lemma.

Lemma 3. *A connected clustered graph whose underlying graph has n vertices and n edges is c-planar.*

Observe that, while any connected graph with n vertices and $n+2$ edges is planar, there exist connected clustered graphs with n vertices and $n+1$ edges that are not c-planar. See Fig. 4.

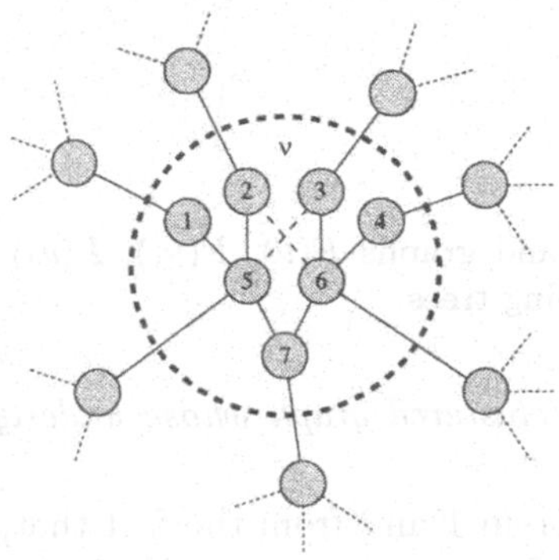

Fig. 4. Inserting edge $(2,6)$ and $(5,3)$ generates a crossing.

The following lemma generalizes Lemma 3.

Lemma 4. *Let $C = (G,T)$ be a connected clustered graph whose underlying graph G is a tree. Let $u_1,\dots,u_k,v_1,\dots,v_k$ be vertices of G such that edges $(u_1,v_1),(u_2,v_2),\dots,(u_k,v_k)$ are not in G. If for each pair $(u_i,v_i),(u_j,v_j)$ $(i \neq j)$, $lca(u_i,v_j) \neq lca(lca(u_i,v_i),lca(u_j,v_j)) \neq lca(u_j,v_i)$, then C remains c-planar after adding $(u_1,v_1),(u_2,v_2),\dots,(u_k,v_k)$.*

Because of Lemma 4, we can do the following. We visit T bottom-up. For each ν of T we check (i) if no edge of $G(\nu) - ST(\nu)$ has been already reinserted and (ii) if $G(\nu) - ST(\nu)$ is not empty. If both conditions hold, then we reinsert an edge of $G(\nu) - ST(\nu)$. We apply the same procedure to all the nodes of T.

Theorem 4. *Let C be a connected clustered graph whose underlying graph has m edges. Algorithm* `SimpleReinsertion` *computes a connected c-planar sub-clustered-graph of C in $O(m)$ time.*

After the construction of C' we reinsert a maximal number of edges of C by testing c-planarity for each edge insertion, using each time the c-planarity testing algorithm in [14]. After the last reinsertion we also compute a c-planar embedding [14].

The following theorem is proved by using Theorems 3, 4, and 2.

Theorem 5. *Let C be a connected clustered graph whose underlying graph has n vertices and m edges. Algorithm* `MaximalcPlanar` *computes a connected maximal c-planar embedded sub-clustered-graph of C in $O(mnc)$ time.*

4 Reinsertion of the Discarded Edges

We describe Algorithm Reinsertion. A technique similar to the one adopted by algorithm Reinsertion is sketched in [20]. Once a maximal connected c-planar embedded sub-clustered-graph $C_{mp} = (G_{mp}, T)$ of $C = (G, T)$ has been computed (where G_{mp} denotes a planar subgraph of G), the remaining edges are reinserted by using a variation of the "classical" technique [5] that is based on computing shortest paths on the dual graph of G_{mp}.

In fact, in the case of clustered graphs it is not possible to apply exactly the same technique that is adopted for graphs. Fig. 5 shows how reinserting an edge (u, v) by following a path on the dual graph of G_{mp} can cause: (i) an edge-region crossing (Fig. 5.a) and/or (ii) more than one crossing between a cluster and one of its incident edges. (Fig. 5.a).

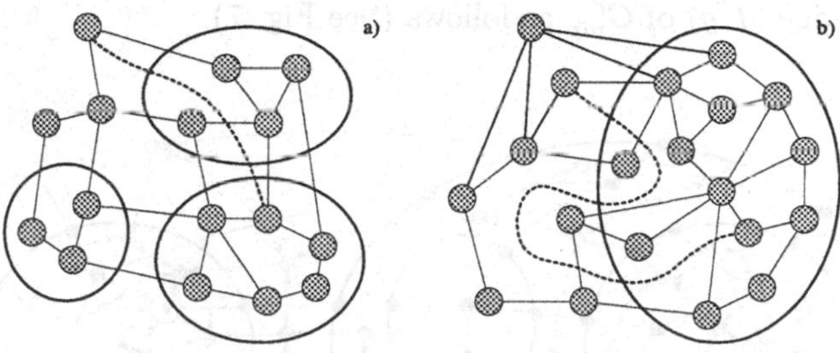

Fig. 5. Examples of wrong insertions: (a) edge-region crossing and (b) more than one crossing between a cluster and one of its incident edges.

In order to avoid the above problems, we compute the shortest paths on a planar embedded graph constructed as follows. Observe that the embedding of C_{mp} induces a planar embedding on G_{mp}.

- We "materialize" the boundary of each cluster. Namely, we augment G_{mp} as follows: (i) for each pair e, ν such that e is an edge of G_{mp} incident on cluster ν, we split e by inserting a *boundary vertex* $v_{e,\nu}$; (ii) for each face we traverse the border counterclockwise and construct a list of boundary vertices; (iii) for each pair of boundary vertices $v_{e_1,\nu}$, $v_{e_2,\nu}$ that belong to the same face f and that are consecutive in the list of f, we insert a *boundary edge* $(v_{e_1,\nu}, v_{e_2,\nu})$. See Fig. 6. Observe that each boundary edge corresponds to one specific cluster and that the boundary edges corresponding to the same cluster are arranged into simple cycles (*boundary cycles*). Boundary cycles and clusters (except for the root of T) are in one-to-one correspondence.
- We compute the planar embedded dual G'_{mp} of the planar graph described in the previous step. Note that each face f of G'_{mp} is inside a certain set of boundary cycles. Such cycles correspond to clusters that are on a rooted path p of T. If the set is empty we *associate* f with the root of T; otherwise, we *associate* f with the lowest node of p.

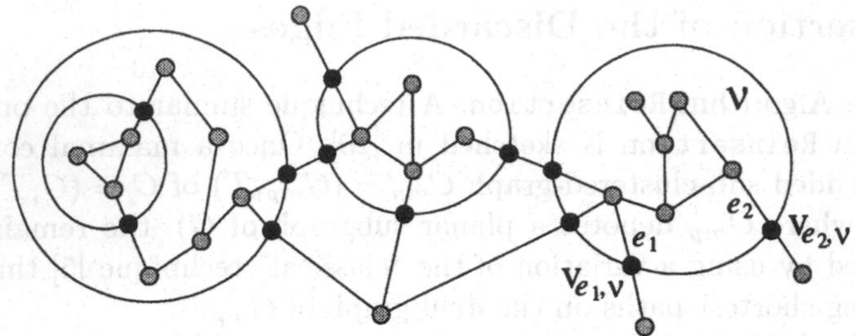

Fig. 6. Insertion of edges and vertices for representing the boundaries of the clusters.

At each edge reinsertion we perform the following algorithm. Let (u, v) be the edge to be reinserted and let $\nu_1, \ldots, \nu_k$ be the nodes of the path of T from u to v (the path contains the allocation node of (u, v)). We orient or temporarily remove each edge (f, g) of G'_{mp} as follows (See Fig. 7).

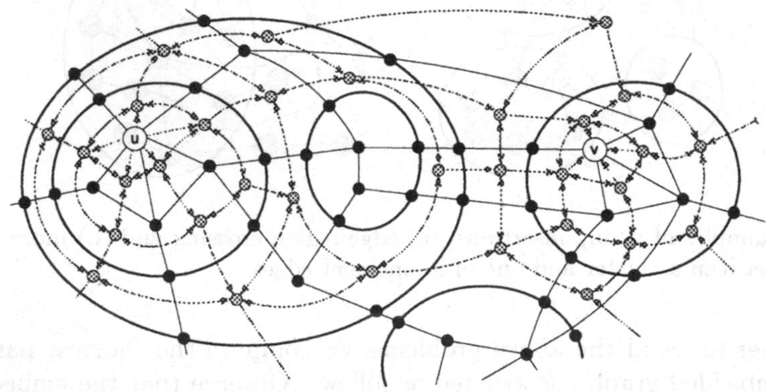

Fig. 7. Orientation and removal of edges of graph G'_{mp}. The edges of G'_{mp} are grey.

Two cases are possible:

1. f is associated with node ν_i and g is associated with node ν_j $(i, j \in \{1, \ldots, k\})$: if $i = j$, then we give a bidirectional orientation to (f, g), else (assume without loss of generality that $i < j$) we orient (f, g) from f to g.
2. either f or g is not associated with a node of $\nu_1, \ldots, \nu_k$: we temporarily remove (f, g).

After orienting the graph and temporarily removing some of its edges, we compute a directed shortest path between the set of faces incident on u and the set of faces incident on v. Edge (u, v) is reinserted into G_{mp} by following the computed shortest path. Dummy vertices representing edge crossings are inserted. The dual graph G'_{mp} is modified accordingly. The temporarily removed edges are restored and the orientation of the edges is removed.

Theorem 6. *Let $C_{mp} = (G_{mp}, T)$ be a maximal connected c-planar embedded sub-clustered-graph of a clustered graph $C = (G, T)$. Let m be the number of edges of G and let m' be the number of edges of $G - G_{mp}$. Starting from C_{mp}, Algorithm* **Reinsertion** *computes a planarization of C with χ dummy vertices (crossings) in $O(m'\chi + m'mc)$ time.*

Proof. Augmenting the embedding of C_{mp} inserting the border edges is done in $O(nc)$ time, where n is the number of vertices of G and c is the number of non-leaf nodes of T. The same amount of time is required to compute the dual graph G'_{mp}. At each edge reinsertion we spend $O(\chi + mc)$ time both for computing the orientation of the dual graph and for computing, with a breadth-first-search, the shortest path. We remark that each reinsertion can originate at most c crossings between the inserted edge and the border edges of the clusters.

In the implementation of Algorithm `Reinsertion` we adopted a slightly different strategy. Although such a strategy does not reduce the time complexity of the algorithm, it has shown positive effects on the execution time of the algorithm.

A heavy step of the algorithm is the one that requires the orientation and the removal of some edges of G'_{mp} at each edge reinsertion. To avoid that, we have used a modified breadth-first-search that is performed directly on G'_{mp}. Suppose that we are reinserting edge (u, v) and that we are visiting face f associated with cluster ν. Also, suppose we are going to traverse edge (f, g). Face g can be visited if and only if the cluster associated with g is either ν or the cluster that immediately follows ν in the path p of T from u to v. Of course, this requires the computation of p. This can be efficiently done by determining the allocation node of (u, v) in constant time exploiting the same variation of the Schieber and Vishkin data structure [22] mentioned in the proof of Theorem 3. A simpler but less efficient alternative would be the one of determining p with a dove-tail bottom-up visit of T starting from u and v.

The following theorem summarizes the time complexity of the whole planarization algorithm.

Theorem 7. *Let $C = (G, T)$ be a clustered-graph. Let n, m, and c be the number of vertices of G, edges of G, and non-leaf nodes of T, respectively. Algorithm* ***ClusteredGraphPlanarizer*** *computes a planarization of C with χ dummy vertices in $O(m\chi + m^2c + mnc)$ time.*

5 A Simple Planarization Algorithm

We now describe a planarization algorithm for a clustered graph $C = (G, T)$ simpler than the one described above. We use it as a reference point in experimenting the effectiveness of our algorithm. We call it `SimplePlanarizer`.

We visit T top-down. We start from the root ν of T and compute graph $F(\nu)$. Then, we apply any planarization algorithm (see, e.g. [5]), in order to compute a planarization of $F(\nu)$. Consider now a child ν_1 of ν. We would like to compute a planarization of $F(\nu_1)$ and to "glue" it into $F(\nu)$. It turns out that the ordering of the edges incident on ν_1 in the planarization of $F(\nu)$ induces a constraint in the planarization of $F(\nu_1)$. Namely, the vertices of $F(\nu_1)$ that are incident on edges of G that are also incident on cluster ν must satisfy two constraints:

– they must stay on the external face of $F(\nu_1)$ and
– while visiting the external face of $F(\nu_1)$ they must appear in the same clockwise order of the edges incident on ν_1 in $F(\nu)$.

The above type of constraints on the topology can be imposed by using the facilities of existing graph drawing libraries (see, e.g. [16]). The same strategy is applied to all the descendants of ν, recursively.

We conclude this section by observing that Algorithm `SimplePlanarizer` works also for the case of non connected clustered graphs.

6 Experimental Study

The algorithms presented in Sections 3, 4, and 5 have been implemented and extensively tested. The implementation of all algorithms have been done by using the C++ language (Visual C++ compiler) and exploiting the basic graph drawing facilities of the `GDToolkit` library [16]. Also, for implementing Algorithm `MaximalcPlanar`, we used the implementation of the c-planarity testing algorithm [14] in the AGD library [1] .

All the experiments have run on a personal computer equipped with an AMD Athlon K7 (500 MHz) and 128 MB of RAM.

We have used two test suites. The first one (*Suite 1*) consists of more than 11,000 graphs with number of vertices ranging from 10 to 100; it has been introduced in [6] and since then has become a widely used test-suite in experimental Graph Drawing. The average density of its graphs is about 1.3.

The second test suite (*Suite 2*) has been conceived to test the algorithms against graphs with higher density. It has been generated with a graph generator that works as follows. For generating a graph the user specifies two values: the number n of vertices and the desired density d. The graph is generated by randomly inserting nd edges on the set of n vertices. The result is kept if it is connected and does not contain multiple edges, otherwise it is discarded and the generation process is repeated. With such a graph generator we have generated 10 groups of graphs. Each group has a fixed density and contains 50 graphs with number of vertices in the range 10 – 100. Densities range from 1.2 to 3.

The graphs of Suites 1 and 2 are just graphs and do not have a cluster structure. In order to perform our experiments, we have augmented them with clusters with the following algorithm. Let G be a graph of a suite. We randomly select a vertex v and compute a breadth-first-search spanning tree of G starting from v. At this point we randomly select a vertex u and perform a breadth-first-search on the spanning tree starting from u and visiting an edge with probability 0.7. The visited set of vertices of G is a cluster. The same algorithm is recursively repeated on the spanning tree using as boundary of the visits the clusters already computed. It is easy to see that the obtained clustered graphs are connected.

They have an average number of clusters that is about 1/4 of the number of vertices. The depth of the obtained inclusion trees is in the range $2-5$. The probability 0.7 of visiting an edge has been chosen after several experiments to keep high the number of generated clusters.

Fig. 8.a shows the average number of edge crossings obtained by Algorithm `SimplePlanarizer` and by Algorithm `ClusteredGraphPlanarizer` on the clustered graphs of Suite 1. Algorithm `ClusteredGraphPlanarizer` outperforms Algorithm `SimplePlanarizer` by about 40%. Another interesting reference point for evaluating the effectiveness of Algorithm `ClusteredGraphPlanarizer` is to compare its behavior against the one of a standard planarizer, used for planarizing the same graphs, considered without clusters. Fig. 8.b shows a comparison with the planarizer of `GDToolkit`. The results are quite encouraging and show how clusters have a limited impact on the number of crossings.

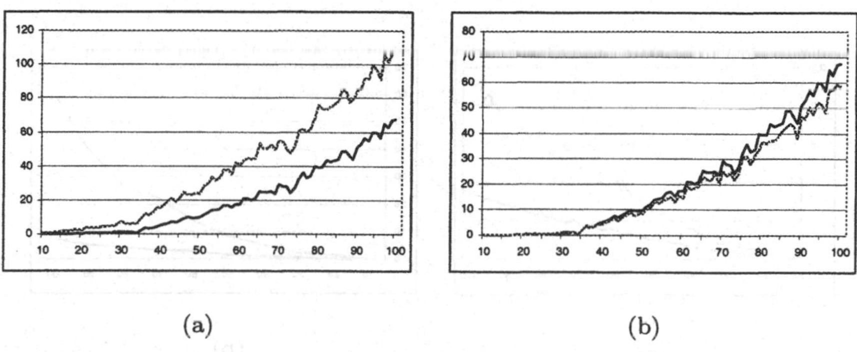

(a) (b)

Fig. 8. Suite 1. (a) Average number of edge crossings obtained by Algorithm `SimplePlanarizer` and by Algorithm `ClusteredGraphPlanarizer` (curve below). (b) Average number of edge crossings obtained by Algorithm `ClusteredGraphPlanarizer` and by the `GDToolkit` planarizer (curve below).

Fig. 9 confirms the good behavior of `ClusteredGraphPlanarizer` also with graphs of higher density. It has been obtained using the graphs with densities 2 and 3 of Suite 2. We omit the graphics obtained with other densities, since they confirm what appears in Fig. 9.

Fig. 10 shows how Algorithm `ClusteredGraphPlanarizer` have time performance that are reasonable for usual applications. Even for the largest graphs of Suite 1 (Suite 2) the time is less than 4 seconds (9 seconds). Algorithm `SimplePlanarizer` is much less efficient. However, the implementation of Algorithm `SimplePlanarizer` did not exploit sophisticated data structures at the level of Algorithm `ClusteredGraphPlanarizer`. We believe that a more careful implementation could reduce the gap between the time performance of the two algorithms.

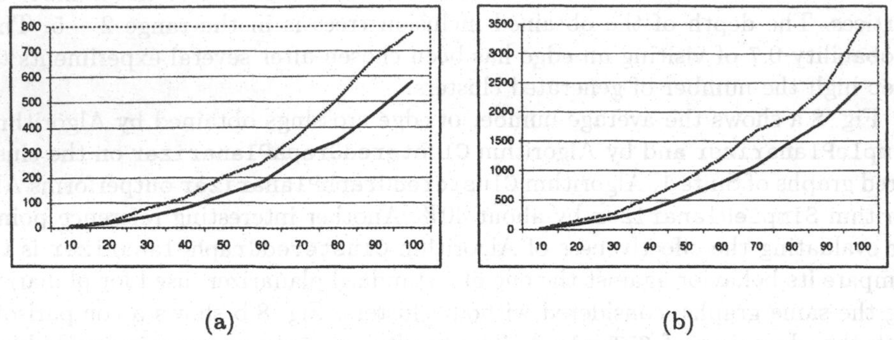

Fig. 9. Suite 2. Average number of edge crossings obtained by Algorithm **SimplePlanarizer** and by Algorithm **ClusteredGraphPlanarizer** (curves below). (a) Density 2. (b) Density 3.

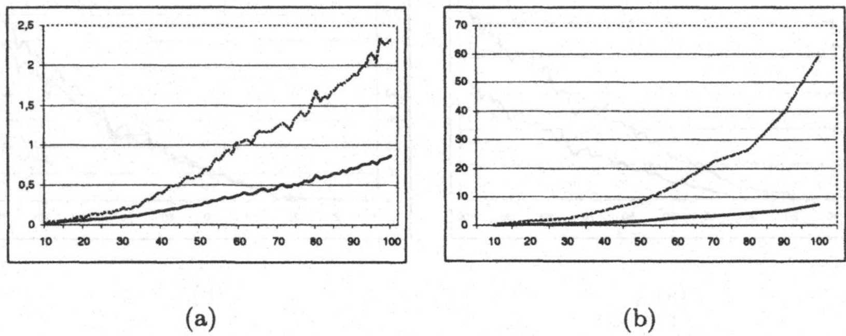

Fig. 10. (a) Suite 1. Average time (seconds) spent by Algorithm **SimplePlanarizer** and by Algorithm **ClusteredGraphPlanarizer** (curve below). (b) Suite 2, density 3. Average time (seconds) spent by by Algorithm **SimplePlanarizer** and by Algorithm **ClusteredGraphPlanarizer** (curve below).

7 A Complete Drawing Algorithm for Clustered Graphs

We have embedded Algorithm **ClusteredGraphPlanarizer** into a complete topology-shape-metrics algorithm for computing drawings of clustered graph within the orthogonal podevsnef [15] drawing convention.

After the planarization has been performed, for computing the shape of the drawing, representing clusters as boxes, we use the same technique adopted in [3, 20]. Namely, we use the variation described in [2] of the min-cost-flow based algorithm in [15]. The boundary cycles representing the borders of the clusters are constrained to have the shape of a rectangle. This can be easily done by using constraints on the flow traversing the edges of those cycles.

We compute the metric of the drawing with the compaction algorithm described in [4]. Fig. 11 shows a drawing computed by the algorithm.

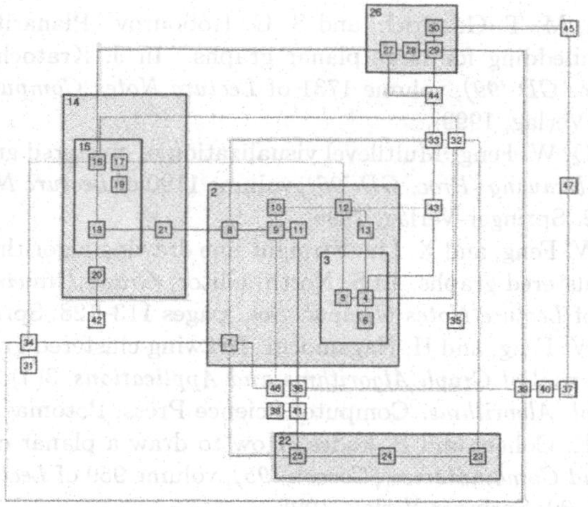

Fig. 11. A clustered graph with 40 vertices and 7 clusters.

8 Open Problems and Future Work

Several problems are still open in the field of clustered graphs. In particular, we are interested in finding planarization algorithms that are more efficient (theoretically and experimentally) than the one described in this paper. Further, the role of connectivity in the c-planarity of clustered graphs is still unclear. Namely, it is unknown whether the c-planarity testing problem for non connected clustered graphs is computationally hard or not. Studies on this matter could open new perspectives for the planarization problem.

References

1. AGD. A library of algorithms for graph drawing. Online. http://www.mpi-sb.mpg.de/AGD/.
2. P. Bertolazzi, G. Di Battista, and W. Didimo. Computing orthogonal drawings with the minimum numbr of bends. *IEEE Transactions on Computers*, 49(8), 2000.
3. U. Brandes, S. Cornelsen, and D. Wagner. How to draw the minimum cuts of a planar graph. In J. Marks, editor, *Graph Drawing (Proc. GD '00)*, volume 1984 of *Lecture Notes Comput. Sci.*, pages 103–114. Springer-Verlag, 2000.
4. G. Di Battista, W. Didimo, M. Patrignani, and M. Pizzonia. Orthogonal and quasi-upward drawings with vertices of arbitrary size. In *Proc. GD '99*, volume 1731 of *LNCS*, pages 297–310, 2000.
5. G. Di Battista, P. Eades, R. Tamassia, and I. G. Tollis. *Graph Drawing.* Prentice Hall, Upper Saddle River, NJ, 1999.
6. G. Di Battista, A. Garg, G. Liotta, R. Tamassia, E. Tassinari, and F. Vargiu. An experimental comparison of four graph drawing algorithms. *Comput. Geom. Theory Appl.*, 7:303–325, 1997.
7. H. N. Djidjev. A linear algorithm for the maximal planar subgraph problem. In *Proc. 4th Workshop Algorithms Data Struct.*, Lecture Notes Comput. Sci. Springer-Verlag, 1995.

8. C. A. Duncan, M. T. Goodrich, and S. G. Kobourov. Planarity-preserving clustering and embedding for large planar graphs. In J. Kratochvil, editor, *Graph Drawing (Proc. GD '99)*, volume 1731 of *Lecture Notes Comput. Sci.*, pages 186–196. Springer-Verlag, 1999.
9. P. Eades and Q. W. Feng. Multilevel visualization of clustered graphs. In S. North, editor, *Graph Drawing (Proc. GD '96)*, volume 1190 of *Lecture Notes Comput. Sci.*, pages 101–112. Springer-Verlag, 1996.
10. P. Eades, Q. W. Feng, and X. Lin. Straight line drawing algorithms for hierarchical graphs and clustered graphs. In S. North, editor, *Graph Drawing (Proc. GD '96)*, volume 1190 of *Lecture Notes Comput. Sci.*, pages 113–128. Springer-Verlag, 1996.
11. P. Eades, Q. W. Feng, and H. Nagamochi. Drawing clustered graphs on an orthogonal grid. *Journal of Graph Algorithms and Applications*, 3(4):3–29, 2000.
12. S. Even. *Graph Algorithms.* Computer Science Press, Potomac, Maryland, 1979.
13. Q. W. Feng, R. Cohen, and P. Eades. How to draw a planar clustered graph. In *Computing and Combinatorics (Cocoon '95)*, volume 959 of *Lecture Notes Comput. Sci.*, pages 21–30. Springer-Verlag, 1995.
14. Q. W. Feng, R. F. Cohen, and P. Eades. Planarity for clustered graphs. In P. Spirakis, editor, *Symposium on Algorithms (Proc. ESA '95)*, volume 979 of *Lecture Notes Comput. Sci.*, pages 213–226. Springer-Verlag, 1995.
15. U. Fößmeier and M. Kaufmann. Drawing high degree graphs with low bend numbers. In F. J. Brandenburg, editor, *Graph Drawing (Proc. GD '95)*, volume 1027 of *Lecture Notes Comput. Sci.*, pages 254–266. Springer-Verlag, 1996.
16. GDToolkit. Graph drawing toolkit. Online. http://www.dia.uniroma3.it/~gdt.
17. M. L. Huang and P. Eades. A fully animated interactive system for clustering and navigating huge graphs. In S. H. Whitesides, editor, *Graph Drawing (Proc. GD '98)*, volume 1547 of *Lecture Notes Comput. Sci.*, pages 374–383. Springer-Verlag, 1998.
18. M. Jünger, E. K. Lee, P. Mutzel, and T. Odenthal. A polyhedral approach to the multi-layer crossing number problem. In G. Di Battista, editor, *Graph Drawing (Proc. GD '97)*, number 1353 in Lecture Notes Comput. Sci., pages 13–24. Springer-Verlag, 1997.
19. M. Jünger and P. Mutzel. Maximum planar subgraphs and nice embeddings: Practical layout tools. *Algorithmica*, 16(1):33–59, 1996. (special issue on Graph Drawing, edited by G. Di Battista and R. Tamassia).
20. D. Lütke-Hüttmann. *Knickminimales Zeichnen 4-planarer Clustergraphen.* Master's thesis, Universität des Saarlandes, 1999.
21. T. Nishizeki and N. Chiba. Planar graphs: Theory and algorithms. *Ann. Discrete Math.*, 32, 1988.
22. B. Schieber and U. Vishkin. On finding lowest common ancestors: Simplification and parallelization. *SIAM J. Comput.*, 17(6):1253–1262, 1988.
23. K. Sugiyama and K. Misue. Visualization of structural information: Automatic drawing of compound digraphs. *IEEE Trans. Softw. Eng.*, 21(4):876–892, 1991.
24. K. Sugiyama, S. Tagawa, and M. Toda. Methods for visual understanding of hierarchical systems. *IEEE Trans. Syst. Man Cybern.*, SMC-11(2):109–125, 1981.

An Algorithm for Finding Large Induced Planar Subgraphs

Keith Edwards[1] and Graham Farr[2,*]

[1] Department of Applied Computing
University of Dundee
Dundee, DD1 4HN
U.K.
kedwards@computing.dundee.ac.uk

[2] School of Computer Science and Software Engineering
Monash University (Clayton Campus)
Clayton
Victoria 3168
Australia
gfarr@csse.monash.edu.au

Abstract. This paper presents an efficient algorithm that finds an induced planar subgraph of at least $3n/(d+1)$ vertices in a graph of n vertices and maximum degree d. This bound is sharp for $d = 3$, in the sense that if $\varepsilon > 3/4$ then there are graphs of maximum degree 3 with no induced planar subgraph of at least εn vertices. Our performance ratios appear to be the best known for small d. For example, when $d = 3$, our performance ratio of at least 3/4 compares with the ratio 1/2 obtained by Halldórsson and Lau. Our algorithm builds up an induced planar subgraph by iteratively adding a new vertex to it, or swapping a vertex in it with one outside it, in such a way that the procedure is guaranteed to stop, and so as to preserve certain properties that allow its performance to be analysed. This work is related to the authors' work on fragmentability of graphs.

1 Introduction

Finding a large planar subgraph of a graph is an important problem in graph drawing [5]. In this paper, an *induced* planar subgraphs is sought, and its size is taken to be the number of its vertices.

Formally, the problem we would like to solve is the following.

MAXIMUM INDUCED PLANAR SUBGRAPH (MIPS)
Input: Graph G, on n vertices.
Output: A largest set of vertices $P \subseteq V(G)$ such that the induced subgraph $\langle P \rangle$ is planar.

Of course, one can also look at the complementary, and computationally equivalent, problem of finding the smallest set R of vertices whose removal leaves

* Some of the work of this paper was done while the second author visited the Department of Applied Computing, University of Dundee, in July–September 2000.

P. Mutzel, M. Jünger, and S. Leipert (Eds.): GD 2001, LNCS 2265, pp. 75–83, 2002.
© Springer-Verlag Berlin Heidelberg 2002

behind a planar graph (i.e., so that $\langle V(G) \setminus R \rangle$ is planar), and we sometimes find this point of view convenient.

Unfortunately, MIPS is NP-hard [18,19], and is also hard to approximate [22]. Halldórsson [11] gives an approximation algorithm that finds an induced planar subgraph of size $\Omega(n^{-1}(\log n/\log\log n)^2)$ times the optimum. For graphs of maximum degree d, Halldórsson and Lau [12] give a linear time algorithm with a performance ratio of $1/\lceil (d+1)/3 \rceil$.

In this paper, we present a polynomial time algorithm that finds an induced planar subgraph of at least $3n/(d+1)$ vertices in a graph of maximum degree d. This bound is sharp for $d = 3$, and some upper bounds are given for $d > 3$. We also ask: for each d, what is the largest $\alpha(d)$ such that an induced planar subgraph of $\alpha(d) \cdot n$ vertices can be found in every graph of maximum degree d? Note that this ratio compares the size of the induced planar subgraph found with that of the whole graph, and so is not a performance ratio, though it certainly implies a performance ratio of at least $3/(d+1)$. In a sense, our result removes the ceiling from the performance ratio of Halldórsson and Lau given above, which for graphs of low d with $d \not\equiv 2 \pmod 3$ is a significant improvement. For example, when $d = 3$, our algorithm has performance ratio at least $3/4$, whereas that of [12] is $1/2$. Also, the induced planar subgraphs found by Halldórsson and Lau's algorithm have maximum degree at most 2, whereas those found by our algorithm are not so restricted in structure.

The algorithm we present is virtually implicit in our proof of [8, Theorem 3.2], though that proof is, on the face of it, non-algorithmic. Our purpose here is to state the algorithm clearly and explicitly to the graph drawing community, and to discuss its properties and some of the questions it raises.

The fact that the algorithm performs well for graphs of bounded degree suggests that it may be useful in practical applications, where graphs are often sparse (although that is not quite the same thing, of course).

We recognise that MIPS has attracted much less attention than the Maximum Planar Subgraph (MPS) problem, in which a planar subgraph in the usual sense (i.e., not necessarily induced) is sought, and its size is taken to be the number of its edges (and its vertex set may as well be the vertex set of the input graph); see, e.g., [15,16,17,20,23]. MPS is known to be NP-hard [21], even for graphs of maximum degree 3 [9]. A number of exact or approximate algorithms have been developed and studied [2,3,4,7,10,15,16,23], and the best known performance ratio is 4/9, due to Călinescu et al. [3]. The literature on MIPS is much smaller. (Liebers [20] reviews MIPS in two pages, while devoting ten to MPS.) Apart from the complexity references cited above, we note work on finding large induced planar subgraphs of graphs of a given genus [6,13,14].

Throughout the paper, n is the number of vertices in the graph(s) under discussion, m is the number of edges, and d is the maximum degree. If $X \subseteq V(G)$ then $k(X)$ denotes the number of components of $\langle X \rangle$. If $X, Y \subseteq V(G)$, then $E(X, Y)$ denotes the set of edges with one endpoint in X and the other in Y. If $v \in V(G)$ and $X \subseteq V(G)$, then $N_X(v)$ denotes the set of those vertices in X that are adjacent to v.

2 The Algorithm

By way of introduction, consider the following easy algorithm:

Algorithm 1. An obvious iterative algorithm

```
1.  Input: Graph G.
2.  P := ∅
    R := V(G)
3.  while ( there exists x0 ∈ R such that
            ⟨P ∪ {x0}⟩ is planar )
    {
            P := P ∪ {x0}
            R := R \ {x0}
    }
```

This algorithm just keeps on adding vertices to P for as long as it is possible to do so, while preserving planarity of $\langle P\rangle$. It is discussed briefly by Liebers [20], who notes that it has complexity $O(nm)$, where $m = |E(G)|$. Observe that, when it stops, each vertex in R must be adjacent to at least two vertices in P (else such a vertex could be added to P by the algorithm). A simple counting argument then shows that the induced planar subgraph so found has at least $2n/(d+2)$ vertices. (Consider $|E(P,R)|$. Count (or rather, bound) it in two ways.) Note that the same lower bound is achieved by an algorithm which replaces the condition in Step 3 of Algorithm 1 by the simpler test

```
( there exists x0 ∈ R such that
  deg_P(x0) ≤ 1  )
```

The modified algorithm may often do worse than Algorithm 1, and the vertex set P found may not even be maximal (subject to planarity of $\langle P\rangle$). Nonetheless, the set P found still has at least $2n/(d+2)$ vertices. It will also be easier to implement, and faster.

The algorithm we now present improves on this basic iterative approach by examining more carefully whether the vertex under consideration, x_0, can usefully be added to P. Sometimes it is advantageous to swap x_0 with some vertex already in P. Care is needed to ensure that an algorithm that does this will eventually stop.

The algorithm is based on the proof of [8, Theorem 3.2].

The algorithm maintains four sets of vertices of the input graph G: P, the set of vertices of the induced planar subgraph constructed so far; $R = V(G) \setminus P$; F, the set of vertices that belong to those components of $\langle P\rangle$ that are trees (i.e., the *f*orest part of $\langle P\rangle$); and $N = P \setminus F$ (*n*on-trees).

Algorithm 2. Finding an induced planar subgraph of at least $3n/(d+1)$ vertices.

```
1.  Input: Graph G.
2.  P := ∅
    N := ∅
    F := ∅
    R := V(G)
```

```
3.  while ( there exists x_0 ∈ R such that
           N_N(x_0) ≤ 1 or (N_N(x_0) = 2 and N_F(x_0) ≤ 1) )
    {
      3.1. if (N_N(x_0) ≤ 1)
              P := P ∪ {x_0}
              R := R \ {x_0}
              update N, F.
      3.2. else if (the two vertices y, z in N_N(x_0) are in different
           components of N)
              P := P ∪ {x_0}
              R := R \ {x_0}
              update N, F.
      3.3. else if (there is a unique y–z path in ⟨N⟩)
              P := P ∪ {x_0}
              R := R \ {x_0}
              update N, F.
      3.4. else
              Let N_0 be the vertex set of the component of N containing y
              and z.
              Find a y–z path Π in ⟨N_0⟩.
                  Let it be y = x_1, x_2, ..., x_t = z.
              x_i := first vertex on Π (when going from y towards z) such that:
                        there is a path Π' in ⟨N_0⟩ from x_i to some later x_j, j > i,
                        with Π' disjoint from Π except at its endpoints.
              P := (P \ {x_i}) ∪ {x_0}
              R := (R \ {x_0}) ∪ {x_i}
              update N, F.
    }
4.  Output: P.
```

Theorem 1 *Algorithm 2 finds an induced planar subgraph of at least* $3n/(d+1)$ *vertices.*

Proof. See [8, Theorem 3.2]. □

In outline, the proof proceeds as follows. In Algorithm 2, it is easy to see that conditions 3.1, 3.2 and 3.3, if satisfied, allow the vertex x_0 to be added to P. A little more effort shows that, whenever step 3.4 is performed, the graph resulting from this step is planar. Theorem 2 below shows that the algorithm finds an induced planar subgraph. When the algorithm stops, each $v \in R$ satisfies either $d_N(v) \geq 3$ or $d_N(v) = 2 \wedge d_F(v) \geq 2$. These together imply $d_P(v) \geq 3 + d_F(v)/2$. If $d \geq 4$, we count the edges of $E(P, R)$ in two ways (once from P, once from R) and derive (at some length) inequalities yielding the claimed bound. If $d = 3$, count $E(N, R)$ instead.

3 Discussion

3.1 Complexity

Theorem 2 *Algorithm 2 has time complexity $O(nm)$.*

Proof. At each step, Algorithm 2 *either*

(i) increases $|P|$, while keeping $\langle P\rangle$ planar, *or*
(ii) decreases $|E(P)|$, while keeping $|P|$ unchanged and maintaining planarity of $\langle P\rangle$, *or*
(iii) decreases $k(P)$, while keeping $|P|$ and $|E(P)|$ unchanged and maintaining planarity of $\langle P\rangle$.

It follows that the algorithm stops eventually. In fact, if $s_{(i)}, s_{(ii)}, s_{(iii)}$ denote the number of steps of Algorithm 2 that do (i), (ii) and (iii) respectively in the above list, and the total number of steps is $s = s_{(i)} + s_{(ii)} + s_{(iii)}$, then we can find a linear bound on s as follows. It can be shown that any step of type (i) increases $|E(P)| + k(P)$ by at most 2, and any step of type (ii) or (iii) decreases $|E(P)| + k(P)$ by exactly 1. Hence

$$2s_{(i)} - s_{(ii)} - s_{(iii)} \geq |E(P)| + k(P).$$

Combining this with the obvious $s_{(i)} = n$ gives $s_{(ii)} + s_{(iii)} \leq 2n$, so that $s \leq 3n$. (In fact, using $|E(P)| \geq |P| - k(P)$ here gives $s \leq 3n - |P|$. Using our bound on $|P|$ (Theorem 1), we obtain $s \leq 3nd/(d+1)$.) The time taken by each step is no worse than linear in m, so the time complexity is $O(nm)$. □

In the light of the superficial similarity of MIPS and MPS, it is interesting to note that Algorithm 2 actually tries (in steps of type (ii)) to *decrease* the number of edges in $\langle P\rangle$ in situations when it cannot increase $|P|$. Of course, this is quite understandable, as graphs with fewer edges should generally be more amenable to the addition of vertices.

3.2 Performance

It is natural to ask whether our lower bound of $3n/(d+1)$ is the best possible, or whether there is some constant $\varepsilon > 3/(d+1)$ such that every graph on n vertices with maximum degree d has an induced planar subgraph of at least εn vertices. Define $\alpha(d)$ by

$$\alpha(d) = \sup\{\varepsilon \mid \text{every } G \text{ such that } \Delta(G) \leq d \text{ has an induced planar subgraph of } \geq \varepsilon n \text{ vertices}\}.$$

We ask, then, for the value of $\alpha(d)$. Theorem 1 shows that

$$\alpha(d) \geq 3/(d+1). \tag{1}$$

On the other hand, the largest induced planar subgraph of K_{d+1} is K_4, so

$$\alpha(d) \leq 4/(d+1). \tag{2}$$

Furthermore, the result [8, Theorem 3.3] implies that

$$\alpha(d) \leq d/(2d-2).$$

This upper bound is better than (2) for $d \leq 5$, and together with (1) it gives us the exact value

$$\alpha(3) = 3/4. \tag{3}$$

Bühler [1] has calculated the minimum size of all maximum induced planar subgraphs of d-regular graphs on n vertices, for several values of n and d. This is the quantity

$$\text{mips}(n,d) = \min\{\text{mips}(G) \mid G \text{ is } d\text{-regular and } |V(G)| = n\},$$

where mips(G) denotes the size of the maximum induced planar subgraph of G. We give her results in the right column of the following table, with our lower bound $3n/(d+1)$ in the third column for comparison. Clearly our bound is not tight, though whether the difference between it and mips(n,d) is just $o(n)$, or is $\Theta(n)$ and so $\alpha(d) > 3/(d+1)$, is not known to us.

d	n	$\lceil 3n/(d+1) \rceil$	mips(n,d) by [1]
3	10	8	8
	12	9	10
	14	11	11
	16	12	13
4	10	6	8
	11	7	9
	12	8	9
	13	8	10
5	12	6	8

3.3 One More Step

In some cases, it is possible to add one more vertex to P after Algorithm 2 stops. This actually allows us to obtain a sharper bound on maximum induced planar subgraph size.

Theorem 3 *If G is a graph with $d = 3$, $n \equiv 0$ or $1 \pmod 4$ and, after Algorithm 2 stops, $|P| = \lceil 3n/4 \rceil$, then any vertex $v \in R$ has the property that $\langle P \cup \{v\} \rangle$ is planar.*

Proof. Suppose $d = 3$. Observe first that the stopping condition of the algorithm implies that R forms an independent set in G, and that each of its vertices have 3 neighbours in P. Furthermore, if G is connected then $\langle P \rangle$ is too, else another step of type (i), (ii) or (iii) would be possible and the algorithm would not have stopped.

Choose any $v \in R$, and put $P' = P \cup \{v\}$ and $R' = R \setminus \{v\}$. Observe that $|E(R')| = 0$, and if $|P| = \lceil 3n/4 \rceil$ then $|R'| = \lfloor n/4 \rfloor - 1$ so that $|E(P', R')| = 3\,|R'| = 3(\lfloor n/4 \rfloor - 1)$. Hence we have

$$\begin{aligned}|E(P')| &= |E| - |E(P',R')| - |E(R')| \\ &\leq 3\lfloor n/2 \rfloor - 3(\lfloor n/4 \rfloor - 1) \\ &< \lceil 3n/4 \rceil + 4,\end{aligned}$$

provided $n \equiv 0$ or $1 \pmod 4$ (for the sake of the final inequality). Now, if $\langle P' \rangle$ were nonplanar, it would contain a subdivision of $K_{3,3}$. Since it is connected, we would have $|E(P')| = |P'| + 3 \geq \lceil 3n/4 \rceil + 4$, a contradiction. Hence $\langle P' \rangle$ is planar. □

So, when $d = 3$ and $n \equiv 0$ or $1 \pmod 4$, we can ensure that $|P| \geq \lceil 3n/4 \rceil + 1$: either Algorithm 2 gives such a P, or, if not, we can add *any* vertex into P to achieve it. We can summarise the achievements of both Algorithm 2 and Theorem 3 by saying that, in combination, they give an induced planar subgraph of at least $\lceil (3n+2)/4 \rceil$ vertices when $d = 3$: for all G on n vertices with $d = 3$,

$$\text{mips}(G) \geq \lceil (3n+2)/4 \rceil. \tag{4}$$

It is intriguing to note that this refined lower bound equals Bühler's calculation of mips$(n, 3)$ for all the values quoted for $d = 3$ above. We find (using Bühler's program) that they are also equal for the odd values $n = 9, 11, 13$, with the caveat that mips$(n, 3)$ is taken to be a minimum over graphs of maximum degree 3 since such G cannot be 3-regular. This raises the question of whether our refined lower bound is actually sharp, for $d = 3$, in the much stronger sense of actually equalling mips$(n, 3)$, and not just differing from it by $o(n)$ (which is what (3) tells us). This looks unlikely, however: a program of Bühler's was used to establish that mips$(18, 3) = 15$, whereas our lower bound is 14.

3.4 Modifications for Maximality

As with the modified version of Algorithm 1, Algorithm 2 does not necessarily find a maximal induced planar subgraph. (Indeed, Theorem 3 gives cases where it never does.) It can easily be extended to do so, in various ways. The main loop control in Step 3 can be replaced by

$$\begin{aligned}&\textbf{while } (\ [\exists x_0 \in R : \langle P \cup \{x_0\} \rangle \text{ is planar}] \ \textbf{ or} \\ &\quad [\exists x_0 \in R, x \in P : (\langle (P \setminus \{x\}) \cup \{x_0\} \rangle \text{ is planar}) \text{ and} \\ &\qquad (\ [|E((P \setminus \{x\}) \cup \{x_0\})| < |E(P)|] \text{ or} \\ &\qquad [(|E((P \setminus \{x\}) \cup \{x_0\})| = |E(P)|) \text{ and} \\ &\qquad (k((P \setminus \{x\}) \cup \{x_0\}) < k(P))\,]\ )\,]\ )\end{aligned}$$

The tests of Steps 3.1–3.3 could be combined and replaced by a single test of whether or not $\langle P \cup \{x_0\} \rangle$ is planar. The resulting algorithm will find a maximal induced planar subgraph, and should in practice produce larger induced planar subgraphs than Algorithm 2. On the other hand, it will be slower, harder to implement from scratch, and the techniques we have discussed here and in [8] do not give a stronger result on its performance.

References

1. S. Bühler, *Planarity of Graphs*, M.Sc. dissertation, University of Dundee, 2000.
2. J. Cai and X. Han and R. E. Tarjan, An $O(m \log n)$-Time Algorithm for the Maximal Planar Subgraph Problem, *SIAM J. Comput.* 22 (1993) 1142–1162.
3. G. Călinescu, C. G. Fernandes, U. Finkler and H. Karloff, A better approximation algorithm for finding planar subgraphs, in: *Proc. 7th Ann. ACM-SIAM Symp. on Discrete Algorithms* (Atlanta, GA, 1996); also: *J. Algorithms* **27** (1998) 269–302.
4. Robert Cimikowski, An analysis of some heuristics for the maximum planar subgraph problem, in: *Proc. 6th Ann. ACM-SIAM Symp. on Discrete Algorithms* (San Francisco, CA, 1995), pp 322–331, ACM, New York, 1995.
5. G. Di Battista, Peter Eades, Roberto Tamassia and I. Tollis, *Graph Drawing: Algorithms for the Visualization of Graphs*, Prentice Hall, 1999.
6. H. N. Djidjev and S. M. Venkatesan, Planarization of graphs embedded on surfaces, in: M. Nagl (ed.), *Proc. 21st Internat. Workshop on Graph-Theoretic Concepts in Computer Science* (WG'95) (Aachen, 1995), Lecture Notes in Comput. Sci., **1017**, Springer, Berlin, 1995, pp. 62–72.
7. M. E. Dyer, L. R. Foulds and A. M. Frieze, Analysis of heuristics for finding a maximum weight planar subgraph, *Eur. J. Operational Res.* **20** (1985) 102–114.
8. K. Edwards and G. Farr, Fragmentability of graphs, *J. Combin. Theory (Ser. B)* **82** (2001) 30–37.
9. L. Faria, C. M. H. de Figueiredo and C. F. X. Mendonça, Splitting number is NP-complete, in: Juraj Hromkovic and Ondrej Sýkora (eds.), *Proc. 24th Internat. Workshop on Graph-Theoretic Concepts in Computer Science (WG '98)* (Smolenice Castle, Slovakia, 18–20 June 1998), Lecture Notes in Comput. Sci., **1517**, Springer, 1998, pp. 285–297.
10. O. Goldschmidt and A. Takvorian, An efficient graph planarization two-phase heuristic, *Networks* **24** (1994) 69–73.
11. Magnús M. Halldórsson, Approximation of weighted independent set and hereditary subset problems, *Journal of Graph Algorithms and Applications* **4** (1) (2000) 1–16.
12. Magnús M. Halldórsson and Hoong Chuin Lau, Low-degree graph partitioning via local search with applications to constraint satisfaction, max cut, and coloring, *Journal of Graph Algorithms and Applications* **1** (3) (1997) 1–13.
13. J. P. Hutchinson, On genus-reducing and planarizing algorithms for embedded graphs, in: R. B. Richter (ed.), *Graphs and Algorithms*, Contemporary Mathematics 89, Amer. Math. Soc., Providence, RI, 1989.
14. Joan P. Hutchinson and Gary L. Miller, On deleting vertices to make a graph of positive genus planar, *Discrete algorithms and complexity* (Kyoto, 1986), *Perspect. Comput.*, 15, pp. 81–98, Academic Press, Boston, MA, 1987.
15. M. Jünger and P. Mutzel, The polyhedral approach to the maximum planar subgraph problem: new chances for related problems, *Graph Drawing '94*, pp. 119–130.
16. M. Jünger and P. Mutzel, Maximum planar subgraphs and nice embeddings: practical layout tools, *Algorithmica* **16** (1996) 33–59.
17. P. C. Kainen, A generalization of the 5-color theorem, *Proc. Amer. Math. Soc.* **45** (1974) 450–453.
18. M. S. Krishnamoorthy and N. Deo, Node-deletion NP-complete problems, *SIAM J. Comput.* **8** (1979) 619–625.
19. J. M. Lewis and M. Yannakakis, The node-deletion problem for hereditary properties is NP-complete, *J. Comput. System Sci.* **20** (1980) 219–230.

20. Annegret Liebers, Planarizing graphs — a survey and annotated bibliography, *Journal of Graph Algorithms and Applications* **5** (1) (2001) 1–74.
21. P. C. Liu and R. C. Geldmacher, On the deletion of nonplanar edges of a graph, Proc. 10th. S-E Conf. on Comb., Graph Theory, and Comp. 1977, *Congressus Numerantium*, No. 24 (1979) 727–738.
22. C. Lund and M. Yannakakis, The approximation of maximum subgraph problems, in: *Proc. 20th Int. Colloquium on Automata, Languages and Programming* (ICALP), Lecture Notes in Comput. Sci., **700**, Springer-Verlag, 1993, pp. 40–51.
23. P. Mutzel, *The Maximum Planar Subgraph Problem*, Ph.D. Thesis, Univ. zu Köln, 1994.

A Characterization of DFS Cotree Critical Graphs

Hubert de Fraysseix[1] and Patrice Ossona de Mendez[2]

[1] UMR 8557, CNRS, Paris, France
hf@ehess.fr, http://www.ehess.fr/centres/cams/person/hf/
[2] UMR 8557, CNRS, Paris, France
pom@ehess.fr, http://www.ehess.fr/centres/cams/person/pom/

Abstract. We give a characterization of DFS-cotree critical graphs which is central to the linear time Kuratowski finding algorithm implemented in PIGALE[1].

1 Introduction

The present paper is a part of the theoretical study underlying a linear time algorithm finding a Kuratowski subdivision in a non planar graph ([1]; see also [7] for another algorithm). It relies on the concept of DFS cotree critical graph, which is an efficient sub-product of DFS-based planarity testing algorithms (such as [4] and [3]). Roughly speaking, a DFS-cotree critical graph is a simple graph of minimum degree 3 having a DFS tree, such that any non-tree (i.e. cotree) edge is *critical*, in the sense that its deletion would lead to a planar graph. A first study of DFS-cotree critical appeared in [2], in which it is proved that a DFS-cotree critical graph either is isomorphic to K_5 or includes a subdivision of $K_{3,3}$ and no subdivision of K_5. We give here a full characterization of DFS-cotree critical graphs: a simple graph is DFS-cotree critical if and only if it is either K_5, or a Möbius pseudo-ladder having a simple path including all the non critical edges (see Figure 1).

2 Preliminary Definitions

A Möbius pseudo-ladder is a natural extension of Möbius ladders allowing triangles. This may be formalized by the following definition.

Definition 1. *A* Möbius pseudo-ladder *is a non planar simple graph, which is the union of a polygon* $(v_1, \ldots, v_n)$ *and chords (called* bars*) such that any two non adjacent bars are interlaced (recall two non adjacent edges* $\{v_i, v_j\}$ *and* $\{v_k, v_l\}$ *are* interlaced *if, in circular order, one finds exactly one of* $\{v_k, v_l\}$ *between* v_i *and* v_j*).*

[1] PIGALE, which stands for "Public Implementation of a Graph Algorithm Library and Editor", is freely available at ftp://pr.cams.ehess.fr/pub/pigale.tar.gz

P. Mutzel, M. Jünger, and S. Leipert (Eds.): GD 2001, LNCS 2265, pp. 84–95, 2002.
© Springer-Verlag Berlin Heidelberg 2002

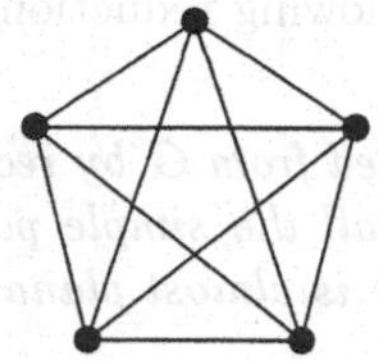

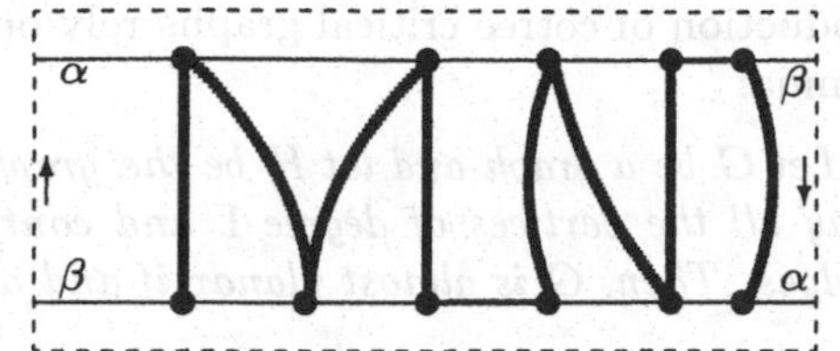

Fig. 1. The DFS-cotree critical graphs are either K_5 or a Möbius pseudo-ladders having all its non critical edges (thickest) included in a single path

This definition means that a Möbius pseudo-ladder may be drawn in the plane as a polygon and internal chords such that any two non adjacent chords cross. In the projective plane, a Möbius pseudo ladder may be drawn without crossing (see Figure 2). Notice that $K_{3,3}$ and K_5 are both Möbius pseudo-ladders.

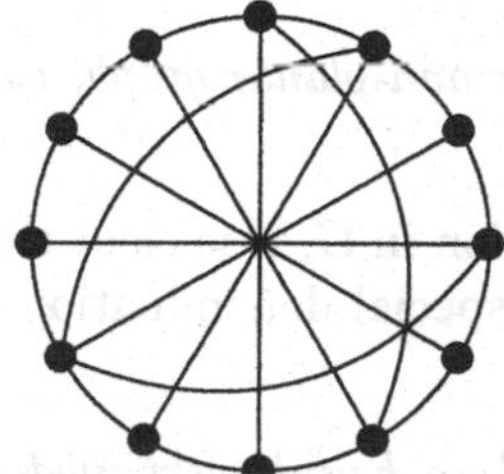

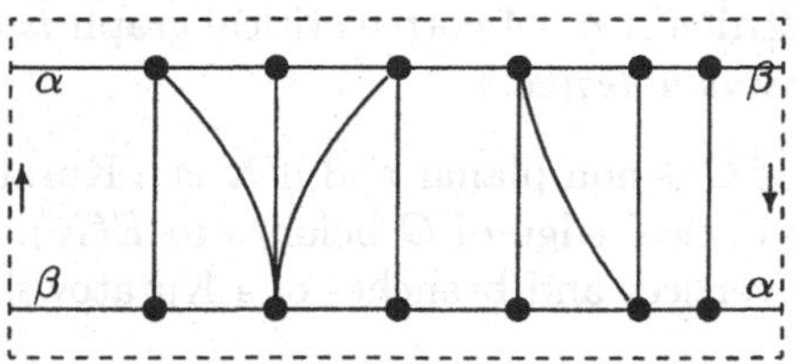

Fig. 2. A Möbius pseudo-ladder on the plane and on the projective plane

Definition 2. *Let G be a graph. An edge $e \in E(G)$ is* critical *if $G - e$ is planar.*

Definition 3. *An* almost planar graph *is a graph, such that the set of non critical edges of G is acyclic.*

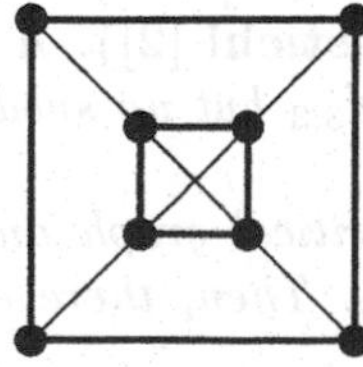

Fig. 3. A non planar graph with bold non-critical edges (notice that the set of non-critical edges is not acyclic and hence the graph is not almost planar)

The introduction of cotree critical graphs rely on the following reduction/ extension lemma:

Lemma 1. *Let G be a graph and let H be the graph obtained from G by recursively deleting all the vertices of degree* 1 *and contracting all the simple paths into single edges. Then, G is almost planar if and only if H is almost planar.*

Proof. First notice that the critical edges of G that remains in H are critical edges of H, according to the commutativity of deletion and contraction operations (for $e \in E(H)$, if $G - e$ is planar so is $H - e$).

Assume G is almost planar. Then, for any simple path P of G, either all the edges of P are critical or they are all non critical. Hence, if H had a cycle of non critical edges, they would define a cycle of non critical edges of G.

Conversely, assume H is almost planar. Adding a vertex of degree 1 does not change the status of the other edges and cannot create a cycle of non critical edges. Similarly, subdividing an edge creates two edges with the same status without changing the status of the other edges and hence cannot create a cycle of non critical edges. □

Definition 4. *A* cotree critical graph *is a almost planar non-planar graph, with minimum degree* 3.

If G is non planar and if K is a Kuratowski subdivision in G, it is clear that any critical edge of G belongs to $E(K)$. This justify a special denomination of the vertices and branches of a Kuratowski subdivision:

Definition 5. *Let G be a non planar graph and let K be a Kuratowski subdivision of G. Then, a vertex is said to be a K-*vertex *(resp. a K-*subvertex*, resp. a K-*exterior vertex*) if it is a vertex of degree at least* 3 *of K (resp. a subdivision vertex of K, resp. a vertex not in K). A K-*branch *is the subdivided path of K between two K-vertices. A K-branch is* critical *if it includes at least one critical edge.*

3 Cotree Critical Graphs

We shall first recall the following result on cotree critical graphs (expressed here with our terminology):

Theorem 1 (Fraysseix, Rosenstiehl [2]). *A cotree critical graph is either a hut or includes a subdivision of $K_{3,3}$ but no subdivision of K_5.*

Lemma 2. *Let G be a cotree critical graph and let K be a Kuratowski subdivision of G isomorphic to $K_{3,3}$. Then, there exists in $E(G) \setminus E(K)$ no path between:*

- *two vertices (K-vertices or K-subvertices) of a same K-branch of K,*
- *two K-subvertices of K-adjacent K-branches of K.*

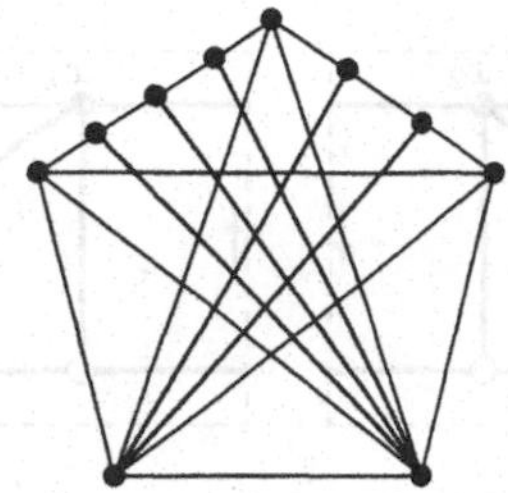

Fig. 4. A hut drawn as a Möbius pseudo-ladder

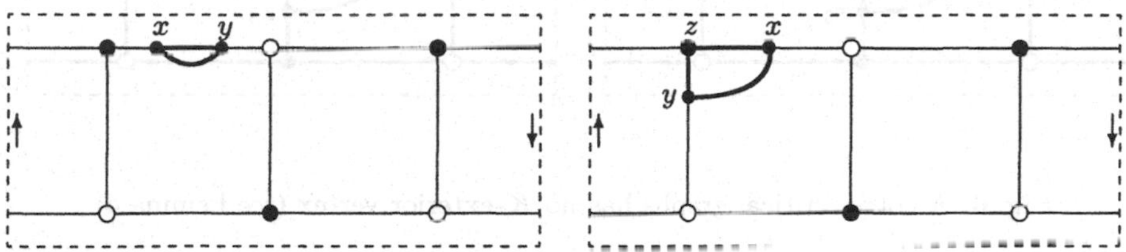

Fig. 5. Forbidden paths in cotree critical graphs (see Lemma 2)

Proof. The two cases are shown Fig 5.

If two vertices x and y (K-vertices or K-subvertices) of a same K-branch of K are joined by a path, both this path and the one linking a and y in K are non critical. Hence G is not cotree critical, a contradiction.

If two K-subvertices x and y of K-adjacent K-branches of K are linked by a path, this path is non critical. Moreover, if z is the K-vertex adjacent to the branches including x and y, both paths from z to x and x to y are non critical. Hence, G includes a non critical cycle, a contradiction. □

Lemma 3. *Every cotree critical graph is 3-connected.*

Proof. Let G be a cotree critical graph. Assume G has a cut-vertex v. Let H_1, H_2 be two induced subgraphs of G having v as their attachment vertex and such that H_1 is non planar. As G has no degree 1 vertex, H_2 includes a cycle. All the edges of this cycle are non critical, a contradiction. Hence, G is 2-connected.

Assume G has an articulation pair $\{v, w\}$ such that there exists at least two graphs H_1, H_2 different from a path having v, w as attachment vertices. As G is non planar, we may choose H_1 in such a way that $H_1 + \{v, w\}$ is a non planar graph (see [6], for instance). As there exists in H_2 two disjoints paths from v to w, no edge of these paths may be critical and H_2 hence include a cycle of non critical edges, a contradiction. □

Lemma 4. *Let G be a cotree critical graph and let K be a Kuratowski subdivision of G. Then, G has no K-exterior vertices, that is: $V(G) = V(K)$.*

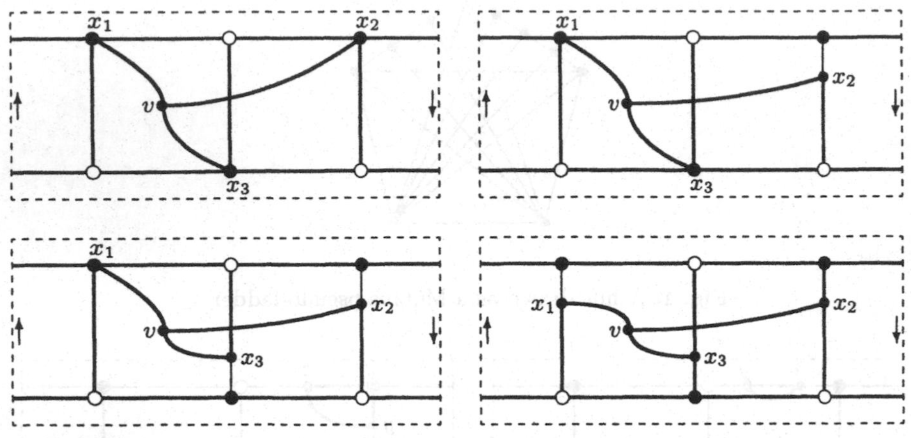

Fig. 6. A cotree critical graphs has no K-exterior vertex (see Lemma 4)

Proof. Assume $V(G) \setminus V(K)$ is non empty and let v be a vertex of G not in K. According to Lemma 3, G is 3-connected. Hence, there exists 3 disjoint paths P_1, P_2, P_3 from v to K. As $K + P_1 + P_2 + P_3$ is non planar partial subgraph of G free of vertices of degree 1, it is a subdivision of a 3-connected graph, according to Lemma 1 and Lemma 3. Thus, the vertices of attachment x_1, x_2, x_3 of P_1, P_2, P_3 in K are all different. According to Lemma 2, no path in $E(G) \setminus E(H)$ may link K-vertices with different colors. Thus, we may assume no white K-vertex belong to $\{x_1, x_2, x_3\}$ and four cases may occur as shown Fig 6. All the four cases show a cycle of non critical edges, a contradiction. □

Corollary 1. *If G is cotree critical, no non critical K-branch may be subdivided, that is: every non critical K-branch is reduced to an edge.*

Proof. If a branch of K is non critical, there exists a $K_{3,3}$ subdivision avoiding it. Hence, the branch is reduced to an edge, according to Lemma 3. □

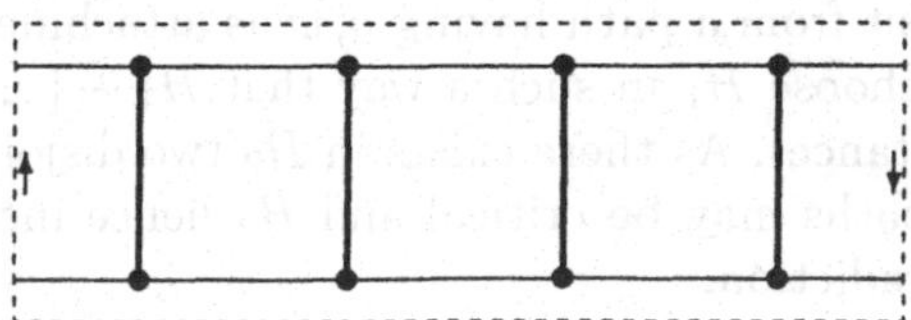

Fig. 7. The 4-bars Möbius ladder M_4 (all bars are non critical)

Let G be a cotree-critical graph obtained by adding an edge linking two subdivision vertices of non adjacent edges of a $K_{3,3}$. This graph is unique up to isomorphism and is the Möbius ladder with 4 non critical bars shown Figure 7.

The same way we have introduced K-vertices, K-subvertices and K-branches relative to a Kuratowski subdivision, we define M-vertices, M-subvertices and M-branches relative to a Möbius ladder subdivision.

Lemma 5. *Let G be a cotree critical graph and $\{x, y, z\}, \{x', y', z'\}$ the K-vertices of a $K_{3,3}$ subdivision in G.*

If $[x, z']$ or $[x', z]$ is a critical K-branch, then all the edges from $]x, z[$ to $]x', z'[$ and the K-branch $[y, y']$ are pairwise interlaced, with respect to the cycle (x, y', z, x', y, z').

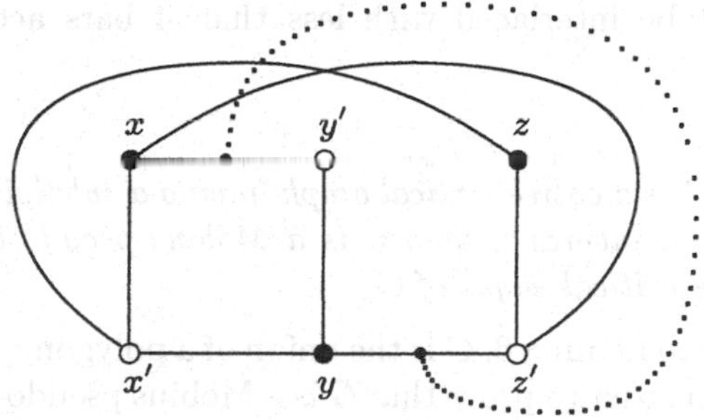

Fig. 8. No edges is allowed from $]x, z[$ to $]x', z'[$ by the "outside" (see Lemma 5)

Proof. The union of the $K_{3,3}$ subdivision and all the edges of G incident to a vertex in $]x, z[$ and a vertex in $]x', z'[$ becomes uniquely embeddable in the plane when removing of the K-branches $[x, z']$ or $[x', z]$. Figure 8 displays the outline of a normal drawing of G is the plane which becomes plane when removing any of the K-branch $[x, z']$ or $[x', z]$. By planarity, given that an edge from $]x, z[$ to $]x', z'[$, if drawn outside, crosses both $[x, z']$ and $[x', z]$, all the edges from $]x, z[$ to $]x', z'[$ and the K-branch $[y, y']$ are drawn inside the cycle (x, y', z, z', y, x') without crossing and thus are interlaced with respect to the cycle (x, y', z, x', y, z'). The result follows. □

Lemma 6. *If G is a cotree critical graph having a subdivision of Möbius ladder M with 4 bars as a subgraph, then it is the union of a polygon γ and non-critical chords. Moreover the 4 bars b_1, b_2, b_3, b_4 of M are chords and any other chord is interlaced with all of b_1, b_2, b_3, b_4 with respect to γ.*

Proof. Let G be a cotree critical graph having a subdivision of Möbius M ladder with 4 bars b_1, b_2, b_3, b_4 as a subgraph. First notice that all the bars of the Möbius ladder are non critical and that, according to Corollary 1, they are hence reduced to edges. According to Lemma 4, M covers all the vertices of G as it includes a

$K_{3,3}$ and hence the polygon γ of the ladder is Hamiltonian. Thus, the remaining edges of G are non critical chords of γ.

Let e be a chord different from b_1, b_2, b_3, b_4.

- Assume e is adjacent to none of b_1, b_2, b_3, b_4.
  Then it can't be interlaced with less than 3 bars, according to Lemma 2, considering the $K_{3,3}$ induced by at least two non interlaced bars. It can't also be interlaced with 3 bars, according to Lemma 5, considering the $K_{3,3}$ induced by the 2 interlaced bars (as $\{x, x'\}, \{z, z'\}$) and one non interlaced bar (as $\{y, y'\}$).
- Assume e is adjacent to b_1 only.
  Then it can't be interlaced with less than 4 bars, according to Lemma 2, considering the $K_{3,3}$ induced by b_1 and two non interlaced bars.
- Assume e is adjacent to b_1 and another bar.
  Then it can't be interlaced with less than 4 bars according to Lemma 2 again.

□

Theorem 2. *If G is a cotree critical graph having a subdivision of Möbius ladder M with 4 bars as a subgraph, then it is a Möbius pseudo-ladder whose polygon γ is the set of the critical edges of G.*

Proof. According to Lemma 6, G is the union of a polygon γ and chords including the 4 bars of M. In order to prove that G is a Möbius pseudo-ladder, it is sufficient to prove that any two non adjacent chords are interlaced with respect to that cycle. We choose to label the 4 bars b_1, b_2, b_3, b_4 of M according to an arbitrary traversal orientation of γ. According to Lemma 6, any chord e is interlaced with all of b_1, b_2, b_3, b_4 and hence its endpoints are traversed between these of two consecutive bars $b_{\alpha(e)}, b_{\beta(e)}$ (with $\beta(e) \equiv \alpha(e) + 1 \pmod 4$), what well defines the α and β functions from the chords different from b_1, b_2, b_3, b_4 to $\{1, 2, 3, 4\}$.

As all the bars are interlaced pairwise and as any chord is interlaced with all of them, we only have to consider two non adjacent chords e, f not in $\{b_1, b_2, b_3, b_4\}$.

- Assume $\alpha(e)$ is different from $\alpha(f)$.
  Then, the edges e and f are interlaced, as f is interlaced with both $b_{\alpha(e)}$ and $b_{\beta(e)}$.
- Assume $\alpha(e)$ is equal to $\alpha(f)$.
  Let b_i, b_j be the bars such that $j \equiv \beta(e) + 1 \equiv \alpha(e) + 2 \equiv i + 3 \pmod 4$. Then, considering the $K_{3,3}$ induced by γ and the bars $b_i, b_{\alpha(e)}, b_j$, it follows from Lemma 6 that e and f are interlaced.

□

4 DFS Cotree Critical Graphs

An interesting special case of cotree critical graphs, the DFS cotree critical graphs, arise when the tree may be obtained using a Depth-First Search, as it happens when computing a cotree critical subgraph using a planarity testing algorithm. Then, the structure of the so obtained DFS cotree critical graphs appears to be quite simple and efficient to exhibit a Kuratowski subdivision.

In this section, we first prove that any DFS cotree graph with sufficiently many vertices includes a Möbius ladder with 4 bars as a subgraph and hence are Möbius pseudo-ladders, according to Theorem 2. We then prove that these Möbius pseudo-ladders may be fully characterized.

Before proceeding, we shall recall some basic properties of DFS trees.

Definition 6. *A* DFS tree *of a connected graph G rooted at $v_0 \in V(G)$ may be recursively defined as follows: If G has no edges, the empty set if a DFS tree of G. Otherwise, let $G_1, \dots, G_k$ the the connected components of $G - v_0$, then a DFS tree of G is the union of the DFS trees $Y_1, \dots, Y_k$ of $G_1, \dots, G_k$ rooted at $v_1, \dots, v_k$ (where $v_1, \dots, v_k$ are neighbors of v_0 in G), and the edges $\{v_0, v_1\}, \dots, \{v_0, v_k\}$.*

Definition 7. *A* DFS cotree-critical graph *G is a cotree critical graph, which non-critical edge set may be extended into a DFS-tree of G.*

Lemma 7. *If G is k-connected ($k \geq 1$) and Y is a DFS tree of G rooted at v_0, then there exists a unique chain from v_0 in Y of length $k-1$.*

Proof. According to Definition 7, the lemma is satisfied for $k = 1$. Assume the lemma is true for $k \geq 1$, and let v_0 be a vertex of a $k+1$-connected graph G. Then, $G - v_0$ has a unique connected component H which is k-connected. A DFS tree Y_G of G will be the union of a DFS tree Y_H of H rooted at a neighbor v_1 of v_0 and the edge $\{v_0, v_1\}$. As there exists, by induction, a unique chain from v_1 in Y_H of length $k-1$, there will exist a unique chain from v_0 in Y_G of length k. □

Corollary 2. *If G is 3-connected and Y is a DFS tree of G rooted at v_0, then v_0 has a unique son which also has a unique son.*

Lemma 8. *Let G be a cotree critical graph and let K be a Kuratowski subdivision of G isomorphic to $K_{3,3}$. Then, two non K-adjacent K-vertices a, b cannot be adjacent to K-subvertices on a same K-branch.*

Proof. The three possible cases are shown Figure 9; in all cases, a cycle of non critical edges exist. □

Lemma 9. *Let G be a cotree critical graph and let K be a Kuratowski subdivision of G isomorphic to $K_{3,3}$. If G has two edges interlaced as shown Figure 10, then G is not DFS cotree critical.*

Proof. Assume G is cotree critical. By case analysis, one easily checks that any edge of G outside $E(K)$ is either incident to a or b. Hence, all the vertices of G incident to at most one non critical edge is adjacent to a vertex incident with at least 3 non critical edges (a or b). According to Corollary 2, the non critical edges cannot be extended to a DFS tree of G, so G is not DFS cotree critical. □

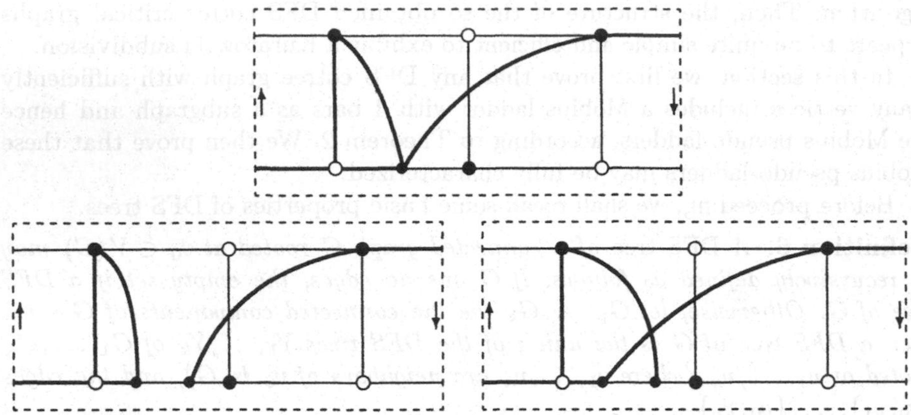

Fig. 9. No two non-adjacent K-vertices may be adjacent to K-subvertices on the same K-branch (see Lemma 8)

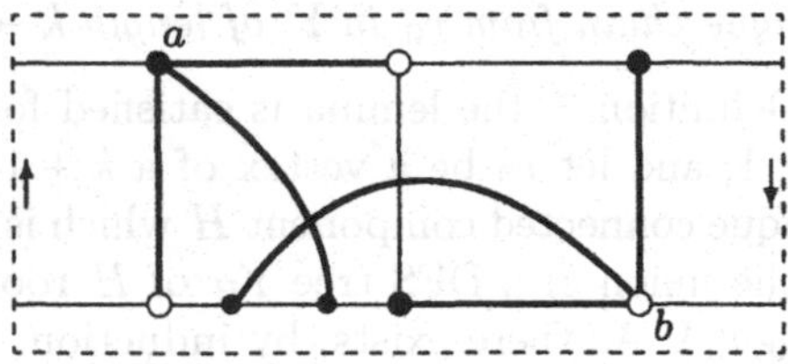

Fig. 10. Case of two adjacent K-vertices adjacent to K-subvertices on the same K-branch by two interlaced edges (see Lemma 9)

Lemma 10. *Let G be a DFS-cotree critical graph and let K be a $K_{3,3}$ subdivision in G. Then, no two edges in $E(G) \setminus E(K)$ may be incident to a same K-vertex.*

Proof. Four cases may occur, as shown Figure 11. By a suitable choice of the Kuratowski subdivision, the last two cases are easily reduced to the two first ones.

- Consider the first case.
  Assume there exists a K-subvertex v between x and y. Then, v is not adjacent to a K-vertex different from a, according to Lemma 8 and Lemma 9. If v were adjacent to another K-subvertex w, the graph would include a Möbius ladder with 4 bars as a subgraph and, according to Theorem 2, would be a Möbius pseudo-ladder in which $\{a, y\}$ and $\{v, w\}$ would be non interlaced chords, a contradiction. Thus, v may not be adjacent to a vertex different from a and we shall assume, without loss of generality, that no K-subvertex exists between x and y.

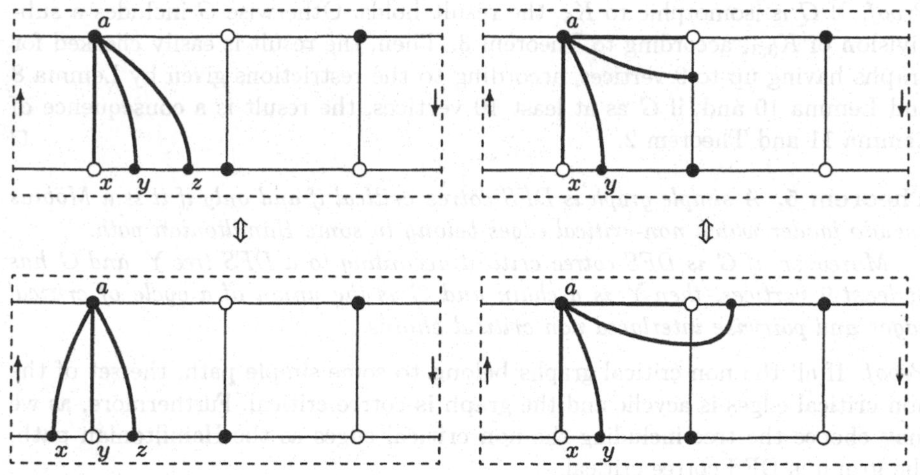

Fig. 11. Cases of Lemma 10

Therefore, if G is DFS-cotree critical, y cannot be a son of a, as it would be adjacent (by the left or by the right) to another son of a. Hence, vertex y has to be the father of the vertex a. As it may not have degree more than one in the DFS tree, it is the root, what contradicts the 3-connexity of G, according to Corollary 2.

– Consider the second case.
We may assume (by symmetry) that the father of a does not belong to the same branch as y and, as previously, that no K-subvertices exists between x and y. As both x and y are adjacent to a by non critical edges, they are both sons of a, the set of non critical edges may not be extended to a DFS tree of G.

□

Lemma 11. *If G is DFS-cotree critical includes a subdivision of $K_{3,3}$ and has at least* 10 *vertices, then G includes a 4-bars Möbius ladder as a subgraph.*

Proof. According to Lemma 10, no two edges in $E(G) \setminus E(K)$ may be incident to a same K-vertex. If two adjacent K-vertices are linked to two K-subvertices, then G includes a 4-bars Möbius ladder as a subgraph. Otherwise, if G has at least 10 vertices, there exists an edge linking two K-subvertices and G includes a 4-bars Möbius ladder as a subgraph. □

Theorem 3 (Fraysseix, Rosenstiehl [2]). *A DFS cotree critical graph is either isomorphic to K_5 or includes a subdivision of $K_{3,3}$ but no subdivision of K_5.*

Theorem 4. *Any DFS-cotree critical graph is a Möbius pseudo-ladder.*

Proof. If G is isomorphic to K_5, the result holds. Otherwise G includes a subdivision of $K_{3,3}$, according to Theorem 3. Then, the result is easily checked for graphs having up to 9 vertices, according to the restrictions given by Lemma 8 and Lemma 10 and, if G as at least 10 vertices, the result is a consequence of Lemma 11 and Theorem 2. □

Theorem 5. *A simple graph is DFS cotree critical if and only if it is a Möbius pseudo ladder which non-critical edges belong to some Hamiltonian path.*

Moreover, if G is DFS cotree critical according to a DFS tree Y and G has at least 9 vertices, then Y is a chain and G is the union of a cycle of critical edges and pairwise interlaced non critical chords.

Proof. If all the non critical graphs belong to some simple path, the set of the non critical edges is acyclic and the graph is cotree critical. Furthermore, as we may choose the tree including the non critical edges as the Hamiltonian path, the graph is DFS cotree critical.

Conversely, assume G is DFS cotree critical. The existence of an Hamiltonian including all the non-critical edges is easily checked for graph having up to 9 vertices. Hence, assume G has at least 10 vertices. According to Theorem 5, G is a Möbius pseudo ladder. By a suitable choice of a Kuratowski subdivision of $K_{3,3}$, it follows from Lemma 10 that no vertex of G may be adjacent to more than 2 non critical edges. Let Y be a DFS tree including all the non critical edges. Assume Y has a vertex v of degree at least 3. Then, one of the cases shown Figure 12 occurs (as v is incident to at most 2 non critical edges) and hence v is adjacent to a terminal w of T. The vertex w cannot be the root, according to Corollary 2. Then, among the two neighbors of w in the polygon of the Möbius pseudo ladder, one is not an ancestor of w, what contradicts the fact that Y is a DFS tree (as the cotree edges of a DFS tree always join a vertex to one of its ancestors). □

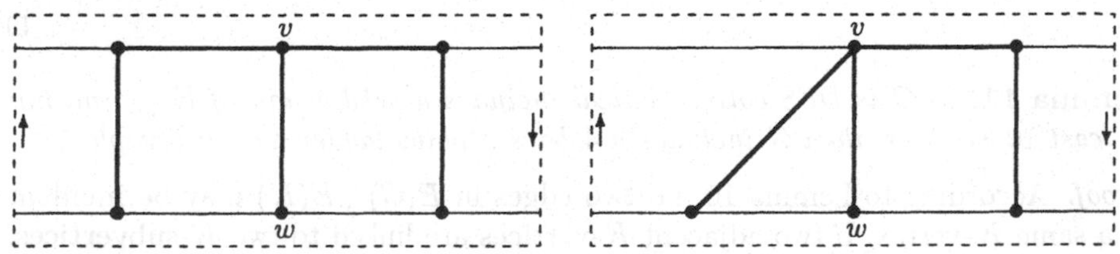

Fig. 12. A vertex of degree at least 3 in the tree is adjacent to a terminal of the tree (see Theorem 5)

References

1. H. de Fraysseix and P. Ossona de Mendez, *An algorithm to find Kuratowski subdivision in DFS cotree critical graphs*, Proceedings of the twelfth australasian workshop on combinatorial algorithms (Edy Try Baskoro, ed.), Institut Teknologi Bandung, Indonesia, 2001.
2. H. de Fraysseix and P. Rosenstiehl, *A discriminatory theorem of Kuratowski subgraphs*, Lecture Notes in Mathematics **1018** (1981), 214–222.
3. ———, *A characterization of planar graphs by Trémaux orders*, Combinatorica **5** (1985), no. 2, 127–135.
4. J.E Hopcroft and R.E. Tarjan, *Efficient planarity testing*, J. Assoc. Comput. Math. **21** (1974), 549–568.
5. K. Kuratowski, *Sur le problème des courbes gauches en topologie*, Fund. Math. **15** (1930), 271–293.
6. W.T. Tutte, *Graph theory*, Encyclopedia of Mathematics and its applications, vol. 21, Addison-Wesley, 1984.
7. S.G. Williamson, *Depth-first search and kuratowski subgraphs*, Journal of the ACM **31** (1984), no. 4, 681–693.

An Improved Lower Bound for Crossing Numbers

Hristo Djidjev[1] and Imrich Vrto[2,*]

[1] Department of Computer Science, Warwick University
Coventry CV4 7AL, United Kingdom
hristo@dcs.warwick.ac.uk

[2] Department of Informatics, Institute of Mathematics
Dúbravská 9, 842 35 Bratislava, Slovak Republic
vrto@savba.sk

Abstract. The crossing number of a graph $G = (V, E)$, denoted by $\text{cr}(G)$, is the smallest number of edge crossings in any drawing of G in the plane. Leighton [14] proved that for any n-vertex graph G of bounded degree, its crossing number satisfies $\text{cr}(G) + n = \Omega(\text{bw}^2(G))$, where $\text{bw}(G)$ is the bisection width of G. The lower bound method was extended for graphs of arbitrary vertex degrees to $\text{cr}(G) + \frac{1}{16}\sum_{v \in G} d_v^2 = \Omega(\text{bw}^2(G))$ in [15,19], where d_v is the degree of any vertex v. We improve this bound by showing that the bisection width can be replaced by a larger parameter - the cutwidth of the graph. Our result also yields an upper bound for the path-width of G in term of its crossing number.

1 Introduction

The crossing number of a graph $G = (V, E)$, denoted by $\text{cr}(G)$, is the smallest number of edge crossings in any drawing of G in the plane. It represents a fundamental measure of non-planarity of graphs and has been studied for more than 40 years. The problem is attractive from practical point of view too. It is known that the aesthetics and readability of graph-like structures (information diagrams, class hierarchies, flowcharts...) heavily depends on the number of crossings [4,16], when the structures are visualized on a 2-dimensional medium. Another natural appearance of the problem is in the design of printed circuit boards and VLSI circuits [14]. The area of a VLSI circuit is strongly related to the crossing number of the underlying graph. The problem is NP-hard [7] and the best theoretical exact and approximation algorithms are in [5,10]. A survey on heuristics is in [3]. Concerning crossing numbers of standard graphs, there are only a few infinite classes of graphs for which exact or tight bounds are known [13]. The main problem is the lack of efficient lower bound methods for estimating the crossing numbers of explicitely given graphs. The survey on known methods is in [17]. One of the powerful methods is based on the bisection

* This research was supported by the VEGA grant No. 02/7007/20 and the DFG grant No. Hr14/5-1.

P. Mutzel, M. Jünger, and S. Leipert (Eds.): GD 2001, LNCS 2265, pp. 96–101, 2002.
© Springer-Verlag Berlin Heidelberg 2002

width concept. The bisection width of a graph G is the minimum number of edges whose removal divides G into two parts having at most $2|V|/3$ vertices each. Leighton [14] proved that in any n-vertex graph G of bounded degree, the crossing number satisfies $\mathrm{cr}(G)+n = \Omega(\mathrm{bw}^2(G))$. The lower bound was extended to

$$\mathrm{cr}(G) + \frac{1}{16}\sum_{v\in V} d_v^2 \geq \frac{1}{40}\mathrm{bw}^2(G)$$

in [15,19], where d_v is the degree of any vertex v. We improve this bound by showing that the bisection width can be replaced by a larger parameter - the cutwidth of the graph, defined as follows. Consider an injection of the vertices of G into points of a horizontal line. Draw the edges above the line using semicircles. Find a vertical line between a pair of consecutive points, which cuts the maximal number of edges. Minimize the maximum over all injections. The minmax value is called the cutwidth of G, denoted by $\mathrm{cw}(G)$. Note that the cutwidth is a standard graph invariant appearing e.g. in the linear VLSI layouts [20], and is related to such a classical topic like the discrete isoperimetric problem [1]. We prove

$$\mathrm{cr}(G) + \frac{1}{16}\sum_{v\in G} d_v^2 \geq \frac{1}{1176}\mathrm{cw}^2(G).$$

Regardless the constant factors, the improvement is evident as $\mathrm{cw}(G) \geq \mathrm{bw}(G)$ and there are connected graphs with $\mathrm{bw}(G) = 1$ but with arbitrarily large cutwidth. If $\mathrm{cw}(G) \approx \mathrm{bw}(G)$ then the bisection lower bound is better up to a constant factor because of the big constant in our estimation. An improvement on it remains an open problem. Anyway, the aim of this note is to show that from the asymptotical point of view, the graph invariant that essentially influences the crossing number is not the bisection width but the cutwidth. The new crossing number lower bound is tight up to a constant factor. E.g., the cutwidth of the n-cube Q_n is $\lceil 2^{n+1}/3\rceil$, see [2], which implies

$$\mathrm{cr}(Q_n) = \Omega(4^n),$$

while it is known that $\mathrm{cr}(Q_n) \geq 4^n/20 - O(n^2 2^n)$, [18] and the best upper bound for the crossing number of the hypercube graph is $(163/1024)4^n$, [6].

Moreover, the additive term $\sum_{v\in G} d_v^2$ can not be removed, since the crossing number of any planar graph is 0 and there exists a planar graph P (e.g. the star) such that $\mathrm{cw}^2(P) = \Omega(\sum_{v\in P} d_v^2)$.

As a byproduct of our result we obtain the following contribution to the topological graph theory. The path-decomposition of a graph G is a sequence $D = X_1, X_2, ..., X_r$ of vertex subsets of G, such that every edge of G has both ends in some set X_i and if a vertex of G occurs in some sets X_i and X_j with $i < j$, then the same vertex occurs in all sets X_k with $i < k < j$. The width of D is the maximum number in any X_i minus 1. The path-width of G, $\mathrm{pw}(G)$, is the minimum width over all path-decompositions of G.

A graph $G = (V, E)$ is k-crossing critical if $\mathrm{cr}(G) = k$ and $\mathrm{cr}(G - e) < \mathrm{cr}(G)$, for all edges $e \in E$. Hliněný [11] proved that $\mathrm{pw(G)} \leq 2^{f(k)}$, where $f(k) =$

$O(k^3 \log k)$. This answers an open question of Geelen et al. [9] whether crossing critical graphs with bounded crossing numbers have bounded path-widths.

From the practical point of view, as pw(G) $\leq n-1$, the above upper bound is useful for $k \leq \sqrt[3]{\log n}$ only, while the maximal order of crossing numbers is n^4. Our result implies that if $\mathrm{cr}(G) = k$, then pw(G) $= O\left(\sqrt{k + \sum_{v\in V} d_v^2}\right)$, without the crossing-criticality assumption.

2 A New Lower Bound

We will make use of the following theorem [8].

Theorem 1. *Let $G = (V, E)$ be a planar graph with non-negative weights on its vertices that sum up to one and every weight is at most 2/3. Let d_v is the degree of any vertex v. Then there exists at most $(\sqrt{3}+\sqrt{2})\sqrt{\sum_{v\in V} d_v^2/2}$ edges whose removal divides G into disjoint subgraphs $G_1 = (V_1, E_1)$ and $G_2 = (V_2, E_2)$ such that the weight of each is at most 2/3.*

Theorem 1 implies an upper bound for the cutwidth of planar graphs which deserves an independent interest.

Theorem 2. *For any planar graph $G = (V, E)$*

$$\mathrm{cw}(G) \leq \frac{6\sqrt{2}+5\sqrt{3}}{2}\sqrt{\sum_{v\in V} d_v^2},$$

where d_v is the degree of any vertex v.

Proof. Apply Theorem 1 to G. Assign weights to vertices:

$$\mathrm{weight}(u) = \frac{d_u^2}{\sum_{v\in V} d_v^2}.$$

1. Assume weight$(u) \leq 2/3$ for all u. By deleting $(\sqrt{3}+\sqrt{2})\sqrt{\sum_{v\in V} d_v^2/2}$ edges we get graphs $G_1 = (V_1, E_1)$ and $G_2 = (V_2, E_2)$ such that for $i = 1, 2$ weight$(V_i) \leq 2/3$, which implies
$$\sum_{v\in V_i} d_v^2 \leq \frac{2}{3}\sum_{v\in V} d_v^2.$$
2. Assume there exists a vertex u such that weight$(u) > 2/3$. By deleting edges adjacent to u we get disjoint subgraphs $G_1 = (V_1, E_1)$ and $G_2 = (V_2, E_2)$, where G_2 is a one vertex graph. We have weight$(V_1) = 1 -$ weight$(u) < 2/3$ and
$$\sum_{v\in V_1} d_v^2 < \frac{2}{3}\sum_{v\in V} d_v^2.$$
The number of edges between G_1 and G_2 is
$$d_u \leq \frac{\sqrt{3}+\sqrt{2}}{2}\sqrt{\sum_{v\in V} d_v^2}.$$

Placing the graphs G_1 and G_2 consecutively on the line and adding the deleted edges we obtain the estimation

$$\mathrm{cw}(G) \le \max\{\mathrm{cw}(G_1), \mathrm{cw}(G_2)\} + \frac{(\sqrt{3}+\sqrt{2})}{2}\sqrt{\sum_{v\in V} d_v^2}.$$

Solving the recurrence we find

$$\mathrm{cw}(G) \le \frac{\sqrt{3}+\sqrt{2}}{2}\sum_{i=0}^{\infty}\left(\frac{2}{3}\right)^{i/2}\sqrt{\sum_{v\in V} d_v^2} = \frac{6\sqrt{2}+5\sqrt{3}}{2}\sqrt{\sum_{v\in V} d_v^2}.$$

□

Our main result is

Theorem 3. *Let $G=(V,E)$ be a graph. Let d_v denote the degree of any vertex v. Then the crossing number of G satisfies*

$$\mathrm{cr}(G) + \frac{1}{16}\sum_{v\in V} d_v^2 \ge \frac{1}{1176}\mathrm{cw}^2(G).$$

Proof. Consider a drawing of G with $\mathrm{cr}(G)$ crossings. Introducing a new vertex at each crossing results in a plane graph H with $\mathrm{cr}(G)+n$ vertices. By Theorem 2 we have

$$\mathrm{cw}(H) \le \frac{6\sqrt{2}+5\sqrt{3}}{2}\sqrt{\sum_{v\in H} d_v^2} = \frac{6\sqrt{2}+5\sqrt{3}}{2}\sqrt{\sum_{v\in G} d_v^2 + 16\mathrm{cr}(G)}.$$

Finally, note that $\mathrm{cw}(G) \le \mathrm{cw}(H)$, which proves the claim. □

This result immediately gives an upper bound for the path-width of G in term of its crossing number as the result of Kinnersley [12] implies that $\mathrm{pw}(G) \le \mathrm{cw}(G)$.

Corollary 1. *Let $G=(V,E)$ be a graph. Then*

$$\mathrm{pw}(G) < 9\sqrt{16\mathrm{cr}(G) + \sum_{v\in V} d_v^2}.$$

3 Final Remarks

We proved a new lower bound formula for estimating the crossing numbers of graphs. The former method was based on the bisection width of graphs. Our method replaces the bisection width by a stronger parameter - the cutwidth. While the bisection width of a connected graph can be just one edge, which implies a trivial lower bound only, the cutwidth based method gives nontrivial lower bounds in most cases. A drawback of the method is the big constant factor in the formula. Currently we are able to reduce it to about 500.

A natural question arises how to find or estimate the cutwidth of a graph. The most frequent approach so far was its estimation from below by the bisection width. This of course degrades the cutwidth method to the bisection method. Provided that $cw(H)$ in known or estimated from below, for some graph H, we can use a well-known relation $cw(G) \geq cw(H)/cg(H,G)$, where $cg(H,G)$ is the congestion of G in H. Another possibility is to use the strong relation of the cutwidth problem to the so called discrete edge isoperimetric problem [1]. Informally, the problem is to find, for a given k, a k-vertex subset of a graph with the smallest "edge boundary". A good solution to the isoperimetric problem provides a good lower bound for the cutwidth.

References

1. Bezrukov, S.L.,: Edge Isoperimetric Problems on Graphs. In: Lovász, L., Gyarfás, A.,Katona, G.O.H., Recski, A., Székely, L. (eds.): Graph Theory and Combinatorial Biology. Bolyai Soc. Mathematical Studies 7. Akadémia Kiadó, Budapest (1999) 157-197
2. Bezrukov, S., Chavez, J.D., Harper, L.H., Röttger, M., Schroeder, U.-P.: The Congestion of n-Cube Layout on a Rectangular Grid. Discrete Mathematics **213** (2000) 13-19
3. Cimikowski, R.: Algorithms for the Fixed Linear Crossing Number Problem. Submitted to Discrete Applied Mathematics
4. Di Battista, G., Eades, P., Tamassia, R., Tollis, I.G.: Graph Drawing: Algorithms for Visualization of Graphs. Prentice Hall (1999)
5. Even, G., Guha, S., Schieber, B.: Improved Approximations of Crossings in Graph Drawing and VLSI Layout Area. In: 32th Annual Symposium on Theory of Computing. ACM Press (2000) 296-305
6. Faria, L., Herrera de Figuerado, C.M.: On Eggleton and Guy conjectured upper bounds for the crossing number of the n-cube. Mathematica Slovaca **50** (2000) 271-287
7. Garey, M. R., and Johnson, D. S.: Crossing Number is NP-complete. SIAM J. Algebraic and Discrete Methods **4** (1983) 312–316
8. Gazit, H., Miller, G.L.: Planar Separators and the Euclidean Norm. In: SIGAL Intl. Symposium on Algorithms. Lecture Notes in Computer Science, Vol. 450. Springer Verlag, Berlin (1990) 338-347
9. Geelen, J. F., Richter, R.B., Salazar, G.: Embedding Graphs on Surfaces. Submitted to J. Combinatorial Theory-B
10. Grohe, M.: Computing Crossing Numbers in Quadratic Time. In: 33rd Annual ACM Symposium on Theory of Computing. ACM Press (2001)
11. Hliněný, P.: Crossing-Critical Graphs and Path-Width. In: 9th Intl. Symposium on Graph Drawing. Lecture Notes in Computer Science, Springer Verlag, Berlin (2001)
12. Kinnersley, N.: The Vertex Separation Number of a Graph Equals its Path-Width. Information Processing Letters **142** (1992) 345-350
13. Liebers, A.: Methods for Planarizing Graphs - a Survey and Annotated Bibliography. J. of Graph Algorithms and Applications **5** (2001) 1-74
14. Leighton, F. T.: Complexity Issues in VLSI, M.I.T. Press, Cambridge (1983)
15. Pach, J., Shahrokhi, F., Szegedy, M.: Applications of Crossing Number. Algorithmica **16** (1996) 111-117

16. Purchase, H.: Which Aestethic has the Greatest Effect on Human Understanding? In: 5th Intl. Symposium on Draph Drawing. Lecture Notes in Computer Science, Vol. 1353. Springer Verlag, Berlin, (1997) 248-261
17. Shahrokhi, F., Sýkora, O., Székely, L.A., Vrťo, I.: Crossing Numbers: Bounds and Applications. In: Bárány, I., Boroczky, K. (eds.): Intuitive Geometry. Bolyai Society Mathematical Studies 6. Akadémia Kiadó, Budapest (1997) 179-206
18. Sýkora, O., Vrťo, I.: On the Crossing Number of the Hypercube and the Cube Connected Cycles. BIT **33** (1993) 232-237
19. Sýkora, O., Vrťo, I.: On VLSI Layouts of the Star Graph and Related Networks. Integration, The VLSI Journal **17** (1994) 83-93
20. Wei-Liang Lin, Amir H. Farrahi, A. H., Sarrafzadeh, M.: On the Power of Logic Resynthesis. SIAM J. Computing **29** (2000) 1257-1289

Crossing-Critical Graphs and Path-Width

Petr Hliněný*

School of Mathematical and Computing Sciences,
Victoria University,
P.O. Box 600, Wellington, New Zealand,
and
Institute for Theoretical Computer Science** (ITI MFF),
Charles University,
Malostranské nám. 25, 118 00 Praha 1, Czech Republic.

(e-mail: hlineny@member.ams.org , fax: +64-4-4635045)

Abstract. The crossing number $\mathrm{cr}(G)$ of a graph G, is the smallest possible number of edge-crossings in a drawing of G in the plane. A graph G is crossing-critical if $\mathrm{cr}(G-e) < \mathrm{cr}(G)$ for all edges e of G. G. Salazar conjectured in 1999 that crossing-critical graphs have path-width bounded by a function of their crossing number, which roughly means that such graphs are made up of small pieces joined in a linear way on small cut-sets. That conjecture was recently proved by the author [9]. Our paper presents that result together with a brief sketch of proof ideas. The main focus of the paper is on presenting a new construction of crossing-critical graphs, which, in particular, gives a nontrivial lower bound on the path-width. Our construction may be interesting also to other areas concerned with the crossing number.

1 Introduction

In this section we informally introduce the problem and our contributions to it. The reader is referred to the next section for formal definition and statements.

We are interested in drawing of (nonplanar) graphs in the plane that have a small number of edge-crossings. There are many practical applications of such drawings, including VLSI design [3], or graph visualization [4,14]. Crossing-number problems are often discussed on Graph Drawing conferences, recently for example [12,18,14].

Determining the crossing number of a graph is a hard problem [6] in general, and the crossing number is not even known exactly for complete or complete bipartite graphs. A lot of work has been done investigating the crossing number of particular graph classes like $C_m \times C_n$, see [15,16,8]. For general graphs, research so far focused mainly on relations of the crossing number to nonstructural graph

* The research was partially supported by a New Zealand Marsden Fund research grant to Geoff Whittle, and by a Czech research grant GAČR 201/99/0242.

** Supported by the Ministry of Education of Czech Republic as project LN00A056.

P. Mutzel, M. Jünger, and S. Leipert (Eds.): GD 2001, LNCS 2265, pp. 102–114, 2002.
© Springer-Verlag Berlin Heidelberg 2002

properties like the number of edges, for example [1,11,13]. On the other hand, crossing-critical graphs play a key role in investigation of structural properties of the crossing number. Our result tries to give some insight to the general structure of crossing-critical graphs, about which is not much known yet.

In Section 2, we state that if G is a k-crossing-critical graph, then G cannot contain a subdivision of a "large in k" binary tree. It is known [17] that the latter condition is equivalent to G having "bounded in k path-width", which roughly means that G is made up of small pieces joined in a linear way on small cut-sets. We also sketch basic proof ideas for this result in Section 3, while the whole proof (which is rather long) can be found in [9].

We mainly focus on constructions of crossing-critical graphs that give good lower bounds on the path-width (in terms of binary trees) in Section 4. Specifically, we present new general classes of k-crossing-critical graphs for $k \geq 3$, and we prove their values of the crossing number. These classes contain graphs with binary trees of heights up to $k + 2$. We think that these classes may be also interesting to other areas concerned with the crossing number.

2 Definitions and Results

We consider finite simple graphs in the paper. We usually speak about actual drawings of graphs instead of abstract graphs here. If $\varrho : [0, 1] \to \mathbb{R}^2$ is a simple continuous function, then $\varrho([0, 1])$ is a *simple curve*, and $\varrho((0, 1))$ is a *simple open curve*.

Definition. A graph G is *drawn* in the plane if the vertices of G are distinct points of $\mathbb{R}^2$, and every edge $e = uv \in E(G)$ is a simple open curve ϱ such that $\varrho(0) = u$, $\varrho(1) = v$. Moreover, it is required that no edge contains a vertex of G, and that no three distinct edges of G share a common point. An *(edge-) crossing* is any point of the drawing that belongs to two distinct edges.
(Notice that our edge as a topological object does not include its endpoints. In particular, when we speak about a crossing, we do not mean a common end of two edges.)

Definition. The *crossing number* $\mathrm{cr}(G)$ of a graph G is the smallest possible number of edge-crossings in a drawing of G in the plane. A graph G is *crossing-critical* if $\mathrm{cr}(G - e) < \mathrm{cr}(G)$ for all edges $e \in E(G)$. A graph G is *k-crossing-critical* if G is crossing-critical and $\mathrm{cr}(G) = k$.

The crossing number stays the same if we consider drawings on the sphere instead of the plane, or if we require piecewise-linear drawings. (However, if we require the edges to be straight segments – so called rectilinear crossing number, we get completely different behavior; but we are not dealing with this concept here.) Also, the crossing number is clearly preserved under subdivisions of edges (although not under contractions). Thus it is not an essential restriction when we consider simple graphs only.

One annoying thing about the crossing number is that there exist other possible definitions of it, and we do not know whether they are all equivalent or not. The *pairwise-crossing number* $\mathrm{cr}_{\mathrm{pair}}$ is defined similarly, but it counts the number of crossing pairs of edges, instead of crossing points. The *odd-crossing number* $\mathrm{cr}_{\mathrm{odd}}$ counts the number of pairs of edges that cross odd number of times only. It clearly follows that $\mathrm{cr}_{\mathrm{odd}}(G) \leq \mathrm{cr}_{\mathrm{pair}}(G) \leq \mathrm{cr}(G)$, and it was proved by Tutte [19] that $\mathrm{cr}_{\mathrm{odd}}(G) = 0$ implies $\mathrm{cr}(G) = 0$. The best known general relation between these crossing numbers is due to Pach and Tóth [13] who proved $\mathrm{cr}(G) \leq 2\,\mathrm{cr}_{\mathrm{odd}}(G)^2$. Our results are formulated for the ordinary crossing number, however, they hold as well for the pairwise-crossing number.

Further we define the path-width of a graph and present its basic properties. A notation $G \restriction X$ is used for the subgraph of G induced by the vertex set X. A *minor* is a graph obtained from a subgraph by contractions of edges.

Definition. A *path decomposition* of a graph G is a sequence of sets $(W_1, W_2, \dots, W_p)$ such that $\bigcup_{1\leq i\leq p} W_i = V(G)$, $\bigcup_{1\leq i\leq p} E(G \restriction W_i) = E(G)$, and $W_i \cap W_k \subseteq W_j$ for all $1 \leq i < j < k \leq p$. The width of a path decomposition is $\max\{|W_i| - 1 : 1 \leq i \leq p\}$. The *path-width* of a graph G, denoted by $\mathrm{pw}(G)$, is the smallest width of a path decomposition of G.

It is known [17] that if G is a minor of H, then $\mathrm{pw}(G) \leq \mathrm{pw}(H)$. A *binary tree* of height h a rooted tree T such that the root has degree 2, all other non-leaf vertices of T have degrees 3, and every leaf of T has distance h from the root. (A binary tree of height h has $2^{h+1} - 1$ vertices.) Since the maximal degree of a binary tree T is 3, a graph H contains T as a minor if and only if H contains T as a subdivision. The important connection between binary trees and path-width was first established by Robertson and Seymour in [17], while the following strengthening is due to [2]:

Theorem 2.1. (Bienstock, Robertson, Seymour, Thomas)
(a) If T is a binary tree of height h, then $\mathrm{pw}(T) \geq \frac{h}{2}$.
(b) If $\mathrm{pw}(G) \geq p$, then G contains any tree on p vertices as a minor.

We look closer at some facts about crossing-critical graphs. By the Kuratowski theorem, there are only two 1-crossing-critical graphs K_5 and $K_{3,3}$, up to subdivisions. On the other hand, an infinite family of 2-crossing-critical graphs with minimal degree more than 2 was found by Kochol in [10]. One may easily observe that every edge-transitive graph is crossing-critical, while the converse is not true, of course.

Ding, Oporowski, Thomas and Vertigan [5] have proved that every 2-crossing-critical graph satisfying certain simple assumptions and having sufficiently many vertices belongs to a well-defined infinite graph class. In particular, these graphs have bounded path-width. Analyzing the structure of other known infinite classes of crossing-critical graphs, G. Salazar formulated the following conjecture, appearing in [7].

Conjecture 2.2. (Salazar, 1999) There exists a function g such that any k-crossing-critical graph has path-width at most $g(k)$.

The paper [7] proves a weaker statement that the tree-width of a crossing-critical graph is bounded. Our Theorem 2.3 [9], together with Theorem 2.1, immediately imply a solution to Salazar's conjecture.

Theorem 2.3. *There exists a function f such that no k-crossing-critical graph contains a subdivision of a (complete) binary tree of height $f(k)$. In particular, $f(k) \leq 6 \cdot (72 \log_2 k + 248) \cdot k^3$.*

Corollary 2.4. *Let f be the function from Theorem 2.3. If $\boldsymbol{G}$ is a k-crossing-critical graph, then the path-width of $\boldsymbol{G}$ is at most $2^{f(k)+1} - 2$.*

Remark. It is important that Theorem 2.3 speaks about crossing-critical graphs, since an arbitrary graph of a fixed crossing number k may contain a binary tree of any height. There is no direct connection between the crossing number and the path-width of a graph without an assumption of being crossing-critical.

A natural question arises about lower bounds on the function f from Theorem 2.3. An easy argument shows that $f(k)$ must grow with k: The complete graph $\boldsymbol{K}_n$ is crossing-critical for $n \geq 5$ with the crossing number growing roughly as $\Theta(n^4)$, and $\boldsymbol{K}_n$ contains a binary tree of height $\lfloor \log_2 n \rfloor - 1$. (In fact, the path-width of K_n is $n - 1$.) However, we are able to provide much better bounds on f as consequences of a general construction presented in Section 4:

Theorem 2.5. *Let f be the function from Theorem 2.3, and $k \geq 3$. Then $f(k) \geq k + 3$, or $f(k) \geq k$ if we consider only simple 3-connected graphs.*

3 Upper Bound Sketch

The whole proof [9] of Theorem 2.3 is quite long, so here we present only an informal short sketch of it. Suppose that $\boldsymbol{G}$ is a graph drawn in the plane with k crossings. The basic idea behind our proof is that if sufficiently many nested edge-disjoint cycles "separate" all crossed edges from some edge e in $\boldsymbol{G}$, then $\boldsymbol{G}$ cannot be crossing-critical since deleting e cannot decrease its crossing number. (This trick was suggested earlier by Salazar in connection with the tree-width of crossing-critical graphs.) Unfortunately, considering sequences of single cycles is not enough to achieve our goal. So we actually work with so called "nesting" and "cutting" sequences in the graph $\boldsymbol{G}$ (see Lemmas 3.1 and 3.2).

Recall that $\boldsymbol{G}$ is a graph drawn in the plane. Informally speaking, a *multicycle* M in $\boldsymbol{G}$ is a collection of (not necessarily disjoint) cycles of $\boldsymbol{G}$ such that no two of these cycles are crossed or nested. (These words implicitly refer to the infinite face of the drawing.) The finite faces bounded by the cycles of M are called the *interior faces* of M. We say that a multicycle M is *nested* in a multicycle M', denoted by $M \preceq M'$, if each interior face of M is contained in some interior face of M'. We say that M is *strictly nested* in M', denoted by $M \lll M'$, if $M \preceq M'$, and if M and M' share at most one vertex. See an illustration in Fig. 1.

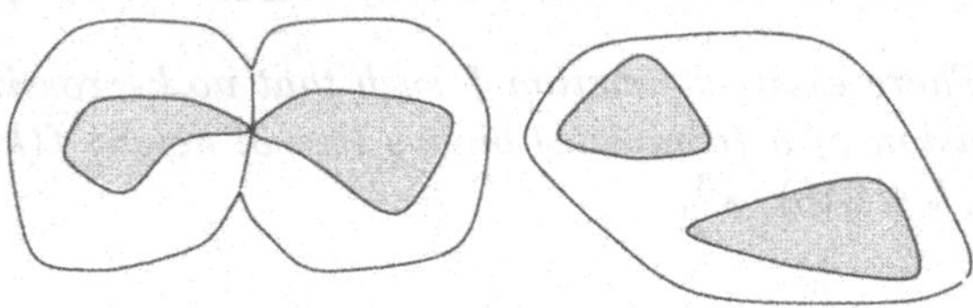

Fig. 1. An example of two strictly nested multicycles $M \lll M'$ (shaded M consists of 4 cycles, and M' consists of 3 cycles).

Let $M_1 \lll M_2 \lll \ldots \lll M_c$ be a sequence of c strictly nested multicycles in the graph G. Suppose that all crossed edges of G are contained in the interior faces of M_1, and that, for each interior face Φ of M_i, $2 \le i \le c$, every component of the subgraph of G drawn inside Φ intersects some cycle of M_{i-1} in Φ. Then $\mathcal{M}_c(G) = (M_1, \ldots, M_c)$ is called a *c-nesting sequence* in G.

Lemma 3.1. *Suppose that there exists a* $(3k-1)$*-nesting sequence in a 2-connected graph* H *drawn in the plane. Then* H *is not k-crossing-critical.*

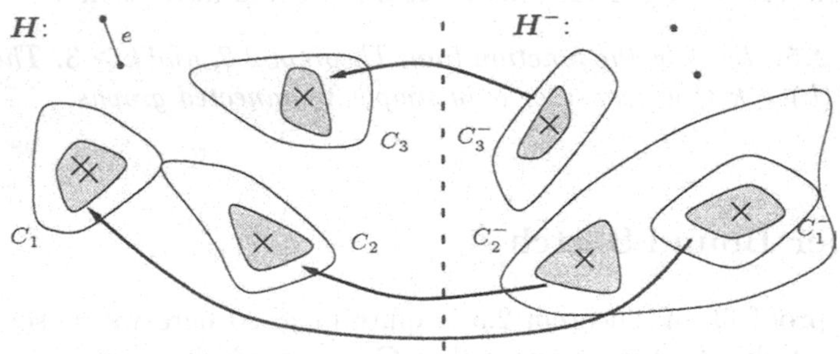

Fig. 2. An illustration to Lemma 3.1; how a better drawing of H is obtained using parts of the drawing $H^- \simeq H - e$ that has fewer than k crossings.

Proof. Let $\mathcal{M}_{3k-1}(H) = (M_1, \ldots, M_{3k-1})$ be a $(3k-1)$-nesting sequence in H. Briefly speaking, our goal is to delete an edge e in the exterior of M_{3k-1}, draw the new graph with fewer crossings, and use pieces of the new drawing to "improve" the drawing H. Notice that M_1 consists of at most k cycles (one for each crossings), and that the number of cycles does not increase in the sequence.

If H is k-crossing-critical, then there exists a drawing H^- of the graph $H-e$ with fewer than k crossings. We denote by $M_1^-, \ldots, M_{3k-1}^-$ the corresponding multicycles in $\boldsymbol{H}^-$. One edge-crossing may involve at most two multicycles, so at least k multicycles of $M_2^-, \ldots, M_{3k-1}^-$ are not crossed in $\boldsymbol{H}^-$. Thus there exists an index $2 \leq i \leq 3k-1$ such that M_i^- is not crossed, and that M_i^-, M_{i-1}^- consist of the same number m of cycles. Let $\boldsymbol{H}_j$, $j = 1, \ldots, m$ be the subgraph of $\boldsymbol{H}$ drawn inside the j-th cycle of M_i, and let $\boldsymbol{H}_j^-$ be the corresponding subgraph in $\boldsymbol{H}^-$. The graphs $\boldsymbol{H}_j \simeq \boldsymbol{H}_j^-$ are connected by the definition. Since the multicycle M_i^- in $\boldsymbol{H}^-$ is not crossed, we may "cut" the subdrawings $\boldsymbol{H}_j^-$ and "paste" them into the interior faces of M_i in $\boldsymbol{H}$ instead of $\boldsymbol{H}_j$, $j = 1, \ldots, m$. Recall that all crossings of $\boldsymbol{H}$ belonged to some $\boldsymbol{H}_j$. Therefore, the new drawing of $\boldsymbol{H}$ has at most as many crossings as $\mathrm{cr}(\boldsymbol{H}^-) < k$, a contradiction. ∎

We say that a sequence $P_1, \ldots, P_q$ of pairwise disjoint paths in a graph $\boldsymbol{G}$ is a *q-cutting sequence* if each set $V(P_i)$ is a cut in $\boldsymbol{G}$ separating $X \cup P_1 \cup \ldots \cup P_{i-1}$ from $P_{i+1} \cup \ldots \cup P_q$, where X is a subgraph formed by all crossed edges of $\boldsymbol{G}$. Similarly as in the previous lemma we prove:

Lemma 3.2. *Suppose that there exists a $4k$-cutting sequence in a 2-connected graph $\boldsymbol{H}$ drawn in the plane. Then $\boldsymbol{H}$ cannot be k-crossing-critical.*

Finally, the lengthy part of the proof of Theorem 2.3 comes in. We want to show that a 2-connected graph with a sufficiently large binary tree contains a long nesting or cutting sequence. Obviously, if our graph $\boldsymbol{H}$ is not 2-connected, we may prove the theorem separately for the blocks of $\boldsymbol{H}$.

Lemma 3.3. *Let $\boldsymbol{H}$ be a 2-connected graph that is drawn in the plane with k crossings. Suppose that $\boldsymbol{H}$ contains a subdivision of a binary tree of height $6 \cdot (72 \log_2 k + 248) \cdot k^3$. Then there exists a $(3k-1)$-nesting sequence or a $4k$-cutting sequence in $\boldsymbol{H}$.*

To prove the lemma, we try to inductively construct a c-nesting sequence in $\boldsymbol{H}$ for $c = 1, 2, \ldots, 3k-1$, such that the multicycles in the sequence satisfy certain rather complicated connectivity property, and that a "large portion" of the subdivision of a binary tree in $\boldsymbol{H}$ stays outside of the sequence. Let us denote by $f'(k) = (72 \log_2 k + 248)k^2$, by $f(k) = 6kf'(k)$, and by $f_i(k) = (6k - 2i - 1)f'(k)$. The first multicycle M_1 of the sequence encloses all crossed edges of $\boldsymbol{H}$, and, at each step c, there is a subdivision $\boldsymbol{U} \subset \boldsymbol{H}$ of a binary tree of height $f_c(k)$ drawn in the infinite face of the last multicycle M_c. For simplicity, say that $\boldsymbol{U}$ actually is a binary tree.

Now we briefly describe a single step of our construction. We divide the binary tree $\boldsymbol{U}$ of height $f_c(k)$ into "layers" of heights $f'(k)$, $f'(k)$, and $f_{c+1}(k)$. (For example, a subtree of $\boldsymbol{U}$ in the "middle layer" has its root at distance $f'(k)$ and its leaves at distance $2f'(k)$ from the root of $\boldsymbol{U}$.)

- First we look whether the leaves of some middle-layer subtree of $\boldsymbol{U}$ are "surrounded" by a common face of $\boldsymbol{H}$. If this happens, then either there is a next multicycle M_{c+1} for our nesting sequence (using part of boundary of the common face), or selected paths of the mentioned subtree form a $4k$-cutting sequence.

- If we are not successful in the previous step, then we argue that most of the middle-layer subtrees are "cut in half" by closed curves in the drawing $\boldsymbol{H}$. If sufficiently many of such curves do not intersect M_c, then they form many graph cycles in $\boldsymbol{H}$. We use the cycles to construct a multicycle M_{c+1} such that some of the bottom-layer subtrees of $\boldsymbol{U}$ of height $f_{c+1}(k)$ stays in the infinite face of M_{c+1}.
- Otherwise, most of middle-layer subtrees are connected by pairwise disjoint paths to vertices of M_c. In such case we apply the above mentioned connectivity property of our sequence (which is specifically tailored to solve this case); and using the connecting paths, we construct another $(3k-1)$-nesting or $4k$-cutting sequence in $\boldsymbol{H}$ straight away.

We skip the details of this proof here.

4 "Crossed-Fence" Construction

Let k be a positive integer. We describe a graph class parametrized by k, and we later prove that the graphs from this class are k-crossing-critical. (The name "fence" for the class was chosen by resemblance of the example from Fig. 3.)

Definition. Let $C_1, C_2, \ldots, C_k$ be a sequence of some k edge-disjoint graph cycles, let $\boldsymbol{F}_0 = C_1 \cup C_2 \cup \ldots \cup C_k$ be a graph, and let $u_1, u_2 \in V(C_1)$, $u_3, u_4 \in V(C_k)$. The 5-tuple $(\boldsymbol{F}_0; u_1, u_2; u_3, u_4)$ is called a *k-fence* if the following conditions (F1-4) are true:

(F1) For $1 \leq i, j \leq k$ and $|i-j| \geq 2$, the cycles C_i, C_j are vertex-disjoint. Moreover, $u_1, u_2 \notin V(C_i)$ for $i > 1$, and $u_3, u_4 \notin V(C_i)$ for $i < k$.

(F2) The graph $\boldsymbol{F}_0 = C_1 \cup \ldots \cup C_k$ is connected and planar.

Let $n = 1, 2$. We define a set $X_n \subset V(\boldsymbol{F}_0)$ recursively as follows: $u_n \in X_n$; and, for $i = 1, 2, \ldots, k-1$ and $j = i+1$, if $x \in X_n \cap V(C_i)$, $x' \in V(C_i) \cap V(C_j)$ are such that there is a path $P \subset C_i$ with ends x, x' internally disjoint from C_j, then we add x' into X_n. We define sets X_n, $n = 3, 4$ analogously for $i = k, k-1, \ldots, 2$ and $j = i - 1$.

(F3) For $n = 1, 2$ (for $n = 3, 4$) and $2 \leq i \leq k-1$ the next holds: if $P \subset C_i$ is a path with both ends in $X_n \cap V(C_i) \cap V(C_{i-1})$ (in $X_n \cap V(C_i) \cap V(C_{i+1})$), then P intersects $V(C_{i+1})$ (P intersects $V(C_{i-1})$).

(F4) The sets X_1, X_2, X_3, X_4 are pairwise disjoint. For $1 \leq i \leq k$; if $v_n \in X_n \cap V(C_i)$, $n = 1, 2, 3, 4$, then the vertices v_1, v_3, v_2, v_4 lie in this cyclic order on the cycle C_i.

Moreover, a graph $\boldsymbol{F}$ is called a *crossed k-fence* if $\boldsymbol{F} = \boldsymbol{F}_0 \cup Q_1 \cup Q_2 \cup Q$ and u_1, u_2, u_3, u_4 are such that the following is true:

(F5) $\boldsymbol{F}_0$ is a graph, $u_1, u_2, u_3, u_4 \in V(\boldsymbol{F}_0)$, and $(\boldsymbol{F}_0; u_1, u_2; u_3, u_4)$ is a k-fence.

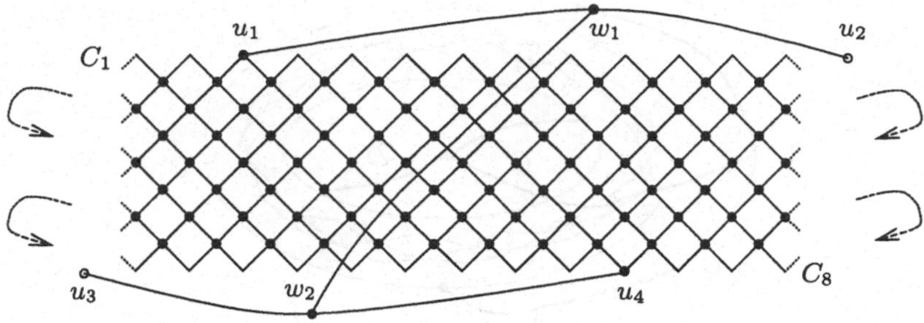

Fig. 3. A basic example of a crossed 8-fence. (The "fence" is winded on a cylinder.)

(F6) Q_1 is a path with ends u_1, u_2 internally disjoint from $\boldsymbol{F}_0$, and Q_2 is a path with ends u_3, u_4 internally disjoint from $\boldsymbol{F}_0 \cup Q_1$. For some internal vertices $w_1 \in V(Q_1), w_2 \in V(Q_2)$ of the paths Q_1, Q_2, the path Q connects w_1, w_2 and is internally disjoint from $\boldsymbol{F}_0 \cup Q_1 \cup Q_2$.

As an illustration to the above definition we present an example of a crossed 8-fence in Fig. 3. We also add some informal comments to the definition: By the definition, the graph $\boldsymbol{F}_0$ is planar. Moreover, it immediately follows from (F1-2) that we may draw $\boldsymbol{F}_0$ without crossings as a "bunch of concentric cycles", i.e. each cycle C_i is a closed curve separating $C_1 \cup \ldots \cup C_{i-1}$ from $C_{i+1} \cup \ldots \cup C_k$. (See also Fig. 4.) Notice that the definition of a fence $(\boldsymbol{F}_0; u_1, u_2; u_3, u_4)$ is symmetric with respect to any one of u_1, u_2, u_3, u_4 (possibly reversing the order of cycles in $\boldsymbol{F}_0$). Notice also that the sets X_n, $n = 1, 2, 3, 4$ intersect all cycles of $\boldsymbol{F}_0$. More properties of a fence are illustrated by two easy lemmas.

Lemma 4.1. *Let $(\boldsymbol{G}_0; u_1, u_2; u_3, u_4)$ be a k-fence, $k \geq 2$, where $\boldsymbol{G}_0 = C_1 \cup C_2 \cup \ldots \cup C_k$. We denote by $\boldsymbol{G}_0' = C_2 \cup C_3 \cup \ldots \cup C_k$, and by u_i', $i = 1, 2$ some vertex of $C_1 \cap C_2$ such that there is a path $P_i \subset C_1$ with ends u_i, u_i' internally disjoint from C_2. Then $(\boldsymbol{G}_0'; u_1', u_2'; u_3, u_4)$ is a $(k-1)$-fence.*

Proof. Let us look at the definition of a fence on page 108. The conditions (F1-2) from the definition are clearly satisfied for $\boldsymbol{G}_0'$. In particular, $u_1', u_2' \notin V(C_i)$ for $i > 2$ since $u_1', u_2' \in V(C_1)$. We denote by X_n', $n = 1, 2, 3, 4$ the sets defined analogously to X_n for $\boldsymbol{G}_0'$. Then $X_n' = X_n \setminus V(C_1)$ for $n = 3, 4$, and $X_n' \subset X_n$ for $n = 1, 2$ since $u_n' \in X_n$ by the definition. So validity of the conditions (F3-4) for $\boldsymbol{G}_0'$ follows easily, and $\boldsymbol{G}_0'$ forms a $(k-1)$-fence. ∎

Lemma 4.2. *Let $\boldsymbol{G}$ be a crossed k-fence, $k \geq 1$. Then* $\mathrm{cr}(\boldsymbol{G}) \leq k$*, and* $\mathrm{cr}(\boldsymbol{G} - e) \leq k - 1$ *for all edges $e \in E(\boldsymbol{G})$.*

Proof. This is an easy proof again, so we only sketch it. (See the scheme in Fig. 4.) We use the notation $\boldsymbol{G} = \boldsymbol{G}_0 \cup Q_1 \cup Q_2 \cup Q$ and $\boldsymbol{G}_0 = C_1 \cup \ldots \cup C_k$ analogously to the definition of a crossed fence. As noted above, $\boldsymbol{G}_0$ can be

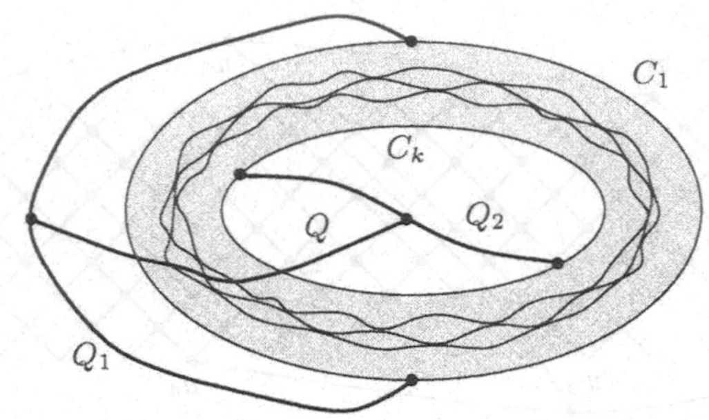

Fig. 4. A generic plane drawing of a crossed k-fence.

drawn without crossings as a "bunch" of concentric cycles. We may add the paths Q_1, Q_2 "outside" and "inside" to $\boldsymbol{G}_0$ again without crossings. Finally, we draw the path Q connecting a vertex of Q_1 to a vertex of Q_2 so that it crosses each of the k cycles C_i of $\boldsymbol{G}_0$ exactly once.

Next, we show how to modify the previous drawing of $\boldsymbol{G}$ to get a drawing of $\boldsymbol{G} - e$ with less than k crossings: If $e \in E(\boldsymbol{G}_0)$, then we may avoid the crossing of Q with the cycle C_j, $e \in E(C_j)$. If $e \in E(Q)$, then $\boldsymbol{G} - e$ is planar. Lastly, if $e \in E(Q_1)$ (which is symmetric to $e \in E(Q_2)$), then we may redraw $(Q \cup Q_1) - e$ so that it does not cross C_1. ∎

Lemma 4.3. *Let $\boldsymbol{G}$ be a crossed k-fence, $k = 1$ or $k \geq 3$. Then* $\mathrm{cr}(\boldsymbol{G}) \geq k$.

Proof. We use induction on k. In the base case $k = 1$, $\boldsymbol{G}$ is a subdivision of the nonplanar graph $K_{3,3}$, and so $\mathrm{cr}(\boldsymbol{G}) = 1$. Unfortunately, our statement is false for $k = 2$; a crossed 2-fence may have crossing number 1. Thus we must avoid referring to that case in the induction. We first present a general inductive step, and then we show how to overcome the exceptional value of 2.

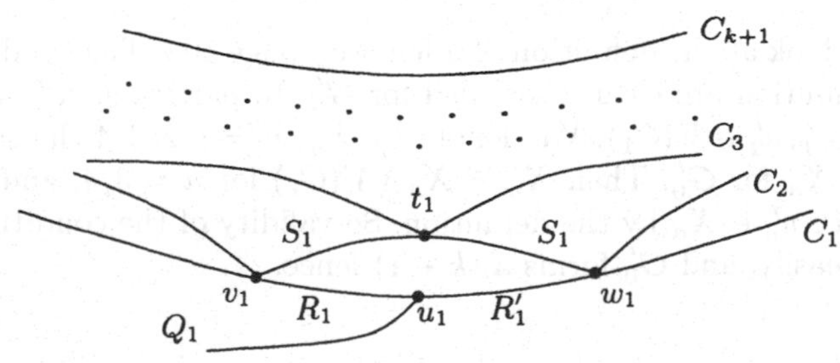

Fig. 5. An illustration to the proof.

Let us have an optimal drawing $\boldsymbol{H}$ of a crossed (k+1)-fence, where $\boldsymbol{H} = \boldsymbol{H}_0 \cup Q_1 \cup Q_2 \cup Q$ and $\boldsymbol{H}_0 = C_1 \cup \ldots \cup C_{k+1}$, $u_1, u_2 \in V(C_1)$, $u_3, u_4 \in V(C_{k+1})$, as in the definition of a crossed fence. By connectivity of $\boldsymbol{H}_0$, there are two (possibly equal for now) vertices $v_1, w_1 \in V(C_1) \cap V(C_2)$ such that the edge-disjoint paths $R_1, R_1' \subset C_1$ connecting u_1 to v_1 and u_1 to w_1, resp., are internally disjoint from C_2. (See Fig. 5.) We define vertices v_2, w_2 and paths $R_2, R_2' \subset C_1$ analogously for u_2. Then $v_1, w_1 \in X_1$ and $v_2, w_2 \in X_2$ by the definition. It follows from (F4) that v_1, w_1, v_2, w_2 are pairwise distinct.

We first assume that some edge $e \in E(C_1)$ is crossed in $\boldsymbol{H}$. Up to symmetry, we may assume that $e \notin E(R_1)$ and $e \notin E(R_2)$. We set $\boldsymbol{H}_0' = C_2 \cup C_3 \cup \ldots \cup C_{k+1}$, $u_1' = v_1$, $u_2' = v_2$. By Lemma 4.1, $(\boldsymbol{H}_0'; u_1', u_2'; u_3, u_4)$ is a k-fence; and hence $\boldsymbol{H}' = \boldsymbol{H}_0' \cup (Q_1 \cup R_1) \cup (Q_2 \cup R_2) \cup Q$ is a crossed k-fence. However, the drawing $\boldsymbol{H}'$ has at least one crossing less than $\boldsymbol{H}$ since $e \notin E(\boldsymbol{H}')$. Therefore, $\mathrm{cr}(\boldsymbol{H}) \geq \mathrm{cr}(\boldsymbol{H}') + 1 \geq k + 1$ if $k \neq 2$.

Second, we assume that no edge of C_1 is crossed in $\boldsymbol{H}$. We denote by $S_1 \subset C_2$ the path with ends v_1, w_1 and disjoint from v_2, w_2; and $S_2 \subset C_2$ with ends v_2, w_2 analogously. By (F3), both paths S_1, S_2 intersect the cycle C_3. Moreover, since $C_3 \cup \ldots \cup C_{k+1} \cup Q_1 \cup Q_2 \cup Q$ is a connected graph, all three paths S_1, S_2, Q_1 are drawn in the same region of C_1 by the Jordan Curve Theorem. (Recall that C_1 is drawn as an uncrossed closed curve.) It follows from the order of the path ends on C_1 that the path Q_1 must cross both paths S_1, S_2, say in edges $e_1 \in E(S_1)$, $e_2 \in E(S_2)$. We denote by $t_n \in V(S_n) \cap V(C_3)$, $n = 1, 2$ vertices such that (F3) there are subpaths $S_n' \subset S_n - e_n$ connecting v_n (or w_n, up to symmetry) to t_n and internally disjoint from C_3. We set $\boldsymbol{H}_0'' = C_3 \cup \ldots \cup C_{k+1}$, $u_1'' = t_1$, $u_2'' = t_2$. Then $(\boldsymbol{H}_0''; u_1'', u_2''; u_3, u_4)$ is a $(k-1)$-fence by double application of Lemma 4.1, and so $\boldsymbol{H}'' = \boldsymbol{H}_0'' \cup (Q_1 \cup R_1 \cup S_1') \cup (Q_2 \cup R_2 \cup S_2') \cup Q$ is a crossed $(k-1)$-fence. Therefore, $\mathrm{cr}(\boldsymbol{H}) \geq \mathrm{cr}(\boldsymbol{H}'') + 2 \geq k + 1$ if $k - 1 \neq 2$.

Finally, we resolve the exceptions left above. Suppose that $k = 2$ in the first case, and that the second case cannot be symmetrically applied (i.e. C_3 is crossed as well). Then we may actually repeat this step twice (for C_1 and C_3 in $\boldsymbol{H}$), and refer to the inductive assumption for $k - 1 = 1$. Suppose that $k = 3$ in the second case, and that the first case cannot be symmetrically applied (i.e. neither C_4 is crossed). Then again, we argue twice in the same way, showing that both cycles C_2, C_3 of $\boldsymbol{H}$ are crossed at least twice each by the paths Q_1, Q_2, resp. Hence $\mathrm{cr}(\boldsymbol{H}) \geq 2 + 2 = 4$ in this case, as desired. ∎

The previous Lemmas 4.2,4.3 immediately imply:

Theorem 4.4. *Let $\boldsymbol{G}$ be a crossed k-fence, $k = 1$ or $k \geq 3$. Then $\boldsymbol{G}$ is a k-crossing-critical graph.* ∎

5 Lower Bounds

In this section we are going to prove Theorem 2.5 by exhibiting crossed fences that contain large binary trees. (The example of a crossed k-fence from Fig. 3 contains a subdivision of a binary tree of height about $\frac{k}{2}$, however, we provide even better constructions now.)

Lemma 5.1. *There exists a graph* $\boldsymbol{H}^k$, $k \geq 1$ *such that* $\boldsymbol{H}^k$ *is a crossed* k*-fence, and that* $\boldsymbol{H}^k$ *contains a binary tree of height* $k+2$.

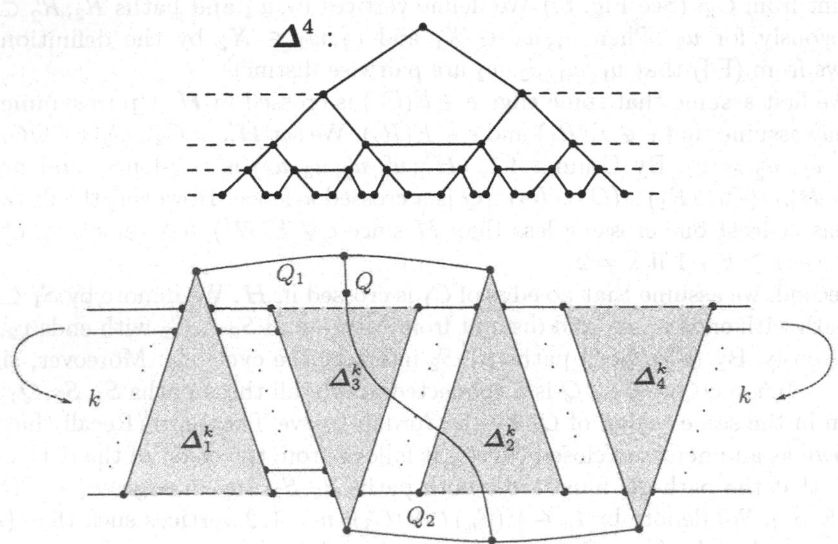

Fig. 6. A scheme of the construction of graph $\boldsymbol{H}^k$ (cf. Lemma 5.1).

Proof. We denote by $\boldsymbol{\Delta}^k$ a graph described as follows: The vertex set of $\boldsymbol{\Delta}^k$ consists of all words starting with a symbol r and appending a sequence of at most k symbols chosen from $0, 1$. Each vertex (word) $\langle x \rangle$ of $\boldsymbol{\Delta}^k$ is adjacent to both $\langle x\,0 \rangle, \langle x\,1 \rangle$ (unless x is longer than k). Moreover, each vertex of pattern $\langle y\,01^i \rangle$, $i \geq 1$ in $\boldsymbol{\Delta}^k$ is adjacent to $\langle y\,10^i \rangle$. (The exponent of a symbol counts repetition of this symbol in the word.) The construction is illustrated by an example of $\boldsymbol{\Delta}^4$ on the top of Fig. 6. Clearly, $\boldsymbol{\Delta}^k$ has a spanning binary tree with the root $\langle r \rangle$.

A graph $\boldsymbol{H}_0^k$ is a disjoint union of four copies $\boldsymbol{\Delta}_1^k, \boldsymbol{\Delta}_2^k, \boldsymbol{\Delta}_3^k, \boldsymbol{\Delta}_4^k$ of $\boldsymbol{\Delta}^k$ joined by edges in the following way: For $i = 1, \ldots, k$, a vertex $\langle r\,0^i \rangle$ of $\boldsymbol{\Delta}_1^k$ is adjacent to a vertex $\langle r\,0^{k+1-i} \rangle$ of $\boldsymbol{\Delta}_4^k$; and a vertex $\langle r\,1^i \rangle$ of $\boldsymbol{\Delta}_1^k$ is adjacent to a vertex $\langle r\,1^{k+1-i} \rangle$ of $\boldsymbol{\Delta}_3^k$. Vertices of $\boldsymbol{\Delta}_2^k$ are analogously adjacent to $\boldsymbol{\Delta}_3^k$ and to $\boldsymbol{\Delta}_4^k$. Fig. 6 shows a scheme of the construction. We claim that $\boldsymbol{H}_0^k$ is a k-fence (see the definition on page 108): It is easy to see the cycles C_i – the vertex set $V(C_i)$ is formed by all words in $\boldsymbol{\Delta}_1^k$, $\boldsymbol{\Delta}_2^k$ of length i or $i+1$, and by all words in $\boldsymbol{\Delta}_3^k$, $\boldsymbol{\Delta}_4^k$ of length $k+2-i$ or $k+1-i$. The vertices u_1, u_2, u_3, u_4 from the definition are the respective roots, and the sets X_1, X_2, X_3, X_4 are the respective vertex sets, of $\boldsymbol{\Delta}_1^k, \boldsymbol{\Delta}_2^k, \boldsymbol{\Delta}_3^k, \boldsymbol{\Delta}_4^k$.

The graph $\boldsymbol{H}^k$ results from $\boldsymbol{H}_0^k$ by adding paths Q_1, Q_2 and Q, each of length two, as required by the definition of a crossed fence. Then $\boldsymbol{H}^k$ has a spanning binary tree of height $k+2$, the root of which is the middle vertex of path Q. ∎

Lemma 5.2. *There exists a simple 3-connected graph $\tilde{\boldsymbol{H}}^k$, $k \geq 1$ such that $\tilde{\boldsymbol{H}}^k$ is a crossed k-fence, and $\tilde{\boldsymbol{H}}^k$ contains a binary tree of height $k-1$.*

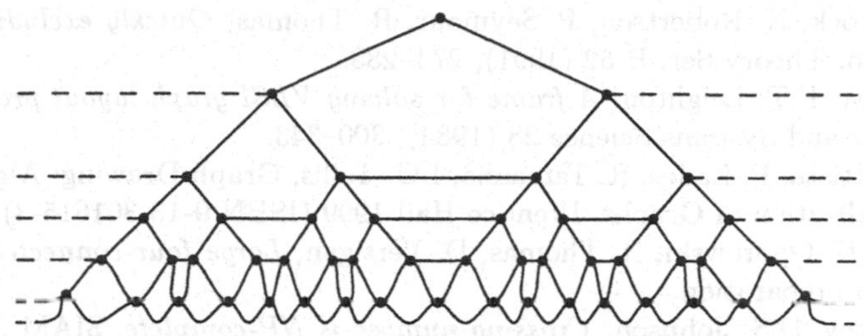

Fig. 7. A modified construction of a graph $\tilde{\boldsymbol{\Delta}}^6$ (cf. Lemma 5.2).

Proof. The construction of $\tilde{\boldsymbol{H}}^k$ is almost the same as the previous construction of $\boldsymbol{H}^k$, but we use copies of a graph $\tilde{\boldsymbol{\Delta}}^k$ instead of $\boldsymbol{\Delta}^k$. Simply speaking, $\tilde{\boldsymbol{\Delta}}^k$ is obtained from $\boldsymbol{\Delta}^{k-2}$ by adding two paths that "form a lace on the bottom vertices", as shown in Fig. 7. Rest follows the scheme in Fig. 6, with a minor variation that the path Q now consists of one edge.

We formally describe the construction as follows: The graph $\tilde{\boldsymbol{\Delta}}^k$ results from $\boldsymbol{\Delta}^{k-1}$ (see in the proof of Lemma 5.1) by contracting all edges of the pattern $\{\langle y\,01^i\rangle, \langle y\,10^i\rangle\}$ where y is a prefix of length $k-1-i$, and by adding all edges of the pattern $\{\langle z\,0\rangle, \langle z\,1\rangle\}$ where z is a prefix of length $k-1$. Then $\tilde{\boldsymbol{H}}_0^k$ is constructed from four copies $\tilde{\boldsymbol{\Delta}}_1^k, \tilde{\boldsymbol{\Delta}}_2^k, \tilde{\boldsymbol{\Delta}}_3^k, \tilde{\boldsymbol{\Delta}}_4^k$ of $\tilde{\boldsymbol{\Delta}}^k$. A vertex $\langle r\,0^i\rangle$, $i = 2, \ldots, k-1$ of $\tilde{\boldsymbol{\Delta}}_1^k$ is adjacent to a vertex $\langle r\,0^{k+1-i}\rangle$ of $\tilde{\boldsymbol{\Delta}}_3^k$, a vertex $\langle r\,0\rangle$ of $\tilde{\boldsymbol{\Delta}}_1^k$ is adjacent to $\langle r\,0^{k-1}\rangle$ of $\tilde{\boldsymbol{\Delta}}_3^k$, and $\langle r\,0^{k-1}\rangle$ of $\tilde{\boldsymbol{\Delta}}_1^k$ is adjacent to $\langle r\,0\rangle$ of $\tilde{\boldsymbol{\Delta}}_3^k$. Analogously, vertices of $\tilde{\boldsymbol{\Delta}}_1^k$ are adjacent to vertices of $\tilde{\boldsymbol{\Delta}}_4^k$, and vertices of $\tilde{\boldsymbol{\Delta}}_2^k$ are adjacent to vertices of $\tilde{\boldsymbol{\Delta}}_3^k$ and of $\tilde{\boldsymbol{\Delta}}_4^k$.

It is a routine work to verify that $\tilde{\boldsymbol{H}}^k$ is a simple 3-connected graph and a crossed k-fence. The largest binary tree in $\tilde{\boldsymbol{H}}^k$ spans $\tilde{\boldsymbol{\Delta}}_1^k \cup \tilde{\boldsymbol{\Delta}}_2^k \cup Q_1$ and it has height $k-1$. ∎

Using Theorem 4.4, the proof of Theorem 2.5 is now finished.

6 Conclusions

We have shown polynomial lower and upper bounds on the height $f(k)$ of a subdivision of a largest binary tree that may be contained in a k-crossing-critical graph. Unfortunately, these bounds are still far apart. We do not make any conjecture about the correct asymptotic for the function $f(k)$ from Theorem 2.3, but we think that it would be closer to the linear lower bound than to the cubic upper bound.

References

1. M. Ajtai, V. Chvátal, M.M. Newborn, E. Szemerédi, *Crossing-free subgraphs.*, Theory and practice of combinatorics, 9–12, North-Holland Math. Stud. 60, North-Holland, Amsterdam-New York, 1982.
2. D. Bienstock, N. Robertson, P. Seymour, R. Thomas, *Quickly excluding a forest*, J. Combin. Theory Ser. B 52 (1991), 274–283.
3. S.N. Bhatt, F.T. Leighton, *A frame for solving VLSI graph layout problems*, J. of Computer and Systems Science 28 (1984), 300–343.
4. G. Di Battista, P. Eades, R. Tamassia, I.G. Tollis, Graph Drawing: Algorithms for the Visualization of Graphs, Prentice Hall 1999 (ISBN 0-13-301615-3).
5. G. Ding, B. Oporowski, R. Thomas, D. Vertigan, *Large four-connected nonplanar graphs*, in preparation.
6. M.R. Garey, D.S. Johnson, *Crossing number is NP-complete*, SIAM J. Algebraic Discrete Methods 4 (1983), 312–316.
7. J. Geelen, B. Richter, G. Salazar, *Embedding grids on surfaces*, manuscript.
8. L.Y. Glebsky, G. Salazar, *The conjecture* $\mathrm{cr}(C_m \times C_n) = (m-2)n$ *is true for all but finitely* n*, for each* m, submitted.
9. P. Hliněný, *Crossing-number critical graphs have bounded path-width*, submitted. `http://www.mcs.vuw.ac.nz/~hlineny/doc/crpath2.ps.gz`
10. M. Kochol, *Construction of crossing-critical graphs*, Discrete Math. 66 (1987), 311–313.
11. F.T. Leighton, Complexity Issues in VLSI, M.I.T. Press, Cambridge, 1983.
12. P. Mutzel, T. Ziegler, *The Constrained Crossing Minimization Problem*, In: Proceedings Graph Drawing '99, Štiřín Castle, Czech Republic, September 1999 (J. Kratochvíl ed.), 175–185; Lecture Notes in Computer Science 1731, Springer Verlag, Berlin 2000 (ISBN 3-540-66904-3).
13. J. Pach, G. Tóth, *Which crossing number is it, anyway?*, Proc. 39th Foundations of Computer Science (1998), IEEE Press 1999, 617–626.
14. H. Purchase, *Which Aesthetics has the Greates Effect on Human Understanding*, In: Proceedings Graph Drawing '97, Rome, Italy, September 18–20 1997 (G. DiBattista ed.), 248–261; Lecture Notes in Computer Science 1353, Springer Verlag, Berlin 1998 (ISBN 3-540-63938-1).
15. R.B. Richter, C. Thomassen, *Intersections of curve systems and the crossing number of* $C_5 \times C_5$, Discrete Comput. Geom. 13 (1995), 149–159.
16. R.B. Richter, G. Salazar, *The crossing number of* $C_6 \times C_n$, Australas. J. Combin. 23 (2001), 135–143.
17. N. Robertson, P. Seymour, *Graph minors I. Excluding a forest*, J. Combin. Theory Ser. B 35 (1983), 39–61.
18. F. Shahrokhi, I. Vrt'o, *On 3-Layer Crossings and Pseudo Arrangements*, In: Proceedings Graph Drawing '99, Štiřín Castle, Czech Republic, September 1999 (J. Kratochvíl ed.), 225–231; Lecture Notes in Computer Science 1731, Springer Verlag, Berlin 2000 (ISBN 3-540-66904-3).
19. W.T. Tutte, *Toward a theory of crossing numbers*, J. Combinatorial Theory 8 (1970), 45–53.

One Sided Crossing Minimization Is NP-Hard for Sparse Graphs*

Xavier Muñoz[1]**, W. Unger[2], and Imrich Vrťo[3],***

[1] Departament de Matemàtica IV, Universitat Politècnica de Catalunya
Jordi Girona 1-3, 08034 Barcelona, Spain
[2] Lehrstuhl für Informatik I, RWTH Aachen
Ahornstrasse 55, 52074 Aachen, Germany
[3] Department of Informatics, Institute of Mathematics, Slovak Academy of Sciences,
Dúbravská 9, 842 35 Bratislava, Slovak Republic

Abstract. The one sided crossing minimization problem consists of placing the vertices of one part of a bipartite graph on prescribed positions on a straight line and finding the positions of the vertices of the second part on a parallel line and drawing the edges as straight lines such that the number of pairwise edge crossings is minimized. This problem represents the basic building block used for drawing hierarchical graphs aesthetically or producing row-based VLSI layouts. Eades and Wormald [3] showed that the problem is NP-hard for dense graphs. Typical graphs of practical interest are usually very sparse. We prove that the problem remains NP-hard even for forests of 4-stars.

1 Introduction

The one sided crossing minimization problem consists of placing of the vertices of one part of a bipartite graph on prescribed positions on a straight line and finding the positions of the vertices of the second part on a parallel line and drawing the edges as straight lines such that the number of pairwise edge crossings is minimized. This task represents the basic building block used for drawing hierarchical graphs [2,9] or producing row-based VLSI layouts [11,13]. Hierarchical graphs are abstractions of the various information schemes, flowcharts, PERT diagrams and relationship-structures from economic and social science. It is known that the aesthetics and readability of such diagrams heavily depends on the number of line crossings. The main aim is to produce such drawings automatically which led to the design of tens of various heuristics and minimization procedures. We only point on a very exhaustive recent comparative survey of the best heuristics [1].

* This work has been partially done with the aid of GRAAL network (Graphs and Algorithms)

** Supported by the Comisión Interministerial de Ciencia y Tecnología (CICYT) under project TIC 2000 1017.

*** This research was partially supported by the DFG grant No. Hr14/5-1 and the VEGA grant No. 02/7007/20.

P. Mutzel, M. Jünger, and S. Leipert (Eds.): GD 2001, LNCS 2265, pp. 115–123, 2002.
© Springer-Verlag Berlin Heidelberg 2002

From theoretical point of view the problem is NP-hard [3]. However the graphs produced by the NP-hardness proof are very dense. More precisely, the graphs contain $n_1n_2/3$ edges where n_1 and n_2 are the numbers of vertices on the first and the second line, respectively. The median heuristic of Eades and Wormald [3] approximates the crossing number in polynomial time within a factor 3. If the maximum degree of the vertices on the free side is 2,3 or 4, then there exists a 2-approximation algorithms [16]. DiBattista et al. [2] proved that any algorithm gives a good approximation for dense graphs but the problem remains the class of sparse graphs. Note that the typical instances of the problem in practice are very sparse graphs [1]. The companion problem - the two sided crossing minimization, when there are no prescribed positions of vertices on either sides is usually solved by iterating the one sided crossing minimization problem. The problem is NP-hard too [5], approximable for regular graphs by a polylogarithmic factor and solvable in polynomial time for trees [12]. This give rise to a natural question, what is the complexity of the one sided crossing minimization problem for sparse graphs or trees. In this paper we prove that the problem remains NP-hard even for forests of 4-stars. The result is in some sense expectable as the related problem, the one sided maximal planar subgraph problem, was proved to be NP–hard for forests of 2-stars [4].

The paper is organized as follows. In Section 2 we give basic notions. In Section 3 we solve the problem for graphs having degrees at most 2 on the free side. Our main result is presented in Section 4 and Section 5 contains some final remarks.

2 Notations

Given a digraph $G = (V, A)$, and given two its spanning subgraphs $H_1 = (V, A_1)$, $H_2 = (V, A_2$, the direct sum $H_1 \cup H_2$ is defined to be the digraph $(V, A_1 \cup A_2)$. Given a digraph G the (multi)digraph nG is obtained from G by replacing each arc by n parallel arcs.

Given a directed graph G, the Feedback Arc Set problem (FAS) consists of finding the minimum number of arcs of A whose removal makes G acyclic. The decision version of the FAS problem is NP–complete [6]. Let $fas(G)$ denote the size of the minimum fedback arc set for G.

A *k–star* S_k is a tree with $(k+1)$ vertices and k leaves. The *center* of the star is the only vertex of S_k which is not a leaf.

A *directed star* is a digraph whose underlying graph is a star and in which all arcs are pointing towards the center or all arcs are pointing away from the center. We will call them *centripetal stars* and *centrifugal stars* respectively.

We will refer to the One Sided Crossing Minimization problem for graphs of maximal degree k on the free side as the OSCM-k problem.

We will focus on bipartite graphs being forests of stars S_k in which leaves are on prescribed positions on the bottom line and centers must be placed on the top line in such a way that the number of crossings is minimum.

The vertices on the bottom (resp. top) line will be denoted by $v_1, v_2, ..., v_m$ (resp. $w_1, w_2, ..., w_n$). The order of $v_1, v_2, ..., v_m$ will be defined by the indexes.

For any two top vertices w_i, w_j, $cr(i, j)$ stands for the number of crossings produced by edges incident to w_i and w_j when w_i is placed to the right of w_j. In the example of Figure 1 we get $cr(2, 3) = 6$.

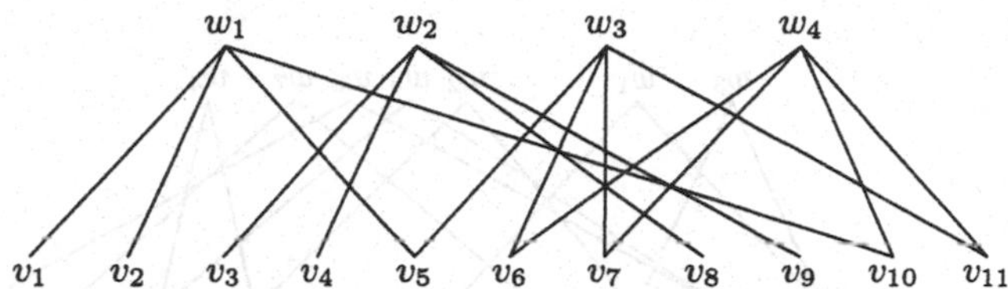

Fig. 1. Example of an input for the OSCM-4

3 Complexity of OSCM-2

In this section we prove that a barycenter algorithm solves the OSCM-2 problem in linear time. Besides its independent interest, the result will be used in proving our main result.

First we give a lower bound on the number of crossings for the OSCM-2 problem. Let G be the input graph. Let w_i (resp. w_j) top vertices with neigbours v_a, v_b with $a < b$ (resp. v_c, v_d with $c < d$). Without loss of generality we assume $a \leqslant c$. The following table give the lower bound on the number of crossings $cr(i, j)$ and $cr(j, i)$.

Situation	$min(cr(i,j), cr(j,i))$	Example in Figure2
$b \leqslant c$	0	$i = 6, j = 8$
$c < b \leqslant d$	1	$i = 1, j = 2$
$d < b \wedge a = c$	1	$i = 1, j = 3$
$d < b \wedge a < c$	2	$i = 6, j = 2$

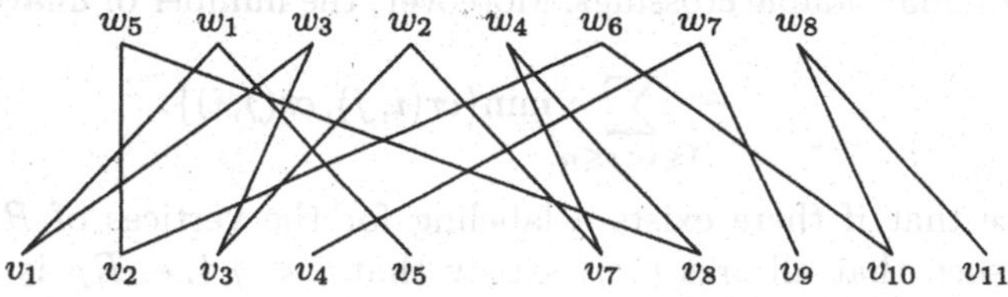

Fig. 2. Example of an input for the OSCM-2

These lower bounds are achieved by placing a node w_i with neighbours v_a and v_b on a unique place on the top line in the position $(a+b)/2 + i/n$ (see for example Figure 3). From this discussion the following corollary is an easy exercise.

Corollary 1. *The OSCM-2 problem can be solved in linear time.*

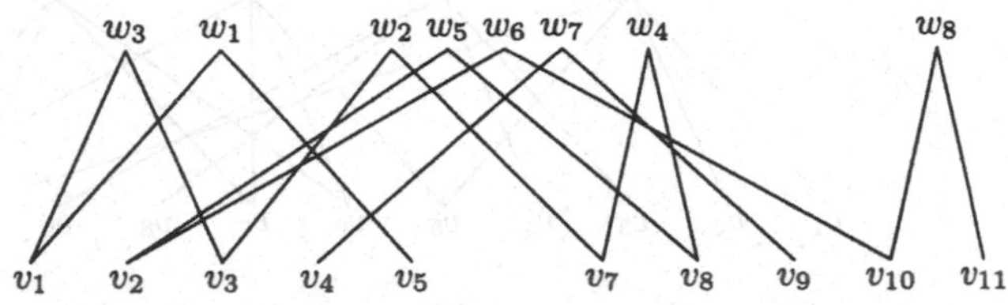

Fig. 3. An optimal solution for the above example

4 Complexity of OSCM-4

In this section we prove that the decision version of the OSCM-4 problem is NP-complete even for forests of 4-stars by a reduction from the decision version of the FAS problem, which immediately implies the NP-hardness of the OSCM-4 problem.

4.1 Restrictions Graph

Definition 1. *Given an instance for the OSCM-k problem with n k-stars labeled* $1, \ldots, n$, *the* restrictions graph *is defined as the (multi)digraph* $R_P = (V, A)$ *with vertex set* $V = \{1, \ldots, n\}$ *corresponding to the n stars of the instance and in which there are* $cr(i,j) - cr(j,i)$ *arcs from i to j,* $i, j \in V$ *iff* $cr(i,j) > cr(j,i)$.

Proposition 1. *Solving the OSCM-k problem for an instance P is equivalent to solving the FAS problem for its restrictions graph* R_P.

Proof. In any ordering of the centers of the stars there will always be a certain number of unavoidable crossings. Moreover, the number of unavoidable crossings will be

$$\sum_{1 \leqslant i < j \leqslant n} \min\{cr(i,j), cr(j,i)\}.$$

Notice that if there exists a labeling for the vertices of R_P with numbers $1, \ldots, n$ such that all arcs (i,j) satisfy that $i < j$ (i.e. R_P is a partial order), then the placing of the centers of the stars in the order given by such a labeling will make the number of crossings to be minimum.

If such a labeling is not possible, then minimizing the number of crossings is equivalent to find a labeling of the vertices of R_P with a minimal number of arcs (i, j) with $i > j$. This is also the same as removing the minimum number of arcs of R_P such that R_P becomes a acyclic.

Since a digraph is compatible with a total order if and only if it is acyclic [14], removing the minimum number of arcs such that R_P becomes acyclic is equivalent to minimizing the number of crossings. Finally observe that if G_P is the graph corresponding to the instance P

$$cr(G_P) = \sum_{1 \leqslant i < j \leqslant n} \min\{cr(i,j), cr(j,i)\} + fas(R_P).$$

□

Let us recall that proving the complexity for OSCM-2 could have be done easily by proving that the restrictions graph is always a partial order (i.e. is always acyclic). We prefered to show a constructive method to make further sections clearer.

Remark 1. Note that the connection between the FAS and the one sided crossing minimization was first observed by Sugiyama et al. [15] by their penalty minimization heuristic. It was also essentially used by Eades and Wormald [3] in their NP-hardness proof for dense graphs. Demetrescu and Finocchi [1] utilized the connection in designing a practical heuristic for the problem.

In the rest of the paper we will show that given any simple digraph G there exists a polynomial time algorithm for finding an instance of the OSCM-4 problem with restrictions graph R_P such that FAS for R_P is equivalent to FAS for G.

4.2 Descent Digraphs

Definition 2. *Given a permutation π on the set $1, 2, \ldots, n$, the* descent digraph *$D_\pi(n)$ is defined as the directed graph with vertices being integers from 1 to n, and in which there is an arc from vertex i to vertex j whenever $i > j$ and $\pi(i) < \pi(j)$.*

The name of descent digraphs comes from the concepts of descents in the context of permutations (see [14] for definition and related results on descents).

Recall the following Lemma:

Lemma 1. *Any forest of directed stars is a descent digraph.*

Proof. It is easy to see that the digraph $D_\pi(n)$ with π being the cyclic permutation $(12\ldots n)$ is a centrifugal star with center n and that $D_\rho(n)$ with ρ being the cyclic permutation $(n(n-1)\ldots 21)$ is a centripetal star with center 1.

Any forest of directed stars can be represented by a product of disjoint cycles of consecutive numbers, and hence it is a descent digraph. □

Notice that given a forest of stars the permutation for the descent digraph is not unique.

4.3 k–Recognizability

Definition 3. *A digraph G is said to be k-recognizable if there is an instance for the OSCM-k problem with restrictions graph being isomorphic to G that can be found in polynomial time.*

Proposition 2. *Let D_π be a descent digraph. Then $2D_\pi$ is 2-recognizable.*

Proof. First of all let us recall that constructing an instance for the OSCM-2 problem with n stars S_2 consists of determining the placement for the $2n$ leaves on a line.

Let us label the two leaves of each S_2 with labels 1, 2 and let us name stars with numbers $1, \dots, n$. That is, each of the $2n$ leaves will be denoted by a pair (i, j), $i \in \{1, 2\}$, $j \in \{1, \dots, n\}$.

If π is a permutation on n symbols, then placing the leaves in the order

$$v_{1,\pi^{-1}(n)} \quad v_{1,\pi^{-1}(n-1)} \quad \cdots \quad v_{1,\pi^{-1}(1)} \quad v_{2,1} \quad v_{2,2} \quad \cdots \quad v_{2,n}$$

produces an instance for the OSCM-2 problem with restrictions graph being $2D_\pi(n)$. In fact, suppose that there is an arc from the vertex i to the vertex j in the descent digraph D_π. It means that $i > j$ and $\pi(i) < \pi(j)$. The inequality $\pi^{-1}(\pi(i)) > \pi^{-1}(\pi(j))$ implies that the left (right) leaf of the star j is placed left from the left (right) leaf of the star i. It means that $c(i, j) = 3$ and $c(j, i) = 1$ and there are 2 arcs from i to j. (See Figure 4.) □

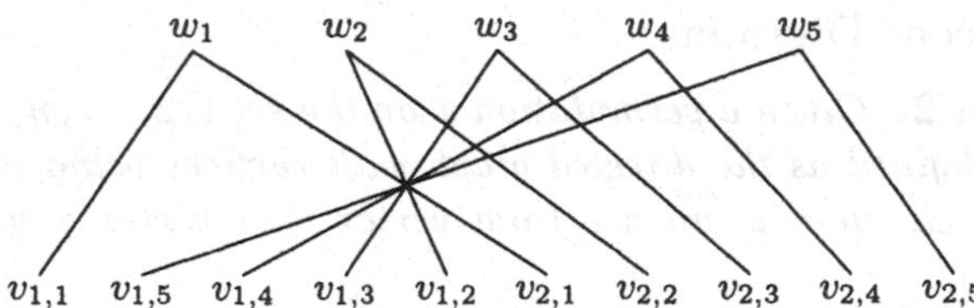

Fig. 4. An instance for the OSCM–2 with the restrictions graph being a centripetal star with 2 parallel arcs isomorphic to $2D_\pi$ with $\pi = (54321)$.

The following lemma is evident

Lemma 2. *Let $\overline{G}$ be a digraph obtained from a digraph G by reversing all arc orientations in G. Then G is k-recognizable iff $\overline{G}$ is k-recognizable.*

Proposition 3. *Let H_0 and H_1 be spanning subgraphs of a digraph G. If H_0 and H_1 are 2–recognizable, then $H_0 \cup H_1$ is 4–recognizable.*

Proof. Place the leaves of the stars corresponding to H_0 and append at the rightmost part the leaves of the stars corresponding to H_1. Identify centers of stars corresponding to the same vertices in H_0 and H_1. The reader can trivially notice that the result is an instance for the OSCM–4 with restrictions graph being $\overline{(H_0 \cup H_1)}$, which by Lemma 2 implies that $H_0 \cup H_1$ is 4–recognizable. Notice that the instance can be generated in linear time. □

4.4 Transformation from FAS

Definition 4. *Given a digraph $G = (V, A)$, the stretch of G, $G' = (V', A')$, is defined as the digraph with vertex set $V' = V \cup A$ and arc set $A' = \{(x, (x, y)) \mid (x, y) \in A\} \cup \{((x, y), y) \mid (x, y) \in A\}$.*

In other words, the stretch of G is defined as the digraph obtained from G by replacing arcs in G by dipaths of length 2.

Proposition 4. *FAS for G is equivalent to FAS for its stretch G'.*

Proof. Let $G = (V, A)$ be a graph and $G' = (V', A')$ be it's stretch.

Assume that $G = (V, F)$ with $F \subset A$ is acyclic, i.e G becomes acyclic by deleting $|A \setminus F|$ arcs. Then $H' = (V', A' \setminus \{((x, y), (x, y) \mid (x, y) \in A \setminus F\})$ is acyclic.

Assume now that $H'' = (V', F')$ with $F' \subset A'$ is acyclic. Then $G' = (V, A \setminus \{(x, y) \mid (x, (x, y)) \in F' \subset A' \vee ((x, y), y) \in F' \subset A'\})$ is acyclic.

In both cases the number of deleted arc stays the same, proving that FAS for G is equivalent to FAS for its stretch G'. □

Proposition 5. *Given any digraph G its stretch G' is the direct sum of two descent digraphs and both digraphs can be identified in polynomial time.*

Proof. Let us consider each vertex $v \in V$) and assign label 1 to the arcs pointing away from it and label 0 to the arcs pointing towards to it. Clearly after completing the labeling for all vertices in V all arcs in A' will have a label (and only one). The subgraph induced by arcs of label 1 is a spanning subgraph of G' consisting of a forest of directed stars and so is the subgraph induced by arcs with label 0. Since forests of stars are descent digraphs, the result is proved. Notice that the digraphs can be identified in linear time. □

Finally consider the following trivial result:

Lemma 3. *FAS for a digraph G is equivalent to FAS for nG.*

Putting together all these results, we have

Theorem 1. *The decision version of the OSCM-4 problem is NP–complete.*

Proof. Given a generic digraph G, construct the stretch G' of G, and descent digraphs H_0, H_1 such that G'= $H_0 \cup H_1$, (Lemma 1 and Proposition 5). By Proposition 2, $2H_0$ and $2H_1$ are both 2-recognizable, and by Proposition 3, $2G'$ is 4-recognizable. Determine an instance I of the OSCM-4 problem with restriction graph isomorphic to $2G'$. By Proposition 1 and Lemma 3, solving I is equivalent to FAS for G' and for G. Finally, let $G_I = (V_0, V_1, E), V_1 = \{1, 2, 3, ..., n\}$ be the graph corresponding to the instance I then

$$cr(G_I) = \sum_{1 \leqslant i < j \leqslant n} \min\{cr(i,j), cr(j,i)\} + 2fas(G).$$

□

5 Conclusions and Further Work

We proved that the one sided crossing minimization problem is NP-hard even for forests of stars of degree 4. So far it was known that the problem is NP-hard for dense graphs. Our result motivates for looking for good approximation algorithms for sparse graphs, which are typical practical instances. As we shown in Section 3, the OSCM-2 problem can be solved in linear time using the barycenter algorithm. It remains to resolve the problem OSCM-3. We conjecture that it is NP-hard.

References

1. Demetrescu, C., Finocchi, I.: Removing Cycles for Minimizing Crossings. J. Experimental Algorithmics. To appear
2. Di Battista, G., Eades, P., Tamassia, R., Tollis, I.G.: Graph Drawing: Algorithms for Visualization of Graphs. Prentice Hall (1999)
3. Eades, P. Wormald, N.C.: Edge Crossings in Drawing Bipartite Graphs. Algorithmica **11** (1994) 379-403
4. Eades, P., Whitesides, S.: Drawing Graphs in Two Layers. Theoretical Computer Science **131** (1994) 361-374
5. Garey, M.R., and Johnson, D.S.: Crossing Number is NP-Complete. SIAM J. Algebraic and Discrete Methods **4** (1983) 312–316
6. Gavril, F.: Some NP-Complete Problems on Graphs. In: 11th Conf. on Information Sciences and Systems. John Hopkins University Press, Baltimore (1977) 91-95
7. Jünger, M., Mutzel, P.: 2-Layer Straightline Crossing Minimization: Performance of Exact and Heuristic Algorithms. J. Graph Algorithms and Algorithms **1** (1999) 11-25
8. Matuszewski, C., Schönfeld, R., Molitor, P.: Using Sifting for *k*-Layer Crossing Minimization. In: 7th Intl. Symp. on Graph Drawing. Lecture Notes in Computer Science, Vol. 1731. Springer-Verlag, Berlin (1999) 217-224
9. Mutzel, P.: Optimization in Leveled Graphs. In: Pardalos, M., Floudas, C.A. (eds.): Encyclopedia of Optimization. Kluwer, Dordrecht (2001)
10. Purchase, H.: Which Aesthetic Has the Greatest Effect on Human Understanding? In: 5th Intl. Symposium on Graph Drawing. Lecture Notes in Computer Science, Vol. 1353. Springer-Verlag, Berlin (1997) 248-261

11. Sechen, C.: VLSI Placement and Global Routing Using Simulated Annealing, Kluwer, Dordrecht (1988)
12. Shahrokhi, F., Sýkora, O., Székely, L.A., Vrt'o, I.: On Bipartite Drawings and the Linear Arrangement Problem. SIAM J. Computing **30** (2000) 1773-1789
13. Stallmann, M., Brglez, F., Ghosh, D.: Heuristics, Experimental Subjects, and Treatment Evaluation in Bigraph Crossing Minimization. J. Experimental Algorithmics. To appear
14. Stanley, R. H.: Enumerative Combinatorics. Wadsworth & Brooks, Monterey (1986)
15. Sugiyama, K., Tagawa, S., Toda, M.: Methods for Visual Understanding of Hierarchical System Structures. IEEE Transactions on Systems, Man, and Cybernetics **SMC–11** (1981) 109-125
16. Yamaguchi, A., Sugimoto, A.: An Approximation Algorithm for the Two-Layered Graph Drawing Problem. In: 5th Annual Intl. Conf. on Computing and Combinatorics. Lecture Notes in Computer Science, Vol. 1627. Springer-Verlag, Berlin (1999) 81-91

Fast Compaction for Orthogonal Drawings with Vertices of Prescribed Size*

Markus Eiglsperger and Michael Kaufmann

Universität Tübingen, Wilhelm-Schickard-Institut für Informatik
{eiglsper/mk}@informatik.uni-tuebingen.de

Abstract. In this paper, we present a new compaction algorithm which computes orthogonal drawings where the size of the vertices is given as input. This is a critical constraint for many practical applications like UML. The algorithm provides a drastic improvement on previous approaches. It has linear worst case running time and experiments show that it performs very well in practice.

1 Introduction

Orthogonal drawings of graphs are extensively used in many application areas. Examples are UML class-diagrams in software engineering or ER-diagrams in database management. One critical property in many of these diagrams is that the vertices of the graph have prescribed size. Take again UML class-diagrams, where the size of the vertices is determined by the contained text.

Traditional algorithms for orthogonal layout do not take prescribed vertex sizes into account. Recently, algorithms based on the topology-shape-metrics approach have been proposed for this problem [1], [11], and [5]. The topology-shape-metrics approach introduced in [3] is a very popular algorithm framework and consists of the three steps planarization, layout and compaction. The lengths in the drawing are assigned in the compaction step. Therefore the prescribed size constraints for vertices affect mainly this step. However, the input of the compaction step, also called the shape of the drawing, must ensure that the prescribed size constraints can be fulfilled. Algorithms based on the Kandinsky framework [7] guarantee this.

The algorithm of Di Battista et. al. [1] creates in a first step a drawing of the graph where the vertices are represented as points and edges overlap if they are adjacent to the same vertex at the same side. Then each vertical and each horizontal grid-line is expanded individually, i.e., each grid-line is replaced by a set of grid-lines such that the vertices can be assigned their prescribed sizes and the edges can be routed without overlap. This expansion is done by solving a Min-Cost-Flow problem for each horizontal and each vertical grid-line. The authors do not give any bounds on the time complexity of their algorithm. They experienced up to 50 seconds computation time on graphs in the Rome graphs test suite [4]. Klau et al. propose in [11] that their compaction approach

* Partially supported by DFG-Grant Ka812/8-1

P. Mutzel, M. Jünger, and S. Leipert (Eds.): GD 2001, LNCS 2265, pp. 124–138, 2002.
© Springer-Verlag Berlin Heidelberg 2002

originating from graph labeling can also be used to solve the compaction problem for drawings with prescribed vertex size. The approach relies on branch-and-cut and has, therefore, exponential worst case running time. The authors give neither a detailed description of how the algorithm can be used in the prescribed vertex size setting nor experimental results.

We will present in this work a new compaction algorithm which constructs orthogonal drawings where vertices have prescribed size. The algorithm requires linear running time which improves significantly on the existing algorithms. The design of the algorithm allows us to also improve the quality of the result in the expense of running time. We also provide the results of empirical tests which demonstrate the effectiveness of the algorithm in practice. Since now, experiments for compaction algorithms have only been performed for 4-graphs [9].

This paper is organized as follows: In section 2, we introduce basic concepts of the topology-shape-metrics approach and the Kandinsky framework. In section 3, we present a new linear time compaction algorithm for 4-graphs. In section 4 we extend this algorithm for the Kandinsky model. In section 5, we show how prescribed vertex sizes can be integrated in this algorithm. Section 6 deals with extensions, implementation issues and concludes with the results of extensive empirical tests.

2 Preliminaries

We assume familiarity with the concept of planarity of graphs. An *embedded planar graph* is a planar graph with a specific circular order of edges around vertices and a specific external face, admitting a planar drawing that respects the given embedding. Unless otherwise specified, the planar graphs we consider are always embedded.

A *planar orthogonal drawing* of a planar graph is a planar drawing that maps each vertex to a point and each edge to a sequence of horizontal and vertical segments. In a *planar orthogonal grid drawing* the vertices and the bends along the edges have integer coordinates. Note that a planar graph admits a planar orthogonal grid drawing if and only if it is a 4-graph, i.e., the vertices of the graph have degree less than or equal to 4. An *orthogonal representation* H is a mapping from the set of faces F of a 4-graph G to clockwise ordered lists of tuples (e_r, a_r, b_r) where e_r is an edge, a_r is the angle formed with the following edge inside the appropriate face, stored as multiple of 90°, and b_r is the list of bends of the edge. Note that in a planar orthogonal drawing, $1 \leq a_r \leq 4$ holds. If there are no bends in H, we call H *simple*.

A *planar orthogonal box drawing* of a planar graph is a planar drawing that maps each vertex to a box and each edge to a sequence of horizontal and vertical segments. In the corresponding grid drawing the center of the boxes and the bends along the edges have integer coordinates. A *quasi-orthogonal representation* Q is defined analogously to orthogonal representation with the difference that 0° angles are allowed. A 0° angle denotes that the following edge is adjacent to the same side of the vertex as the preceding edge. Note that quasi-orthogonal representations are not related to *quasi-orthogonal drawings* as described in [10].

Different drawing conventions have been proposed for planar orthogonal box drawings. We will concentrate on the so-called Kandinsky-models. All Kandinsky-models impose the following constraints on the drawing which we call Kandinsky-properties: the *bend-or-end property* and the *non-empty face property*. The bend-or-end property is defined as follows: Let $G=(V,E)$ be an embedded graph and Γ be an planar orthogonal box drawing of G. Let e_1 and e_2 be two edges adjacent to the same side of a vertex v, e_1 following e_2 in the embedding. Let f be the face to which e_1 and e_2 are adjacent. Then either e_1 must have a first bend with a 270° angle in f or e_2 must have a first bend with 270° angle in f. The non-empty face condition forbids some degenerated cases for triangles in the graph. See [7] for a detailed description of the Kandinsky-properties.

There are several variations of Kandinsky-models: In the original version all vertices were represented by squares of equal size, arranged on a coarse vertex grid [7]. In the *big-node model* [8], the size of the vertices is determined by the number of edges attached to the different sides of the vertex. We also consider the *point model*, in which the vertices are represented by points. As pointed out above, these drawings might not be valid orthogonal drawings since edge overlap may occur. In [1] the *podavsnef*-model is introduced in which vertices have prescribed size. We refer to this model as the *prescribed-size* Kandinsky-model in this work.

We assume that the (quasi-)orthogonal representations, which stem from the second phase of the topology-shape-metrics algorithm have only a linear number of bends. The algorithm works also for cases where the representations have more bends, but the time bounds are no longer valid. The above assumption is justified, since to our knowledge, all algorithms which create (quasi-)orthogonal representations follow this assumption.

3 Compaction of 4-Graphs

In this section, we treat the *compaction problem for orthogonal drawings* and propose a linear time algorithm for it. Our algorithm is a combination of the exact compaction algorithm [12] and the rectangular decomposition technique [15]. Rectangular decomposition already leads directly to a linear time algorithm, but the insight that we gain in combining the two techniques is the base for the following sections. The problem is stated as follows:

Problem 1: Given an embedded 4-graph G with orthogonal representation H, find a drawing Γ of G with orthogonal representation H.

Note that every orthogonal representation can be reduced to a simple orthogonal representation by replacing the bends in H by dummy vertices. We therefore assume for the rest of the section that H is simple. We first review the exact compaction algorithm and then derive our algorithm from it.

3.1 The Approach of Klau/Mutzel

Given an embedded graph G with orthogonal representation H. We calculate from G and H a mapping $dir : E \to \{r, u\}$ and an orientation of G such that there exists an orthogonal drawing Γ of G with orthogonal representation H where the directed edge e points upward if $dir(e) = u$ and points to the right if $dir(e) = r$. We can calculate this in linear time. Note that this assignment is not unique, but the set of solutions are rotations of each other. Note also that for each edge, there are two entries in H. In one of these entries, the edge orientation is the same as the traversal direction of the face and in one of these entries the traversal direction is the opposite of the edge orientation. From this fact we can derive in which absolute direction $d \in \{up, down, left, right\}$ an edge is traversed in a face-traversal and append this information directly to the tuple in H. We assume for the rest of the section that G is directed according to the orientation stated above.

We denote with $G_r = (V, E_r)$ the subgraph of G which contains only horizontal edges, i.e., $E_r = \{e \in E : dir(e) = r\}$ and with $G_u = (V, E_u)$ the subgraph of G which contains only vertical edges, i. e., $E_u = \{e \in E : dir(e) = u\}$.

The connected components of G_r, resp. G_u, are directed paths and form a line in a drawing of G with orthogonal representation H. We denote with S_r, resp. S_u, the set of connected components of G_r, resp. G_u, and call the elements of it *horizontal segments*, resp. *vertical segments*. We say that two segments are *adjacent* if they share a point. We denote with $\alpha(s)$ the start node of the path which forms a segment s and with $\omega(s)$, the endpoint of this path. Every edge e is contained in exactly one segment $seg(e)$ and each node v is contained in exactly one vertical segment $vert(v)$ and one horizontal segment $hor(v)$.

The concept of *constraint graph* describes the ordering of the segments of one type. The edge sets of the constraint graphs $D_u = \{S_r, A_u\}$, resp. $D_r = \{S_u, A_r\}$, are defined as:

$$A_u = \{(hor(v), hor(w)) : (v, w) \in E_u\}, \text{ and}$$
$$A_r = \{(vert(v), vert(w)) : (v, w) \in E_r\}$$

The *shape description* $\mathcal{S} = (D_r, D_u)$ combines two constraint graphs.

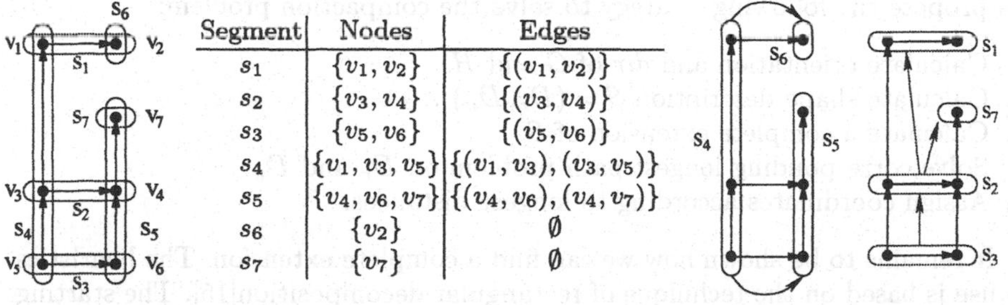

Segment	Nodes	Edges
s_1	$\{v_1, v_2\}$	$\{(v_1, v_2)\}$
s_2	$\{v_3, v_4\}$	$\{(v_3, v_4)\}$
s_3	$\{v_5, v_6\}$	$\{(v_5, v_6)\}$
s_4	$\{v_1, v_3, v_5\}$	$\{(v_1, v_3), (v_3, v_5)\}$
s_5	$\{v_4, v_6, v_7\}$	$\{(v_4, v_6), (v_4, v_7)\}$
s_6	$\{v_2\}$	$\emptyset$
s_7	$\{v_7\}$	$\emptyset$

Fig. 1. Examples of segments and the corresponding constraint graph.

Definition 1. *Let s_r be a vertical segment and s_u be a horizontal segment which are not adjacent. We call s_r and s_u to be* separated *if one of the following conditions hold:*

$$\textit{1. } s_u \xrightarrow{*}_{D_r} ver(\alpha(s_r)) \qquad \textit{2. } ver(\omega(s_r)) \xrightarrow{*}_{D_r} s_u$$
$$\textit{3. } s_r \xrightarrow{*}_{D_u} hor(\alpha(s_u)) \qquad \textit{4. } hor(\omega(s_u)) \xrightarrow{*}_{D_u} s_r$$

A shape description is called *complete* if every pair of segments with opposite direction is separated.

We define the linear program (LP) as:

$$x_b - x_a \geq 1 \ \forall (a,b) \in A_r$$
$$y_b - y_a \geq 1 \ \forall (a,b) \in A_u$$
$$x_s \geq 0 \ \forall s \in S_u$$
$$y_s \geq 0 \ \forall s \in S_r$$

Lemma 1. *[12] If S is complete a feasible solution (x, y) of the linear program (LP) induces an orthogonal drawing $\Gamma : V \to \mathbb{N} \times \mathbb{N}$ with $\Gamma(v) = (x_{vert(v)}, y_{hor(v)})$ of G with orthogonal representation H.*

We can find a feasible solution of (LP) in linear time by solving a longest path problem [13]. When the shape description is not complete, there may be feasible solutions of the (LP) which induce non-valid orthogonal drawings. In [12], it is shown that there always exists a superset of S which is a complete shape description. We call this superset S_{ext} a *complete extension* of S. Klau/Mutzel [12] propose a branch and cut algorithm which finds the complete extension such that the resulting drawing has minimal edge length. This approach has the disadvantage that it has exponential worst-case running time. We cannot hope to do better when we try to optimize the edge length, since this problem is NP-hard[14]. If we drop the goal to achieve optimality, we can get a much faster algorithm by searching a complete shape extension by heuristics.

3.2 A Fast Heuristic

We propose the following strategy to solve the compaction problem:

1. Calculate orientation and *dir* of G and H.
2. Calculate shape description $S = (D_r, D_u)$.
3. Calculate a complete extension of S.
4. Solve corresponding longest path problem on D_r and D_u.
5. Assign coordinates according to longest distances.

It remains to be shown how we can find a complete extension. The heuristics we use is based on the technique of rectangular decomposition[15]. The starting point for the rectangular decomposition strategy is the observation that if all faces of the graphs are rectangles, we can easily solve the compaction problem by applying longest path or network flow algorithms to it. The idea is to subdivide those faces which are not rectangular into rectangles and then solve the problem on this subdivision. This induces a valid embedding on the original graph. What remains is to perform this subdivision efficiently, which can be done by searching certain patterns of angles on the face. We denote 90° angles on a face with a

'0' and 270° angles with a '1'. Every time we find the pattern 100, we cut a rectangle from the face and continue to search the pattern on the remaining face. See Fig. 2 for an illustration. We terminate if there are no more patterns in any face. Using a list, rectangle decomposition can be done in linear time.

Function `init-list`(*Face f*)

```
List l ← ε;
for each (e = (v, w), a, ε, d) in f do
    // Let d' be the direction obtained by rotating d by 90°.;
    if a = 1 then append (0, seg(e), d) to l;
    if a = 3 then append (1, seg(e), d) to l;
    if a = 4 and e ∈ E_u then append (1, seg(e), d), (1, hor(w), d') to l;
    if a = 4 and e ∈ E_r then append (1, seg(e), d), (1, ver(w), d') to l;
end
return l
```

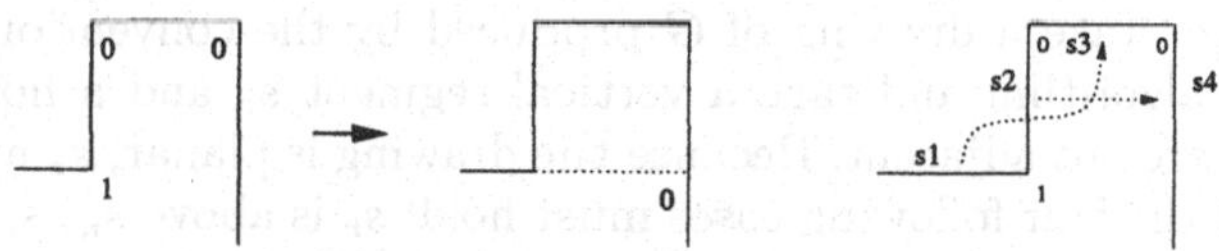

Fig. 2. Decomposition of a face into a rectangle and a remaining face.

From this algorithm we can directly derive a completion heuristic. Assume that we are in the situation illustrated in Fig. 2. Instead of introducing a dummy node and a dummy edge in the graph, we simply add edges to the constraint graph. In the case above, we insert the edge (s_1, s_3) in D_u and the edge (s_2, s_4) in D_r. We handle the other three cases symmetrically. The function `define-box` describes the four cases:

Function `define-box`(*Shape description S, Direction d, Segments* s_1, s_2, s_3, s_4)

```
Let S = ((S_u, A_r)(S_r, A_u));
If d = up then A_u ← A_u ∪ (s2, s4), A_r ← A_r ∪ (s3, s1);
If d = down then A_u ← A_u ∪ (s4, s2), A_r ← A_r ∪ (s1, s3);
If d = left then A_u ← A_u ∪ (s1, s3), A_r ← A_r ∪ (s2, s4);
If d = right then A_u ← A_u ∪ (s3, s1), A_r ← A_r ∪ (s4, s2);
```

The completion algorithm executes on every face first `init-list` and then `decompose`.

Lemma 2. *With the above algorithm, we can calculate a complete shape extension of size $O(n)$ in linear time.*

```
Function decompose(Shape description , List l)
while size(l) > 4 do
    // denote with t_i = (a_i, s_i, d_i) the i-th tuple in l;
    if (a_1 = 1) and (a_2 = 0) and (a_3 = 0) then
        define-box(S, d_1, s_1, s_2, s_3, s_4);
        replace t_1 with (0, s_1, d_1);
        remove t_2 and t_3 from l;
    else
        move t_1 to the rear of l
    end
end
define-box(S, d_1, s_1, s_2, s_3, s_4);
```

Proof. We first show that the complete shape extension has size $O(n)$. The initial shape description has linear size by Euler's formula. Since the rectangle decomposition introduces $O(n)$ rectangles and we insert two edges into it per rectangle, the complete shape extension has linear size. The linear running time follows immediately from this fact, too. It remains to be shown that the extension is complete. Take a drawing of G produced by the conventional rectangle decomposition algorithm and take a vertical segment s_r and a horizontal segment s_u which are not adjacent. Because the drawing is planar, s_r and s_u do not cross, so one of the four following cases must hold: s_r is above s_u, s_r is below s_u, s_r is left of s_u or s_r is right of s_u. Assume w.l.o.g. that s_u is above s_r and that s_u is not to the left of s_r. The other cases are symmetric. Since s_u is above s_r, $s' = hor(\alpha(s_u))$ is also above s_r. Assume that one of the following cases holds:

1. There is a $s_v \in S_u$ such that the intersection of the projection of s_v and s_u on the y-axis is non-empty and there is a path from $vert(\omega(s_r))$ to s_u in D_r.
2. There is a segment $s_h \in S_r$ such that the intersection of the projection of s_h and s' on the x-axis is non-empty and there is a path from s_r to s_h in D_u.

If the assumption is true, we are done. To see this, assume that the second case holds. Then just take a line parallel to the y-axis with x-coordinate in the intersection of the projections. From the rectangles which are intersected by the parallel line, we can now easily construct a path from s_h to s' in D_u.

We give a constructive proof that either s_v or s_h exists. Start at segment s_r and go to the lowest rectangle to the right of it, if such a rectangle exists. In this case, go from this rectangle to the leftmost rectangle above. Iterate until a segment is found which induces an intersection of the projections. Because we proceed monotonically increasing in x- and y-coordinates, such a segment must exist for monotonicity reasons. The existence of the path follows from how we traverse the rectangles.

Theorem 1. *The above algorithm solves the Problem 1 in linear time.*

4 Compaction in the Kandinsky Point Model

Before we present the compaction algorithm for prescribed vertex sizes in the next section, we take a step in between and consider the compaction problem for the Kandinsky point model, which is presented in this section.

Problem 2: Given an embedded graph G with simple quasi-orthogonal representation Q having the Kandinsky properties. Find a drawing Γ of G with quasi-orthogonal representation Q, in which the vertices of G are represented as points.

We assume as in the previous section that G is oriented, marked with direction tags and that bends are replaced by dummy vertices.

In a quasi-orthogonal representation there may be more than one edge adjacent to a side of a vertex. As a consequence a segment does not represent a directed path of edges with the same direction tags as in the previous section. See Fig.3 for such a segment.

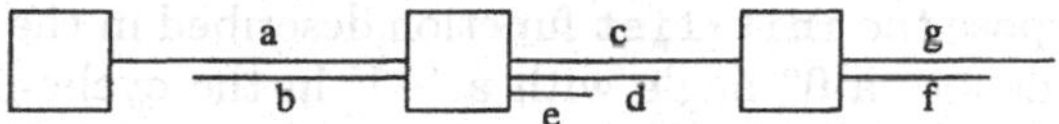

Fig. 3. A segment in a quasi-orthogonal representation, where vertices are denoted as rectangles.

We therefore refine the segment definition and introduce the concept of *sub-segment*. We call two edges (u, v) and (v, w) a *right-join* (*left-join*) if they have the same direction and between them in the cyclic order (reverse cyclic order) there are only edges with different directions. In Fig. 3, edges b and e form a right-join and edges a and c a left-join.

Definition 2. *Let p be a directed path consisting of edges with the same direction tag. The path p is a* sub-segment *if every two consecutive edges in the path are a right-join or every two consecutive edges in the path are a left-join or if p consists of one single edge. The path p is a* maximal sub-segment *if there is no sub-segment containing it.*

Because we will only consider maximal sub-segments, we will omit the word 'maximal' for the rest of this work. Note that the terms sub-segment and maximal-sub-segment have a different meaning as in [12]. The sub-segments of the segment shown in Fig. 3 are (a, c, g),(b, e),(d) and (c, f). An edge is part of at most two sub-segments. As for segments, we distinguish between vertical sub-segments and horizontal sub-segments. A vertex may be adjacent to an arbitrary number of sub-segments. If s is a sub-segment, we denote with $seg(s)$ the segment which contains s. Note that α and ω are no longer well defined on segments, but are well-defined for sub-segments. We define $\mathcal{S} = (D_r, D_u)$ and (LP) analogous to the previous section.

Definition 3. *Let s_r be a vertical sub-segment and s_u be a horizontal sub-segment which are not adjacent. We call s_r and s_u* separated *if one of the following conditions hold:*

1. $seg(s_u) \xrightarrow{*}_{D_r} seg(\alpha(s_r))$
2. $seg(\omega(s_r)) \xrightarrow{*}_{D_r} seg(s_u)$
3. $seg(s_r) \xrightarrow{*}_{D_u} seg(\alpha(s_u))$
4. $seg(\omega(s_u)) \xrightarrow{*}_{D_u} seg(s_r)$

We call a shape description *complete* if every pair of sub-segments with opposite direction is separated unless the overlap of these two segments cannot be avoided because of the representation of vertices as points.

With the same argumentation as in Lemma 1, it follows that if S is complete, a feasible solution (x, y) of (LP) induces an orthogonal drawing $\Gamma : V \rightarrow \mathbb{N} \times \mathbb{N}$ with $\Gamma(v) = (x_{vert(v)}, y_{hor(v)})$ of G with quasi-orthogonal representation Q. As in the previous section a feasible solution of (LP) can be found with a longest path algorithm in linear time. It remains to be shown how we can calculate a complete shape extension for S. We again use rectangle decomposition to perform this, but this time we need two rounds.

Let f be a face in Q. In the first round, we eliminate all 0° angles from f. We modify for this purpose the `init-list` function described in the previous section in two points: we denote a 0° angle with a '-1' in the cyclic-list and we store the sub-segments rather than the segments in the list entries. Then, we perform a modified version of the `decompose` function on this list. The first modification is that we search for patterns described in Fig. 4. The choice of rule (c) or (d) depends on some technical constraints, see [6] for a detailed description. The `define-box` function takes sub-segments instead of segments as arguments. It connects the segments in the constraint graph which correspond to the sub-segments unless it would connect a segment with itself. We terminate when we have traversed the entire cyclic list. Let l' denote the resulting cyclic list. In the

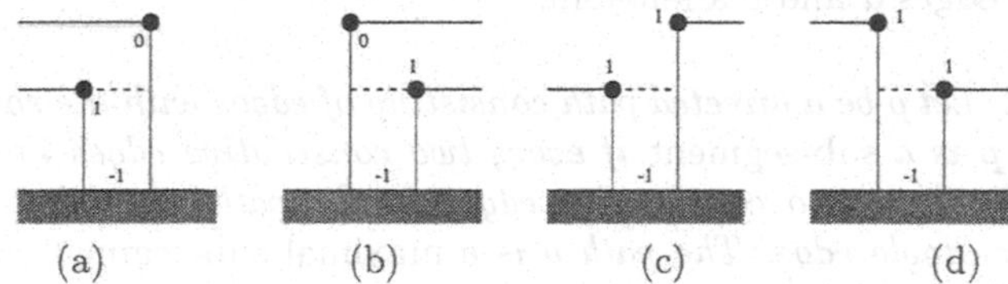

Fig. 4. Rules to eliminate zero degree angles

second round, we simply perform `decompose` on l'.

From the correctness and time bounds proven in the previous section, together with the fact that the traditional version of rectangle decomposition described in [6] is correct, the following theorem follows:

Theorem 2. *The above algorithm solves Problem 2 in linear time.*

5 Compaction in the Prescribed-Size Kandinsky Model

In this section we provide a compaction algorithm for the prescribed-size Kandinsky-model. We denote with *width*(v) the prescribed width of a vertex in G and with *height*(v) the prescribed height. We assume that the number of edges adjacent to the top/bottom side of a vertex v is less than $width(v)+1$ and the number of edges adjacent to the left/right side is less than $height(v)+1$.

Problem 3: Given an embedded graph G with quasi-orthogonal representation Q having the Kandinsky properties. Find a drawing Γ of G with orthogonal representation Q, in which each vertex v of G is represented by a box of the size $(width(v), height(v))$.

We assume as in the previous section that G is oriented, marked with direction tags and that bends are replaced by dummy vertices.

The algorithm proceeds as follows: It first replaces each non-dummy vertices by a rectangular face. The result of this transformation is a 4-graph with a simple orthogonal representation. It creates for this orthogonal representation a complete shape description which is compatible with the vertex size constraints. Finally, it calculates the drawing from the complete extension. We now give a detailed description of the algorithm.

5.1 Simplification

Let $B \subset V$ denote the dummy vertices representing bends. The elements in B have zero size. In a first step, the vertices with non-zero size are replaced by rectangular faces. Let v be such a vertex. For each edge adjacent to v, a new node $p(e, v)$ is created which represents the port of e on v. Also, four corner nodes $nw(v)$, $ne(v)$, $sw(v)$ and $se(v)$ are created. See Fig. 5 for an example. Each node face has four adjacent *node-segments*: the top segment $t(v)$, the bottom segment $b(v)$, the left segment $l(v)$, and the right segment $r(v)$. The result of this

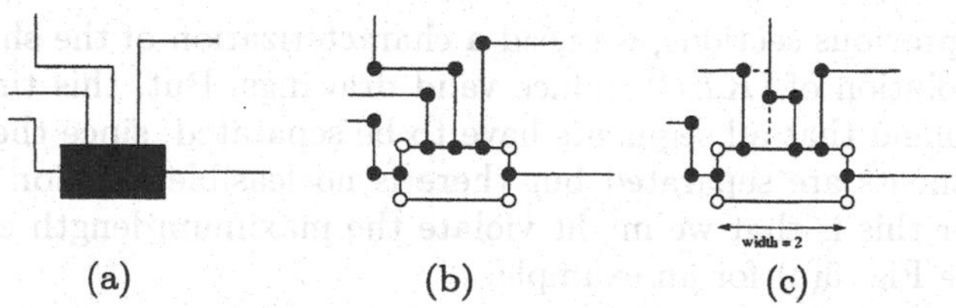

Fig. 5. Transformation of a vertex into a face and an invalid compaction.

simplification step is a 4-graph $G_S = (V_S, E_S)$ with orthogonal representation H_S. The mapping $simple : E \to E_S$ maps every edge in E to the corresponding edge in E_S. G_S has $4|V|+2|E|+|B|$ nodes. Since G and G_S are planar it follows with Eulers formula that the above transformation causes a constant blow up.

5.2 Problem Description

We now try to find a drawing of G_S which induces a valid drawing on G. We call these drawings *valid drawings* of G_S. There are two differences between a valid drawing of G_S and a valid drawing of a 4-graph described in Sec. 3:

1. The edges adjacent to a corner node may have zero length.
2. The segments of the node-faces have prescribed distance.

We extend the algorithm of the previous section to match these requirements. This is done by refining the constraint graph definition of the previous section by adding a length function to the constraint graphs, and introducing auxiliary edges which denote the vertex size. These refinings are similar to the ones in [11]. We define the edge-set of the constraint graph D'_u as $A'_u = A_u \cup N_u^+ \cup N_u^-$, with

$$N_u^+ = \{(b(v), t(v)) : v \in V\}, \text{ and } N_u^- = \{(t(v), b(v)) : v \in V\}$$

The constraint graph D'_r is defined analogously, and $S' = (D'_r, D'_u)$ is the corresponding shape description. Additionally, we define the length function length : $A'_u \cup A'_r \rightarrow Z$ in the following way:

$$\text{length(e)} = \begin{cases} 0 & \text{if } e \in A_u, \text{e is induced by an edge adj. to a corner} \\ \text{height}(v) & \text{if } e \in N_u^+ \\ -\text{height}(v) & \text{if } e \in N_u^- \\ 1 & \text{otherwise} \end{cases}$$

The values of length for A'_r are defined analogously. This leads to the Kandinsky linear program (KLP):

$$\begin{aligned} x_b - x_a &\geq \text{length}(e)\ \forall e = (a,b) \in A'_u \\ y_b - y_a &\geq \text{length}(e)\ \forall e = (a,b) \in A'_r \\ x_s &\geq 0\ \forall s \in S_u \\ y_s &\geq 0\ \forall s \in S_u \end{aligned}$$

As in the previous sections, we need a characterization of the shape descriptions whose solution of (KLP) induce valid drawings. But, this time, it is not enough to demand that all segments have to be separated, since there are cases where all segments are separated but there is no feasible solution for (KLP). The reason for this is that we might violate the maximum length condition for node faces, see Fig. 5(c) for an example.

Definition 4. *A shape description S' is* length-complete *if S' is complete and every cycle in the constraint graphs of S' has non-positive length.*

Lemma 3. *Let S' be a complete shape description. (KLP) has a feasible solution if and only if every cycle in the constraint graphs of S' has non-positive length.*

Proof. It is shown in [12] that if a constraint graph in S has a positive length cycle, then (KLP) is not feasible. We can transform (KLP) to a *system of difference constraints* by multiplying each constraint by -1. This transforms every positive length cycle in a negative length cycle and vice versa. It holds now that a system of difference constraints is feasible if and only if it has no negative length cycle, see, i.e., [2] for a proof.

We cannot apply the longest path algorithms of the previous sections to find a feasible solution of (KLP) because there are cycles and negative edges in the constraint graphs. In the remainder of this section, we will present an algorithm which finds a feasible solution of (KLP) in linear time.

A segment $s \in Q$ induces a set of segments $simple(s)$ in H_S. Note that every negative length edge in a constraint graph connects two segments in H_S which belong to the same segment in Q. Our algorithm considers one segment of Q at a time. Let s be a vertical segment in Q. A critical role is played by the edges in s which are not adjacent to bends. We denote these edges by $nb(s)$ and the corresponding edges in H_S by $nb_S(s)$. Because of the bend or end property the edges in $nb(s)$ form a path in s. Let $e = (v, w) \in nb(s)$, $e_S = simple(e)$. We denote with $d(v, e)$ the minimal distance from $left(v)$ to e in D_r. The same holds for w and $d(w, e)$. This value can be calculated in amortized constant time from the shape. Given an x-value $\bar{x}$ for one arbitrary edge $e_S \in nb_S(s)$, where $e = (v, w)$ is the corresponding edge in Q. We define $x(left(v)) = x(e_S) - d(e_S, v)$ and $x(left(w)) = x(e_S) - d(e_S, w)$. Let e'_S be the following edge of e in the path $nb(s)$, and e''_S the preceding edge. We define $x(e'_S) = x(left(v)) + d(e'_S, v)$ and $x(e''_S) = x(left(w)) + d(e'_S, w)$. We proceed recursively until x is determined for all left sides of nodes and all edges in nb_S. Given $e \in nb_S(s)$ and $\bar{x} = 0$ and the rest of the x-values calculated as described above. Then we let $\text{off}(e, s) = \min_{e' \in nb_S(s)}\{x(e')\}$.

The algorithm to calculate x works now as follows: First, we determine the vertical segments of Q and calculate a topological numbering $s_1, \ldots, s_k$ of them. We denote with $U_i \subseteq S'_u, 0 \le i < k$ the set of vertical segments in H_S whose corresponding segment in Q is s_i. We then calculate a topological numbering on the vertical segments of H_S where the U_i appear in an interval. For every i, we do the following: We perform the standard DAG algorithm for longest paths on U_i. Then we search the edge $e \in nb_S(s_i)$ with maximal x-value. We add $\text{off}(e, s_i)$ to this x-value and apply the above algorithm to determine the x-values of all other segments in s_i. This algorithm has linear running time.

Lemma 4. *A feasible solution of (KLP) can be found in time $O(n)$.*

5.3 Rectangular Decomposition

We will again use the rectangular decomposition method to obtain a length-complete shape extension.

Definition 5. *Let $v \in V \setminus B$ and $d \in \{up, down, left, right\}$. Then $seg(v,d)$ is defined the following way:*

$$seg(v,d) = \begin{cases} left(v) & \text{if } d = up \\ right(v) & \text{if } d = down \\ top(v) & \text{if } d = right \\ bottom(v) & \text{if } d = left \end{cases}$$

The equivalent to a sub-segment in Q is a *meta-segment* in H_S.

Definition 6. *Let $s = e_0, e1, \ldots, e_r$ be a sub-segment, with $e_0 = (v_0, v_1)$, $e_2 = (v_1, v_2), \ldots e_r = (v_r, v_{r+1})$ and $d \in \{up, down, left, right\}$. The corresponding meta-segment $meta(s,d)$ is defined as the set:*

$$\{seg(v_0,d), simple(e_0), seg(v_1,d), simple(e_1), \ldots, simple(e_r), seg(v_{r+1},d)\}$$

On every face f we perform rectangle decomposition in the following way: We first execute the first round of the decomposition algorithm of section 4 on f and obtain a list of three-tuples l. Then we insert for each tuple $t = (a,s,d)$ one *meta-segment node* $mn(s)$ in the corresponding constraint graph which represents the boundaries of the meta-segment $meta(s,d)$ at f. If the face is on the right side of $mn(s)$ we create for every segment s' in $meta(s,d)$ an edge $(mn(s), s')$, if the face is on the left side we create edges $(s', mn(s))$. The created edges have zero length. Then we separate adjacent meta-segments by introducing edges of length 1 between one meta-segment node and one segment node. This is illustrated in Fig. 6(b). As a last step we apply function **decompose** on l, where we use a special version of function **define-box** which connects the appropriate meta-segment nodes in the constraint graphs as illustrated in Fig. 6(a).

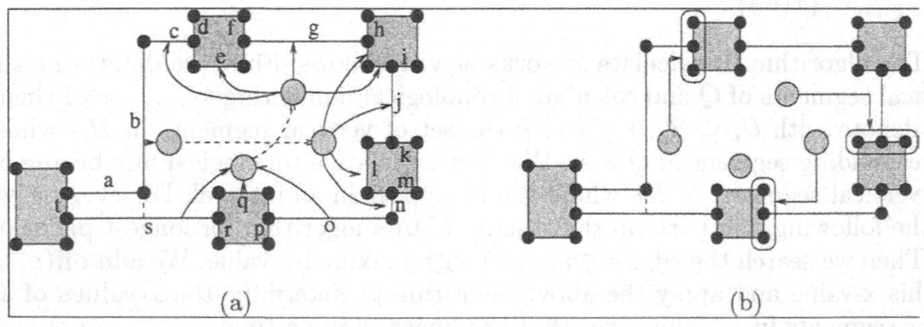

Fig. 6. A face with meta-segments $(a), (b), (c,e,g,i), (h,j,l,m), (o,q,s), (t)$.

Lemma 5. *The algorithm described above makes the shape-description length-complete. The shape description has linear size.*

Proof. (Sketch) We introduce at most $2|E|$ meta-segment nodes. Every segment in H_S is contained in at most two meta-segments and is therefore connected to a meta-segment node at most twice. From this it follows that H_S has linear size. All segments which take part in a decomposition step are separated. With a similar argument as in Lemma 2 then follows that all non-adjacent segments in Q are separated.

Theorem 3. *The above algorithm solves Problem 3 in linear time.*

6 Practical Issues and Experiments

We assumed in the previous section that vertices have sufficient width resp. height to connect all edges adjacent to a certain side. But this assumption may be not satisfied by the layout algorithm which provides the input for the compaction. In this case, we define in the simplification step the distances between certain edges adjacent to the same side of a vertex as zero such that the size constraint can be fulfilled. With a slight modification the non-bending edge on a vertex side can be placed in the middle of the side which yields nice drawings.

We performed no empirical comparison with [1], but we are convinced that the quality of the drawings with respect to the drawing area is as least as good as in [1]. The edge length may be worse, but to further improve the quality of the drawing, we can solve the occurring linear programs with a Min-Cost Flow algorithm instead of using longest path algorithms. Of course this increases the running time. The solution obtained by using the Min-Cost Flow algorithm has minimal edge length with respect to the given complete shape extension (see, e.g., [12]). To improve an existing drawing, we can apply one-dimensional compaction, known from VLSI-design (see, e.g., [13]), as postprocessing. We can reuse for this case the compaction algorithm; we only have to replace the rectangle decomposition step by a separation function based on the visibility of segments in the input drawing.

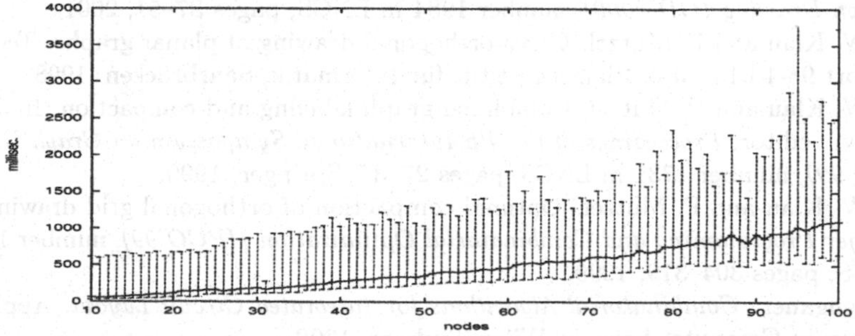

Fig. 7. Average running time of the algorithm on the Rome graphs.

We tested the algorithm on the so called Rome graphs test suite [4]. We assigned to each vertex a random width and height between zero and ten. The input of the compaction algorithm was generated by an implementation of the Kandinsky bend-minimizing algorithm [7]. The running time of the algorithm was always below 4 seconds using the virtual machine provided in the JDK 1.2.2 of Sun Microsystems, with a 64MB memory limit on a Sun Ultra 5 with 333 MHz. The average running time of the algorithm is shown in Fig. 7.

References

1. G. Di Battista, W. Didimo, Maurizo Patrignani, and Maurizio Pizzonia. Orthogonal and quasi-upward drawings with vertices of prescribed size. In J. Kratochvil, editor, *Proceedings of the 7th International Symposium on Graph Drawing (GD'99)*, volume 1731 of *LNCS*, pages 297–310. Springer, 1999.
2. T.H. Cormen, C.E. Leiserson, and R.L. Rivest. *Introduction to Algorithms.* McGraw-Hill, 1990.
3. G. Di Battista, P. Eades, R. Tamassia, and I. G. Tollis. *Graph Drawing: Algorithms for the Visualization of Graphs.* Prentice Hall, 1999.
4. G. Di Battista, A. Garg, , G .Liotta, R. Tamassia, E. Tassinari, and F. Vargiu. An experimental comparison of four graph drawing algortihms. *Comput. Geom. Theory Appl.*, 7:303–325, 1997.
5. M. Eiglsperger. Constraints im Kandinsky-Algorithmus. Master's thesis, Universität Tübingen, 1999.
6. U. Fößmeier. *Orthogonale Visualisierungstechniken für Graphen.* PhD thesis, Eberhard-Karls-Universität zu Tübingen, 1997.
7. U. Fößmeier and M. Kaufmann. Drawing high degree graphs with low bend numbers. In F. J. Brandenburg, editor, *Proceedings of the 3rd International Symposium on Graph Drawing (GD'95)*, volume 1027 of *LNCS*, pages 254–266. Springer, 1996.
8. U. Fößmeier and M. Kaufmann. Algorithms and area bounds for nonplanar orthogonal drawings. In G. Di Battista, editor, *Proceedings of the 5th International Symposium on Graph Drawing (GD'97)*, volume 1353 of *LNCS*, pages 134–145. Springer, 1997.
9. K. Klein G. W. Klau and P. Mutzel. An experimental comparison of orthogonal compaction algorithms. In *Proceedings of the 8th International Symposium on Graph Drawing (GD'2000)*, number 1984 in LNCS, pages 37–51, 2001.
10. G. W. Klau and P. Mutzel. Quasi-orthogonal drawing of planar graphs. Technical Report 98-1-013, Max-Planck-Institut für Informatik, Saarbrücken, 1998.
11. G. W. Klau and P. Mutzel. Combining graph labeling and compaction. In J. Kratochvil, editor, *Proceedings of the 7th International Symposium on Graph Drawing (GD'99)*, number 1731 in LNCS, pages 27–37. Springer, 1999.
12. G. W. Klau and P. Mutzel. Optimal compaction of orthogonal grid drawings. In *Integer Programming and Combinatorial Optimization (IPCO'99)*, number 1610 in LNCS, pages 304–319, 1999.
13. T. Lengauer. *Combinatorial Algorithms for Integrated Circuit Layout.* Applicable Theory in Computer Science. Wiley-Teubner, 1990.
14. M. Patrignani. On the complexity of orthogonal compaction. *Computational Geometry: Theory and Applications*, 19(1):47–67, 2001.
15. R. Tamassia. On embedding a graph in the grid with the minimum number of bends. *SIAM Journal on Computing,* 16(3):421–444, 1987.

Labeling Heuristics for Orthogonal Drawings*

Carla Binucci, Walter Didimo, Giuseppe Liotta, and Maddalena Nonato

Università di Perugia ({binucci,didimo,liotta,nonato}@diei.unipg.it).

Abstract. This paper studies the problem of computing an orthogonal drawing of a graph with labels along the edges. Labels are not allowed to overlap with each other or with edges to which they are not assigned. The optimization goal is area minimization. We provide a unified framework that allows to easily design edge labeling heuristics. By using the framework we implemented and experimentally compared several heuristics. The best performing heuristics have been embedded in the topology-shape-metrics approach.

1 Introduction

The labeling placement problem has a long tradition in the computational geometry, computational cartography, and geographic information system communities. Several papers have been published in the literature presenting algorithms that receive as input a drawing Γ of a graph together with a set of text or symbol labels for the vertices and/or edges and produce as output a labeled drawing with some "good readability" qualities. For example, if Γ is a city map, then the streets must be easy to identify by their names.

Kakoulis and Tollis [12] define three basic requirements that a good labeling of a drawing should have: (i) For any label it must be easy to identify the edge or vertex to which the label is assigned. (ii) A label assigned to a vertex or edge cannot overlap other labels, vertices or edges. (iii) A label must be placed in the best possible position among all acceptable positions.

Depending on the application domain, the above three requirements are expressed in terms of different geometric constraints and optimization goals. Unfortunately, most labeling problems have been proved to be NP-hard in general even in their simplest forms where the given drawing consists of just a set of distinct points or a set of distinct straight-lines; efficient heuristics and polynomial solutions for restricted versions of these problems have been designed. These results have in common that the problem constraints do not allow to change the geometry of the drawing that must be labeled. The interested reader is referred to the on-line bibliography by Wolff and Strijk [17] for references on the subject.

In recent years the labeling problem has been receiving increasing attention also in the graph drawing community (see e.g. [2,6,13,10,11,15,14]). As several

* Research supported in part by the CNR Project "Geometria Computazionale Robusta con Applicazioni alla Grafica ed al CAD", the project "Algorithms for Large Data Sets: Science and Engineering" of the Italian Ministry of University and Scientific and Technological Research (MURST 40%).

P. Mutzel, M. Jünger, and S. Leipert (Eds.): GD 2001, LNCS 2265, pp. 139–153, 2002.
© Springer-Verlag Berlin Heidelberg 2002

authors remark (see e.g. [2,12,13]), the problem allows more flexibility in the graph drawing context where one can change the geometry of a drawing so to free up space for label insertions. In this area a challenging research direction is that of designing algorithms whose input is a graph G with a set of labels for its vertices or edges and whose output is a labeled drawing of G. Namely, most graph drawing algorithms have very poor support for computing labeled drawings and there is a clear need of integrating effective labeling strategies within graph drawing techniques in order to enlarge their range of applications.

For example, the well-known and widely used topology-shape-metrics approach for computing orthogonal drawings of planar graphs is not equipped with effective technology for inserting labels along the edges. One possible solution is to model the labels as dummy vertices and to apply the topology-shape-metrics approach that computes an orthogonal drawing where the dummy vertices are constrained to have fixed size [3]. However, it is not clear how to guarantee in this framework that the resulting labeled drawing has good readability qualities. Namely, the described strategy does not allow any control on choosing the best position for a label along an edge; if for example the graph to be drawn is large and optimizing the area of the drawing is a critical issue, choosing a segment or another for placing a label on an edge may deeply affect the readability of the output. The drawing on the right-hand side of Figure 1 is computed with a random segment selection strategy.

A pioneering work aiming at integrating labeling techniques with the topology-shape-metrics approach is due to Klau and Mutzel [13]. They study the problem of computing a grid drawing of an orthogonal representation with labeled vertices and minimum total edge length. Klau and Mutzel show an elegant ILP formulation of the problem and present the first branch-and-cut based algorithm that combines compaction and labeling techniques.

The present paper makes a further step in the direction defined by Klau and Mutzel by investigating one of the questions that they leave as open. Namely we focus on integrating the topology-shape-metrics approach with algorithms for edge labeling. A precise description of the problem we deal with is as follows: Let G be a planar graph, let H be an orthogonal representation of G, and let L be a set of labels for the edges of G, where each edge is associated with at most one label. We want to compute an orthogonal grid drawing Γ of G such that the edges of Γ are labeled and have the shape defined in H. A label is modeled as an axis parallel rectangle of given integer width and height.

Our geometric constraints and optimization goals are as follows: (i) A label is drawn with one of its sides as a proper subset of a segment of the corresponding edge (the label is "glued" to the edge, so the assignment is unambiguous); (ii) Each label λ associated with a segment s cannot overlap any other label, vertex, or segment except s; (iii) A good placement for our labels is a placement that minimizes the area of the drawing.

We observe that finding an optimal solution for our problem can be too expensive in practice, since the instance where every edge has no associated label coincides with the well known NP-hard compaction problem [16] for orthogo-

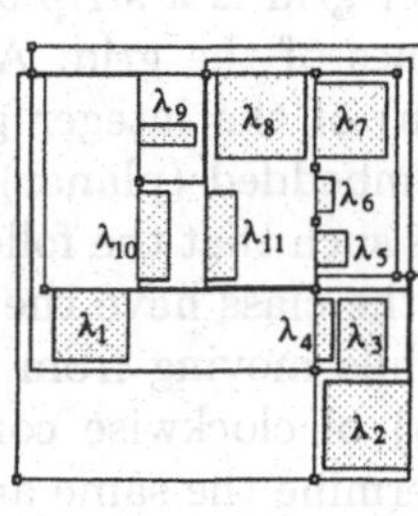

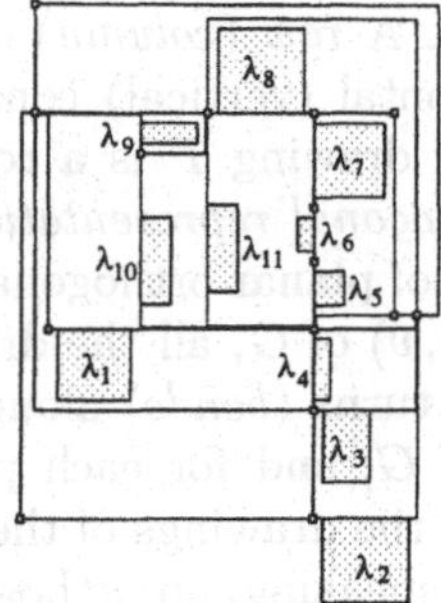

Fig. 1. Two different labeled drawings of the same labeled orthogonal representation. The one the right is computed by choosing randomly the segment and the direction of each label.The one on the left is computed by choosing the segment and the direction of each label with one of our heuristics.

nal representations. We investigate the edge labeling problem with an approach different from that followed by Klau and Mutzel for vertex labeling: Instead of looking for an ILP formulation, we implement simple and robust heuristics and experimentally compare their performances. A more detailed description of our results is given below:

- We define a general framework for constructing heuristics to solve the edge labeling problem for orthogonal representations. The proposed framework is based on a greedy strategy that computes a drawing by inserting a label at a time. The optimization goal is area minimization.
- By using the above framework we designed different algorithms for the edge labeling problem and experimentally compared their performances. We used techniques such as local search and greedy randomized adaptive search procedures (GRASP) [7] to investigate the effectiveness of our heuristics in practice.
- We embedded the best performing heuristics within the topology-shape-metrics approach. Namely, the implementation of these heuristics has enriched the topology-shape-metric technology of the GDToolkit library [9] for computing orthogonal drawings. We designed and implemented two new algorithms, called `Fast Labeler` and `Slow Labeler` that receive as input an edge labeled graph (with vertices of unbounded degree) and produce a compact orthogonal drawing of G. The drawing on the left-hand-side of Figure 1 is computed with the `Slow Labeler` and is the same graph as the one on the right-hand side.

2 Preliminaries

We assume familiarity with basic definitions on graph connectivity, graph planarity, and graph drawing [4]. An *orthogonal (grid) drawing* of a graph is a drawing such that the vertices are represented as points of an integer grid and the edges are represented as chains of horizontal and vertical segments on the integer grid lines. A *row* (*column*) of an integer grid is a strip of the plane between two horizontal (vertical) consecutive lines of the grid. A *row* (*column*) of an orthogonal drawing Γ is a row (column) of the integer grid that intersects Γ. An *orthogonal representation* of an embedded (planar) graph G is an equivalence class of planar orthogonal drawings such that the following holds: (i) For each edge (u, v) of G, all the drawings of the class have the same sequence of left and right turns (*bends*) along (u, v), while moving from u to v. (ii) For each vertex v of G, and for each pair $\{e_1, e_2\}$ of clockwise consecutive edges incident on v, all the drawings of the class determine the same angle between e_1 and e_2. Roughly speaking, an orthogonal representation defines a class of planar orthogonal drawings that may differ only for the length of the segments of the edges.

One of the most popular techniques for computing an orthogonal drawing of a graph G is the so called *topology-shape-metrics* approach [18,19,4]. It consists of three consecutive steps: (i) `Planarization`: An embedding of G is computed, possibly adding dummy vertices to replace crossings. (ii) `Orthogonalization`: An orthogonal representation H of G is computed within the previously computed embedding. (iii) `Compaction`: A final geometry for H is determined. Namely, coordinates are assigned to vertices and bends of H. The distinct phases of the topology-shape-metrics approach have been extensively studied in the literature [4].

Since each vertex of a planar orthogonal drawing is a grid point and since planarity does not allow distinct edges to overlap, each vertex can have at most degree equal to four. Of course, this is a severe limitation for most applications. In order to orthogonally draw graphs of arbitrary vertex degree, different drawing conventions have been introduced in the literature. Here we refer to as the *podevsnef* (planar orthogonal drawing with equal vertex size and not empty faces) drawing convention, defined by Fößmeier and Kaufmann [8]. A *podevsnef* drawing (see Figures 2 (a) and 2 (b)) is an orthogonal drawing such that: (i) Segments representing edges cannot cross, with the exception that two segments that are incident on the same vertex may partially overlap. Observe that the angle between such segments has zero degree. Roughly speaking, a podevsnef drawing is "almost" planar: it is planar everywhere but in the possible overlap of segments incident on the same vertex. Observe in Figure 2 (b) the overlap of segments incident on vertices 1, 2, and 3. (ii) All the polygons representing the faces have area strictly greater than zero.

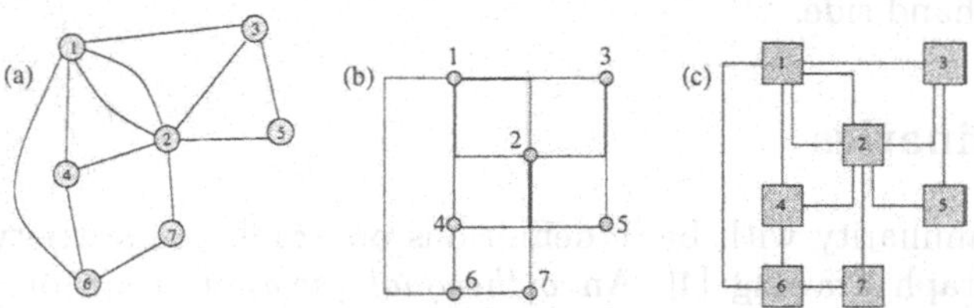

Fig. 2. (a) A planar graph and (b) one of its podevsnef drawings; (c) A more effective visualization of the podevsnef drawing in (b).

Podevsnef drawings are usually visualized representing vertices as boxes with equal size and representing two overlapping segments as two very near segments. See Figure 2 (c). Podevsnef drawings generalize the concept of orthogonal representation, allowing angles between two edges incident to the same vertex to have a zero degree value. The consequence of the assumption that the polygons representing the faces have area strictly greater than zero is that the angles have specific constraints. Namely, each zero degrees angle is in correspondence with exactly one bend [8]. An orthogonal representation corresponding to the above definition is a *podevsnef orthogonal representation*. In a podevsnef orthogonal representation we allow label assigment only to segments that do not overlap (note that each edge has a non-overlapping portion of segment). From now on we shall use the term orthogonal drawing and orthogonal representation for denoting podevsnef orthogonal drawings and podevsnef orthogonal representations, respectively. Also, when we talk about a segment we always refer to a segment (or portion of a segment) of an orthogonal representation that does not overlap with any other segment.

We conclude this section with some geometric definitions that will be used from here on to describe the different label insertion strategies. Let Γ be an orthogonal drawing, and let s be a horizontal (vertical) segment of Γ with endpoints a and b. Assume that a is to the left of b if s is horizontal and that a is below b if s is vertical. Let R be an axis aligned rectangle such that one of the sides of R is a subset of s. We call *position* of R with respect to s the distance between R and a. If such a distance is an integer number the position of R with respect to s is an *integer position*. If R lies on the left-hand side of s while moving from a, we say that R has *the left direction* with respect to s; otherwise R has *the right direction* with respect to s. R is an *empty rectangle* if the interior of R does not contain any point of Γ.

3 A Unified Framework for Greedy Labelers

In this section we describe a general greedy strategy for the edge labeling problem for orthogonal drawings.

3.1 The Greedy Labeler

Let H be an orthogonal representation and let L be a set of labels for the edges of H. We call our general strategy `Algorithm Greedy Labeler`. The algorithm is based on a greedy approach. It first computes a drawing of H with no labels, and then it performs $|L|$ steps. At each step a new label is selected and inserted in the current drawing. The insertion is performed by possibly stretching some of the edges in order to avoid intersections between any two labels or between a label and an edge. Let $\lambda \in L$ be the label of and edge e of H such that λ has not been inserted in the drawing yet. We associate λ with a pair $< place(\lambda), cost(\lambda) >$ such that: (i) $place(\lambda)$ is the *drawing placement* of λ. It is a triplet $< s, d, p >$, where s is the segment of e on which λ will be drawn, d specifies the direction (left or right) of λ with respect to s, and p defines the integer position of λ

with respect to s. (ii) *cost*(λ) is the *drawing cost* of λ. It measures the "price" that must be paid when inserting λ in the current drawing with the constraints defined by its drawing placement. Different definitions of this price give rise to different greedy heuristics. For example, *cost*(λ) can measure how much the area of the current drawing is increased in order to accommodate λ, or it can also look ahead and estimate whether inserting λ can lower the cost of some next insertions.

The insertion of λ in the current drawing is performed by invoking a suitable `Insert Label Procedure` that receives as input λ and the current labeled drawing Γ and computes a new labeled drawing containing all labels in Γ plus λ. The basic idea is to enlarge Γ by inserting a minimal number of columns and/or rows that are needed to draw λ without overlaps. Since the insertion of a new label can change the geometry of the current drawing, it may be needed to update the values of the drawing placement and cost for some of the remaining labels that have not been inserted yet. Namely, there are two main implications of the insertion of λ in the drawing: (i) Inserting rows or columns in the drawing causes additional free area in some faces. This can imply the reduction of the cost of some labels to be drawn yet. (ii) Drawing λ in a face f reduces the free area in f and this can lead to an increase of the drawing cost of some other labels, for example those that are going to be drawn inside f. In both cases the drawing cost of some of the remaining labels must be updated after the insertion of λ. Let λ' be one of those labels whose drawing cost and placement is affected by the insertion of λ; `Algorithm Greedy Labeler` invokes a `Cost Assignment Procedure` that receives as input λ' and the current drawing and computes as output the values $< place(\lambda'), cost(\lambda') >$. A detailed description of `Algorithm Greedy Labeler` is given below.

`Algorithm Greedy Labeler`

input: An orthogonal representation H and a set L of labels for the edges of H.
output: A labeled drawing Γ of H.

`Step 1: Preprocessing Phase.`
- Compute a grid drawing Γ_0 of H with no labels
- Copy all labels in a set Q
- **for each** label $\lambda \in Q$ execute `Cost Assignment Procedure`(λ, Γ_0)

`Step 2: Labeling Phase.`
- **for** $i := 1$ **to** $|L|$ perform the following three steps

  `Step 2.1: Label Selection.`
  - **let** $\lambda \in Q$ be the label such that *cost*(λ) is minimum
  - Remove λ from Q

  `Step 2.2: Label Insertion.` Compute a new drawing Γ_i by executing `Insert Label Procedure` (λ, Γ_{i-1})

  `Step 2.3: Cost Update.`
  - **for each** label $\lambda' \in Q$ that may be drawn in some of the faces whose free area is changed after the insertion of λ, execute `Cost Assignment Procedure`(λ, Γ_i)
- **set** $\Gamma = \Gamma_{|L|}$

`Step 1` (the `Preprocessing Phase`) can be accomplished by means of a standard technique for compacting (podevsnef) orthogonal representations [8, 1]. `Step 2` strongly relies on the `Cost Assignment Procedure(`λ`,`Γ_i`)` and on the `Insert Label Procedure (`λ`,`Γ_{i-1}`)` which are the subject of the next two subsections.

3.2 The Cost Assignment Procedure

The `Cost Assignment Procedure` receives as input the label λ of an edge e and a drawing Γ where e is not labeled yet. For each segment s of e, the procedure moves λ along s and computes a pair $< place_s(\lambda), cost_s(\lambda) >$. $place_s(\lambda)$ is a triplet $< s, d, p >$ where p and d are the integer position and direction of λ with respect to s that have minimum cost; this minimum cost is stored in $cost_s(\lambda)$. Pair $< place_s(\lambda), cost_s(\lambda) >$ is computed by means of a suitable *drawing cost function* that we denote as $CF(\lambda, s, \Gamma)$ and that is the kernel of the `Cost Assignment Procedure`. Finally, the output of the `Cost Assignment Procedure` is the pair $< place(\lambda), cost(\lambda) >$ corresponding to the minimum value of $cost_s(\lambda)$ over all segments s of e.

`Cost Assignment Procedure(`λ`,`Γ`)`

- **let** e be the edge of Γ where λ must be drawn
- **set** $place(\lambda) =< nil, nil, nil >$
- **set** $cost(\lambda) = +\infty$
- **for all** segments s of e
  - **set** $< place_s(\lambda), cost_s(\lambda) >= CF(\lambda, s, \Gamma)$
  - **if** $(cost_s(\lambda) < cost(\lambda))$ **then**
    set $< place(\lambda), cost(\lambda) >=< place_s(\lambda), cost_s(\lambda) >$
- **return** $< place(\lambda), cost(\lambda) >$

We are now ready to provide more details about $CF(\lambda, s, \Gamma)$. For concreteness, we assume that s is vertical (the case where s is horizontal is handled analogously). Let h_λ denote the height of λ and let h_s denote the height of s. We distinguish between two cases.
Case 1: $h_\lambda + 2 \leq h_s$ (this implies that λ can be inserted in Γ on s without stretching s). In this case $CF(\lambda, s, \Gamma)$ analyzes each integer position of λ with respect to s and for each such position it considers the two possible directions (left and right) of λ with respect to s. For a pair $< p, d >$, where p is a given integer position and d is a given direction of λ with respect to s, $CF(\lambda, s, \Gamma)$ executes the following two tasks:

`Task 1:` It evaluates a minimal number of rows and columns that must be added in order to properly draw λ in Γ. Let r and c be the computed number of rows and columns, respectively.

`Task 2:` It computes a cost C that depends on r, c, Γ and the dimensions of λ. If C is lower than $cost_s(\lambda)$, we set $cost_s(\lambda) = C$ and $place_s(\lambda) =< s, d, p + 1 >$. Different heuristics for the labeling problem can be designed within our greedy framework and by changing the definition of C (see also Section 4).

Task 1 applies the following rule (refer to Figure 3). Let h_λ and w_λ be the height and the width of λ, respectively. We compute the largest empty rectangle R such that the position of R with respect to s is p, the direction of R with respect to s is d, and the width of R is equal to $w_\lambda + 1$. Let h_R be the height of R. There are two cases to consider:

- $h_R \geq h_\lambda + 2$. In this case λ can be drawn inside R, in position $p + 1$ with respect to s, and without inserting any rows or columns to the drawing. Hence, r and c are both set to be 0. See for example in Figure 3 (a), where d is the right direction.
- $h_R < h_\lambda + 2$. In this case we insert only rows or only columns in the drawing in order to place λ in position $p + 1$ with respect to s (see for example Figure 3 (b)). The number c of columns is defined as follows. Assume that d is the right direction (the reasoning is symmetric for the case that d is the left direction) and let f be the face of Γ on the right-hand side of s. Let s_d be the closest segment (not touching s) on the boundary of f on the right-hand side of s that has one point in common with the boundary of R; let δ be the distance between s and s_d. See for example Figure 3 (b). Then c is set to $c = w_\lambda + 1 - \delta$. As for the value of r, we distinguish between the case that R has 0 height and the case that R has positive height. In the first case, r is set to be $r = +\infty$. In the second case, r is set to be $r = h_\lambda + 2 - h_R$. If $min(c, r) = r$ then we set $c = 0$, else we set $r = 0$.

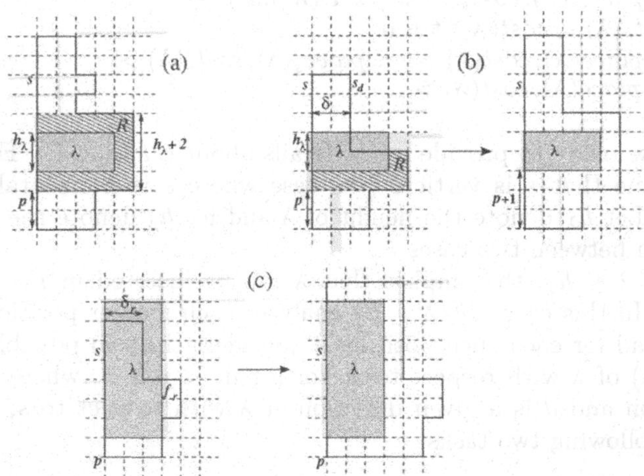

Fig. 3. (a) Label λ can be placed in position $p + 1$ without inserting rows or columns (b) Label λ can be placed in position $p + 1$ adding two rows or three columns. It is inserted by adding two rows. (c) The height of the label plus two is larger than the length of the segment. To place the label we insert two rows and two columns.

Case 2: $h_\lambda + 2 > h_s$ (this implies that s must be stretched in order to support λ). $CF(\lambda, s, \Gamma)$ executes the following operations (see Figure 3 (c)):

- It sets $p = 0$.
- It sets $r = h_\lambda + 2 - h_s$. Also, let f_l be the face of Γ on the left-hand side of s, and let f_r be the face of Γ on the right-hand side of s. Let $c_l = max\{0, w_\lambda + 1 - \delta_l\}$, where δ_l is the minimum distance between s and a segment (not touching s) on the boundary of f_l that is on the left-hand side of s. Let $c_r = max\{0, w_\lambda + 1 - \delta_r\}$, where δ_r is the minimum distance between s and a segment (do not touching s) on the boundary of f_r that is on the right-hand side of s. Let $c = min(c_l, c_r)$. If $c = c_l$ it is set $d = d_l$, else it is set $d = d_r$.
- It computes a cost C that depends on r, c, Γ and the dimensions of λ. It sets $cost_s(\lambda) = C$ and $place_s(\lambda) =< s, d, p+1 >$.

3.3 The Insert Label Procedure

The `Insert Label Procedure(`λ`,`Γ`)` draws label λ in drawing Γ. Label λ is placed in Γ according to $place(\lambda) =< s, d, p >$. The number of rows and columns to be added to Γ so to avoid overlaps is computed as described for $CF(\lambda, s, \Gamma)$. Actually, in order to save computation time one can store the number of rows and columns computed by $CF(\lambda, s, \Gamma)$ into two variables $r(\lambda)$ and $c(\lambda)$ and pass these information to the `Insert Label Procedure(`λ`,`Γ`)`.

We provide some details about the insertion strategy assuming that s is vertical and that d is the right direction (see Figure 3 (b) and Figure 3 (c)). The other cases are handled similarly. The value of position p with respect to s defines a unique point along segment s; let l_h and l_v be the horizontal line and the vertical line through this point, respectively. Let H_u be the closed half-space above l_h and let H_r be the open half-space to the right of l_v. The following tasks are executed: (i) $r(\lambda)$ rows are inserted in Γ by translating of $r(\lambda)$ units to the North direction all points of Γ that lie in H_u (ii) $c(\lambda)$ columns are inserted in Γ by translating of $c(\lambda)$ units to the right direction all points of Γ that lie in H_r. (iii) λ is drawn in position p and direction d with respect to s. Once inserted, λ will be treated by the greedy algorithm as a new face whose segments cannot be stretched.

The following theorem summarizes the complexity of `Algorithm Greedy Labeler` in the worst case. Such a complexity is obtained by considering the case in which all the costs of the remaining labels have to be updated at each new greedy step and by considering the complexity of function $CF()$. We omit the proof due to space limitations, but we observe that our experiments show better time performance in practice than the one theoretically estimated. This can be justified by a more careful analysis of the practical situations. In Section 4 we give some intuitions that are behind such a behavior.

Theorem 1. *Let H be an orthogonal representation of a planar graph with n vertices, and let L be a set of labels for the edges. There exists a general greedy algorithm that computes a labeled drawing Γ of H in $O(|L|nT_L)$ time, where T_L is the total length of the labeled edges of Γ.*

4 Experimental Comparison of Different Labeling Heuristics

We designed three basic heuristics for edge labeling within the framework of `Algorithm Greedy Labeler`. Such heuristics differ for the definition of cost adopted by the `Assignment Procedure` where selecting the label with highest insertion priority. We use the notation of the previous sections and denote with h_λ and w_λ the height and the width of λ, respectively. Also, r and c denote the number of rows and columns computed by $CF(\lambda, s, \Gamma)$. We denote with Δ_A the area increase (measured in terms of grid points in the bounding box of the drawing) of Γ implied by inserting r rows and c columns in Γ with the `Insert Label Procedure`. The three heuristics are as follows.

`Delta-area`: The label with highest insertion priority is the one that causes the minimum increase of area. Therefore, $C = \Delta_A$.

`Max-size-delta-area`: The label with highest insertion priority is the one with maximum area. If two labels have the same area, the label that implies the minimum increase of the area is chosen. Therefore $C = \Delta_A - K(h_\lambda+1)(w_\lambda+1)$ where K is a constant such that $K >> \Delta_A$.

`Max-ratio-delta-area`: The label with highest insertion priority is the one with maximum aspect ratio. If two labels have the same aspect ratio, it is chosen the one that causes the minimum increase of the area. Therefore $C = \Delta_A - Kmax\{(h_\lambda+1)/(w_\lambda+1), (w_\lambda+1)/(h_\lambda+1)\}$, where K is a constant such that $K >> \Delta_A$.

We implemented and experimentally compared the performances of the above three heuristics. The implementation uses a PC-Pentium III (800 MHz and 256 MB RAM), Linux RedHat 6.2 O.S., gnu g++ compiler, and the GDToolkit library [9]. The experimental analysis measured the following quantities:CPU time, area, total edge length, and screen ratio of the labeled drawings.

The test suite for the experiments is a variant of the real-world graphs extensively used in the graph drawing community for experimental analysis (the so-called "Rome-graphs") [5]. Namely, for each graph of this set we computed an orthogonal representation H with a variation of the technique in [8] and randomly generated labels to be assigned to the edges. The assignment of a label to an edge is done by a coin flip (i.e. one edge gets a label with probability 0.5). For each label λ, h_λ and w_λ are in the range of integer values $0-5$ and are defined at random with uniform probability distribution (a label whose height and width are both 0 is a grid point). The resulting test suite consists of about eleven thousand labeled graphs that are grouped into families according to their number of vertices. The number of vertices ranges from 10 to 100.

Concerning the screen ratio the three heuristics have very similar behaviour. The charts of the total edge length have the same behaviour as those of the area, although the differences between the three heuristics are less remarkable for total edge length. For reasons of space, we omit the charts on screen ratio and total edge length and show only a subset of those about area and CPU time. All our experimental results have the property that the curves relative to different heuristics almost never overlap and are always monotonically increasing. Since differences of $5 - 10\%$ may appear not very readable in small figures, we shall often display only charts relative to a subset of the test suite while talking about the performance of the heuristics over the whole test set.

Figure 4 (c) shows the CPU time performance (average values) of the three heuristics. We observe that they look fast in practice since they never require more than two seconds even for the largest graph instances of the test suite. In the implementation of `Algorithm Greedy Labeler` we used the Fibonacci heap data structure to implement Q, that supports selections and updates in constant and logarithmic amortized time. The complexity of updating the cost of a label λ depends on the computation of function $CF(\lambda, s, \Gamma)$. Although in the worst case this computation can require evaluating a number of positions for λ that is equal to the length of s, in many cases it evaluates only a small number of positions. In fact, if λ is so "big" that s must be stretched to accommodate it, just one integer position for λ is taken into account (see **Case 2** of Section 3.2). Conversely, if λ is "small" $CF(\lambda, s, \Gamma)$ usually finds in a few steps a position for λ that do not require to insert rows or columns and then stops its computation. Further, consider that portions of segments that overlap are discarded.

Figure 4 (a) compares the (average) areas of the drawings computed by the three heuristics for graphs with number of vertices in the range $80 - 100$. Heuristics `max-ratio-delta-area` is the best performing and improves the solutions of heuristic `delta-area` by at least an $4-5\%$ factor; there are also cases in which such an improvement is much greater. In our opinion this result is a consequence of the fact that `max-ratio-delta-area` gives higher priority to "skinny" labels, i.e. those that have high aspect ratio. Inserting a skinny label often implies inserting extra rows and/or columns and enlarging the size of several faces in the current drawing. As a consequence, other labels can be inserted without any further insertions of rows and columns. Also, a skinny label λ occupies only a small portion of the face f in which it is drawn; hence inserting λ does not increase too much the cost of other labels that have to be placed in f. On the other extreme, `delta-area` does not take into account in any way the impact that inserting a label in a face can have on further label insertions. The behavior of `max-ratio-delta-area` is in-between these two extremes: it draws first those labels that have large area but this is somehow not properly reflecting our rows/columns insertion strategy that, when possible, stretches the drawing only in one direction for each label insertion. In order to test the robustness of our techniques and further verify the above results we repeated the experiments on the same graphs but where label dimensions vary in the range $5-10$. Figure 4 (b) shows the same trend of performance as the one of Figure 4 (a) and it even emphasizes the difference between `delta-area` and `max-ratio-delta-area` for some graph instances.

Motivated by the above experiments we ran several new experiments to further investigate the effectiveness of the `max-ratio-delta-area` heuristic. Because of the NP-hardness of the problem, estimating how much the feasible solutions of the `max-ratio-delta-area` heuristic are far from optimum is not an easy task. Nevertheless, we executed four experiments that provide evidence of the good behavior of the heuristic. Let Γ be the labeled drawing computed by max-ratio-delta-area for a labeled graph G and let S be the solution space, i.e. the set of all possible labeled orthogonal drawings of G. Our experiments pick other feasible solutions in S and compare them with Γ.

We first compared the areas of drawings computed by max-ratio-delta-area with the area of the drawings computed by a trivial greedy heuristic, called random, that chooses randomly both the ordering in which labels will be inserted and the segment and the drawing placement of each label. Figure 4 (d) shows that max-ratio-delta-area outperforms random by about 20% for most graphs instances.

A second experiment iteratively executes random until a labeled drawing with area less than or equal to the area of the output of max-ratio-delta-area is found. Figure 4 (e) shows the CPU time (in seconds and logarithmic scale) spent by random to find a solution less than or equal to the one computed by max-ratio-delta-area for the same graph. In order to execute such an experimentation in a reasonable time, we performed it over a subset of graphs of the test suite, with maximum running time of 30 minutes for each graph.

A third experiment uses local search to explore the neighborhood of Γ in S. We visit the neighborhood of Γ searching for better solutions. Clearly, setting the neighborhood to be searched involves a tradeoff between solution quality (the larger the neighborhood the better the solution that can be found) and running time (larger neighborhoods require longer time to be searched). Our experiment is as follows: For each label λ assigned to a segment s, we tried to flip λ around s while preserving the directions of the other labels and ran again max-ratio-delta-area with the constraint that the labels are inserted in the same ordering as the one used for constructing Γ. If a better solution is found (i.e. for the flipping of some of the labels) we move on with a new solution and iterate the search, otherwise it is stopped. Figures 4 (f) and (g) show our experimental results about the area and CPU time.

In order to overcome the inner limit of local search strategies that get trapped in a local optimum, we ran a fourth experiment that includes our local search within a greedy randomized adaptive search procedure (GRASP)[7]. A GRASP repeatedly starts the search from different solution points in the feasible region. Such points are not selected at random as in classical multi-start methods, but are computed by perturbing the criterion of the max-ratio-delta-area heuristic when selecting the best current label. Our GRASP is based on choosing at random the label whose cost lies in the range $[\chi_m, \chi_m + (\chi_M - \chi_m)\alpha]$ where α is a real in the interval $[0, 1]$ and χ_m and χ_M are the minimum and the maximum costs of the labels considered in the selection step. We then apply a local search approach to each one of the different solutions and keep the best solution among the computed ones. On each graph we performed 50 GRASP iterations. Figures 4 (h) and (i) show the area and the CPU time experimental results for $\alpha = 0.1$ and $\alpha = 0.3$.

Both the local search approach and the GRASP show that solutions with smaller area than those computed by the max-ratio-delta-area heuristic can be obtained by paying a relatively high price in terms of computation time. The experimental data show that the obtained area improvement is for most case not larger than $7 - 11\%$.

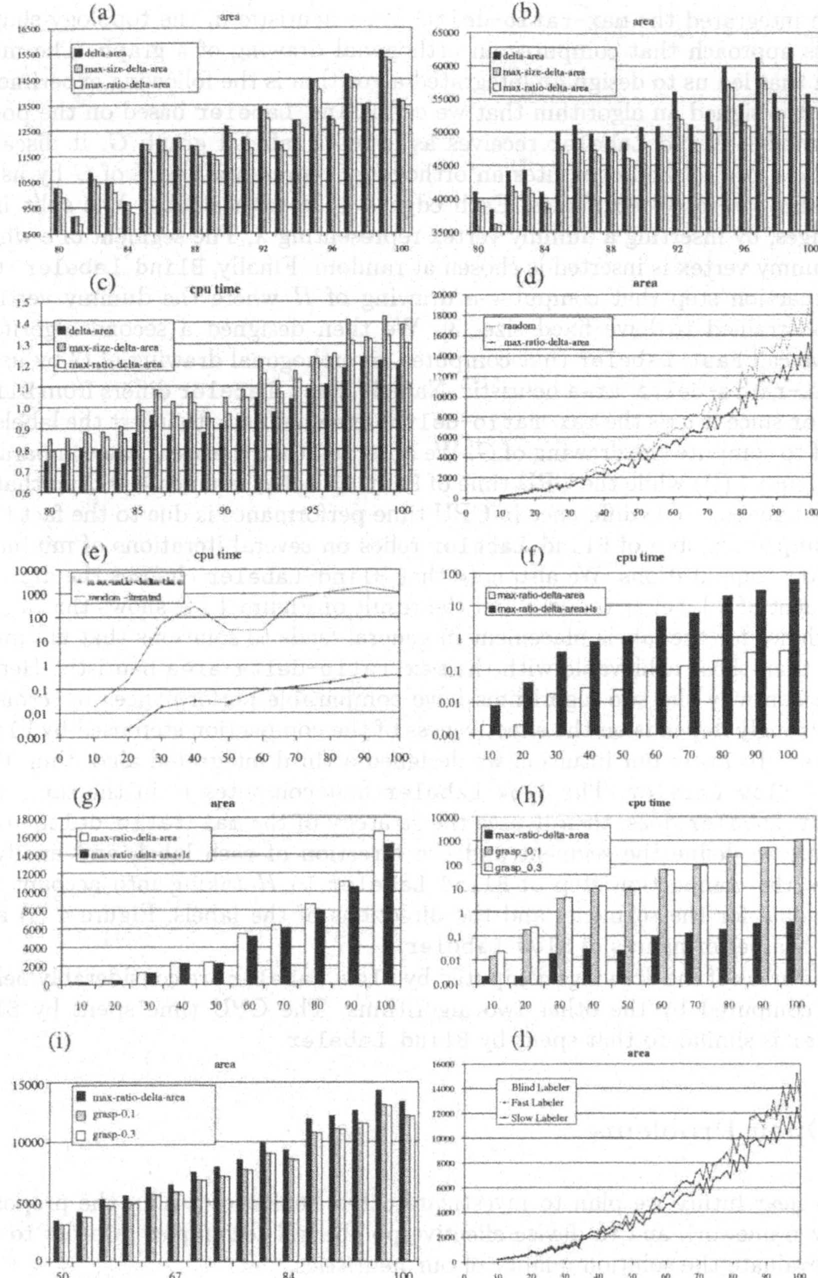

Fig. 4. Charts of the experiments. The x-axis represents the number of vertices. The y-axis reports average values.

We integrated the `max-ratio-delta-area` heuristic in the topology-shape-metrics approach that computes an orthogonal drawing of a graph. The main reason that led us to design an integrated algorithm is the following experiment. We first designed an algorithm that we call `Blind Labeler` based on the podevsnef model. `Blind Labeler` receives as input a labeled graph G, it discards the labels form G and computes an orthogonal representation H of G by using minimum cost flow techniques. Each edge $e \in H$ with a label λ is split into two edges, by inserting a dummy vertex representing λ. The segment of e where the dummy vertex is inserted is chosen at random. Finally, `Blind Labeler` runs a compaction step that computes a drawing of H where the dummy vertices are constrained to have fixed size [3]. We then designed a second algorithm that we call `Fast Labeler` that computes an orthogonal drawing of G by using the `max-ratio-delta-area` heuristic. Namely, `Fast Labeler` differs from `Blind Labeler` since it uses the `max-ratio-delta-area` heuristic to insert the labels in H and to compute the drawing of G. We observed that the areas are comparable (see Figure 4 (l)) while the CPU time of `Fast Labeler` is much less than that of `Blind Labeler`. The difference in CPU time performance is due to the fact that the compaction step of `Blind Labeler` relies on several iterations of minimum cost flow computations. We also note that `Blind-Labeler` chooses the drawing placement of a label at random and the result of Figure 4 (d) shows that a random choice for the labels placement in general leads to solutions that are much worse than those achievable with the `max-ratio-delta-area` heuristic. Hence, the reason why the two algorithms have comparable performances in terms of area strongly depends on the effectiveness of the compaction step used by `Blind Labeler`. To foster our intuition we designed a third integrated algorithm, that we call `Slow Labeler`. The `Slow Labeler` first computes H in the same way as `Fast Labeler` does, then it uses the strategy of the `max-ratio-delta-area` heuristic to define the segment and the direction of each label, and finally it applies the compaction step of `Blind Labeler` to H taking into account the constraints for the segments and the directions of the labels. Figure 4 (l) also shows the performances of `Slow Labeler`.

The areas of the drawings computed by `Slow Labeler` are considerably below those computed by the other two algorithms. The CPU time spent by `Slow Labeler` is similar to that spent by `Blind Labeler`.

5 Open Problems

In the near future we plan to investigate other heuristics within the proposed greedy framework and to devise effective polyhedral techniques in order to exactly evaluate the solution quality of our heuristics.

References

1. P. Bertolazzi, G. Di Battista, and W. Didimo. Computing orthogonal drawings with the minimum numbr of bends. *IEEE Transactions on Computers*, 49(8), 2000.

2. R. Castello, R. Milli, and I. Tollis. An algorithmic framework for visualizing statecharts. In *Proc. GD '00*, volume 1984 of *LNCS*, pages 139–149, 2001.
3. G. Di Battista, W. Didimo, M. Patrignani, and M. Pizzonia. Orthogonal and quasi-upward drawings with vertices of arbitrary size. In *Proc. GD '99*, volume 1731 of *LNCS*, pages 297–310, 2000.
4. G. Di Battista, P. Eades, R. Tamassia, and I. G. Tollis. *Graph Drawing.* Prentice Hall, Upper Saddle River, NJ, 1999.
5. G. Di Battista, A. Garg, G. Liotta, R. Tamassia, E. Tassinari, and F. Vargiu. An experimental comparison of four graph drawing algorithms. *Comput. Geom. Theory Appl.*, 7:303–325, 1997.
6. U. Dogrusoz, K. G. Kakoulis, B. Madden, and I. G. Tollis. Edge labeling in the graph layout toolkit. In *Proc. GD '98*, volume 1547 of *LNCS*, pages 356–363, 1999.
7. T. A. Feo and M. G. C. Resende. Greedy randomized adaptative search procedure. *Journal of Global Optimization*, 6:109–133, 1995.
8. U. Fößmeier and M. Kaufmann. Drawing high degree graphs with low bend numbers. In *Proc. GD '95*, volume 1027 of *LNCS*, pages 254–266, 1996.
9. GDToolkit. Graph drawing toolkit. On line. http://www.dia.uniroma3.it/~gdt.
10. K. G. Kakoulis and I. G. Tollis. On the edge label placement problem. In *Proc. GD '96*, volume 1190 of *LNCS*, pages 241–256, 1997.
11. K. G. Kakoulis and I. G. Tollis. An algorithm for labeling edges of hierarchical drawings. In *Proc. GD '97*, volume 1353 of *LNCS*, pages 169–180, 1998.
12. K. G. Kakoulis and I. G. Tollis. On the complexity of the edge label placement problem. *Comput. Geom. Theory Appl.*, 18:1–17, 2001.
13. G. W. Klau and P. Mutzel. Combining graph labeling and compaction. In *Proc. GD '99*, volume 1731 of *LNCS*, pages 27–37, 2000.
14. G. W. Klau and P. Mutzel. Optimal labelling of point features in the slider model. In *Proc. COCOON '00*, volume 1858 of *LNCS*, pages 340–350, 2000.
15. S. Nakano, T. Nishizeki, T. Tokuyama, and S. Watanabe. Labeling points with rectangles of various shapes. In *Proc. GD '00*, volume 1984 of *LNCS*, pages 91–102, 2001.
16. M. Patrignani. On the complexity of orthogonal compaction. *Comput. Geom. Theory Appl.*, 19:47–67, 2001.
17. T. Strijk and A. Wolff. The map-labeling bibliography. http://www.math-inf.uni-greifswald.de/map-labeling/bibliography/.
18. R. Tamassia. On embedding a graph in the grid with the minimum number of ben ds. *SIAM J. Comput.*, 16(3):421–444, 1987.
19. R. Tamassia, G. Di Battista, and C. Batini. Automatic graph drawing and readability of diagrams. *IEEE Trans. Syst. Man Cybern.*, SMC-18(1):61–79, 1988.

Untangling a Polygon

János Pach and Gábor Tardos

Rényi Institute of Mathematics
Hungarian Academy of Sciences
pach@cims.nyu.edu, tardos@renyi.hu

Abstract. The following problem was raised by M. Watanabe. Let P be a self-intersecting closed polygon with n vertices in general position. How manys steps does it take to untangle P, i.e., to turn it into a simple polygon, if in each step we can arbitrarily relocate one of its vertices. It is shown that in some cases one has to move all but at most $O((n \log n)^{2/3})$ vertices. On the other hand, every polygon P can be untangled in at most $n - \Omega(\sqrt{n})$ steps. Some related questions are also considered.

1 Introduction

Suppose we have a self-intersecting closed polygon P on the screen of our computer, whose vertices are $p_1, p_2, \ldots, p_n$ in this order, and no three vertices are collinear. We are allowed to modify P so that in each step we can grab a vertex and move it to an arbitrary new position. (For simplicity, we assume that the screen is very large, so we are not limited by its size.) At the 5th Czech-Slovak Symposium on Combinatorics in Prague in 1998, Mamoru Watanabe asked the following question. Is it true that every polygon P can be *untangled*, i.e., turned into a noncrossing polygon, in at most εn steps, for some absolute constant $\varepsilon < 1$?

The aim of this note is to answer this question in the negative.

Given another closed polygon Q with vertices $q_1, q_2, \ldots, q_n$ (in this order), let $f(P, Q)$ denote the number of "common points" of P and Q, i.e., the number of indices i, for which $q_i = p_i$. Let $f(P)$ denote the largest number of points that can be kept fixed when we untangle P. Using our notation,

$$f(P) = \max_Q f(P, Q),$$

where the maximum is taken over all noncrossing closed polygons with n vertices. See Fig. 1.

It is easy to see that every polygon P can be untangled in at most $n - \sqrt{n}$ moves. That is, we have

Proposition 1. *For every polygon P with n vertices, we have $f(P) > \sqrt{n}$.*

Proof. Assume without loss of generality that p_n is a vertex of the convex hull of $\{p_1, p_2, \ldots, p_n\}$, and let $p_{\sigma(1)}, p_{\sigma(2)}, \ldots, p_{\sigma(n-1)}$ be the other points of P, listed in clockwise order of visibility around p_n. According to a wellknown lemma of Erdős

P. Mutzel, M. Jünger, and S. Leipert (Eds.): GD 2001, LNCS 2265, pp. 154–161, 2002.
© Springer-Verlag Berlin Heidelberg 2002

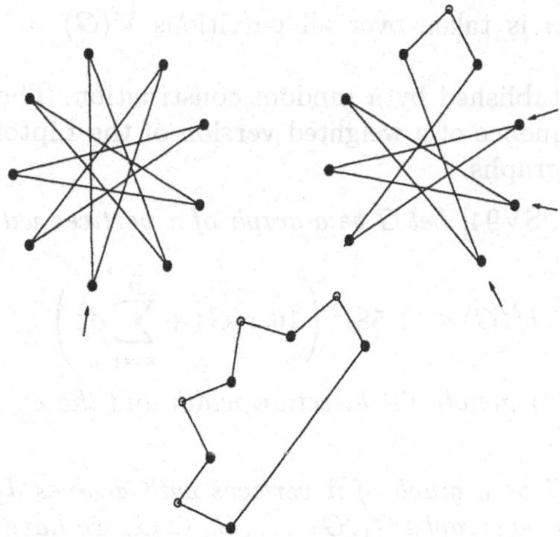

Fig. 1. For a star-polygon P with n vertices, we have $f(P) = \frac{n+1}{2}$

and Szekeres [ES35], every sequence of length k has a monotone subsequence of length $\lceil\sqrt{k}\rceil$. Therefore, there is a sequence $1 \leq i_1 < i_2 < i_3 < \ldots \leq n-1$ of length $\lceil\sqrt{n-1}\rceil$ such that either $\sigma(i_1) < \sigma(i_2) < \sigma(i_3) < \ldots$ or $\sigma(i_1) > \sigma(i_2) > \sigma(i_3) > \ldots$ is true. In either case, the points $p_n, p_{\sigma(i_1)}, p_{\sigma(i_2)}, \ldots$ induce a noncrossing closed polygon Q_0. Let Q denote the n-gon obtained from Q_0 by subdividing its sides with as many points as necessary, to achieve that the index of every point $p_n, p_{\sigma(i_1)}, p_{\sigma(i_2)}, \ldots$ be the same in P as in Q. Clearly, we have $f(P,Q) \geq \lceil\sqrt{n-1}\rceil + 1 > \sqrt{n}$, as required. □

Our main result can now be formulated as follows.

Theorem 2. *For every sufficiently large n, there exists a closed polygon P with n vertices, which cannot be untangled in fewer than $n - c(n\log n)^{2/3}$ moves. That is, we have $f(P) \leq c(n\log n)^{2/3}$, where c is a constant.*

Let G be a graph with vertex set $V(G)$ and edge set $E(G)$, respectively. A *drawing* of G is a representation of G in the plane such that every vertex corresponds to a point, and every edge is represented by a Jordan arc connecting the corresponding two points without passing through any vertex other than its endpoints. Two edges are said to *cross* each other if they have an interior point in common. The *crossing number* $\mathrm{cr}(G)$ of G is defined as the minimum number of crossing pairs of arcs in a drawing of G.

For any partition of the vertex set of G into two disjoint parts, V_1 and V_2, let $E(V_1, V_2) \subseteq E(G)$ denote the set of edges with one endpoint in V_1 and the other in V_2. Define the *bisection width* of G as

$$b(G) = \min |E(V_1, V_2)|,$$

where the minimum is taken over all partitions $V(G) = V_1 \cup V_2$ such that $|V_1|, |V_2| \leq 2n/3$.

Theorem 2 is established by a random construction. The proof is based on the following consequence of a weighted version of the Lipton-Tarjan separator theorem for planar graphs.

Lemma 3. [PSS94],[SV94] *Let G be a graph of n vertices with degrees $d_1, d_2, \ldots, d_n$. Then*

$$b^2(G) \leq (1.58)^2 \left(16\mathrm{cr}(G) + \sum_{k=1}^{n} d_k^2\right),$$

where $b(G)$ and $\mathrm{cr}(G)$ denote the bisection width and the crossing number of G, respectively.

Corollary 4. *Let G be a graph of n vertices with degrees $d_1, d_2, \ldots, d_n$. Then, for any edge disjoint subgraphs $G_1, G_2, \ldots, G_j \subseteq G$, we have*

$$\sum_{i=1}^{j} b(G_i) \leq 1.58 j^{1/2} \left(16\mathrm{cr}(G) + \sum_{k=1}^{n} d_k^2\right)^{1/2}.$$

Let d_{ik} denote the degree of the k-th vertex in G_i. Corollary 4 immediately follows from Lemma 3. Indeed, applying Lemma 3 to each G_i separately and adding up the resulting inequalities, we obtain

$$\sum_{i=1}^{j} b^2(G_i) \leq (1.58)^2 \left(16 \sum_{i=1}^{j} \mathrm{cr}(G_i) + \sum_{i=1}^{j} \sum_{k=1}^{n} d_{ik}^2\right)$$
$$\leq (1.58)^2 \left(16\mathrm{cr}(G) + \sum_{k=1}^{n} d_k^2\right).$$

Therefore, we have

$$\left(\sum_{i=1}^{j} b(G_i)\right)^2 \leq j \sum_{i=1}^{j} b^2(G_i) \leq (1.58)^2 j \left(16\mathrm{cr}(G) + \sum_{k=1}^{n} d_k^2\right),$$

as required.

2 Proof of Theorem 2

We start with an auxiliary lemma.

Lemma 5. *Let H_b denote the graph (cycle) defined on the vertex set $V = \{1, 2, \ldots, t\}$, whose edges are $(1,2), (2,3), \ldots, (t,1)$. Let H_r be a randomly selected Hamilton cycle on the same vertex set, i.e., let*

$$E(H_r) = \{(\sigma(1), \sigma(2)), (\sigma(2), \sigma(3)), \ldots, (\sigma(t), \sigma(1))\},$$

where σ is a random permutation of V.

Then, for every $s < t$ and K, the probability that the crossing number of $H = H_b \cup H_r$ is at most K satisfies

$$\text{Prob}[\text{cr}(H) \le K] \le \binom{t}{D}^2 \left(\frac{3t}{s}\right)^D \frac{s^{t-D}}{(t-D)!},$$

where $D = \lfloor 71\sqrt{t(K+t)/s} \rfloor$.

Proof. We refer to the edges of H_b and H_r as *black* and *red* edges, respectively.

Suppose that $\text{cr}(H) \le K$. The degree of every vertex in H is at most 4, so it follows from Corollary 4 that the sum of the bisection widths of any j pairwise (edge) disjoint subgraphs of H is at most

$$2 \cdot 1.58 j^{1/2}(16K + 16t)^{1/2} < 13 j^{1/2}(K+t)^{1/2}. \tag{1}$$

We multiplied the upper bound given in Corollary 4 by a factor of 2, because any two vertices of H may be connected by 2 edges: one black and one red.

Let s be a positive integer. By deleting relatively few edges, decompose H into connected components of sizes smaller than s as follows. In the first step, delete $b(H)$ edges such that H falls into two components, each having at most $\frac{2}{3}|V(H)| = \frac{2}{3}t$ vertices. As long as there is a component $H' \subset H$ whose size is at least s, by the removal of $b(H')$ edges, *cut* it into two smaller components, each of size at most $(2/3)|V(H')|$. When there are no such components left, stop.

Let $\mathcal{H}$ denote the family of all components arising at *any* level of the above procedure (e.g., we have $H \in \mathcal{H}$), and let $\mathcal{H}_0 \subseteq \mathcal{H}$ consist of the components of the *final* decomposition. That is, $\mathcal{H}_0$ contains all elements of $\mathcal{H}$, whose sizes are smaller than s. On the other hand, the size of every element of $\mathcal{H}_0$ is at least $s/3$, so we have that

$$|\mathcal{H}_0| \le \frac{t}{s/3} = \frac{3t}{s}, \tag{2}$$

and the total number of cuts is $|\mathcal{H}_0| - 1 < 3t/s$.

Define the *order* of any element $H' \in \mathcal{H}$ as the largest integer k, for which there is a chain

$$H_0 \subsetneqq H_1 \subsetneqq \ldots \subsetneqq H_k \tag{3}$$

in $\mathcal{H}$ such that $H_0 \in \mathcal{H}_0$ and $H_k = H'$. Thus, $\mathcal{H}_0$ is the set of elements of $\mathcal{H}$ of order 0. For any k, let $\mathcal{H}_k$ denote the set of all elements of $\mathcal{H}$ of order k.

For a fixed $k \ge 1$, the elements of $\mathcal{H}_k$ are pairwise (vertex) disjoint. Recall that in a chain (3) we have $|V(H_1)| \ge s$ and the ratio of the sizes of any two consecutive members is at least 3/2. Therefore, the number of vertices in any element of $\mathcal{H}_k$ is at least $(3/2)^{k-1}s$, which in turn implies that for $k \ge 1$

$$j_k := |\mathcal{H}_k| \le \frac{t}{(3/2)^{k-1}s} = \frac{(2/3)^{k-1}t}{s}.$$

Applying (1) to the subgraphs (components) in $\mathcal{H}_k$, we obtain that the total number of edges removed, when they are first subdivided during our procedure, is at most

$$13\left(\frac{2}{3}\right)^{\frac{k-1}{2}} \left(\frac{t}{s}\right)^{1/2} (K+t)^{1/2}.$$

Summing up over all k, we conclude that the total number of edges deleted during the whole procedure does not exceed

$$D = \left\lfloor 71\sqrt{\frac{t(K+t)}{s}} \right\rfloor .$$

Consequently, if we want to give an upper bound on Prob[cr$(H) \leq K$], it is sufficient to bound the probability that H can be decomposed into sets of size smaller than s by the deletion of precisely D black and precisely D red edges. In what follows, we estimate this probability.

The D black edges that are deleted can be chosen in $\binom{t}{D}$ different ways. The remaining black edges form D paths. The vertex set of any subgraph in $\mathcal{H}$ is the union of the vertex sets of a few of these paths. By (2), there are at most $(3t/s)^D$ possibilities for the partition of V induced by $\mathcal{H}$, once the deleted black edges are chosen.

We consider the red Hamiltonian cycle to be picked with an *orientation.* There are $\binom{t}{D}$ different ways how to pick the *starting points* of the D deleted red edges.

The probability that a randomly selected red Hamiltonian cycle "respects" a fixed partition of V into parts of size smaller than s, except for the edges originating at a fixed set of size D, is at most $s^{t-D}/(t-D)!$. Indeed, when we start drawing H_r randomly at a point, and we reach a vertex x which is not the starting point of a deleted red edge, then the probability that the endpoint of the red edge starting at this point belongs to the part of the partition which contains x is less than s divided by the number of vertices in V not yet visited by the initial portion of H_r. Summarizing: the probability that H can be decomposed into sets of size smaller than s by the deletion of D black and D red edges is at most

$$\binom{t}{D}^2 \left(\frac{3t}{s}\right)^D \frac{s^{t-D}}{(t-D)!},$$

and the lemma follows. □

Now we are in a position to establish Theorem 2.

Consider a regular n-gon and let $p_1, p_2, \ldots, p_n$ be a *random permutation* of its vertices. Let P denote the closed polygon obtained by connecting the p_i-s in this order. We claim that with high probability $f(P) \leq c(n \log n)^{2/3}$, where c is a constant.

For any positive integer t, we have $f(P) \geq t$ if and only if there is a t-element subset $T \subseteq \{p_1, p_2, \ldots, p_n\}$ such that there is a noncrossing closed polygon Q with vertices $q_1, q_2, \ldots, q_n$, in this order, with $q_i = p_i$ whenever $p_i \in T$.

To estimate the probability of this event for a fixed t-element set T, define two Hamilton cycles, H_b and H_r, on the vertex set T as follows. Let H_b consist of all edges of the convex hull of T. These edges are called *black.* A vertex $p_i \in T$ is connected to another vertex $p_j \in T$ by an edge of H_r, if p_i and p_j are consequtive in the cyclic order induced on T by the random permutation. That is, if $i < j$, then there is no index k with $i < k < j$ such that and $p_k \in T$ or there is no index k with $k < i$ or $j < k$ with $p_k \in T$. The edges in H_r are said to be *red*. Let $H = H_b \cup H_r$.

Suppose now that there is a noncrossing closed polygon Q with vertices $q_1, q_2, \ldots, q_n$, such $q_i = p_i$ whenever $p_i \in T$. By slightly changing the positions of its vertices not belonging to T, if necessary, we may achieve that the no three vertices of Q are collinear.

Consider the drawing of H, in which every vertex is represented by itself, every black edge is represented by a straight line segment, and every red edge by the corresponding portion of Q. In this drawing, there is no crossing between edges of the same color. Since every edge of Q can cross the black cycle (the boundary of the convex hull of T) in at most two points, we obtain the the number of crossings, and hence $\text{cr}(H)$, are at most $2n$. Thus, we have

$$\text{Prob}[f(P) \geq t] \leq \binom{n}{t} \text{Prob}[\text{cr}(H) \leq 2n].$$

Notice that any fixed set T uniquely determines H_b, but H_r is a uniformly distributed random Hamiltonian cycle on T determined by the random permutation $p_1, \ldots, p_n$. After substituting $t = 150(n \log n)^{2/3}$ and applying Lemma 5 with $K = 2n$ and $s = 101n^{1/3} \log^{4/3} n$, Theorem 2 follows by computation:

$$\text{Prob}[f(P) \geq t] \leq \binom{n}{t} \binom{t}{D}^2 \left(\frac{3t}{s}\right)^D \frac{s^{t-D}}{(t-D)!}$$

Here $D = \lfloor 71\sqrt{t(K+t)/s} \rfloor < t/\log n$ and hence we get

$$\text{Prob}[f(P) \geq t] \leq 2^{-t}.$$

3 Related Problems and Remarks

1. Proposition 1 (with a weaker constant) also follows from the main result in [PW98]: Every planar graph with m vertices admits a crossing-free drawing in the plane such that its vertices are mapped into arbitrarily prespecified points and each of its edges are represented by a polygonal curve with fewer than Cm bends, where C is a constant. (Apply this result to the cycle with vertices $p_{\lceil\sqrt{Cn}\rceil}, p_{\lceil 2\sqrt{Cn}\rceil}, p_{\lceil 3\sqrt{Cn}\rceil}, \ldots$, where each of these points has to be mapped into itself.)

2. It is easy to see that any polygon of n vertices and only one crossing pair of edges can be untangled in $\lceil n/4 \rceil$ moves. Indeed, deleting the two crossing edges, the polygon falls into two disjoint paths. Let $p_1, \ldots, p_m$, $m \leq n/2$ denote the vertices of one of these paths in their natural order. We move p_1 close to the crossing of the two deleted edges and we move p_i close to p_{m+2-i} for $i = 2, \ldots, \lceil m/2 \rceil$. One can do this in a way to obtain a simple polypon. Figure 2 shows an example of a polygon with a single crossing that cannot be untangled with $o(n)$ moves and it seems that one cannot untangle it moving substantially fewer than $n/4$ vertices.

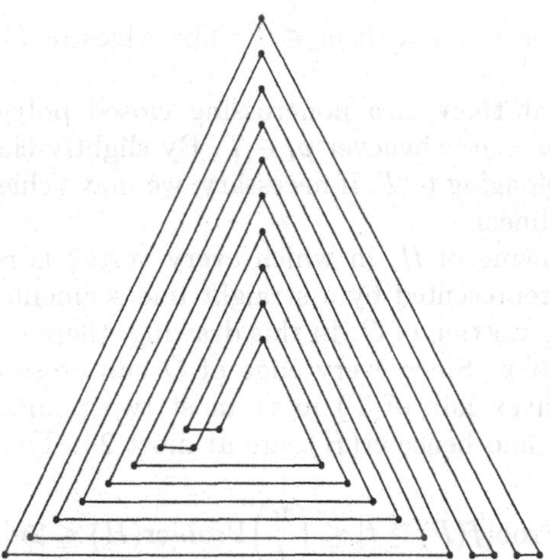

Fig. 2. A polygon with one crossing which cannot be untangled in few moves.

Obviously, if a polygon has c crossings, then it can be untangled without moving the vertices of its longest crossing-free section, whose length is at least $\lceil n/(2c) \rceil$. This bound is naturally far from being optimal.

More generally, we can raise the following

Problem 1. Let P be a polygon of n vertices with the property that every edge of P crosses at most k other edges. Is it true that P can be untangled so that at least $c_k n$ vertices remain fixed, for a suitable constant $c_k > 0$ depending only on k?

The answer to this question is in the affirmative in the special case when every edge e of the polygon is disjoint from all other edges, whose distances from e along P are larger than a constant k.

3. One can ask similar questions for straight-line drawings of *planar graphs* rather than closed polygons. Now we are allowed to relocate any vertex, keeping all of its connections straight. Our goal is to get rid of all crossings, moving as few vertices as possible.

Problem 2. Let P be a (not necessarily crossing-free) straight-line drawing of a planar graph with n vertices. Can P be untangled leaving n^{ε} vertices fixed, for an absolute constant $\varepsilon > 0$?

References

[ES35] P. Erdős and G. Szekeres, A combinatorial problem in geometry, *Compositio Mathematica* **2** (1935), 463–470.

[PSS94] J. Pach, F. Shahrokhi, and M. Szegedy, Applications of the crossing number, *Proc. 10th ACM Symposium on Computational Geometry,* 1994, 198–202. Also in: *Algorithmica* **16** (1996), 111–117.

[PW98] J. Pach and R. Wenger, Embedding planar graphs with fixed vertex locations, in: *Graph Drawing '98 (Sue Whitesides, ed.), Lecture Notes in Computer Science* **1547**, Springer-Verlag, Berlin, 1998, 263–274.

[SV94] O. Sýkora and I. Vrťo, On VLSI layouts of the star graph and related networks, *Integration, The VLSI Journal* **17** (1994), 83-93.

Drawing with Fat Edges*

Christian A. Duncan[1], Alon Efrat[2], Stephen G. Kobourov[2], and Carola Wenk[3]**

[1] Department of Computer Science
University of Miami
Coral Gables, FL 33124
duncan@cs.miami.edu

[2] Department of Computer Science
University of Arizona
Tucson, AZ 85721
{alon,kobourov}@cs.arizona.edu

[3] Institut für Informatik
Freie Universität Berlin
Berlin, Germany
wenk@inf.fu-berlin.de

Abstract. In this paper, we introduce the problem of drawing with "fat" edges. Traditionally, graph drawing algorithms represent vertices as circles and edges as closed curves connecting the vertices. In this paper we consider the problem of drawing graphs with edges of variable thickness. The thickness of an edge is often used as a visualization cue, to indicate importance, or to convey some additional information. We present a model for drawing with fat edges and a corresponding polynomial time algorithm that uses the model. We focus on a restricted class of graphs that occur in VLSI wire routing and show how to extend the algorithm to general planar graphs. We show how to take an arbitrary wire routing and convert it into a homotopic equivalent routing such that the distance between any two wires is maximized. Moreover, the routing uses the minimum length wires. Maximizing the distance between wires is equivalent to finding the drawing in which the edges are drawn as thick as possible. To the best of our knowledge this is the first algorithm that finds the maximal distance between any two wires and allows for wires of variable thickness. The previous best known result for the corresponding decision problem with unit wire thickness is the algorithm of Gao *et al.*, which runs in $O(kn^2 \log(kn))$ time and uses $O(kn^2)$ space, where n is the number of wires and k is the maximum of the input and output complexities. The running time of our algorithm is $O(kn + n^3)$ and the space required is $O(k+n)$. The algorithm generalizes naturally to general planar graphs as well.

* This is a report of ongoing research. The full proofs and new results will be maintained in the full version of the paper, which is available at
`http://www.cs.arizona.edu/~alon/papers/fatedges.ps.gz`

** Supported by Deutsche Forschungsgemeinschaft, grant AL 253/4-3.

P. Mutzel, M. Jünger, and S. Leipert (Eds.): GD 2001, LNCS 2265, pp. 162–177, 2002.
© Springer-Verlag Berlin Heidelberg 2002

1 Introduction

In the area of graph drawing many algorithms have been developed for the classic problem of visualizing graphs in 2D and 3D. If the underlying graph is weighted, the edge weight information is typically displayed as a label near the edge. It seems natural to assign to the edges a width (or thickness) proportional to their weights. If the weights are in a large range then a logarithmic scale may be used. Surprisingly, there does not seem to be any previous work on drawing graphs with edges of varying thickness.

Some related work has been done in addressing a classic VLSI problem, the continuous homotopic routing problem (CHRP) [2,9]. For the CHRP problem, we need to route wires with fixed terminals among fixed obstacles when a sketch of the wires is given, i.e., each wire is given a specified homotopy class. If the wiring sketch is not given or the terminals are not fixed, the problem is NP-hard [10,14,15]. In the CHRP problem we are given a wiring layout and some constant ϵ and we want to find if this wiring can be continuously transformed to a new wiring in which the wire separation is at least ϵ. In this setting the graph is given a fixed planar embedding and the maximum degree is 1. It is easy to see that the CHRP problem can be rephrased as the following graph drawing problem: does there exist a planar drawing in which all the wires can be drawn with thickness ϵ?

In this paper we address a more general optimization problem which we call the Fat Edge Drawing (FED) problem: given a planar weighted graph G with maximum degree 1 and an embedding for G, find a planar drawing such that all the edges are drawn as thick as possible and proportional to the corresponding edge weights. The FED problem is a generalization of the CHRP problem since it allows for edges of different weights and solves the maximization problem, rather than the decision problem. The General Fat Edge Drawing Problem (GFED) is the FED problem without the maximum degree condition. We present an algorithm for the FED problem which easily generalizes to an algorithm for the GFED problem.

1.1 Previous Work

Some of the early work on continuous homotopic routing was done by Cole and Siegel [2] and Leiserson and Maley [9]. They show that in L_∞ norm a solution can be found in $O(k^3 \log n)$ time and $O(k^3)$ space, where n is the number of wires and k is the maximum of the input and output complexities of the wiring. Maley [11] shows how to extend the distance metric to arbitrary polygonal distance functions (including Euclidean distance) and presents a $O(k^4 \log n)$ time and $O(k^4)$ space algorithm. Note that k can be arbitrarily larger than n. In fact, it is easy to construct examples in which k is arbitrarily large. More surprisingly, even after shortest paths have been computed for each wire, k can be as large as $k = \Omega(2^n)$, see Figure 1. The best result so far is due to Gao *et al.* [4] who present a $O(kn^2 \log(kn))$ time and $O(kn^2)$ space algorithm.

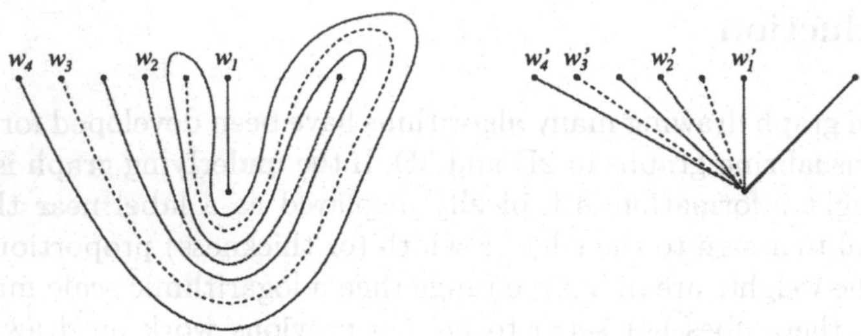

Fig. 1. An example with exponential complexity: $k = \Omega(2^n)$: On the left is the initial wiring sketch, and on the right is the wiring after the shortest paths have been computed. The number of edge segments in the shortest paths w_1', w_2', w_3', w_4' is 1, 2, 4, 8. In general, wire w_i' has 2^{i-1} edge segments. Note that on the right many edge segments are parallel.

Similar work has been done for a special class of grid graphs: finite subgraphs of the planar rectangular grid. The first algorithms for such restricted versions of the problem are presented by Mehlhorn and Kaufmann [6,7], and by Schrijver [16,17]. Work in this area is related to the river routing problem in VLSI chips [2,3,12,14].

A related problem was considered by Pach and Wenger [13]. They address the problem of laying out a planar graph at predefined locations in the plane. Given a planar graph on n vertices and a point set with n points, we want to draw the graph subject to the condition that each vertex v_i is mapped to a point p_i. This can be done with at most $O(n)$ bends per edge using the algorithm of Pach and Wenger [13]. In the wire routing setting, this implies that $k = \theta(n^2)$.

As a part of our algorithm we use a geometric shortest path algorithm. For triangulated polygons, the Euclidean shortest path between two points can be computed in linear time using the algorithms of Chazelle [1] or Lee and Preparata [8]. The latter algorithm is known as the funnel algorithm and it can be extended to river routing [2,4,9]. In our setting, the shortest paths can be found in optimal $O(nk)$ time using the algorithm of Hershberger and Snoeyink [5].

1.2 Our Results

We show how to solve the FED problem in $O(nk+n^3)$ time and $O(n+k)$ space. We describe the algorithm in the tradition of the homotopic wire routing where n is the number of wires and k is the maximum of the initial and final complexities of all the paths. We also show how to extend the FED algorithm to an algorithm for the GFED problem with the same time and space bounds.

Our FED algorithm solves a more general problem than the CHRP problem and is an improvement in both space and time complexity over the best known algorithm for the CHRP problem.

2 Continuous Homotopic Routing

In this paper we use some basic definitions from Gao et al. [4]. Let $W = \{w_1, w_2, \ldots, w_n\}$ be a set of *wires* (also called *paths*). *Paths* are non-intersecting planar continuous curves. Such a collection of wires is called *collisionfree.* Let $T = \{w_i(0), w_i(1) : 1 \leq i \leq n\}$ be the set of *terminals* (endpoints of the wires) of W, $|T| = 2n$. Let $V = T \cup O$ be the set of *vertices,* where O is a set of at most $O(n)$ point *obstacles* disjoint from W.

Let $p, q : [0,1] \longrightarrow \mathbb{R}^2$ be two continuous curves parameterized by arc-length. Then p and q are homotopic with respect to a set $V \subseteq \mathbb{R}^2$ if there exists a continuous function $h : [0,1] \times [0,1] \rightarrow \mathbb{R}^2$ with the following three properties:

1. $h(0,t) = p(t)$ and $h(1,t) = q(t)$, for $0 \leq t \leq 1$
2. $h(\lambda, 0) = p(0) = q(0)$ and $h(\lambda, 1) = p(1) = q(1)$ for $0 \leq \lambda \leq 1$
3. $h(\lambda, t) \notin V$ for $0 \leq \lambda \leq 1, 0 < t < 1$

We call a set of paths $P = \{p_1, p_2, \ldots, p_n\}$ a *homotopic shift* of a set of wires $W = \{w_1, w_2, \ldots, w_n\}$ if the following two conditions are met:

1. p_i is homotopic to w_i with respect to $V, 1 \leq i \leq n$
2. P is collisionfree, i.e., no two paths in P intersect

Let for each path p a weight $\omega_p > 0$ be given. Let the weighted distance between two paths p and q be the minimum weighted distance between all point pairs u and v with $u \in p$ and $v \in q$, where the weighted distance between u and v is the Euclidean distance multiplied by $2/(\omega_p + \omega_q)$. Note that if all weights have unit weight this yields the Euclidean distance. Then the separation, $s(P)$, of a set of paths P is defined to be the minimum weighted distance between any two paths $p_i, p_j \in P$.

In this paper we address the following Fat Edge Drawing (FED) problem: Given a set $W = \{w_1, w_2, \ldots, w_n\}$ of wires with terminal set T and obstacle set O. Furthermore, for each wire w_i let an associated weight $\omega_i > 0$ be given. Find a homotopic shift $P = \{p_1, p_2, \ldots, p_n\}$ of W with maximum separation, i.e., $s(P) \geq s(Q)$ for all homotopic shifts Q of W.

Note that the maximum separation for a set of paths P directly yields the maximum thickness with which the paths in P can be drawn without intersections. Indeed, if the weighted distance between two paths p and q is $2\epsilon/(\omega_p + \omega_q)$, then p can be drawn with thickness $\omega_p\epsilon/(\omega_p + \omega_q)$, q can be drawn with thickness $\omega_q\epsilon/(\omega_p + \omega_q)$, and p and q are disjoint. Computing a homotopic shift for W, T, and O with maximum separation thus corresponds to drawing a planar graph, with vertex degrees at most 1, with edges as thick as possible.

3 Algorithm Overview

We are now ready to outline the general technique for finding the continuous homotopic routing of maximum separation. We begin with the initial set of wires W and compute for each $w \in W$ the shortest path w' homotopic to w, which yields a set of shortest paths W' that is a homotopic shift of W. This can be done in $O(nk)$ time using a result from Hershberger and Snoeyink [5]. The idea of the shortest path computation is to triangulate the region using the $O(n)$ terminals and obstacles, yielding a triangulation of size $O(n)$. A shortest path w' homotopic to a given path w can be computed in time $O(C_w + \Delta_w)$, where C_w is the complexity of w and Δ_w is the number of times w intersects a triangulation edge which is $O(nC_w)$. Since $\sum_{w \in W} C_w = k$, the total time taken to compute all shortest paths becomes $O(k + nk) = O(nk)$. The storage required by this algorithm is $O(n + k)$.

From W' we compute a wire routing with maximum separation by applying the following kinetic approach: We let the wires simultaneously grow in thickness over time, in a speed proportional to the individual weight of each wire. Throughout this growth process, the wires remain as short as possible, and the homotopy between wires is preserved. As a result, the line segments of the original polygonal paths are deformed into two types of curves. We borrow the names for these curves from plumbing jargon: *straights* and *elbows*. *Straights* are rectangular regions and *elbows* are formed by the arc of an annulus with two given radii, Figure 2. We examine elbows and straights in depth in Section 4.

We introduce a compact routing data structure which allows us to represent the thick wires, i.e., the straights and elbows, and to maintain the growth process efficiently. From the beginning to the end of the growth process, the algorithm that maintains the data structure requires $O(n^3)$ time and $O(n^2 + k)$ space. In order to achieve this time bound independent of the complexity of the wires k (which can be as large as $\Omega(2^n)$), we group wires together. Note that in W' many of the wires may travel in parallel, see Figure 1 for an example. We take advantage of this and group such parallel wires into *bundles* and maintain *straight bundles* and *elbow bundles* instead of single straights and elbows in our compact routing structure.

In Theorem 2 we prove that the number of bundles stored in this data structure is only $O(n)$ and in Theorem 3 we prove that the space required is $O(n+k)$. After the growth process stops we reconstruct the set of maximally separated wires which consist of straight line segments and circular arcs.

4 Compact Routing Structure

4.1 Straights, Elbows, and Bundles

Definition 1. The wires in the routing are broken into connected sequences of fat edge segments called *elbows* and *straights*, as described below (see Figure 2). The thickness of each such segment f is determined by $\omega_f t$, where ω_f is the weight of the segment that corresponds to the weight of its initial wire, and $t > 0$ represents the current time frame. Let $V := T \cup O$ be the set of vertices.

- An **elbow** segment, e, associated with a vertex v and two straight segments from v to u and from v to w; $u, v, w \in V$; is a connected region of the plane, formed by a piece of an annulus centered at v and having thickness

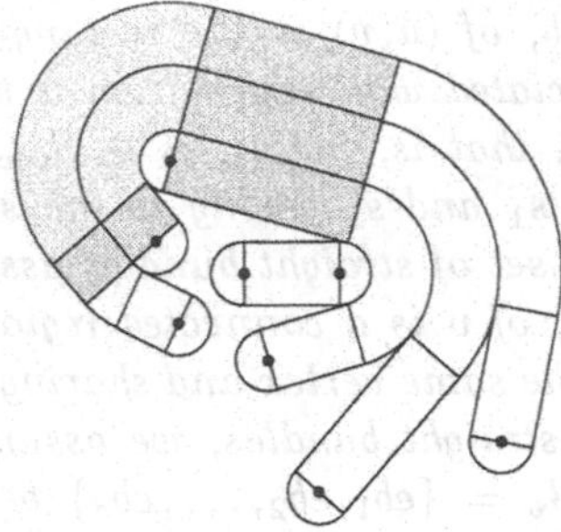

Fig. 2. An example of wires composed of straights and elbows. The shaded regions are bundles.

$w_f t$. More formally $e = \{p \in \mathbb{R}^2 : r_1 \leq ||p - v|| \leq r_2$ with $r_2 - r_1 = w_f t$ and the orientation of the segment $\overline{pv} \in [\theta_1, \theta_2]\}$. The elbow e maintains the property that for any point $p \in e$ the line segment $\overline{pv}$ intersects only other elbows associated with vertex v. Note that the half disk and the disk centered around v are also elbows. For notation, let us call these regions **terminal elbows**. Let $E_v = \{e_1, e_2, \ldots e_d\}$ be the set of elbows associated with v.

- A **straight** segment, s, associated with a segment f from $u \in V$ to $v \in V$ is a rectangular region (fully) connecting two elbows, $e_v \in E_v$ and $e_u \in E_u$. The thickness of the segment is equal to $w_f t$. Let $S_{uv} = \{s_1, s_2, \ldots s_d\}$ be the set of straight segments associated with edges between u and v.

Note that every non-terminal elbow has two associated straight segments. Every terminal elbow has at most one straight segment, the one emanating from the corresponding vertex. Throughout the algorithm the following disjointness property is maintained:

Property 1. The only way two segments can intersect is along their boundary while the interiors are disjoint. The only possible intersections are:

- Two straight segments intersect only if they are associated with the same two vertices (u, v).
- Two elbow segments intersect only if they have the same associated vertex.
- A straight segment s intersects an elbow segment e if and only if e is one of s's two connecting segments.

The total number of elbow and straight segments depends on k which can be very large, see Figure 1. In order to reduce this complexity, straights and elbows are bundled and only the bundles are manipulated.

Definition 2. *Straight bundles are made of straight edge segments and elbow bundles are made of elbow segments as described below:*

- *A* **straight bundle**, *sb, of (u, v) is the rectangular region formed by the union of straights associated with edges from u to v. We assume that the bundles are maximized, that is, for $s_1, s_2 \in S_{u,v}$, the straights s_1 and s_2 intersect if and only if s_1 and s_2 belong to the same bundle. Let $SB_{uv} = \{sb_1, sb_2, \ldots sb_d\}$ be the set of straight bundles associated with u and v.*
- *An* **elbow bundle**, *eb, of v is a connected region formed by the union of elbows associated with the same vertex and sharing the same straight bundles on both ends. As with straight bundles, we assume that the elbow bundles are maximized. Let $EB_v = \{eb_1, eb_2, \ldots, eb_d\}$ be the set of elbow bundles associated with v. A terminal elbow has at most one straight segment and belongs to its own elbow bundle.*

For any vertex v there may be many elbow bundles. It is not hard to show that for any pair of vertices (u, v) there may be at most 3 straight bundles between them — one below both vertices, one above both vertices, and a "diagonal" bundle, that is below one vertex and above the other.

Lemma 1. *If the number of straight bundles in our compact routing structure is m then there are at most $O(m)$ elbow bundles.*

Proof Sketch: This proof relies on the fact that the bundles are "ordered" on both sides of a group of elbows. As one elbow bundle ends, one straight bundle must also end on one of the two sides of the elbow groups. □

4.2 Compact Routing Structure $\mathcal{S}$

Given the two types of segments and bundles we show how they can be stored and updated in a manageable data structure.

Definition 3. *The compact routing structure $\mathcal{S}$ represents a fat-wire routing as follows:*

- *Terminals and obstacles are stored with pointers to their terminal elbows*
- *Elbow and straight bundles are stored along with their weights*
- *Elbow and straight bundles store the ordered list of wires they represent*
- *A straight bundle sb stores:*
  - *two linked lists of left and right adjacent elbow bundles*
  - *the number of straight segments it represents*
  - *its weight (the sum of the weights of each of its straight segments)*
- *An elbow bundle eb stores:*
  - *the set of elbow bundles adjacent to it (both above and below)*
  - *the two adjacent straight bundles*
  - *the number of elbow segments that it represents*
  - *its weight (the sum of the weights of each of its elbow segments)*
  - *its* layered *weight (the sum of the weights of each of the elbow segments between the elbow bundle and the vertex center, not counting its own weight)*

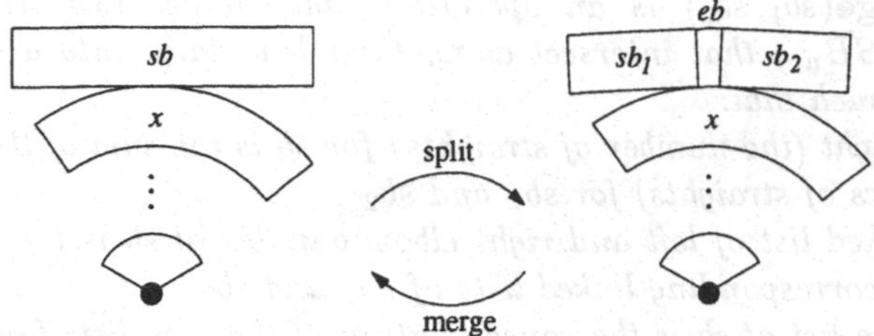

Fig. 3. Split and merge operations. The split operation splits the straight bundle at an elbow bundle into a straight-elbow-straight bundle sequence. The merge operation merges a sequence of straight-elbow-straight bundles into a single straight bundle.

After the homotopic shortest paths have been computed, we need to find out which bundles participate in the compact routing structure and what is the order of the wires inside each bundle. The next lemma describes this process.

Lemma 2. *Given a set of shortest paths W' the compact routing structure $\mathcal{S}$ can be initialized in $O(n^2 + k)$ time.*

Due to lack of space the proof is omitted from this extended abstract.

4.3 Maintaining $\mathcal{S}$

Lemma 3. *We can maintain the following operations in constant time:*

- `report`(*sb* **or** *eb*) *is an operation that returns a description of the bundle at the current time frame*
- `split`(sb, x) *(see Figure 3) is an operation that splits a straight bundle sb into three connected bundles sb_1, eb, and sb_2 such that:*
  - *x is an elbow bundle that initially intersects sb at the boundary*
  - *sb_1 is adjacent to all elbow bundles on the left portion of sb*
  - *sb_2 is adjacent to all elbow bundles on the right portion of sb*
  - *eb is adjacent to sb_1 and sb_2 and is associated with the same vertex as x*
  - *The wire lists, weights, and number of segments for the three new bundles are the same as the wire list, weight, and number of segments for sb*
- `merge`(sb_1, eb, sb_2) *(see Figure 3) is an operation that merges two straight bundles and an elbow bundle into one straight bundle sb such that:*
  - *eb is adjacent to sb_1 and sb_2*
  - *eb has arc length 0, implying that sb_1 intersects sb_2*
  - *sb represents the region $sb_1 \cup sb_2$ and is adjacent to the left elbow bundles of sb_1 and the right elbow bundles of sb_2*
  - *The wire lists, weights and number of segments for all the bundles are the same*

- **bundlemerge**(sb_1, sb_2) *is an operation that merges two straight bundles* $sb_1, sb_2 \in SB_{uv}$*, that intersect along their boundary, into a single straight bundle* sb *such that:*
  - *the weight (the number of straights) for* sb *is the sum of the weights (the numbers of straights) for* sb_1 *and* sb_2
  - *the linked list of left and right elbow bundles of* sb *is the concatenation of the corresponding linked lists of* sb_1 *and* sb_2
  - *the wire list of* sb *is the concatenation of the wire lists for* sb_1 *and* sb_2

  If the left (resp. right) elbow bundles of sb *are adjacent to the same straight bundle to their left (resp. right), then those elbow bundles get merged into a single elbow bundle* eb*. The wire list, weight, layered weight, and number of elbows* eb *represents follow from the corresponding entries in the elbow bundles being merged.*
- **bundlesplit**(sb, eb) *is an operation that splits a straight bundle* $sb \in SB_{uv}$ *at the adjacent elbow bundle* eb *into two straight bundles* $sb_1, sb_2 \in SB_{uv}$*, that intersect along their boundary, such that:*
  - eb *is the first elbow bundle in the left or right adjacency list of* sb_2
  - *the linked lists of left and right elbow bundles of* sb *are the concatenation of the corresponding linked lists of* sb_1 *and* sb_2
  - *the wire list of* sb *is the concatenation of the wire lists for* sb_1 *and* sb_2
  - *the weight (the number of straights) for* sb_2 *is the sum of the weights (the number) of the adjacent elbow bundles*
  - *the weight (the number of straights) for* sb *is the sum of the weights (the numbers of straights) for* sb_1 *and* sb_2

  If sb *is adjacent (on the side not adjacent to* eb*) to exactly one elbow bundle, then this elbow bundle also gets split into two elbow bundles. The weights, layered weights, and numbers of elbows those two elbow bundles represent follow from the entries in* sb_1 *and* sb_2*.*

Lemma 4. *Given* $\mathcal{S}$ *we can uncompress the bundles of* $\mathcal{S}$ *into straight segments and elbow segments in time* $O(k)$ *where* k *is the final complexity of the paths.*

Proof Sketch: The idea is to go through the structure using the wire identities in the wire list and greedily unzip the bundles segment by segment starting at any terminal and progressing along its path. The paths obtained in this way consist of straight line segments and circular arcs. □

4.4 Bounding the Number of Bundles in $\mathcal{S}$

In this section we argue that our compact routing structure $\mathcal{S}$ contains $O(n)$ bundles in total. We do this by showing that $\mathcal{S}$ is (nearly) a planar graph implying the stated size.

Definition 4. *A* **planar fat embedding** *of a graph* $G = (V, E)$ *is an embedding of* G *with the following properties:*

- *Every vertex $v \in V$ is represented by a simply connected closed region P_v.*
- *Every edge $f = (u, v) \in E$ is represented by a simply connected closed region $P_f = P_{uv}$.*
- *For any pair of vertices, $u, v \in V$, the two associated regions are disjoint, $P_u \cap P_v = \emptyset$.*
- *For any pair of non-incident edges, $f, g \in E$, the two associated regions are disjoint, $P_f \cap P_g = \emptyset$.*
- *For any pair of incident edges, $f, g \in E$, the interiors of the two associated regions are disjoint, while their boundaries might intersect: $P_f \cap P_g \subseteq \delta P_f$.*
- *For any edge $f \in E$ and vertex $v \in V$, $P_f \cap P_v \neq \emptyset$ if and only if v is incident to f, i.e. f is an edge between v and some other vertex.*

In other words, all vertices and edges are represented by simply connected regions. All such regions are disjoint except for between edges and vertices that share endpoints.

We now show that a planar fat embedding is indeed a planar embedding.

Theorem 1. *A graph $G = (V, E)$ is planar if and only if it can be represented by a planar fat embedding.*

Proof Sketch: One direction is straightforward: if G is planar then there exists a regular planar embedding, which is a planar fat embedding.

To show the other direction, we need to show how to map a planar fat embedding $\mathcal{E}$ into a regular planar embedding $\mathcal{E}'$. We do this by showing how to embed the graph using points for vertices and paths for edges. First, for any vertex $v \in V$, we place the vertex at any point $p_v \in P_v$. Next for any edge $f = (u, v) \in E$, we route the edge along the shortest path from p_u to p_v that lies completely within $P_u \cup P_f \cup P_v$. Notice since P_f must intersect both P_u and P_v the region is connected and a path exists.

The new embedding is planar since no two vertices can share the same point, and if two edge paths intersect the two edges must share a vertex endpoint in common. In fact, since the paths used are the shortest possible, if two edges intersect they touch and remain touching until they reach the shared endpoint. This implies that the embedding has no crossings and so G must be planar. □

To show that the number of bundles in our compact routing structure is $O(n)$, we describe how it represents a planar fat embedding which implies that it also represents a planar graph that has $O(n)$ size, where n is the number of vertices.

Theorem 2. *Let $\mathcal{S}$ be a compact routing structure for a set of wires $W = \{w_1, w_2, \ldots, w_n\}$ with $O(n)$ terminal and obstacle vertices. The number of bundles stored in $\mathcal{S}$ is $O(n)$. The total storage required for $\mathcal{S}$ is $O(k)$.*

Proof Sketch: We show how the representation in $\mathcal{S}$ can be mapped to a planar fat embedding of a graph $G = (V, E)$. Let V be the set of terminals and obstacles. For each vertex $v \in V$ we let P_v be the simply connected region formed by the

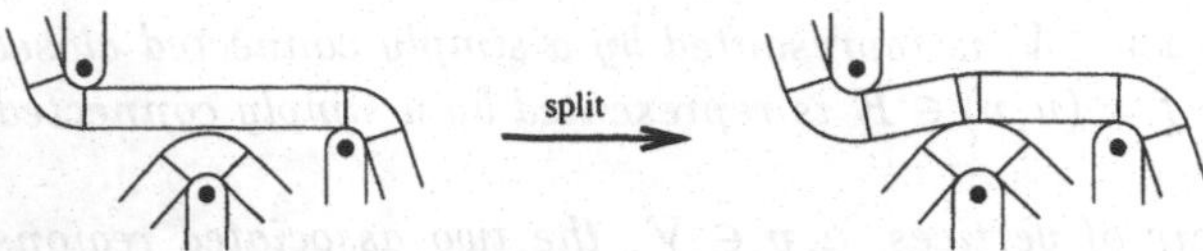

Fig. 4. A split event. The elbow bundle pushes the straight bundle up such that it gets replaced by a straight-elbow-straight bundle sequence.

regions in EB_v (the elbow bundles of v). For each pair of vertices $u, v \in V$ let us consider the set of straight bundles SB_{uv}. We know that $|SB_{uv}| \leq 3$. If $|SB_{uv}| > 0$ we define $f := (u, v) \in E$ and let $P_{uv} = P_f$ be one of the three bundle regions, namely sb_1.

These regions form a planar fat embedding for $G = (V, E)$. It follows from Theorem 1 that G is planar. Therefore, $|E|$ is $O(|V|)$ and the total number of bundles stored in $\mathcal{S}$ is $O(n)$.

Since besides the lists of wires $\mathcal{S}$ stores only a constant amount of information for each bundle, the total storage space for $\mathcal{S}$ is $O(k + n) = O(k)$. □

5 Algorithm

The algorithm uses the following steps:

1. Compute the set of shortest paths W' from the given wire set W
2. Initialize the compact routing data structure $\mathcal{S}$ with W'
3. Thicken the wires in W' (maintaining $\mathcal{S}$) until maximum possible thickness
4. Extract paths from the bundles in $\mathcal{S}$

We argued in Section 3 that the computation of the shortest paths can be done in time $O(nk)$ with $O(n+k)$ space complexity. In Section 4 we argued that the initialization of the compact routing data structure, i.e., the bundling of the edges, can be done in $O(k+n^2)$ time. The final extraction of the paths from the elbow and straight bundles in the compact routing structure for the maximum possible thickness can be done in time $O(k)$, see Section 4. In the remainder of this section we concentrate on step 3, the wire thickening process, and show that it can be done in $O(n^3)$ time and $O(k + n)$ space.

Let $t \geq 0$ be a general thickness parameter which we also refer to as *time*. At time frame t we assign to each wire $w' \in W'$ with weight ω' the thickness $\omega' t$. Starting at $t = 0$ with all wires in W' having thickness 0, we let t monotonically grow, such that the thicknesses of the wires also grow monotonically, and we maintain the invariant that the wires are as short as possible. As the wires grow three types of events can happen: *Split events*, *merge events*, and *stop events*. In a split event (see Figure 4) an elbow bundle touches and then bends a straight bundle, such that this straight bundle gets replaced by a straight-elbow-straight bundle sequence. In a merge event (see Figure 5) an elbow bundle

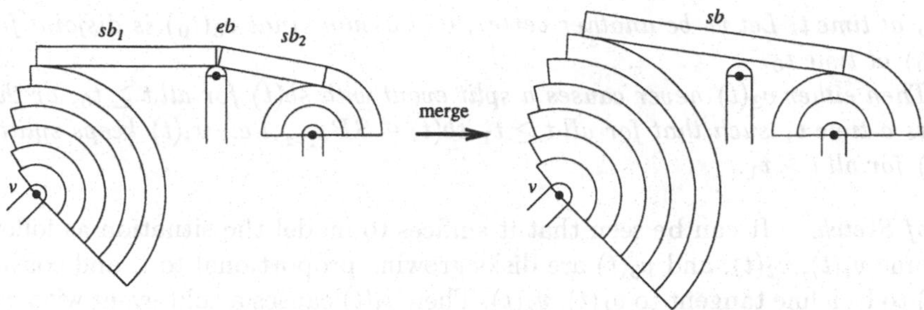

Fig. 5. A merge event. Vertex v pushes the left straight bundle sb_1 up such that the elbow bundle eb in the middle disappears. The straight-elbow-straight bundle sequence $sb_1 - eb - sb_2$ gets replaced by a single straight bundle sb.

straightens, such that the corresponding straight-elbow-straight bundle sequence gets replaced by a single straight bundle. In a stop event two elbow bundles touch each other, which means that at this time the growth process stops, because two elbow bundles cannot bend or push each other away anymore.

We construct a priority queue of events, with the key being the time at which the events occur. For increasing time we update the compact routing data structure and the event queue successively for each event. We obtain the events we store in the queue by considering for each bundle the next event it will cause independent of other bundles:

- For each straight bundle we store the time at which it hits the next elbow bundle (not taking any other straight bundles into account).
- For each elbow bundle we store the time at which it hits the next straight bundle or elbow bundle (not taking any other bundles into account).
- For each elbow bundle we store the time when it gets straightened (only taking the two incident straight bundles into account).

The first item corresponds to a split event, the second to a split or a stop event, and the third to a merge event. Note that the time at which an elbow bundle gets straightened as well as the time at which two bundles hit can be computed in constant time. Thus, for a fixed bundle the bundle it will hit next can be found in $O(n)$ time. We initialize the event queue by inserting the next event for each bundle, which takes $O(n + \log n)$ time per bundle, thus $O(n^2)$ in total.

Lemma 5. *Merge or split events are done in $O(n)$ time.*

Due to lack of space the proof is omitted from this extended abstract.

Lemma 6. *Let $sb(t) \in SB_{v_1 v_2}$ be a straight bundle defined by the vertices $v_1(t)$ and $v_2(t)$, where t is some time frame and $v(t)$ denotes the union of elbow bundles*

EB_v at time t. Let v_3 be another vertex, and assume that $v_3(t_0)$ is disjoint from $sb(t_0)$ at time t_0.

Then either $v_3(t)$ never causes a split event with $sb(t)$ for all $t \geq t_0$, or there exists a time t_1 such that for all $t \geq t_1$ $sb(t) \notin SB_{v_1 v_2}$, i.e., $v_3(t)$ keeps splitting $sb(t)$ for all $t \geq t_1$.

Proof Sketch: It can be seen that it suffices to model the situation as follows: Assume $v_1(t)$, $v_2(t)$, and $v_3(t)$ are disks growing proportional to t, and consider $sb(t)$ to be a line tangent to $v_1(t)$, $v_2(t)$. Then $v_3(t)$ causes a split event with $sb(t)$ iff $v_3(t)$ is tangent to $sb(t)$. In order to find the time t and the corresponding line $sb(t)$ at which this event happens, we need to solve a system of three equations in three variables. Assuming the three vertices are in general position this system has a unique solution. Details are omitted from this extended abstract. □

Although there are examples with $\Omega(n)$ vertices that are each involved in $\Omega(n)$ events with a single wire, we can show that the total number of events that occur during the growth process is bounded by $O(n^2)$.

Lemma 7. *The total number of events that the structure goes through is $O(n^2)$.*

Proof. Consider the case of a split event between an elbow bundle around a vertex v and a straight bundle sb defined by two vertices u and w. WLOG let sb lie above both u and w. Once sb gets split there will never again be a straight bundle between u and w touching both u and v from above, because the bundles grow thicker monotonically in time. This follows from Lemma 6. Since there are only $O(n^2)$ possible straight bundles this proves the claim for split events.

For merge events the argument is similar: Let the two straight bundles about to merge be sb_1 and sb_2. Let sb_1 and sb_2 be defined by (u, v) and (v, w), respectively. Let u be the vertex with many elbows around it that pushes the straight bundle sb_1 between u and v up. Then, either sb_1 or sb_2 can never occur again. Indeed, once u has lifted sb_1 up above u the only way to make sb_2 touch v again is for a vertex to push sb_1 down, but this again destroys sb_1 as a straight bundle forever. Similarly sb_2 could also be destroyed in order to make sb_1 appear again. In any case, either sb_1 or sb_2 will never appear again. And since there are only $O(n^2)$ possible straight bundles this proves the claim for merge events.

Since there are $O(n)$ different elbow bundles the number of stop events is clearly $O(n^2)$. Since the processing of each split or merge event introduces only a constant number of new events, the number of invalid events in the event queue, i.e., events that refer to non-existing straight or elbow bundles, is also $O(n^2)$. Thus the total number of events in the queue is $O(n^2)$. □

The growth process stops at the first stop event. From Lemma 7 we know that this will happen after at most $O(n^2)$ events. Each event can be processed and the data structure maintained in $O(n)$ time according to Lemma 5, which yields a total runtime of $O(n^3)$. We need $O(k)$ space for the compact routing data structure and $O(n^2)$ space to store the events, for a total of $O(n^2 + k)$. This directly yields our main theorem:

Theorem 3. *The continuous homotopic wire routing with maximum separation can be computed in $O(n^3 + nk)$ time and $O(n + k)$ space.*

Proof. From Lemma 5 and Lemma 7 follows the result with $O(n^2 + k)$ space complexity. We can reduce the space needed for the algorithm to $O(n+k)$ without increasing the asymptotic running time. This is done as follows: After computing the compact routing structure, the only place where superlinear space is needed is in the priority queue, used for discovering future events. Instead of keeping all future events, we divide the execution of the algorithm into $O(n)$ *phases*, where in each phase $O(n)$ events happen. The priority queue then contains only the next $O(n)$ events, so only $O(n)$ space is needed. After each phase, we find the next $O(n)$ events by checking each bundle against each other bundle (which is done in $O(n^2)$ time), so the total time spent for finding future events remains $O(n^3)$, as before. □

6 General Planar Graphs

We presented a $O(nk+n^3)$ time algorithm for the FED problem. Since the FED problem restricts us to graphs with maximum degree 1, we would like to extend it to general graphs which yields the GFED problem. It is easy to extend our algorithm to an algorithm for general planar graphs. Recall the GFED problem: given a weighted planar graph (not necessarily of degree 1) and an embedding for it, find a planar drawing with the edges drawn as thick as possible with thicknesses proportional to the edge weights. We can modify our algorithm as follows: Let each vertex grow at a rate proportional to its degree. In this setting our modified algorithm will find the optimum solution. However, the solution may not be optimal in the sense that some vertices may occupy more space than they need, thus causing the algorithm to terminate earlier, see Figure 6.

This problem can be addressed by allowing vertices to have a variable rate of growth as follows. Each vertex is the smallest circle such that its adjacent edges do not overlap outside that circle. Note that the largest diameter circle needed for a vertex of degree i is i. The problem with this approach is that the angles of the adjacent edges change dynamically throughout the algorithm and hence would require updates at every event for every elbow.

7 Open Problems

Some of the open problems related to the FED and GFED problems include:

- Is $k = \Omega(2^n)$ the worst case complexity for the FED problem or can it be worse than that?
- Can the FED algorithm be extended to non-point obstacles?
- What is a "good" model for the GFED problem?
- Can the FED algorithm be modified to an algorithm for the GFED problem with vertices growing at varying rates?

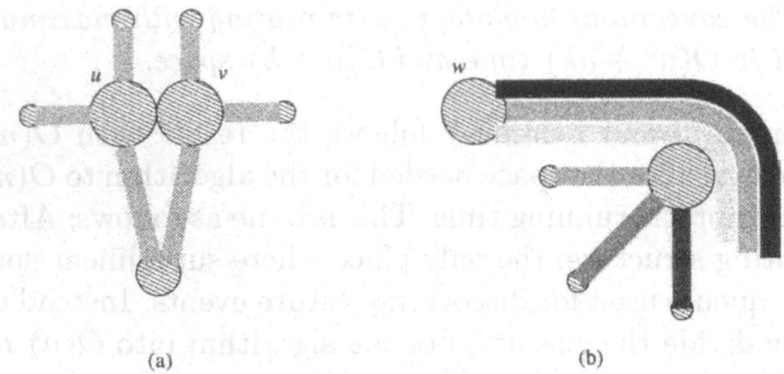

Fig. 6. The FED algorithm applied to general planar graphs. In both graphs shown the max degree is 3 and all vertices have uniform weights. Vertices of different degree have different sizes, more precisely, a vertex of degree i has diameter i. (a) The algorithm stopped because vertices u and v touched. However, vertices u and v need not be that large since their edges "fan out." (b) In this case vertex w has to be large since its edges do not fan out.

Acknowledgements. We would like to thank Jack Snoeyink for introducing us to the homotopic routing problem. We thank Cesim Erten and Jack Snoeyink for many fruitful discussions.

References

1. B. Chazelle. A theorem on polygon cutting with applications. In *23th Annual Symposium on Foundations of Computer Science*, pages 339–349, Los Alamitos, Ca., USA, Nov. 1982. IEEE Computer Society Press.
2. R. Cole and A. Siegel. River routing every which way, but loose. In *25th Annual Symposium on Foundations of Computer Science*, pages 65–73, Los Angeles, Ca., USA, Oct. 1984. IEEE Computer Society Press.
3. D. Dolev, K. Karplus, A. Siegel, A. Strong, and J. D. Ullman. Optimal wiring between rectangles. In *Conference Proceedings of the Thirteenth Annual ACM Symposium on Theory of Computation*, pages 312–317, Milwaukee, Wisconsin, 11–13 May 1981.
4. S. Gao, M. Jerrum, M. Kaufmann, K. Mehlhorn, W. Rülling, and C. Storb. On continuous homotopic one layer routing. In *Proceedings of the Fourth Annual Symposium on Computational Geometry (Urbana-Champaign, IL, June 6–8, 1988)*, pages 392–402, New York, 1988. ACM, ACM Press.
5. Hershberger and Snoeyink. Computing minimum length paths of a given homotopy class. *CGTA: Computational Geometry: Theory and Applications*, 4, 1994.
6. Kaufmann and Mehlhorn. On local routing of two-terminal nets. *JCTB: Journal of Combinatorial Theory, Series B*, 55, 1992.
7. M. Kaufmann and K. Mehlhorn. Routing through a generalized switchbox. *Journal of Algorithms*, 7(4):510–531, Dec. 1986.
8. D. T. Lee and F. P. Preparata. Euclidean Shortest Paths in the Presence of Rectilinear Barriers. *Networks*, 14(3):393–410, 1984.
9. C. E. Leiserson and F. M. Maley. Algorithms for routing and testing routability of planar VLSI layouts. In *Proceedings of the Seventeenth Annual ACM Symposium on Theory of Computing*, pages 69–78, Providence, Rhode Island, 6–8 May 1985.
10. C. E. Leiserson and R. Y. Pinter. Optimal placement for river routing. *SIAM Journal on Computing*, 12(3):447–462, Aug. 1983.

11. F. M. Maley. *Single-Layer Wire Routing.* PhD thesis, Massachusetts Institute of Technology, 1987.
12. A. Mirzaian. River routing in VLSI. *Journal of Computer and System Sciences*, 34(1):43–54, Feb. 1987.
13. J. Pach and R. Wenger. Embedding planar graphs at fixed vertex locations. In *Proc. 6th Int. Symp. Graph Drawing (GD '98)*, pages 263–274, 1998.
14. R. Pinter. River-routing: Methodology and analysis, 1983.
15. D. Richards. Complexity of single layer routing. *IEEE Transactions on Computers*, 33:286–288, 1984.
16. Schrijver. Disjoint homotopic paths and trees in a planar graph. *Discrete & Computational Geometry*, 6, 1991.
17. A. Schrijver. Edge-disjoint homotopic paths in straight-line planar graphs. *SIAM Journal on Discrete Mathematics*, 4(1):130–138, Feb. 1991.

Detecting Symmetries by Branch & Cut*

Christoph Buchheim and Michael Jünger

Universität zu Köln, Institut für Informatik,
Pohligstraße 1, 50969 Köln, Germany
{buchheim,mjuenger}@informatik.uni-koeln.de

Abstract. We present a new approach for detecting automorphisms and symmetries of an arbitrary graph based on branch & cut. We derive an IP-model for this problem and have a first look on cutting planes and primal heuristics. The algorithm was implemented within the ABACUS-framework; its experimental runtimes are promising.

1 Introduction

The display of symmetries is one of the most desirable properties of a graph drawing [7]. Each recognizable symmetry reduces the complexity of the drawing for the human viewer, see Fig. 1. Furthermore, the existence of a symmetry can be an important structural property of the data being displayed.

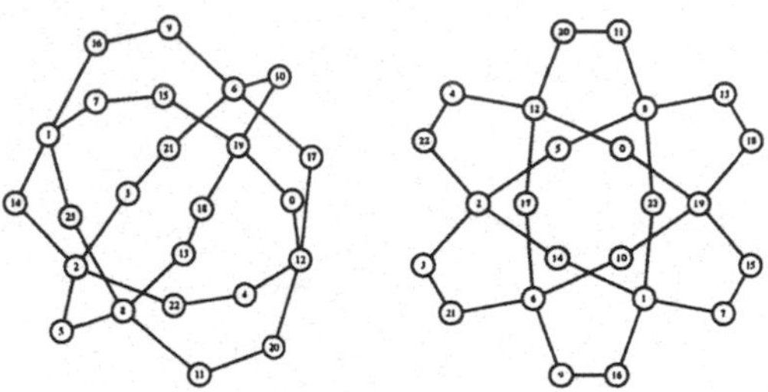

Fig. 1. Two drawings of the same abstract graph, both computed by a spring embedder. On the left, the initial drawing was chosen randomly; on the right, the displayed symmetries have been computed explicitly before

Since detecting symmetries of an arbitrary graph is NP-hard [6], most algorithms are either restricted to special classes of graphs [4] or have an implicit tendency to display symmetries while failing to find them in general [2,3]. Explicit algorithms consist of two steps: First, compute an abstract symmetry

* This work was partially supported by the IST Programme of the EU under contract number IST-1999-14186 (ALCOM-FT).

P. Mutzel, M. Jünger, and S. Leipert (Eds.): GD 2001, LNCS 2265, pp. 178–188, 2002.
© Springer-Verlag Berlin Heidelberg 2002

of the graph, i.e., an automorphism that can be displayed by some drawing of the graph in the plane; second, compute such a drawing with regard to the usual aesthetical requirements like a small number of edge crossings or edge bends. In this article, we focus on the first step, presenting an algorithm that can be applied to every graph and surely finds its best symmetry, i.e., a symmetry of maximum order. The algorithm uses branch & cut to solve an integer programming (IP) formulation of the symmetry detection problem. By the NP-hardness of the problem, one cannot expect polynomial runtime, but the experimental runtimes are promising.

In the next section, we list the required definitions and facts concerning automorphisms and symmetries. In Sect. 3, we derive the IP-model for symmetry detection in arbitrary graphs. In Sect. 4, we explain the technique of node labeling that improves our algorithm at many points. The components of the branch & cut-approach are presented in Sect. 5. After discussing experimental runtimes in Sect. 6 and possible extensions in Sect. 7, we summarize in Sect. 8 what has been achieved and what can still be done.

2 Preliminaries

Throughout this article, a *graph* is a simple undirected graph $G = (V, E)$ with $n = |V|$, see Sect. 7 for other types of graphs.

An *automorphism* of G is a permutation π of V with $(i, j) \in E$ if and only if $(\pi(i), \pi(j)) \in E$ for all $i, j \in V$. The set of automorphisms of G forms a group with respect to composition, denoted by $\text{Aut}(G)$. The *order* of $\pi \in \text{Aut}(G)$ is $\text{ord}(\pi) = \min\{k \in \mathbf{N} \mid \pi^k = \text{id}_V\}$, where id_V denotes the identity permutation of V. For a node $i \in V$, the set $\text{orb}_\pi(i) = \{\pi^k(i) \mid k \in \mathbf{N}\}$ is the *π-orbit* of i. Finally, the *fixed* nodes are those in $\text{Fix}(\pi) = \{i \in V \mid \pi(i) = i\}$.

A *reflection* of G is an automorphism $\pi \in \text{Aut}(G)$ with $\pi^2 = \text{id}_V$, i.e., an automorphism of order 1 or 2. For $k \in \{1, \ldots, n\}$, a *k-rotation* of G is an automorphism $\pi \in \text{Aut}(G)$ such that $|\text{orb}_\pi(i)| \in \{1, k\}$ for all $i \in V$ and $|\text{Fix}(\pi)| \leq 1$ if $k \neq 1$. Observe that each 2-rotation is a reflection, but not vice versa, and that the identity id_V is both a reflection and a 1-rotation.

If there exists a drawing of G in the plane and an isometry of the plane that maps nodes to nodes and edges to edges, this drawing induces an automorphism of G. Any automorphism induced like this is called a *geometric automorphism* or a *symmetry* of G. We have the following characterization, for a proof see [2]:

Lemma 1. *An automorphism of a graph G is a symmetry if and only if it is a rotation or a reflection.*

Obviously, the symmetries displayed by a single drawing of G form a subgroup of $\text{Aut}(G)$, but the set of all symmetries of G is not closed under composition. The following lemma characterizes the sets of symmetries that can be displayed together, for a proof see [2] again:

Lemma 2. *A set of automorphisms can be displayed by a single drawing of G if and only if it generates a subgroup of Aut(G) that is generated by a single symmetry or by a rotation π_1 and a reflection π_2 satisfying $\pi_2\pi_1 = \pi_1^{-1}\pi_2$.*

3 The IP-Model

In this section, we derive an integer linear program (ILP) modeling the set of symmetries of the given graph $G = (V, E)$. First observe that permutations of V can be described by the following program:

$$\begin{array}{rll} x_{ij} \in \{0,1\} & \text{for all } i,j \in V & \\ \sum_{j \in V} x_{ij} = 1 & \text{for all } i \in V & (1) \\ \sum_{i \in V} x_{ij} = 1 & \text{for all } j \in V\,. & \end{array}$$

A value of 1 for a variable x_{ij} is interpreted as mapping node i to node j. The equations of the first type make sure that each node i is mapped to exactly one node j, so that the variables induce a function $V \to V$. By the equations of the second type, this function is bijective.

Observe that by relaxing the integrality constraints in (1) we get the assignment polytope. All vertices of this polytope have only integer components. Unfortunately, this does not remain true if the model is restricted to automorphisms. It is easy to see that an automorphism of G is a permutation of V such that the corresponding matrix $X = (x_{ij})$ commutes with the adjacency matrix A_G of G. Hence automorphisms are singled out by adding the n^2 equations $A_G X = X A_G$ to (1).

Next, we model the characterization of Lemma 1 to restrict the search to symmetries. For this, we introduce new variables $l_k \in \{0,1\}$ for $k \in \{1,\dots,n\}$, where a value of 1 for l_k means an order of k for the symmetry induced by the variables x_{ij}. Besides requiring $\sum_{k=1}^{n} l_k = 1$, we first have to ensure that all orbits have length one or k if $l_k = 1$. For this, let $C = (i_1, \dots, i_p)$ be a cycle in V with $p \geq 2$, and define $x(C) = \sum_{r=1}^{p} x_{i_r i_{r+1}}$ with $i_{p+1} = i_1$. Then

$$x(C) \leq p - 1 + l_p \qquad (2)$$

is valid for any symmetry, since we always have $x(C) \leq p$, and $x(C) = p$ implies that C is an orbit of length $p \geq 2$, hence $l_p = 1$. The set of all inequalities of type (2) guarantees that all non-trivial orbits have the size given by the variables l_k. Since there are exponentially many cycles of V, we have to add these inequalities in the separation phase, see Sect. 5.3.

Finally, we have to deal with the fixed nodes. If $l_k = 1$ for $k \geq 3$, the number $\sum_{i \in V} x_{ii}$ of fixed nodes is at most one by definition of a rotation; if $l_2 = 1$, it is at most $n - 2$. Summing up, the following ILP characterizes symmetries:

$$\begin{array}{rll} x_{ij} \in \{0,1\} & \text{for all } i,j \in V & \\ l_k \in \{0,1\} & \text{for all } k \in \{1,\dots,n\} & \\ \sum_{j \in V} x_{ij} = 1 & \text{for all } i \in V & \\ \sum_{i \in V} x_{ij} = 1 & \text{for all } j \in V & \\ A_G X = X A_G & & (3) \\ \sum_{k=1}^{n} l_k = 1 & & \\ x(C) \leq |C| - 1 + l_{|C|} & \text{for all cycles } C \text{ of } V \text{ with } |C| \geq 2 & \\ \sum_{i \in V} x_{ii} \leq (n-1) l_1 + (n-3) l_2 + 1\,. & & \end{array}$$

It remains to specify an objective function that selects a "best" symmetry. Our goal is to maximize the order of the symmetry first; within the symmetries of maximum order, we want to minimize the number of fixed nodes. Both is realized by the following objective function:

$$\max\ (n+1)\sum_{k=1}^{n} kl_k - \sum_{i\in V} x_{ii}\ .$$

4 Node Labelings

Our approach makes extensive use of *labelings.* A labeling of a graph $G = (V, E)$ is a coloring $c\colon V^2 \to \mathbf{N}$, such that for each automorphism $\pi \in \mathrm{Aut}(G)$ and all nodes $i, j \in V$ we have $c(i, j) = c(\pi(i), \pi(j))$. We define a corresponding *node labeling* $c\colon V \to \mathbf{N}$ by $c(i) = c(i, i)$. For reasons to be explained later, we always prefer a fine labeling, i.e., one with many different colors, to a coarse labeling with less colors. For every graph G, there is a finest node labeling, called the *automorphism partitioning* of G, but its computation is isomorphism complete [8]. Nevertheless, there are heuristics finding the automorphism partitioning in most cases. We use the algorithm of Bastert [1] to compute a labeling once before starting our branch & cut-process. This algorithm needs $O(n^3 \log n)$ time and $O(n^2)$ space. Observe that we always have a trivial labeling of G by defining

$$c(i,j) = \begin{cases} 0 & \text{if } i \neq j \text{ and } (i,j) \notin E \\ 1 & \text{if } i \neq j \text{ and } (i,j) \in E \\ 2 & \text{if } i = j\ . \end{cases}$$

In the following, we always assume that the computed labeling is at least as fine as the trivial one.

One reason for using node labelings is the following: If $c(i) \neq c(j)$ for two nodes $i, j \in V$, no automorphism will map i to j, by definition of c. Hence the variable x_{ij} in (3) will always be zero, so that we can leave it out from the beginning. This often reduces the size of the ILP immensely. If the automorphism group is trivial and the node labeling assigns a different color to each node, all x-variables except for the x_{ii} can be deleted.

Furthermore, we may be able to omit many of the variables l_k: For a color $c_0 \in \mathbf{N}$, consider the set $P = \{i \in V \mid c(i) = c_0\}$ and let $p = |P|$. Any symmetry π of G with $\mathrm{ord}(\pi) \geq 3$ has at most one fixed node, and all other orbits have length $\mathrm{ord}(\pi)$. By definition, we know that $\pi(P) = P$ for any $\pi \in \mathrm{Aut}(G)$, hence the set $P \setminus \mathrm{Fix}(\pi)$ can be divided into orbits of length $\mathrm{ord}(\pi)$, so that $\mathrm{ord}(\pi)$ must divide $|P \setminus \mathrm{Fix}(\pi)| \in \{p-1, p\}$. For all $k \geq 3$ neither dividing p nor $p-1$, we can leave out l_k, and this holds for any $c_0 \in \mathbf{N}$.

5 The Branch & Cut-Approach

Branch & Cut is a powerful method for solving hard combinatorial optimization problems. Given an ILP-formulation of the problem, branch & cut proceeds as follows: First, the integrality constraints of the ILP are ignored and the remaining *LP-relaxation* is solved. If the computed optimal solution is integer, the algorithm stops, since this is an optimal solution for the ILP, too. Otherwise, there must be a linear inequality that is valid for the original ILP but violated by the computed fractional solution. If such a *cutting plane* can be found by some *separation* algorithm, it is added to the LP-relaxation as a new constraint, and the new LP is solved again. This is the *cutting phase.* If no more cutting planes can be (or shall be) computed, one has to *branch*: A variable x with a fractional LP-value $\overline{x}$ is selected and two subproblems are created. In the first, the constraint $x \leq \lfloor \overline{x} \rfloor$ is added, in the second, $x \geq \lceil \overline{x} \rceil$ is added. Both problems are solved recursively.

Another ingredient is *primal heuristics.* With the help of the fractional LP-solution, these try to find a feasible solution for the ILP. If the LP-solution in a subproblem has an objective value smaller – when maximizing – than the objective value of some feasible solution, this subproblem cannot contain the optimal solution and can thus be deleted.

In this section, we first describe two kinds of cutting planes that are very easy to handle, see Sect. 5.1. In Sects. 5.2 and 5.3, we present more sophisticated classes of cutting planes; the first one is valid for every automorphism, the second one is in particular useful to reject non-geometric automorphisms. In Sect. 5.4, we present primal heuristics.

5.1 Simple Inequalities

The valid inequalities presented in this section are easy to handle; they can be added to ILP (3) from the beginning, since their number is at most n. Hence no separation is necessary. As in Sect. 4, consider $P = \{i \in V \mid c(i) = c_0\}$ for $c_0 \in \mathbf{N}$ and assume $p = |P| \geq 2$. Let π be any symmetry of G with $\operatorname{ord}(\pi) \geq 3$. In Sect. 4, we argued that $\operatorname{ord}(\pi)$ must divide p or $p-1$, since by definition we have $\pi(P) = P$. In the first case, we have $P \cap \operatorname{Fix}(\pi) = \emptyset$, hence we derive

$$\sum_{i \in P} x_{ii} \leq p(l_1 + l_2) + \sum_{k \geq 3,\, k \mid (p-1)} l_k \,.$$

Furthermore, we know that at least ($p \bmod \operatorname{ord}(\pi)$) nodes of P are fixed, since the non-fixed nodes are divided into orbits of length $\operatorname{ord}(\pi)$, hence

$$\sum_{i \in P} x_{ii} \geq \sum_{k=1}^{n} (p \bmod k) l_k$$

is a valid inequality for (3).

5.2 Homomorphism Inequalities

The only constraints in (3) depending on the edges of the graph are the equations $A_G X = X A_G$. In this section, we present a class of cutting planes that focus on the edges, too, and that is equivalent to $A_G X = X A_G$ for integer solutions. Nevertheless, this is not true for fractional solutions. In our branch & cut-algorithm, the best strategy is to use both types of constraints.

Let c be the labeling of G again. Up to now, we only considered the node colors $c(i)$. In the following, we make use of all colors $c(i,j)$. Let $I \subseteq V^2$ be a set of node-pairs such that for all $(i,j),(k,l) \in I$ we have $c(i,k) \neq c(j,l)$. By definition, no automorphism can map both i to j and k to l in this case. Hence the following *homomorphism inequality* is valid for ILP (3):

$$\sum_{(i,j)\in I} x_{ij} \leq 1\,. \tag{4}$$

We now explain how to find valid constraints of type (4) that are violated by the current LP-solution $\overline{x}_{ij}$. This is a special independent set problem: Consider the graph $G_c = (V^2, E_c)$ with $((i,j),(k,l)) \in E_c$ if and only if $c(i,k) = c(j,l)$. Assign the weight $\overline{x}_{ij}$ to the node (i,j) of G_c. Then (4) is valid if and only if I is an independent set in G_c, and it is violated if and only if the total weight of I is greater than 1. We can thus use any maximum weight independent set heuristic for the heuristical separation of homomorphism inequalities.

Observe that a fine labeling is favorable again: The finer the labeling, the sparser the graph G_c, the larger the feasible sets I, the more restrictive the corresponding inequalities (4). In general, the homomorphism inequalities are no facets of the symmetry polytope. Nevertheless, they perform well, even if the underlying labeling is the trivial one. Leaving out the homomorphism inequalities increases runtime significantly.

5.3 Decomposition Inequalities

The inequalities presented in this section improve and generalize the constraints of type (2) in Sect. 3. They aim at eliminating non-geometric automorphisms and are based on the fact that all non-trivial orbits of a symmetry have the same length.

In the general case, let $P \subseteq V$ be any set of nodes of G and $I \subseteq P^2$. Let $p = |P|$. We can consider $H = (P, I)$ as a directed graph with loops. A *cycle decomposition* of H is a set of node-disjoint directed cycles in H, such that every node is contained in one of these cycles. Now define $K(H)$ as the set of all $k \in \{1, \ldots, n\}$ such that there exists a cycle decomposition of H with all cycles of length one or k and, for $k \geq 3$, at most one cycle of length one. Using these definitions, we have the following *decomposition inequality*:

$$\sum_{(i,j)\in I} x_{ij} \leq p - 1 + \sum_{k\in K(P,I)} l_k\,. \tag{5}$$

Indeed, the sum on the left hand side is at most p. If it is equal to p, the symmetry π induced by the x_{ij} satisfies $\pi(P) = P$. Hence by definition of $K(P,I)$ we must have $l_k = 1$ for some $k \in K(P,I)$.

The constraints of type (2) are special decomposition inequalities: Consider a cycle $C = (i_1, \ldots, i_p)$ of V again. For the ease of exposition, we will view C as a function by defining $C(i_r) = i_{r+1}$ for $r < p$ and $C(i_p) = i_1$. If we set $P = \{i_1, \ldots, i_p\}$ and $I = \{(i,j) \in P^2 \mid C(i) = j\}$, we have $K(P,I) = \{p\}$, so in this case (5) becomes (2). But, without generalization, these constraints perform poorly. We will illustrate this by an example. Assume that the current LP-solution has an orbit of length five, say $C = (1,2,3,4,5)$, but l_5 is zero. The constraint $x(C) \leq 4 + l_5$ will not allow this; after adding it to the LP, the additional value of one often escapes to the cycle $C^2 = (1,3,5,2,4)$ in the following sense: In the next LP-solution, we get $x_{i,C(i)} = \frac{4}{5}$ and $x_{i,C^2(i)} = \frac{1}{5}$ for $i \in \{1,2,3,4,5\}$, which can be prevented neither by $x(C) \leq 4 + l_5$ nor by $x(C^2) \leq 4 + l_5$.

To solve this problem, we aim at combining several constraints of type (2) in one constraint of type (5). We will give two examples of this strategy. First, consider $I = \{(i,j) \in P^2 \mid i \neq j\}$. Then we have $k \in K(P,I)$ if and only if $k \geq 2$ and $k \mid p$, hence

$$\sum_{i,j \in P,\ i \neq j} x_{ij} \leq p - 1 + \sum_{k \geq 2,\ k \mid p} l_k$$

is a valid inequality. Second, let $e, f \in \{1, \ldots, p\}$ with $e < f$ and $\gcd(f-e,p) = 1$. For $I = \{(i,j) \in P^2 \mid C^e(i) = j \text{ or } C^f(i) = j\}$, i.e., for the combination of C^e and C^f, one can show $K(P,I) = \{p/\gcd(e,p), p/\gcd(f,p)\}$. In particular, if p is odd and $p \neq 1$, we get $x(C) + x(C^2) \leq p - 1 + l_p$ as a valid inequality.

In general, it is difficult – even for very restricted classes of graphs (P,I) – to develop an algorithm for computing $K(P,I)$, let alone a simple formula as above. We are convinced that the algorithm can be improved at this point by finding other classes of graphs that can be treated efficiently.

We separate all tractable types of decomposition inequalities by the same straightforward heuristical scheme. We first compute a permutation π of V in the following way: At the beginning, the nodes $\pi(i)$ and $\pi^{-1}(i)$ are undefined for all $i \in V$. Then we traverse all pairs $(i,j) \in V^2$ by descending LP-value of x_{ij}. For the pair (i,j), we set $\pi(i) = j$ and $\pi^{-1}(j) = i$ if $\pi(i)$ and $\pi^{-1}(j)$ are undefined. This yields the permutation π. Now for every π-orbit $C = (i_1, \ldots, i_p)$, we check if the constraint (2) or some improved constraint of type (5) is violated.

5.4 Primal Heuristics

Finding symmetries heuristically is difficult, since a local variation of the graph usually changes the set of its symmetries completely, so that straightforward greedy heuristics do not work. The following lemma helps (see Sect. 5.2 for the definition of the graph G_c):

Lemma 3. *Each maximum clique in G_c has n nodes. There is a one-to-one correspondence between the set of maximum cliques in G_c and $Aut(G)$.*

We skip the simple proof; we just mention that the automorphism corresponding to a maximum clique $Q \subseteq V^2$ is given by $\pi(i) = j$ for all $(i,j) \in Q$.

Lemma 3 tells us that we can use a heuristic for the maximum clique problem in G_c and hope that the computed clique has n nodes so that it corresponds to an automorphism of G. If we are lucky, this automorphism is even a symmetry. Using this algorithm as a stand-alone heuristic is not very effective, since in most cases, the computed clique will either not be maximum or correspond to the trivial automorphism of G. In the branch & cut-setting, we apply the primal heuristics whenever the LP-solver comes up with a fractional solution $\overline{x}_{ij}$, hoping that this solution will guide us to a nearby integer solution. A good way to use this knowledge in our maximum clique heuristic is to assign the weight $\overline{x}_{ij}$ to the node $(i, j) \in G_c$ (as we did in Sect. 5.2) and to search for a maximum *weight* clique in G_c.

We also use another primal heuristic: We consider the permutation π of V computed for the separation of decomposition inequalities described in Sect. 5.3, then we check if this permutation induces a symmetry of G. This heuristic is very fast; the additional runtime for checking is linear.

6 Experimental Runtimes

The branch & cut-algorithm presented in the previous section has been implemented in C++ using ABACUS [5] with CPLEX. As mentioned before, the problem being solved is NP-hard, so that no polynomial time bound can be given. Instead, we present experimental runtimes in this section. All instances have been solved on a Sun UltraSPARC-II (296 MHz) machine. We are still working on improving the algorithm at several points and presume that the runtimes will decrease in the future.

Since we do not have test instances from applications, we have to create random instances. This is delicate: Using random graphs is not useful, since symmetries are rare, most graphs do not have any non-trivial automorphism at all. These graphs are usually very easy to handle. But if the user of our algorithm would not expect his graphs to have symmetries with a reasonable probability, he probably would not use it.

Given a fixed number n of nodes, we therefore create graphs with symmetries of any feasible order, i.e., of any order $k \in \{1, \ldots, n\}$ such that $k \mid n$ or $k \mid (n-1)$. To create a symmetry of a given order k, we first choose a number f of fixed nodes, where $f \leq 1$ for $k \geq 3$ and $k \mid (n - f)$. Next, we randomly choose a permutation of $V = \{1, \ldots, n\}$ with f fixed elements and all other orbits of length k. This permutation induces a permutation π of the node-pairs in V^2. For each π-orbit $I \subseteq V^2$, we add an edge between i and j either for all $(i, j) \in I$ or for none. The resulting simple graph has a symmetry of order k by construction, but observe that it may also have a better symmetry by chance. For $k = 1$, the resulting graph is just a random graph with a random number of edges between zero and $n(n-1)/2$.

The influence of order on runtime is shown in Table 1. Here, we created 10 000 graphs with 50 nodes each by the method explained above and sorted the results by the order of the computed symmetry. We display average and maximum cpu-time (in seconds), average and maximum number of generated subproblems, and average and maximum number of solved LPs. Observe that runtime increases with order, at least for high orders. In Table 2, we display the results for $n \leq 80$, where for each n we created 1000 test instances by the method explained above, with equally many symmetries of each feasible order.

Table 1. Runtimes for $n = 50$, sorted by order

Order	Time (sec.) avg	Time (sec.) max	#Subproblems avg	#Subproblems max	#LPs avg	#LPs max
1	1.74	1.95	1.00	1	1.00	1
2	2.12	2.82	1.00	1	1.00	1
5	2.10	2.48	1.00	1	1.00	1
7	2.09	2.46	1.00	1	1.00	1
10	2.19	5.85	1.00	1	2.40	21
25	2.44	7.72	1.00	1	1.25	12
49	3.93	27.02	1.00	1	2.23	15
50	8.02	83.27	1.00	1	4.86	34

Table 2. Runtimes for $1 \leq n \leq 80$

n	Time (sec.) avg	Time (sec.) max	#Subproblems avg	#Subproblems max	#LPs avg	#LPs max
1–10	0.05	30.43	1.51	197	2.52	328
11–20	1.48	2756.89	2.00	707	4.49	1096
21–30	4.00	6976.54	1.09	139	2.49	244
31–40	1.38	720.96	1.00	1	2.03	78
41–50	2.47	143.86	1.00	3	2.02	51
51–60	4.36	147.17	1.00	1	2.00	42
61–70	7.00	208.94	1.00	1	1.94	39
71–80	10.32	763.62	1.00	1	2.01	63

These tables show that on average the branch & cut-algorithm creates very few subproblems. Even the number of LPs to be solved is very small in general. Furthermore, the average number of subproblems and LPs surprisingly decreases for larger graphs. This is due to the following fact: The larger the graph is, the lower is the probability to get some "extra structure" by chance, i.e., some automorphism or similar structure different from the symmetry created explicitly. This extra structure may distract our algorithm. Particularly hard to handle are graphs with many automorphisms but few symmetries. For this reason, we also created graphs with not necessarily geometric automorphisms. For a given number n of vertices, we computed a random permutation of $V = \{1, \ldots, n\}$ and added the edges exactly as in the last step of the symmetry creation. For these instances, the results are significantly worse, see Table 3. We checked 100 instances for each $n \leq 30$.

Table 3. Runtimes for hard instances

n	Time (sec.) avg	Time (sec.) max	#Subproblems avg	#Subproblems max	#LPs avg	#LPs max
1–5	0.01	0.08	1.23	7	1.55	17
6–10	0.06	1.19	2.01	41	3.75	52
11–15	0.83	193.67	5.40	909	10.51	1172
16–20	4.08	991.59	4.02	433	12.24	1157
21–25	25.99	3533.09	10.60	2523	32.08	7453
26–30	99.42	9197.55	7.68	1295	24.37	2147

This shows that the most important improvement to be achieved is a better rejection of non-geometric automorphisms, for example by better decomposition inequalities, see Sect. 5.3. If we search for arbitrary automorphisms instead of symmetries, the runtimes are much shorter for the same instances.

Finally, we applied the symmetry detection algorithm to all 11 523 graphs of the "Rome Library" [9]. Here, the number of nodes ranges from 10 to 100. We observed an average runtime of 4.26 seconds, the average number of subproblems and LPs was 1.06 and 1.11, respectively. The maximum runtime was 19.29 seconds (for `grafo11203.100` with 100 nodes); the algorithm needed at most three subproblems and five LPs. Nine graphs have a 3-rotation and 7 845 of the remaining graphs have a reflection, with many fixed nodes in general.

7 Extensions

The algorithm presented in this article can be applied to more general types of graphs. In the most general case, we have a colored graph, i.e., a set V of nodes and a coloring $c\colon V^2 \to \mathbb{N}$. Then an automorphism of $G = (V, c)$ is defined as a permutation π of V such that $c(i,j) = c(\pi(i), \pi(j))$ for all $i, j \in V$. By this, we can also model multigraphs (by defining $c(i,j)$ as the number of edges between the nodes i and j) and directed graphs (the coloring c does not have to be symmetric). The number of loops at node i can be stored in $c(i,i)$. Now define the adjacency matrix of G by $A_G = (c(i,j))_{i,j}$, and observe that labelings can be defined and computed for general colored graphs as well as for simple graphs (the coloring c itself is a labeling of G). Then our algorithm carries over without modification.

Another extension concerns Lemma 2 in Sect. 2. After computing the best symmetry π_1 as explained above, we can compute a second symmetry π_2 that can be displayed simultaneously with π_1 without being generated by π_1 – if such a symmetry exists – by the same procedure, except for a slight modification of the ILP (3). Using Lemma 2, one can show that it is necessary and sufficient to enforce $\pi_2^2 = \mathrm{id}_V$, $\pi_2\pi_1 = \pi_1^{-1}\pi_2$, and, if $\mathrm{ord}(\pi_1)$ is even, that $\pi_2 \neq \pi_1^e$ for $e = \mathrm{ord}(\pi_1)/2$. These conditions translate to

$$\begin{aligned} x_{ij} &= x_{ji} && \text{for all } i, j \in V \\ x_{\pi_1(i),j} &= x_{i,\pi_1(j)} && \text{for all } i, j \in V \\ \textstyle\sum_{i \in V} x_{i,\pi_1^e(i)} &\leq n - 2 && \text{if } e = \operatorname{ord}(\pi_1)/2 \text{ is integer.} \end{aligned}$$

Finally, we can modify ILP (3) to solve other problems related to symmetry and automorphism detection, such as graph isomorphism or the greatest common subgraph problem. The necessary adjustments of (3) are easy to derive, but to get a fast algorithm, it is indispensable to find new and special cutting planes for each of these problems.

8 Conclusion and Outlook

This presentation should be considered as a first step towards a polyhedral approach to the symmetry detection problem in arbitrary graphs. At the current state, most graphs can be processed fast, but in some cases, the runtimes are exorbitant. We are convinced that the algorithm can be improved sharply, especially by finding new types of cutting planes. For this, deeper insight into the structure of the polytopes describing automorphisms or symmetries is needed.

References

1. O. Bastert. New ideas for canonically computing graph algebras. Technical Report TUM-M9803, Technische Universität München, Fakultät für Mathematik, 1998.
2. P. Eades and X. Lin. Spring algorithms and symmetry. *Theoretical Computer Science*, 240(2):379–405, 2000.
3. H. de Fraysseix. An heuristic for graph symmetry detection. In J. Kratochvíl, editor, *Graph Drawing '99*, volume 1731 of *Lecture Notes in Computer Science*, pages 276–285. Springer-Verlag, 1999.
4. S.-H. Hong, P. Eades, and S.-H. Lee. Finding planar geometric automorphisms in planar graphs. In K.-Y. Chwa et al., editors, *Algorithms and computation. 9th international symposium, ISAAC '98*, volume 1533 of *Lecture Notes in Computer Science*, pages 277–286. Springer-Verlag, 1998.
5. M. Jünger and S. Thienel. The ABACUS system for branch-and-cut-and-price-algorithms in integer programming and combinatorial optimization. *Software – Practice & Experience*, 30(11):1325–1352, 2000.
6. J. Manning. Computational complexity of geometric symmetry detection in graphs. In *Great Lakes Computer Science Conference*, volume 507 of *Lecture Notes in Computer Science*, pages 1–7. Springer-Verlag, 1990.
7. H. Purchase. Which aesthetic has the greatest effect on human understanding? In Giuseppe Di Battista, editor, *Graph Drawing '97*, volume 1353 of *Lecture Notes in Computer Science*, pages 248–261. Springer-Verlag, 1997.
8. R. C. Read and D. G. Corneil. The graph isomorphism disease. *Journal of Graph Theory*, 1:339–363, 1977.
9. The Rome library of undirected graphs. Available at: `www.inf.uniroma3.it/people/gdb/wp12/undirected-1.tar.gz`.

Drawing Graphs Symmetrically in Three Dimensions*

Seok-Hee Hong

Basser Department of Computer Science, University of Sydney, Australia.
shhong@cs.usyd.edu.au

Abstract. In this paper, we investigate symmetric graph drawing in three dimensions. We show that the problem of drawing a graph with a maximum number of symmetries in three dimensions is NP-hard. Then we present a polynomial time algorithm for finding maximum number of three dimensional symmetries in planar graphs.

1 Introduction

Symmetry is one of the most important aesthetic criteria for Graph Drawing. It clearly reveals the structure of a graph. Symmetric graph drawing has been investigated by a number of authors [2,4,5,6,7,8,9,10,13,16,17,18]. However, most of previous work on symmetric graph drawing has mainly focused on two dimensions [2,4,5,6,7,8,13,16,17,18].

Symmetry in three dimensions is much richer than that in two dimensions. For example, a maximal symmetric drawing of the icosahedron in two dimensions shows 6 symmetries. However, the maximal symmetric drawing of the icosahedron in three dimensions shows 120 symmetries.

In this paper, we investigate symmetric graph drawing in three dimensions. First, we show that the problem of drawing a graph with a maximum number of symmetries in three dimensions is NP-hard. Then we present a polynomial time algorithm for finding maximum number of three dimensional symmetries in planar graphs.

To draw a graph symmetrically in three dimensions, there are two steps: first find the three dimensional symmetries, then construct a drawing which displays these symmetries. The first step is the more difficult one. This paper concentrates on the first step.

This paper is organized as follows. In the next section, we explain necessary background. We show that the problem of drawing a graph with a maximum number of symmetries in three dimensions is NP-hard in Section 3. In Section 4, Section 5, Section 6, and Section 7, we present an algorithm for finding maximum number of three dimensional symmetries of planar graphs. Section 8 concludes.

* In this extended abstract, many proofs are omitted. This research has been supported by a grant from the Australian Research Council. Three dimensional drawings are available from `http://www.cs.usyd.edu.au/~shhong/research7.htm`.

P. Mutzel, M. Jünger, and S. Leipert (Eds.): GD 2001, LNCS 2265, pp. 189–204, 2002.
© Springer-Verlag Berlin Heidelberg 2002

2 Symmetric Graph Drawing in Three Dimensions

In this section, we explain the types of three dimensional symmetry and describe our model for drawing graphs symmetrically in three dimensions.

2.1 Symmetries in Three Dimensions

Symmetry in three dimensions is richer and more complex than symmetry in two dimensions. The types of symmetry in three dimensions can be classified as *rotation*, *reflection*, *inversion* and *rotary reflection* [19]. A rotational symmetry in three dimensions is a rotation about an *axis* and a reflectional symmetry in three dimensions is a reflection in a *plane*. Inversion is a reflection in a *point*. Rotary reflection is a composition of a rotation and a reflection.

A *finite rotation group* in three dimensions is one of following three types: a *cyclic group* (C_n), a *dihedral group* (D_n) and the rotation group of one of the *Platonic solids* [19]. There are only five regular Platonic solids, the *tetrahedron*, the *cube*, the *octahedron*, the *dodecahedron* and the *icosahedron*.

The *full symmetry group* of a finite object in three dimensions is much more complex than the rotation group. The complete list of all possible symmetry groups in three dimensions can be found in [19].

2.2 Symmetric Drawing in Three Dimensions

For the purpose of drawing graphs in three dimensions, we require three *non-degeneracy* conditions: no two vertices are located at the same point, no two edges overlap, and no vertex lies on an edge with which it is not incident.

A symmetry of a three dimensional drawing D of a graph G induces a permutation of the vertices, which is an *automorphism* of the graph; this automorphism is *displayed* by the symmetry. We say that the automorphism is a *three dimensional symmetry* of a graph G.

The symmetry group of a three dimensional graph drawing D of a graph G induces a subgroup of the automorphism group of G. The subgroup P is a *three dimensional symmetry group* if there is a three dimensional drawing D of G which displays every element of P as a symmetry of the drawing.

3 General Graphs

In this section, we show that the problem of drawing a graph with a maximum number of symmetries in three dimensions is NP-hard.

The aim of this paper is to investigate graph drawings in three dimensions that show a maximum number of symmetries. For this purpose, we need to find a three dimensional symmetry group which has maximum size. We now define our main problem.

Three Dimensional Symmetry Group Problem (3DSGP)
Input: A graph G.
Output: A three dimensional symmetry group P of G which has maximum size.

In general, $3DSGP$ is NP-hard; the following subproblem is NP-complete.

Three Dimensional Size 3 Symmetry Group Problem (3D3SGP)
Input: A graph G.
Question: Does G have a three dimensional symmetry group of size at least 3?

Theorem 1. *The problem* $3D3SGP$ *is NP-complete, and the problem* $3DSGP$ *is NP-hard.*

To show that $3D3SGP$ is NP-complete, we consider the following problems. A vertex is *fixed* by a given three dimensional symmetry if it is mapped onto itself by that symmetry.

1. $3DROT$: Does G have a drawing D in three dimensions which shows any nontrivial rotational symmetry?
2. $3DROT_0$: Does G have a drawing D in three dimensions which shows any nontrivial rotational symmetry with no fixed vertex?
3. $3DREF$: Does G have a drawing D in three dimensions which shows any reflectional symmetry?
4. $3DREF_0$: Does G have a drawing D in three dimensions which shows any reflectional symmetry with no fixed vertex?
5. $3DINV$: Does G have a drawing D in three dimensions which shows an inversion?
6. $3DINV_0$: Does G have a drawing D in three dimensions which shows an inversion with no fixed vertex?
7. $3DINV_1$: Does G have a drawing D in three dimensions which shows an inversion with one fixed vertex?

We now show that each of these problems, except $3DREF$ is NP-complete. For $3DREF$, note that any drawing of a graph in a plane displays reflectional symmetry in three dimensions.

First we consider a three dimensional symmetry with no fixed vertex.

Theorem 2. $3DREF_0$, $3DINV_0$ *and* $3DROT_0$ *are NP-complete.*

Proof. This immediately follows from [14] that the detection of *fixed point free* automorphism is NP-complete.

We now consider the problem of three dimensional symmetry with fixed vertex.

Theorem 3. $3DINV$, $3DINV_1$ *and* $3DROT$ *are NP-complete.*

Proof. We can reduce these problems to the corresponding problem in two dimensions, which is NP-complete [18].

To prove the general complexity results in Theorem 1, note that any symmetry group of size greater than 2 must include a symmetry that is either a rotation or an inversion. Thus Theorem 1 follows from Theorem 3.

However, if we restrict the problem into subclasses of planar graphs such as trees and series-parallel digraphs, then $3DSGP$ is solvable in linear time [9, 10]. For drawings of planar graphs in three dimensions, we require two further properties. First, two nonadjacent edges should not intersect, that is, we do not allow edge crossings. Second, the drawing should be *linkless*, that is, no two undirected cycles form a nontrivial link. In the next section, we show that $3DSGP$ is solvable in polynomial time for planar graphs.

4 Planar Graphs

In this section, we present an algorithm for finding three dimensional symmetries in planar graphs. We use connectivity to divide the problem into cases.

1. G is triconnected.
2. G is biconnected.
3. G is one-connected.

Each case relies on the result of the previous case. The following theorem summarizes the result of this section.

Theorem 4. *There is a polynomial time algorithm which computes a maximum size three dimensional symmetry group of planar graphs.*

The proof of this theorem is outlined in the remainder of this paper. The first case is the simplest and described in Section 5. The biconnected case and the one-connected case are more difficult and described in Section 6 and Section 7 respectively.

5 The Triconnected Case

If G is a triconnected planar graph, then we simply use automorphism to find three dimensional symmetry. This is based on the following theorem.

Theorem 5. *[1,15] Every triconnected planar graph G can be realized as the 1-skeleton of a convex polytope P in $\mathbb{R}^3$ such that all automorphisms of G are induced by isometries of P.*

From Theorem 5, we can compute three dimensional symmetry group of triconnected planar graph G from the automorphism group of G. Automorphism of triconnected planar graphs can be computed in linear time [12]. However we need $O(n^2)$ space to represent the automorphism group.

6 The Biconnected Case

If the input graph G is biconnected, then we break it into *triconnected components* in a way that is suitable for the task. The overall algorithm is composed of three steps.

```
Algorithm 3DBiconnected_Planar
```

1. Construct the SPQR-tree T_1 of G, and root T_1 at its center.
2. *Reduction*: For each level i of T_1 (from the lowest level to the root level)
   a) For each leaf node on level i, compute labels.
   b) For each leaf node on level i, label the corresponding virtual edge of the parent node with the labels.
   c) Remove the leaf nodes on level i.
3. Compute a maximum size three dimensional symmetry group at the labeled center.

We briefly sketch the idea of the algorithm. The algorithm begins by constructing the SPQR-tree for the input biconnected planar graph. Then we use a kind of "reduction" [11]. The operation traverses the SPQR-tree from the leaf nodes to the center level by level. First it computes the labels for the leaf nodes. The labels are a pair of integers, a list of integers and boolean values that capture some information of the symmetry of the leaf nodes. Then it labels the corresponding virtual edge in the parent node and delete each leaf node. This reduction process stops at the root (the center of the SPQR-tree). The center may be a node or an edge. Using the information encoded on the labels, we compute a maximum size three dimensional symmetry group at the center.

We now explain each step of the algorithm.

6.1 SPQR-Tree

We briefly review the definition of the SPQR-tree. For details, see [3].

The SPQR-tree represents a decomposition of a biconnected planar graph into triconnected components. There are four types of nodes in the SPQR-tree T_1 and each node v in T_1 is associated with a graph which is called as the *skeleton* of v (*skeleton*(v)). The node types and their skeletons are:

1. Q-node: The skeleton consists of two vertices which are connected by two multiple edges.
2. S-node: The skeleton is a simple cycle with at least 3 vertices.
3. P-node: The skeleton consists of two vertices connected by at least 3 edges.
4. R-node: The skeleton is a triconnected graph with at least 4 vertices.

In fact, we use slightly different version of the SPQR-tree. We use the SPQR-tree without Q nodes. Also we root the SPQR-tree T_1 at its center. The SPQR-tree is unique for each biconnected planar graph. Let v be a node in T_1 and u is a parent node of v. The graph *skeleton*(u) has one common *virtual edge* with *skeleton*(v), which is called as a *virtual edge* of v.

6.2 Reduction Process

The reduction process takes the SPQR-tree of a biconnected graph. The SPQR-tree T_1 is rooted at the center, based on the following theorem.

Theorem 6. *The center of the SPQR-tree is fixed by a three dimensional symmetry group of a biconnected planar graph.*

Reduction proceeds from the nodes of maximum level toward the center, deleting each leaf node v at the lowest level at each iteration.

The reduction process clearly does not decrease the three dimensional symmetry group of the original graph. This is not enough; we need to also ensure that the three dimensional symmetry group is not increased by reduction. This is the role of the labels. As a leaf v is deleted, the algorithm labels the virtual edge e of v in $skeleton(u)$ where u is a parent of v. Roughly speaking, they encode information about the deleted leaf to ensure that every three dimensional symmetry of the labeled reduced graph extends to a three dimensional symmetry of the original graph.

The reduction process stops when it reaches the root.

6.3 The Labels and Labeling Algorithms

We now define the labels which play very important role in the reduction process.

Our aim is to label the virtual edges so that we can compute a maximum size three dimensional symmetry group of the original graph by computing a maximum size three dimensional symmetry group of the labeled center.

At each stage of the reduction process, labels for the deleted leaf nodes are needed. This is because the reduced graph overestimates a three dimensional symmetry group of the original graph. By checking labels whether they preserve the three dimensional symmetry or not, we can compute a three dimensional symmetry group of the original graph exactly from the reduced graph.

Let v be an internal node of T_1. We say that a virtual edge e of $skeleton(v)$ is a *parent (child) virtual edge* if e corresponds to a virtual edge of u which is a parent (child) node of v. We define a *parent separation pair* $s = (s_1, s_2)$ of v as the two endpoints of a parent virtual edge e.

When we compute the labels of v, we need to delete the parent virtual edge e from $skeleton(v)$. We denote the resulting graph by $skeleton^-(v)$.

Suppose that nodes $v_1, v_2, \ldots v_k$ of the SPQR-tree T_1 are deleted at one iteration of the reduction process. These nodes correspond to virtual edges $e_1, e_2, \ldots, e_k$ in the level above the current level. For each e_i, we need to compute the following labels.

1. isomorphism code: a pair $Iso(e_i)$ of integers.
2. rotation code: a list L_{e_i} indicating the size of possible rotation groups of $skeleton^-(v_i)$ that fixes the parent separation pair.
3. reflection code:

a) $Ref_{swap}(e_i)$: a boolean label indicating whether $skeleton^{-}(v_i)$ has a reflectional symmetry that *swaps* the parent separation pair.
b) $Ref_{fix}(e_i)$: a boolean label indicating whether $skeleton^{-}(v_i)$ has a reflectional symmetry that *fixes* the parent separation pair.

4. inversion code: a boolean label $Inv(e_i)$ indicating whether $skeleton^{-}(v_i)$ has an inversion that swaps the parent separation pair.

Note that we need these labels when the virtual edge is fixed by a three dimensional symmetry of the parent node. We now describe each labeling algorithm.

Computation of an Isomorphism Code. The isomorphism code $Iso(c)$ consists of a pair of integers. This is because the $skeleton(v)$ has an orientation with respect to the parent separation pair [11]. The isomorphism code can be computed in linear time using a planar graph isomorphism algorithm [12].

Computation of a Rotation Code. In addition to the isomorphism code, we attach further information about a rotational symmetry of $skeleton^{-}(v)$ which fixes the parent separation pair. The algorithm computes a list L_e, the size of the possible rotation groups. In fact, each element of L_e consists of a pair of integers: one which indicates the size of the rotation group and the other which indicates whether the rotational symmetry has a fixed edge.

Note that the rotational symmetry should respect the isomorphism code of the child virtual edge. Further, if the rotational symmetry fixes a child virtual edge, then we need to compute the intersection of the rotation groups. We now state the algorithm.

Algorithm Compute_Rotation_Code

1. Compute the list L_e of rotation groups of $skeleton^{-}(v)$ which fixes the parent separation pair and respects the isomorphism codes of child virtual edges.
2. For each element ρ of L_e, if there is a child virtual edge e_j in $skeleton^{-}(v)$ which is fixed by ρ, then compute the intersection of the rotation groups of L_e and L_{e_j}.

Step1 uses the triconnected case in Section 5 if $skeleton^{-}(v)$ is triconnected. Otherwise it uses `3DBiconnected_Planar` recursively. Step 2 can be computed in linear time using a bit array representation [9].

Computation of a Reflection Code. Further, we need information about a reflectional symmetry. This can be divided into two cases: *swaps* the parent separation pair or *fixes* the parent separation pair. First we describe an algorithm for $Ref_{swap}(e)$.

Note that the reflectional symmetry should respect the isomorphism code of the child virtual edge. Further, if the reflectional symmetry fixes a child virtual edge, then we need to test its label. We now state the algorithm.

Algorithm Compute_Reflection_Code_Swap

1. Test $skeleton^{-}(v)$ whether it has a reflectional symmetry α which swaps the parent separation pair and respects the isomorphism codes of child virtual edges.
2. If α exists, then
   a) For each child virtual edge e_j that is fixed by α, check followings:
      i. if α fixes the endpoints of e_j, then $Ref_{fix}(e_j) = true$.
      ii. if α swaps the endpoints of e_j, then $Ref_{swap}(e_j) = true$.
   b) If one of these properties fails,
      then $Ref_{swap}(e) := false$; else $Ref_{swap}(e) := true$.
   else $Ref_{swap}(e) := false$.

An algorithm for computing $Ref_{fix}(e)$ is very similar to the algorithm for computing $Ref_{swap}(e)$. We omit this algorithm from this extended abstract.

Computation of an Inversion Code. Further, we need information about an inversion which swaps the parent separation pair. The algorithm is similar to the case of reflectional symmetry.

Algorithm Compute_Inversion_Code

1. Test $skeleton^{-}(v)$ whether it has an inversion β which swaps the parent separation pair and respects the isomorphism codes of child virtual edges.
2. If β exists, then
   If for each child virtual edge e_j that is fixed by β, $Inv(e_j) = true$,
   then $Inv(e) := true$; else $Inv(e) := false$.
   else $Inv(e) := false$.

Note that these labeling algorithms are for R-nodes. When v is a P-node, then we use similar algorithms to the case of parallel compositions in series parallel digraphs [10]. When v is an S-node, then we use similar algorithms to the case of series compositions in series parallel digraphs [10]. We omit these algorithms from this extended abstract.

Based on each labeling algorithm, now we are ready to find three dimensional symmetry at the center.

6.4 Finding Three Dimensional Symmetry at the Center

In this section, we briefly describe how to compute a maximum size three dimensional symmetry group at the labeled center.

Note that the center of the SPQR-tree may be a node c or an edge e. If the center is a node c, then we can further divide into three cases by its type. If c is a R-node, then we use the triconnected case in Section 5 to compute a three dimensional symmetry group. See Figure 1 (a) for example. We construct a three dimensional drawing of $skeleton(c)$ and then replace each child virtual edge e_j by a drawing of $skeleton^{-}(v_j)$. We repeat this process recursively. Note that we place a drawing of $skeleton^{-}(v_j)$ on a plane to maximize symmetry.

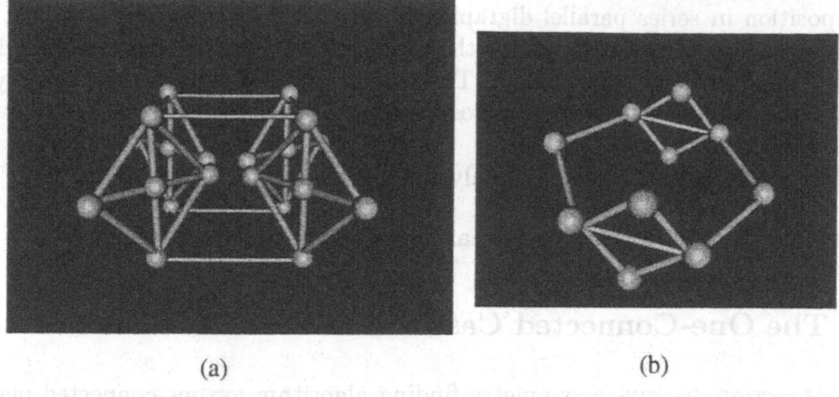

(a) (b)

Fig. 1. *Example of (a) R-node and (b) S-node.*

If c is a P-node, then we use similar algorithm to the case of a parallel composition in series parallel digraphs [10]. See Figure 2 (a) for example. If c is an S-node, then we use similar algorithm to the case of labeled cycle. See Figure 1 (b). We omit this algorithm from this extended abstract.

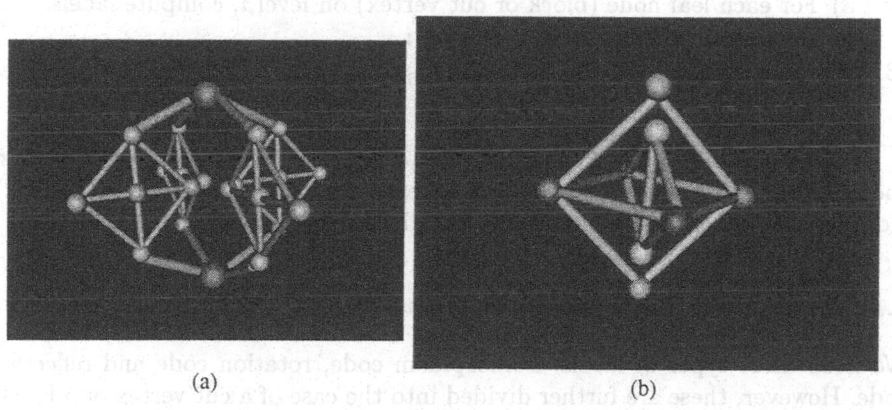

(a) (b)

Fig. 2. *Example of (a) P-node and (b) Special case.*

However, there may exist some other node v which is fixed by a three dimensional symmetry group. See Figure 2 (b). We call this special case as *enclosing case.* Thus to find a maximum size three dimensional symmetry group at the center c, we compute these two cases and then find the maximum.

If the center is an edge, then we find the maximum among three cases: parallel composition, reduction composition and enclosing composition. Parallel composition means that we construct a drawing with two labeled edges such as a parallel composition in series parallel digraphs. Reduction composition means that we compute labels of one node u and then delete u by labeling the corresponding virtual edge e of the other node v. Then we compute a three dimensional symmetry group at v. Enclosing composition means that we construct a drawing such as the special case.

We conclude this section by analyzing the time complexity.

Theorem 7. `3DBiconnected_Planar` *takes* $O(n^2)$ *time.*

7 The One-Connected Case

In this section, we give a symmetry finding algorithm for one-connected planar graph G. The algorithm is similar to the biconnected case: we use reduction approach. We also use algorithm `3DBiconnected_Planar` as a subroutine.

The method proceeds from the leaves of the *block-cut vertex tree* (BC-tree) to the center; we may regard the BC-tree as rooted at the center. The overall algorithm is thus composed of three steps.

Algorithm `3DOneconnected_Planar`

1. Construct the BC-tree T_2 of G, and root T_2 at its center.
2. *Reduction*: For each level i of T_2 (from the lowest level to the root level)
   a) For each leaf node (block or cut vertex) on level i, compute labels.
   b) Remove all the leaf nodes on level i.
3. Compute a maximum size three dimensional symmetry group at the labeled center.

The reduction process is similar to the biconnected case. In this case we take the BC-tree and then compute labels at each leaf node (block or cut vertex). However, the labels are different. We now define the labels.

7.1 The Labels and Labeling Algorithms

We need three types of labels: isomorphism code, rotation code and reflection code. However, these are further divided into the case of a cut vertex or a block. Let B_i represent a block and c_i represent a cut vertex.

1. isomorphism code : an integer $Iso_B(B_i)$ (or $Iso_C(c_i)$).
2. rotation code : a list L_{B_i} (or L_{c_i}) indicating the size of possible rotation groups of B_i (or c_i) which fixes the parent node.
3. reflection code : an integer $Ref_B(B_i)$ (or $Ref_C(c_i)$) indicating whether B_i (or c_i) has a reflectional symmetry which fixes the parent node.

Note that we need these labels when the block or cut vertex is fixed by a three dimensional symmetry of the parent node. We now describe each labeling algorithm.

Computation of an Isomorphism Code of a Block. Suppose that $B_1, B_2, \dots, B_m$ are the blocks on the lowest level and $p_1, p_2, \dots, p_m$ are the parent cut vertices for the blocks. We compute isomorphism code $Iso_B(B_i)$ for each B_i using a planar graph isomorphism algorithm which takes linear time [11, 12]. Note that the isomorphism should respect the isomorphism code of the child cut vertex. We now describe the algorithm.

Algorithm Compute_Iso_B

for each $B_i, i = 1, 2, \dots, m$,

if there is an isomorphism α between B_i and B_j such that,

a) $\alpha(p_i) = p_j$.

b) for each cut vertex c_k of B_i,

i. $\alpha(c_k)$ is a cut vertex.

ii. $Iso_C(c_k) = Iso_C(\alpha(c_k))$.

then assign isomorphism code such that $Iso_B(B_i) = Iso_B(B_j)$.

Computation of a Rotation Code of a Block. A rotation code of a block B consists of a list L_B which represents the size of possible rotation groups of B which fixes the parent cut vertex p of B. In fact, each element of L_B consists of a pair of integers: one which indicates the size of the rotation group and the other which indicates whether the rotational symmetry has a fixed edge which is adjacent to p. We need this information when the block is fixed by a rotational symmetry of the parent node. Thus we compute the list of the size of possible rotation groups which fixes the parent cut vertex p of B.

Note that the rotational symmetry should respect the isomorphism code of the child virtual edge. Further, if the rotational symmetry has a fixed child cut vertex c_j, then we need to compute the intersection of the rotation group of B and the rotation group of c_j. Let $c_1, c_2, \dots, c_k$ be the child cut vertices of B. We now describe the algorithm.

Algorithm Compute_Rot_B

1. Compute the list L_B of the size of the rotation groups of B which fixes the parent cut vertex p and respect the isomorphism code of child cut vertex, together with information about the fixed edge which is adjacent to p.
2. For each element ρ of L_B, if there is a child cut vertex c_j of B whose $\rho(c_j) = c_j$, then
   a) Let f be the number of fixed edges in B which is adjacent to c_j.
   b) Compute L'_{c_j} from the list L_{c_j} of the size of rotation groups of c_j which has at most $2 - f$ fixed child blocks with a fixed edge adjacent to c_j.
   c) Compute the intersection of L_B and L'_{c_j}.

At step 1, we use algorithm `3DBiconnected_Planar` in Section 6 to compute the rotation group of B.

Computation of a Reflection Code of a Block. The label $Ref_B(B)$ represents whether the block B has a reflectional symmetry which fixes the parent cut vertex p. Let $c_1, c_2, \ldots, c_k$ be the child cut vertices of B.

In fact, the algorithm computes a ternary value for $Ref_B(B)$. The interpretation of $Ref_B(B)$ is:

1. $Ref_B(B) = -1$: B has no reflectional symmetry which fixes p.
2. $Ref_B(B) = 1$: B has a reflectional symmetry which has a fixed edge adjacent to p.
3. $Ref_B(B) = 0$: B has a reflectional symmetry which has no fixed edge adjacent to p.

First we find a reflectional symmetry α of B which fixes the parent cut vertex using **3DBiconnected_Planar**. Then we check whether each fixed child cut vertex c_j preserves the reflectional symmetry. For this purpose, we need some information about the reflection code $Ref_C(c_j)$. The interpretation of values of $Ref_C(c_j)$ is:

1. $Ref_C(c_j) = 0$: c_j does not preserve α.
2. $Ref_C(c_j) = 1$: c_j preserves α.

Finally we assign the value, depending on the fixed edge which is adjacent to p. We now state the algorithm.

Algorithm Compute_Ref_B

1. Test B whether it has a reflectional symmetry α such that
   a) $\alpha(p) = p$.
   b) for each child cut vertex c_j of B,
      i. $\alpha(c_j)$ is a child cut vertex.
      ii. If $\alpha(c_j) = c_k$, then $Iso_C(c_j) = Iso_C(\alpha(c_k))$.
      iii. If $\alpha(c_j) = c_j$, then $Ref_C(c_j) = 1$.
2. If α exists, then
   If there is a fixed edge which is adjacent to p,
   then $Ref_B(B) := 1$; else $Ref_B(B) := 0$.
   else $Ref_B(B) := -1$.

Computation of an Isomorphism Code of a Cut Vertex. Suppose that $c_1, c_2, \ldots, c_k$ are the cut vertices on the lowest level. We compute $Iso_C(c_i)$ for each c_i, $i = 1, 2, \ldots, k$, which represents an isomorphism code of c_i. More specifically, $Iso_C(c_i) = Iso_C(c_j)$ if and only if the subgraph which is rooted at c_i is isomorphic to the subgraph which is rooted at c_j. We now state the algorithm.

Algorithm Compute_Iso_C

1. For each c_i:
   a) Let $B_{i1}, B_{i2}, \ldots, B_{im}$ be the child blocks of c_i.
   b) $s(c_i) := (Iso_B(B_{i1}), Iso_B(B_{i2}), \ldots, Iso_B(B_{im}))$.
   c) Sort $s(c_i)$.

2. Let Q be the list of $s(c_i)$, $i = 1, 2, \ldots, k$.
3. Sort Q lexicographically.
4. For each c_i, compute $Iso_C(c_i)$ as follows: Assign the integer 1 to c_i whose $s(c_i)$ is the first distinct tuple of the sorted sequence Q. Assign the integer 2 to c_j whose $s(c_j)$ is the second distinct tuple, and so on.

Computation of a Rotation Code of a Cut Vertex. The rotation code consists of a list L_c which represents the size of the possible rotation groups of the cut vertex c. In fact, each element of L_c consists of a pair of integers: one which indicates the size of the rotation group and the other which indicates the number of fixed child blocks which has a fixed edge adjacent to c.

Let B_p be the parent block of c. We use L_c when c is fixed by a rotational symmetry of B_p. When we compute L_{B_p}, we need to compute the intersection of two rotation groups.

Suppose that ρ is the rotational symmetry of B_p which fixes c. In B_p, c may have 0, 1, or 2 fixed edges. Thus, we compute the rotation group of c with three cases: 0, 1, or 2 fixed blocks which has a fixed edge adjacent to c.

Let $B_1, B_2, \ldots, B_m$ be the child blocks of c. To compute L_c, we need to compute the intersection with the rotation group of the fixed child block B_j. We use L_{B_j} for this purpose. The list L_{B_j} represents the size of the possible rotation groups of B_j that fixes c, with some information about the fixed edge which is adjacent to c.

The algorithm is an adaptation of the *pyramid* case of the tree algorithm in three dimensions [9]. Note that at most two blocks with a fixed edge adjacent to c can be fixed by the rotational symmetry. Further we can fix more blocks outside the fixed block if it does not have a fixed edge which is adjacent to c. For each fixed block, we need to compute the intersection of the rotation groups. We can compute the intersection of the rotation groups in linear time using the method in [9]. We omit this algorithm from this extended abstract.

Computation of a Reflection Code of a Cut Vertex. The label $Ref_C(c)$ represents whether a cut vertex c preserves a reflectional symmetry which fixes the parent block. Let B_p be the parent block of c and $B_1, B_2, \ldots, B_m$ be the child blocks of c. Suppose that α is a reflectional symmetry which fixes B_p. We use $Ref_C(c)$ to decide whether c preserves α of B_p. More specifically, this indicates that whether $B_1, B_2, \ldots, B_m$ can be attached to c, preserving α.

In fact, we assign an integer to represent whether it preserves α. $Ref_C(c) = 1$ if c preserves α. Otherwise $Ref_C(c) = 0$.

To compute $Ref_C(c)$, we use $Ref_B(B_j)$. The label $Ref_B(B_j)$ indicates that whether B_j has a reflectional symmetry which fixes c. Further, it indicates that whether there is a fixed edge adjacent to c. We now state the algorithm.

`Algorithm Compute_Ref_C`

1. Partition $B_1, B_2, \ldots, B_m$ into isomorphism classes X_l using $Iso_B(B_j)$.
2. Let $y_l := |X_l|$.

3. If all y_l are even, then $Ref_C(c) := 1$.
4. If there is an y_l which is odd and $Ref_B(B_j) = -1$ for $B_j \in X_l$, then $Ref_C(c) := 0$.
   else $Ref_C(c) := 1$.

7.2 Finding Three Dimensional Symmetry at the Center

In this section, we briefly describe how to compute a maximum size three dimensional symmetry group at the center. We can compute a maximum size three dimensional symmetry group of the whole graph by computing a maximum size three dimensional symmetry group at the labeled center. This is based on the following theorem.

Theorem 8. *The center of the BC-tree is fixed by a three dimensional symmetry group of an one-connected planar graph.*

The center may be a block or a cut vertex. If the center is a block B, then we use algorithm `3DBiconnected_Planar` in Section 6. If there is a fixed cut vertex c, then we need to check the labels of c. See Figure 3 (a) for example.

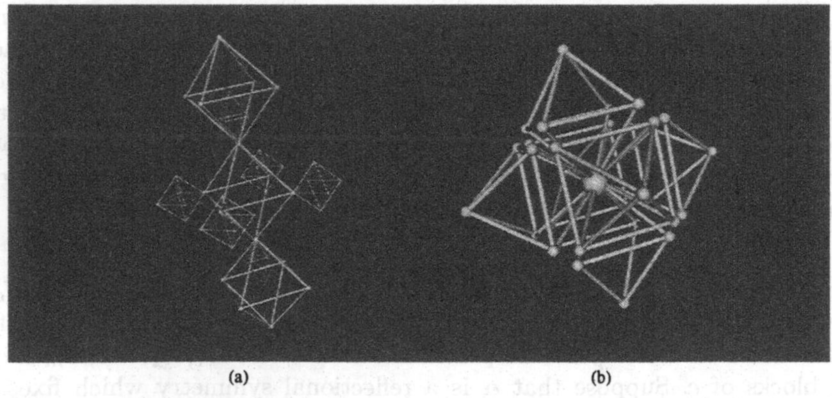

(a) (b)

Fig. 3. *Example of (a) B-center and (b) c-center.*

If the center is a cut vertex c, then we use a similar method that was used in the case of trees [9]. If there is a fixed block B, then we need to check the labels of B. See Figure 3 (b). We omit this algorithm from this extended abstract.

We conclude this section by analyzing the time complexity.

Theorem 9. `3DOneconnected_Planar` *takes $O(n^2)$ time.*

8 Conclusion

In this paper, we show that the problem of drawing a graph with a maximum number of symmetries in three dimensions is NP-hard. Then we present a polynomial time algorithm for finding maximum number of three dimensional symmetries in planar graphs. As a further work, we would like draw general graphs symmetrically in three dimensions.

References

1. L. Babai, Automorphism Groups, Isomorphism, and Reconstruction, Chapter 27 of *Handbook of Combinatorics*, Volume 2, (Ed. Graham, Groetschel and Lovasz), Elsevier Science, 1995.
2. S. Bachl, Isomorphic Subgraphs, *Graph Drawing*, Lecture Notes in Computer Science 1731, (Ed. J. Kratochvil), pp. 286-296, Springer Verlag, 1999.
3. G. Di Battista and R. Tamassia, On-Line Maintenance of Triconnected Components with SPQR-Trees, *Algorithmica* 15, pp. 302-318, 1996.
4. H. Chen, H. Lu and H. Yen, On Maximum Symmetric Subgraphs, *Graph Drawing*, Lecture Notes in Computer Science 1984, (Ed. J. Marks), pp. 372-383, Springer Verlag, 2001.
5. P. Eades and X. Lin, Spring Algorithms and Symmetry, *Theoretical Computer Science*, Vol. 240 No.2, pp. 379-405, 2000.
6. H. Fraysseix, An Heuristic for Graph Symmetry Detection, *Graph Drawing,* Lecture Notes in Computer Science 1731, (Ed. J. Kratochvil), pp. 276-285, Springer Verlag, 1999.
7. S. Hong, P. Eades and S. Lee, Drawing Series Parallel Digraphs Symmetrically,*Computational Geometry: Theory and Applicatons* Vol. 17, Issue 3-4, pp. 165-188, 2000.
8. S. Hong, P. Eades and S. Lee, An Algorithm for Finding Geometric Automorphisms in Planar Graphs, *Algorithms and Computation,* Lecture Notes in Computer Science 1533, (Ed. Chwa and Ibarra), pp. 277-286, Springer Verlag, 1998.
9. S. Hong and P. Eades, An Algorithms for Finding Three Dimensional Symmetry in Trees, *Graph Drawing*, Lecture Notes in Computer Science 1984, (Ed. J. Marks), pp. 360-371, Springer Verlag, 2001.
10. S. Hong and P. Eades, An Algorithms for Finding Three Dimensional Symmetry in Series Parallel Digraphs, *Algorithms and Computation*, Lecture Notes in Computer Science 1969, (Ed. D. T. Lee and S. Teng), pp. 266-277, Springer Verlag, 2000.
11. J. E. Hopcroft and R. E. Tarjan, Isomorphism of Planar Graphs, *Complexity of Computer Computations,* R. E. Miller and J. W. Thatcher, eds., Plenum Press, New York, pp. 131-151, 1972.
12. J. E. Hopcroft and J. K. Wong, Linear Time Algorithm for Isomorphism of Planar Graphs, *Proceedings of the Sixth Annual ACM Symposium on Theory of Computing,* pp. 172-184, 1974.
13. R. J. Lipton, S. C. North and J. S. Sandberg, A Method for Drawing Graphs, In *Proc. ACM Symposium on Computational Geometry*, pp. 153-160, ACM, 1985.
14. A. Lubiw, Some NP-Complete Problems similar to Graph Isomorphism, *SIAM Journal on Computing* 10(1), pp. 11-21, 1981.
15. P. Mani, Automorphismen von Polyedrischen Graphen, *Math. Annalen*, 192, pp. 279-303, 1971.

16. J. Manning and M. J. Atallah, Fast Detection and Display of Symmetry in Trees, *Congressus Numerantium* 64, pp. 159-169, 1988.
17. J. Manning and M. J. Atallah, Fast Detection and Display of Symmetry in Outerplanar Graphs, *Discrete Applied Mathematics* 39, pp. 13-35, 1992.
18. J. Manning, *Geometric Symmetry in Graphs*, Ph.D. Thesis, Purdue Univ., 1990.
19. G. E. Martin, *Transformation Geometry, an Introduction to Symmetry*, Springer, New York, 1982.

User Hints for Directed Graph Drawing

Hugo A.D. do Nascimento* and Peter Eades**

Basser Department of Computer Science,
The University of Sydney, Australia
{hadn,peter}@cs.usyd.edu.au

Abstract. This paper investigates an interactive approach where users can help a system to produce nice drawings of directed graphs by giving hints to graph drawing algorithms. Hints can be three kinds of operations: focus on a specific part of the drawing that needs improvement, insertion of layout constraints, and manual changes of the drawing. These hints help the system to escape from local minima, reduce the size of the solution space to be explored, and input domain knowledge. The overall aim is to produce high quality drawings. We present a system based on this approach and a pilot study involving human tests.

1 Introduction

Many graph drawing methods have been developed to produce drawings of graphs to satisfy aesthetic criteria, such as showing few edge crossings and monotone edge direction (eg. upward or downward). The optimization problems inherent in graph drawing are mostly NP-hard. The aesthetics may also conflict, that is, there is no optimum solution for two criteria simultaneously. As a consequence, graph drawing methods are mainly heuristics that work reasonably fast, but may result in poor quality drawings. Even amongst papers in Graph Drawing [1,2,3,4], one can find drawings that are produced in a few seconds, but present many edge crossings, edge bends, no symmetry, etc. In application areas such as Software Engineering, the drawings are often worse; see [5]. For instance, drawings of the Unix System Family tree that appear in many papers about directed graph drawings show at least one edge crossing [2]. It is interesting to note that this graph is upward planar.

There are several approaches for dealing with the weakness of automatic graph drawing methods. The most popular one is to apply the method for generating an initial drawing, and then improve the drawing manually. This is perhaps the most common way of creating a winning drawing for the *Graph Drawing Contest* [6]. The automatic method solves a great part of the problem by finding approximate positions for the vertices, and then the user produces the final layout by attending to other problems (including problems driven by domain knowledge) that the heuristics do not solve. In many cases, the user can easily

* Lecturer of the Inst. of Informatics, UFG-Brazil. PhD scholarship of CAPES-Brazil.
** Supported by the Australian Research Council.

P. Mutzel, M. Jünger, and S. Leipert (Eds.): GD 2001, LNCS 2265, pp. 205–219, 2002.
© Springer-Verlag Berlin Heidelberg 2002

recognize part of the drawing that needs to be improved and a way of improving it, as long as a good initial drawing is provided.

An alternative approach is to develop better (and more complex) algorithms that consider several rules about how to draw edges and vertices, or to use meta-heuristics that explore the space of possible layout solutions. It is not clear, however, that better algorithms and methods can ever eliminate the possibility of having the user doing some post-processing. This is because there are always some instances for which sufficiently fast algorithms do not produce the best solution, or the space of solutions is too large to be adequately searched in a reasonable amount of time. Moreover, it is common to have a large number of good equivalent layouts for the same graph, where the decision of which one to be taken is subjective, or domain dependent. Even when some subjective aspects can be modeled as objective functions and constraints in flexible algorithms [7,8,9], it is difficult to ensure that all user preferences are considered, and that they imply no ambiguity by leading to a single "optimum" solution. It may also be difficult to assure that the algorithm will find this solution. Thus, in the most extreme situation, the user is still important for validating the result produced by automatic methods or for selecting between a number of good drawings.

In this paper we investigate an approach for drawing directed graphs that considers the user as a *fundamental* element of the graph drawing process. The user plays a very important role by cooperating with an optimization method to produce good quality drawings. Basically, the user provides *hints* to an optimization method, helping it to escape from local minima, to converge much faster to better results, or to clarify subjective and domain dependent aspects.

We apply our user-hints based approach to the problem of drawing directed graphs resulting in an interactive system. Directed graphs were chosen since they appear in several real applications and involve many difficult graph drawing problems. We use the popular method of Sugiyama et al. [2] as our main optimization method. A human pilot study of our system is done in order to analyze the contribution of user hints to the quality of graph drawings. This paper is organized as follows. Section 2 describes some related research regarding the contribution of users to optimization processes. Section 3 presents our approach based on user hints, and Section 4 describes how hints can be incorporated to the Sugiyama method. An interactive system that supports hints is presented in Section 5. Section 6 discusses our experiments with the system and the results obtained. Finally, Section 7 draws our conclusions about interactive graph drawing based on user hints and offers future research directions.

2 Related Work

Due to the high complexity of optimization problems and the constant demand for solutions close to the optimum, some new research has been done in the direction of allowing users to cooperate with automated optimization methods.

A good example of such studies is the work presented by Anderson et al in [10], where the authors introduce a cooperative paradigm called *Human-Guided*

Simple Search (or *HuGSS* for short). HuGSS divides an optimization process into two main subtasks carried out by different entities: the computer is responsible only for finding local minima using a simple hill-climb search; the user is responsible for escaping from the minima leading the search to better solutions. Some visualization techniques help to identify parts of a solution that are promising for improvement. Then the user can either manually change the solution or focus the search on the identified parts. The focus consists of setting a high search priority for a part of the solution and/or defining how deep the search will be executed. The HuGSS paradigm was applied to the problem of capacitated vehicle routing with time windows, and it showed that a simple hill-climb algorithm could be significantly enhanced by user interaction. The same idea was used for the graph clustering problem [11].

Another interesting work on user interaction appears in [12]. It consists of an interactive constraint-based system for graph drawing using force-directed placement [13]. The system works as follows: a graph is modeled as an energy system with springs linking all pairs of vertices, and a method for quadratic optimization is set to continuously compute a layout that correspond to a state of minimal energy. While the optimization method is running, the user can incrementally add constraints to the model, in order to confirm the drawing to his or her desires. Constraints are called VOFs (*Visual Organization Features*) and include a variety of layout aspects such as: showing two vertices close to each other, showing an edge as an orthogonal line, etc. VOFs are implemented as extra springs and added to the original energy model. The system solves constraints searching for a new state that minimizes the energy of the entire set of springs (composed by the original springs of the graph plus the constraint springs). The user may manually move some vertices to help the optimization method to escape from local minima. This characterizes a very general and flexible approach for solving layout constraints that are added during run time.

User guidance can also occur in a much softer way, not by setting algorithms to do specific tasks, but indicating whether or not they are on the right path. This approach was adopted by Branke et al [14,15] using Genetic Algorithm (GA) for drawing general graphs. In their system a genetic algorithm tries to minimize a weighted function of seven aesthetic criteria (minimum number of crossings, high angular resolutions, many symmetries, etc.) as it is commonly done by other meta-heuristics for Graph Drawing [7,8]. However, instead of executing the meta-heuristic for a long period of time until the drawing has converged, the system stops every few iterations and gets feedback from the user. Basically, it shows the best eight drawings currently produced (one drawing for each one of the seven aesthetic criteria, and the best drawing according to the weighted function), and asks the user for scores between 0 to 9 for each drawing. Then the system uses those scores to adjust the weight of each aesthetic criterion in the weighted function. For instance, giving high scores to drawings with few edges crossings contributes to an increasing of the relative weight assigned to the edge crossing criterion in the system. At the end, the system "knows" the relative importance of each criterion that describes the user's aesthetics for a specific drawing.

All these pieces of research show, in different ways, an emerging trend for having users guiding optimization processes. Our approach is close to the HuGSS paradigm in the sense that users are more active and provide pieces of information directly to the optimization method. However, we also consider the possibility of adding constraints to the problem in runtime and use other optimization methods than just hill-climbing ones. The next sections describe our approach in details.

3 Hints for Directed Graph Drawing

We use *user hints*, or just *hints*, to refer to the information provided by users to optimization methods. A hint should help the system to escape from local minima, to accelerate the optimization process, or to solve ambiguity in cases where there is more than one feasible solution. In the context of this paper, hints help graph drawing algorithms to search for high quality graph drawings according to a set of aesthetic criteria.

3.1 Types of User Hints

We consider three kinds of hints for directed graph drawing:

- **Focus.** The aim of focus is to reduce the space of solutions to be explored by a search method. In general, after running a graph drawing algorithm on the whole graph we get a reasonably good drawing, with some areas that do not satisfy the aesthetic criteria. By identifying this, the user can focus the drawing algorithms again on the areas with poor quality in order to improve them. This means that the algorithms will redraw only the focused areas of the graph. The layout of vertices in the non-focused areas is not changed.
- **Layout Constraints.** Layout constraints are useful for helping the system to fix bad quality aspects of a drawing, or for removing ambiguity about where to draw some vertices. We have adopted two kinds of layout constraints, *Top-Down* and *Left-Right*. The *Top-Down* constraint defines an *above-relation* between two vertices u and v, such that u has to appear somewhere above v in the drawing. Similarly, the *Left-Right* constraint defines an *on-the-left-relation* between two vertices u and v.
- **Manual Changes.** Other drawing aspects that are not easily controlled by focus and layout constraints can be fixed by manual changes. The user does manual changes only on vertices by moving them to a different position of the drawing. Changes on edges can be done by moving their related vertices. The mechanism of manual changes is already commonly used in graph drawing activities as a part of a post-processing and fine-tuning step. However, here we have a much more powerful tool, since changes in a drawing may drive the system out of a local minimum to a better solution.

3.2 A Framework for Giving User Hints

An interactive framework for giving hints is shown in Figure 1. Arrows with a capital label represent the action of giving a hint. Note that all kinds of hints are direct or indirect input to the optimization method. The drawing activity occurs as iterative steps with the user providing hints to the optimization method. Then the method produces a new drawing of the graph by taking into consideration information about the current drawing and the hints. These steps repeat until the user is happy with the drawing.

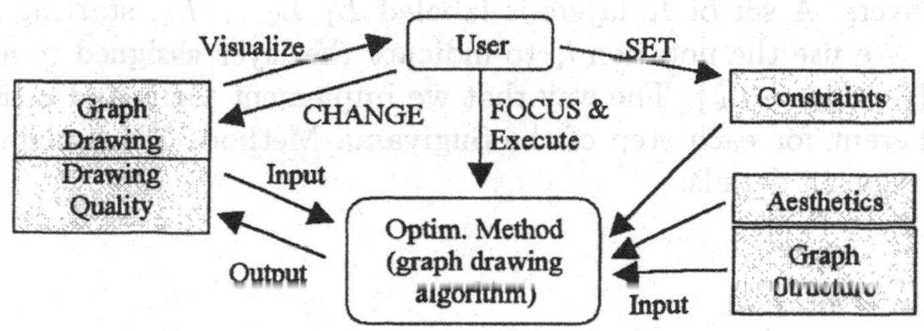

Fig. 1. Interactive framework user hints.

3.3 Implementing Focus and Constraints in the Sugiyama Method

For the purpose of drawing directed graphs, we use an optimization method based on the Sugiyama method [2]. The Sugiyama method draws a graph on a hierarchical set of horizontal lines called layers. The method consists of four steps that in general involve NP-hard problems [1,2]:

1. *Cycle Removal*: it reverses some edges of the graph in order to make it acyclic.
2. *Layer Assignment*: vertices are assigned to layers such that edges show a uniform orientation as much as possible. When an edge intersects one or more layers, it is replaced by a set of small edges (u, n_1),(n_1, n_2),... (n_i, n_{i+1}),(n_k, v), where n_i, $i = 1, 2 \ldots k$ are dummy vertices inserted on each intersection point, and u and v are real vertices.
3. *Crossing Reduction* : vertices are sorted in each layer in order to reduce the number of crossings.
4. *Horizontal Coordinate Assignment*: the X- coordinate of each vertex is computed making edges as straight as possible, and reducing bends and the width of the drawing. All edges changed in step 1 are also reversed to their original orientation.

We preserve the general structure of the Sugiyama method and adjust each step to support focus and layout constraints. Focus has two effects: it limits the action of the graph drawing algorithms to the focused vertices and it defines special constraints that "freeze" the non-focused vertices. Thus, given a graph $G = (V, E)$, we focus on a selected set $A \subseteq V$, by running the Sugiyama method only on A. The X,Y coordinates of the vertices in $V - A$ are kept fixed. On the other hand, layout constraints are modeled either as extra edges added to the graph or as normal constraints that impose an ordering to vertices. Layout constraints can be defined only for real vertices. Some similar kinds of constraints for the Sugiyama method are investigated in [16] and [17].

For simplicity, in the rest of this paper we use the term *selected vertex* meaning a vertex in the selected set A, and *fixed vertex* for a vertex in $V - A$. The drawing is constructed on a grid of integer coordinates. The rows of the grid represent layers. A set of K layers is labeled $L_1, L_2, \ldots L_k$, starting from bottom to up. We use the notation l_v to indicate the layer assigned to a vertex v, with $l_v \in \{L_1, L_2, \ldots L_k\}$. The way that we implement focus and constraints is slightly different for each step of the Sugiyama Method. We explain now this implementation in details.

3.4 Cycle Removal

In the Cycle Removal step, the focus mechanism has no effect. Cycles involving selected vertices and fixed vertices are treated equally (Left-Right constraints also do not affect the drawing). Top-Down layout constraints, however, cause a great impact on the final result of this step.

As an example of how important layout constraints are, consider a graph composed of a ring with four vertices, a, b, c and d. There are four basic ways of drawing this ring such that the number of downward edges is maximum (optimal). These drawings are shown in the Figure 2(a) to (d). Without constraints, the four drawings are equivalent according to the number of downward edges. However, if the user prefers to have the vertex a drawn above the vertex c, and inserts a Top-Down constraint on a and c, this operation reduces the number of optimal solutions to only two, shown in Figures 2(a) and (b). If the user inserts another Top- Down constraint, now on b and d, this lets us with a single optimal solution, Figure 2(a). The user can go even further by inserting another Top-Down constraint to have c drawn above b. In this case, the only valid solution is to reverse the edge (b, c) and (d, a), Figure 2(e), since the constraints defines a precise order: a above c, c above b, and b above d.

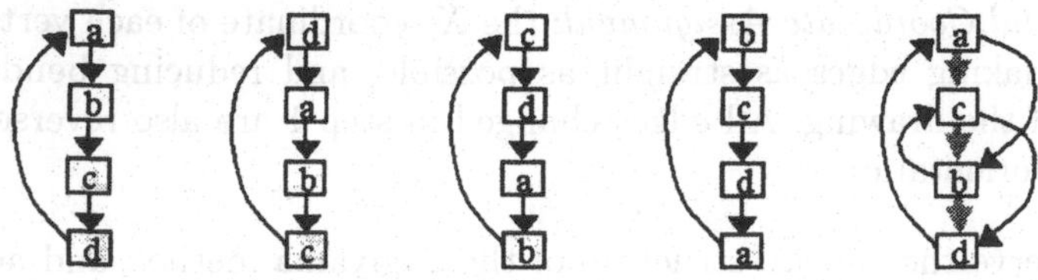

Fig. 2. Different ways of drawing a ring. Thick arrows represent constraints.

In summary, layout constraints can be used not only to reduce the number of feasible solutions, but also to force the system to considerer a specific solution. Note that all layout constraints, Top-Down and Left-Right, involve a pair of vertices. Therefore, layout constraints are modeled as special directed edges, that we call *constraint edges*. Constraint edges can be freely inserted into the system, providing that they do not make a cycle.

Considering the effect of layout constraints, we developed a new approach for the Cycle Removal step. Let $G = (V, E)$ be the graph to be drawn and L the set of Top-Down constraints. First, we construct a new graph G'' by merging L with G. Whenever an original edge and a constraint edge (excluding orientation) connect the same pair of vertices, we remove the original edge and leave the constraint. The merge procedure can be formalized as: $G'' = (V, E'')$, where $E'' = L \cup \{(u,v) \in E | (u,v) and (v,u) notin L\}$. If the resulting graph G'' is acyclic then the problem was solved. Otherwise, a method for the Feedback Arc Set problem is applied to this graph, but it is set to reverse only original edges. For instance, by merging the ring from the previous example with the set of constraints in Figure 2(e) causes the edge (b, c) to be removed. Then, the next step is to reverse some edges in G'' in order to break cycles. The algorithm for this task can reverse any edge, except (a, c), (c, b) and (b, d), which represent constraint. The optimal solution would be to reverse only (d, a).

We modify the *Greedy-Cycle-Removal* heuristic in [1] in order to solve the Feedback Arc Set problem with constraint edges; see Figure 3. The modification, highlighted in bold, is minor, however, it assures that only original edges are reversed. The advantage of using this algorithm is that it is simple and runs in linear time. The algorithm works by removing vertices from the graph and adding them either to a list S_l or to a list S_r. Finally, S_l is concatenated with S_r to form S. The list S provides a sequence of the vertices of G''. Then, all edges $(u, v) \in E'$ with v appearing before u in S are reversed, resulting in a acyclic graph.

Let G_c be a copy of G'.
1. Initialize both S_l and S_r to be empty lists.
2. while G_c is not empty do
   (a) while G_c contains a sink do
   Choose a sink u, remove it from G_c, and prepend it to S_r. (Isolated vertices are also considered sinks at this stage.)
   (b) while G_c contains a source do
   Choose a source u, remove it from G_c, and append it to S_l.
   (c) if G_c is not empty then
   Choose a vertex u, such that **there is no constraint edge (v,u) for any vertex v left in G_c**, and the difference *outdeg(u)-indeg(u)* is maximum; remove u from G_c and append it to S_l.
3. Concatenate S_l with S_r to form S.

Fig. 3. Modified version of the Greedy-Cycle-Removal heuristic presented in [1].

3.5 Layer Assignment

The Layer Assignment is executed for the graph $G'' = (V, E'')$ produced by the previous step. Focus is considered here, and it is implemented by modifying a layering algorithm for not changing the coordinates of fixed vertices. Note that edges with fixed vertices in both ends may not affect the layering algorithm, so they can be removed for saving time. This approach is presented in Figure 4.

1. Let $G'=(V,E')$ be the graph resulting from the previous step, and $A \subseteq V$ the set of selected vertices. Remove all edges (u,v) from E' with u and v fixed vertices (belonging to V-A).
2. Apply a Layering Algorithm on G'. The algorithm, however, should be modified for not changing the coordinates of the fixed vertices.

Fig. 4. Approach for Layer Assignment with constraints and focus.

We use the Longest Path Layering heuristic [18] to construct a layering of G''. This algorithm results in drawings that are in general too wide; however, it can be easily modified to handle focus, and it runs in linear time. The original version of this heuristic places all sinks in the bottom layer, L_1. Then each remain vertex u is placed in layer L_{p+1} where the longest path from u to a sink has length p. In our modified version we consider that all fixed vertices are sinks with predefined layers assigned to them, even though they may have outgoing edges. This implies that each non-sink vertex u will be placed in layer L_{p+q}, where p is the length of the longest path from u to a sink, q is the label of the layer where the sink is, and $p+q$ is maximal. Note that our layering algorithm may violate Top-Down constraints in a special case where they conflict with focus. Consider that there is a chain of directed edges $(v_1, v_2), (v_2, v_3), \ldots, (v_{k-1}, v_k)$, where v_1 is a fixed vertex, $v_2 \ldots v_{k-1}$ are selected vertices, v_k is a sink (or a fixed vertex), and $l_{v_1} - l_{v_k} < (k-1)$, with l_{v_1} and l_{v_k} the layers assigned to v_1 and v_k respectively. If (v_1, v_2) is a constraint edge, then this constraint will be violated since v_2 will be assigned to a layer above v_1. All other edges (v_{i-1}, v_i), $i = 3, \ldots, k$ will point downward.

We developed a solution for the case where v_k is a normal sink. It consists of adding a post-processing step that uses the previously computed layering to shift down some vertices. Basically, for each vertex $u \in V$ taken in the topological order we assign u to a new layer $l_u = min(l_v - 1$: for all vertices v such that there is an edge $(v, u) \in E'')$; if there is no edge $(v, u) \in E''$ for any $v \in V$, u is kept in its current layer. The revised algorithm still runs in linear time. It moves all vertices $v_2, \ldots v_{k-1}$ $k + l_{v_k} - l_{v_1} - 1$ layers down. Unfortunately, the problem persists for the case where v_k is a fixed vertex: the vertex v_{k-1} will be assigned to a layer below vertex v_k. However, this is a problem due rather to a conflict between focus and layout constraints than to the layering algorithm itself. Our approach favors focus against layout constraints.

3.6 Crossing Reduction

In the next step of the Sugiyama method, we use the original graph $G = (V, E)$ as well as the layering defined by the previous step. A version of the barycenter algorithm [2] is applied to handle focus and Left-Right constraints. This version adjusts only the X-coordinate of selected vertices and solves constraints during the processing. A general description of the algorithm is shown in Figure 5.

Let $L_1, L_2, \ldots, L_K$ be the set of K layers defined by the Layer Assignment step.

Repeat until the number of edge crossings is minimal

1. **For $i \leftarrow K$-1 to 1 do**
   a. **For each selected vertex u in layer L_i, move u to its barycenter position according to its adjacent vertices in layer L_{i+1}. If u has no adjacents in L_{i+1}, its actual position is preserved.**
   b. **FixConstraints (i).**
2. **For $i \leftarrow 2$ to K do**
   a. **For each selected vertex u in layer L_i, move u to its barycenter position according to its adjacent vertices in layer L_{i-1}. If u has no adjacents in L_{i-1}, its actual position is preserved.**
   b. **FixConstraints (i).**

Fig. 5. Barycenter algorithm for Crossing Reduction with constraints and focus.

The algorithm uses a heuristic called *FixConstraints(i)* that reorganizes the vertices in layer L_i. This heuristic constructs a set S composed of all selected vertices in L_i that overlap (have the same X-coordinate), have a non-integer X-coordinate or do not satisfy one or more Left-Right constraints. Then the algorithm moves each vertex in S to an empty integer position in L_i where the number of unsatisfied constraints is minimized. If there are many equally possible empty positions in the layer, the algorithm chooses the one closest to the vertex"s current location. Left-Right constraints may involve vertices in different layers. In this case, all constraint edges that have at least one vertex sitting on layer L_i will be analyzed.

Note that our implementation of FixConstraints may not result in satisfaction of all Left-Right constraints, since it analyses locally the layers, and it demands the existence of empty positions for moving vertices. Some vertices may also be fixed, forbidding constraints to be solved. Nevertheless, we expect the heuristic to solve many constraints when applied several times in the barycenter algorithm.

3.7 Horizontal Coordinate Assignment

Finally, the last step of Sugiyama method, the Horizontal Coordinate Assignment, is not explicitly included in our approach. This is because the barycenter algorithm combined with FixConstraints already assigns X-coordinates that do not produce many bends or long edges. Moreover, the algorithm in Figure 5 can be re-applied for improving the horizontal coordinate assignment by focusing only on vertices that cause bends or long edges.

4 The GDHints System

We implemented the Sugiyama steps described in the previous section into an interactive system, called GDHints[1]. A snapshot of the system is shown in Figure 6. The system includes:

- a user interface, by which the user can select vertices for focus, add and delete constraints or perform manual changes;
- graph drawing functions for layering (cycle removal and layer assignment) and ordering (for crossing reduction); and
- displays of quality metrics of drawings.

User-System Cooperation and Quality Feedback. The system and the user work together for improving a drawing of a graph. The drawing is improved when its new layout is better than the previous one in the following priority of aesthetic criteria: (1) number of upward edges, (2) number of edge crossings, (3) number of dummy nodes, (4) number of edge bends and (5) drawing area. At the beginning of the processing, the layering and the ordering functions are automatically executed for a new graph in order to produce an initial drawing. Then the user can call these functions again for redrawing selected parts of the graph. The system evaluates the quality of every new drawings and automatically saves the best drawing generated so far. At any time, the user can return to the best drawing or can force the system to accept the current drawing as the best one.

The system provides useful feedback for the user"s actions. This includes colors for highlighting bad quality aspects of the drawing, and sound and animation events for calling the user"s attention whenever a new solution better than the current best one is found.

5 Pilot Study

An initial study with human experiments was done for validating our approach. Five users took part in the experiments. All of them have a background in Computer Science and in Graph Drawing. The users also had a 30-minute introduction about the system, before starting the experiments.

The study involved three kinds of experiments: $E1$ (constraints only), $E2$ (constraints + focus) and $E3$ (constraints + focus + manual changes).

These experiments were done using six graphs, which details are shown in Table 1. Graphs $G3$, $G4$, $G5$ and $G6$ are from [2,3,4]. In total 90 experiments

[1] Our system can be downloaded from *www.cs.usyd.edu.au/~visual/systems/gdhints*

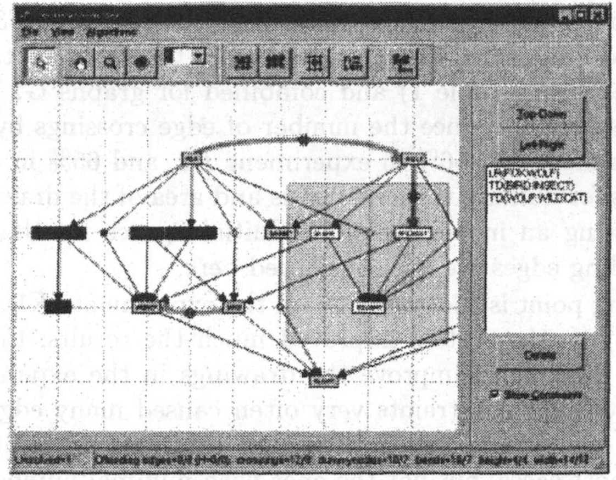

Fig. 6. The interactive system based on user hints

were done (5 users x 3 types of experiments x 6 graphs). We started with small graphs, so that the users could improve their skills in giving hints smoothly. Constraints were allowed in all experiments since they are an advanced feature in interactive graph drawing and we wanted to test them as much as possible. On the other hand, manual changes are very intuitive (the users could tend to use mainly this option). For this reason, we considered manual changes only in experiment $E3$.

The users" actions were recorded into history files for further analysis. After the experiments we also got subjective feedback from the users. The first 15 experiments (related to graph $G1$) were not included in the average analysis of the system, since we considered the users were still learning how to use the system during that time.

Table 2 shows the minimum, the maximum and the average values of the aesthetic criteria for the best drawings produced by the 5 users.

Compared to the initial solutions described in Table 1, the number of offending edges was not improved much. This is because the layering algorithm already produces a result very close to the optimum. On the other hand, there was a significant reduction of the number of edge crossings. The experiments where not all users could improve crossings were the ones based only on constraints. When focus and manual changes were allowed, all five users produced drawings with lesser crossings. Note that the number of crossings could still be smaller than the minimum presented here for some graphs, but this may result in a worse solution, with more offending edges. Regarding the numbers of dummy vertices and bends, and the area of the drawings, they were higher than the initial figures for almost all experiments. This shows that such aesthetics are in general inversely proportional to the improvement of edge crossings. Some examples of drawings produced by the users can be seen in the Appendix. Other drawings are available in our web-page.

We present relative results for experiments $E1$, $E2$ and $E3$ in Table 3. It contains, in percentage, the average values from the previous table divided by the initial values (from Table 1) and combined for graphs $G2$ to $G6$. We can see that the users could reduce the number of edge crossings by about 20% on average in experiment $E1$, 56% in experiment $E2$, and 65% in experiment $E3$. The percentages for dummy vertices, bends and area of the drawings are greater than 100% showing an increasing of the initial figures for these criteria. The values for offending edges are not considered here.

An interesting point is the gap between the experiments $E1$ and $E2$. Adding the focus facility to the system improved much the results. In fact, the users affirmed it was difficult to improve the drawings in the experiment $E1$, since adding new Left-Right constraints very often caused many edge crossings. We concluded that the system was able to find solutions that satisfy Left-Right constraints in most cases, but not the ones with minimal number of crossings.

The usage (percentage of user interactions represented by mouse clicks) of the main operations in the system is presented in Table 4. The row *Total* is the overall results for all experiments. The column *Other* includes align to grid operations, zoom in, zoom out and other actions.

Table 4 shows that constraints played a less important role in the optimization processing compared to focus (*select operations*) and manual changes (*move operations*). Ordering was the most significant operation, and it was executed for improving drawings after giving hints to the system. The users also pointed out that the option for returning to the best solution was very important. They used this facility in a simple search approach for escaping from local minima: performing changes on the drawing and executing the ordering function several times for improving the solution; after some iterations, the user presses a button to recover the new best solution found.

Table 1. Graphs used for the experiments.

Name	|V|	|E|	Quality of the initial drawings (produced by the system using the Sugiyama method)				
			Offending edges	Crossings	Dummy Vertices	Bends	Area
G1 – Waleska	10	20	2	9	17	13	72
G2 – Klayer	18	24	0	5	0	4	24
G3 – Ecosystem	15	26	0	16	7	7	48
G4 – Csyntax	34	46	4	12	39	19	182
G5 – Unix Sys. Family Tree	41	49	0	4	24	9	132
G6 - Dynworld	43	69	6	70	144	56	360

A final result obtained from the experiments regards processing time. The users spent on average 14 minutes on each experiment. However, just 10% of this time was used by the system for doing some processing. In the other 90%, that

Table 2. Quality of the best drawings produced by the users for all graphs.

Graph	Exp	Offending Edges		Crossings		Dummy Vertices		Bends		Area	
		Min-Max	Av	Min-Max	Av	Min-Max	Av	Min-Max	Av	Min-Max	Av
G1	E1	2-2	2	5-7	6	17-30	19.6	12-22	14.6	72-99	79.2
	E2	2-2	2	4-5	4.2	17-17	17	11-14	12.2	72-90	77.4
	E3	2-2	2	3-4	3.8	17-26	18.8	8-13	10.2	40-81	58.4
G2	E1	0-0	0	2-5	3.8	0-8	2.8	0-8	2.4	24-48	33.2
	E2	0-0	0	1-2	1.4	8-11	9	4-10	7.4	42-77	55
	E3	0-0	0	1-2	1.2	6-14	10.2	4-13	7.8	30-80	52.8
G3	E1	0-0	0	9-14	10	7-16	8.8	7-14	8.4	48-60	51.2
	E2	0-0	0	5-13	9	7-57	27.2	5-20	13.2	48-122	85.2
	E3	0-0	0	5-8	6.4	15-37	24.6	10-26	16.8	60-119	84
G4	E1	3-4	3.4	8-12	10.8	39-100	63.8	15-46	28.8	182-440	267.2
	E2	3-4	3.4	6-9	7	38-107	61.6	17-35	21.6	168-288	219.2
	E3	3-4	3.4	4-10	6.4	36-88	56.4	14-34	23.4	168-270	223.6
G5	E1	0-0	0	3-4	3.6	24-24	24	7-12	8.6	132-176	151.8
	E2	0-0	0	0-2	0.4	25-32	28.8	8-17	11.6	132-168	152.8
	E3	0-0	0	0-0	0	25-30	26.8	7-14	9.6	132-156	147.8
G6	E1	6-6	6	52-65	58.4	144-212	167.8	40-55	50.4	360-522	423
	E2	6-6	6	35-60	47.4	144-167	154.6	39-76	58	384-420	404
	E3	6-6	6	35-46	40.6	162-250	193.6	34-94	67.8	384-621	456

Table 3. Overall results of the experiments compared to the quality of the initial drawings.

Experiment	Crossings	Dummy Vertices	Bends	Area
	Av	Av	Av	Av
E1	80.3%	126%	103.4%	124.8%
E2	44.0%	195%	143.9%	151.0%
E3	35.0%	185%	157.1%	151.1%

Table 4. Usage of the main operations.

Exp.	Select	Move	Layering	Ordering	Return to Best Sol.	Add Constr.	Delete Constr.	Other
Total	10.6%	10.7%	9.7%	60.2%	2.2%	4.3%	1.3%	1%
E1	0%	0%	10.7%	76.3%	2.4%	7.7%	2.5%	0.4%
E2	16.2%	0%	11.3%	64.6%	1.9%	4.5%	1.1%	0.4%
E3	15.5%	33.3%	6.9%	39.1%	2.4%	0.6%	0.1%	2.1%

we call idle time, the system was stopped, waiting for the user to do some action. During that time the user was thinking about what kind of hint to give to the system. This indicates that there is much CPU power left for improving the cooperation between the system and the user in our approach. We envision a more collaborative framework where the system may work in background improving the results.

6 Conclusion

User hints, particularly focus and manual changes helped an optimization process based on the Sugiyama method to improve drawings of directed graphs. Some effort was demanded in order to adjust the traditional graph drawing method to allowing user interaction. However, the system with this facility seems more attractive and powerful than just a simple manual post-processing of the drawings. We are now improving our system based on the results obtained from the pilot study. The constraint mechanism is being revised to include new constraints that can represent the users desires better. We are also investigating the use of better optimization methods such as meta-heuristics and exact methods. For such methods, hints can provide a good way of reducing the space of solution. Moreover, we believe that focus and constraints can be implemented in a more straightforward way in those methods.

References

1. G. Di Battista, P. Eades, R. Tamassia, and I. G. Tollis. Graph drawing: algorithms for the visualization of graphs. New Jersey: Prentice-Hall, 1999.
2. K. Sugiyama, S. Tagawa, and M. Toda. Methods for visual understanding of hierarchical systems. IEEE Trans. Syst. Man Cybern., SMC- 11(2):109-125, 1981.
3. R. Tamassia, G. Di Battista, and C. Batini. Automatic graph drawing and readability of diagrams. IEEE Trans. Syst. Man Cybern., SMC- 18(1):61-79, 1988.
4. K. Sugiyama and K. Misue. Visualization of structural information: Automatic drawing of compound digraphs. IEEE Trans. Softw. Eng., 21(4):876–892, 1991.
5. Proceedings of the 22nd Annual Conference on Software Engineering (ICSE 2000). Limerick, Ireland, June 4-11, 2000.
6. Graph Drawing Contest. In Proceedings of the 8th International Symposium on Graph Drawing, GD 2000, Colonial Williamsburg, VA, USA, September 20-23, 2000. ISBN: 3-540-41554-8.
7. R. Davidson and D. Harel. Drawing graphs nicely using simulated annealing. Technical report, Department of Applied Mathematics and Computer Science, The Weizmann Institute of Science, Rehovot, 1989.
8. J. Utech, J. Branke, H. Schmeck, and P. Eades. An evolutionary algorithm for drawing directed graphs. In Proc. of The International Conference on Imaging Science, Systems, and Technology (CISST'98), pp. 154-160, Las Vegas, Nevada: CSREA Press, July 6-9, 1998.

9. H. A. D. do Nascimento, P. Eades and C. F. Xavier de Mendonca Neto. A multiagent approach using A-Teams for Graph Drawing. In Proceedings of the 9th International Conference on Intelligent Systems, Louisville, Kentucky - USA, June 15-16, 2000, pag 39-42.
10. D. Andersen, M. Andersen, M. Lesh, J. Marks, B. Mirtich, D. Ratajczac and K. Ryall, Human guided simple search, to appear in the proceedings of the annual conference of the American Association for Artificial Intelligent, 2000.
11. N. Lesh, J. Marks, and M. Patrigname. Interactive Partitioning. Graph Drawing Conference, 2000.
12. K. Ryall, J. Marks, S. Shieber. An interactive constraint-based system for drawing graphs. In Proc. of the ACM Symposium on User interface Software and Technology (UIST' 97), pages 97-104, Oct. 1997, Banff, Alberta.
13. T. M. J. Fruchterman and E. M. Reingold. Graph drawing by force-directed placement. Software-Practice and Experience, vol. 21, no. 11, 1129-1164, 1991.
14. K. Dauner. "Ein interaktiver Genetischer Algorithmus fur das Zeichnen von Graphen".("Interactive genetic algorithm for graph drawing"). Diplomarbeit (Master Thesis), Institute AIFB, University of Karlsruhe, 76128 Karlsruhe, Germany, 1997.
15. A. E. Jacobsen. "Interaktion und Lernverfahren beim Zeichnen von Graphen mit Hilfe evolutionarer Algorithmen" ("Interaction and learning methods for graph layouts with the help of evolutionary algorithms"). Diplomarbeit (Master Thesis), Institute AIFB, University of Karlsruhe, 76128 Karlsruhe, Germany, 2001.
16. E. Koutsofios and S. North. Drawings graphs with dot. Technical Report, AT&T Bell Laboratoties, Murray Hill, NJ, USA, Sep 1991.
17. K.-F. Böhringer and F. N. Paulisch. Using constraints to achieve stability in automatic graph layout algorithms. Conference proceedings on Empowering people: Human factors in computing system: special issue of the SIGCHI Bulletin, pages 43 - 51,1990.
18. P. Eades. A Heuristics for Graph Drawing. Congr. Numer., 42, 149-160, 1984.

Appendix – Examples of Drawings Produced by the Users

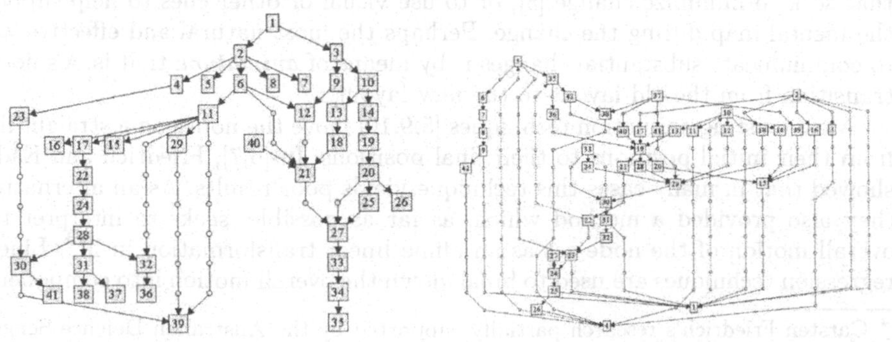

(a) Unix System Family (G5) (b) Forrester's World Dynamics Graph (G6)

Graph Drawing in Motion II

Carsten Friedrich[1]* and Michael E. Houle[2]**

[1] Basser Department of Computer Science
The University of Sydney
Australia
carsten@cs.usyd.edu.au
[2] IBM Research
Tokyo Research Laboratory
Japan
meh@cs.usyd.edu.au

Abstract. Enabling the user of a graph drawing system to preserve the mental map between two different layouts of a graph is a major problem. Whenever a layout in a graph drawing system is modified, the mental map of the user must be preserved. One way in which the user can be helped in understanding a change of layout is through animation of the change. In this paper, we present clustering-based strategies for identifying groups of nodes sharing a common, simple motion from initial layout to final layout. Transformation of these groups is then handled separately in order to generate a smooth animation.

1 Introduction

In many applications where graphs are used to convey relational information, user and application actions can result in drastic changes in structure and layout. Preserving the so-called 'mental map' during these changes has been identified as crucial to the usability of a system [2]. There are two possible approaches to maintaining the user's mental map: either to develop graph drawing algorithms that seek to minimize change [3], or to use visual or other cues to help support the mental map during the change. Perhaps the most natural and effective way to communicate substantial changes is by means of *animation*; that is, a smooth transition from the old layout to the new layout.

Most existing animation techniques [5,9,10] move the nodes on a straight line from their initial positions to their final positions. In [6,7], Friedrich and Eades showed that in many cases this technique yields poor results. As an alternative, they also provided a method which, as far as possible, seeks to interpret the overall motion of the node set as an affine linear transformation in $\mathbb{R}^2$. Linear regression techniques are used to break down the overall motion into components

* Carsten Friedrich's research partially supported by the Australian Defence Science and Technology Organization (DSTO).

** Michael Houle is on leave from the Basser Department of Computer Science, University of Sydney.

P. Mutzel, M. Jünger, and S. Leipert (Eds.): GD 2001, LNCS 2265, pp. 220–231, 2002.
© Springer-Verlag Berlin Heidelberg 2002

due to *scaling*, *rotation*, *shear*, *flip*, and *translation*. These components are then interpolated in parallel and recombined to generate frames for the animation.

Although this method works well when the motion of the nodes is more or less uniform, it performs poorly when several subgraphs have very different underlying motions, as is generally the case when only part of a graph layout is updated. The method attempts to compute an average over all motions of individual nodes, and as a result can decide upon a motion that is not well suited to all parts of the graph. Figure 1 on page 221 shows an example where computing the best overall affine linear transformation produces a poor result[1].

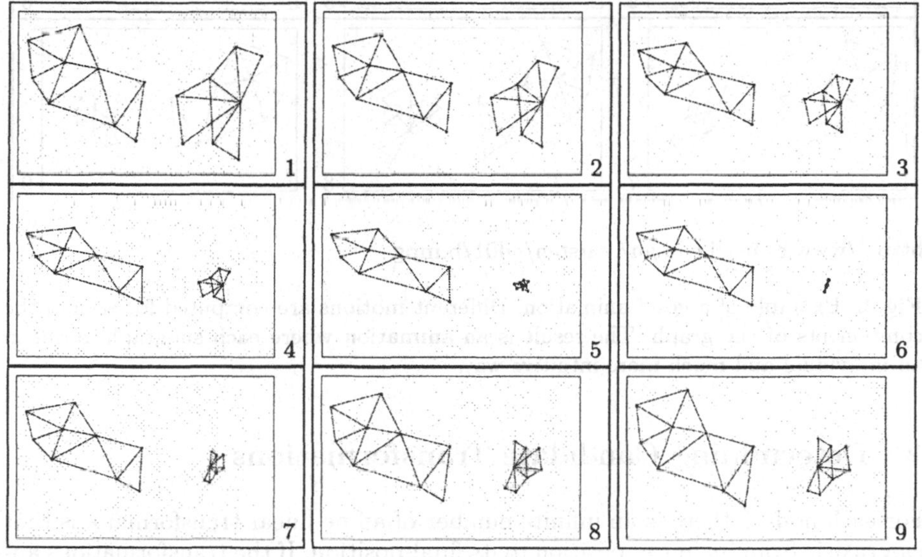

http://www.cs.usyd.edu.au/ carsten/gd01/a.mpg

Fig. 1. Example of a bad animation, starting at Frame 1 and ending at Frame 9. Although only the right-hand component of the graph should move, the calculation and application of an average movement causes distortion in the left-hand component.

In this paper, we use the approach of [7] as the basis of several clustering heuristics, which seek to identify subgraphs which share a similar, structured motion. Applying different transformations to the subgraphs identified allows for the partial update of graph layouts. Figure 2 shows the result of applying the method introduced in Section 3 to the situation of Figure 1.

[1] All examples, in the form of mpeg videos, and a free MPEG player for Windows-NT/95/98 (as well as references to free players for Unix) are available from http://www.cs.usyd.edu.au/ carsten/gd01/.

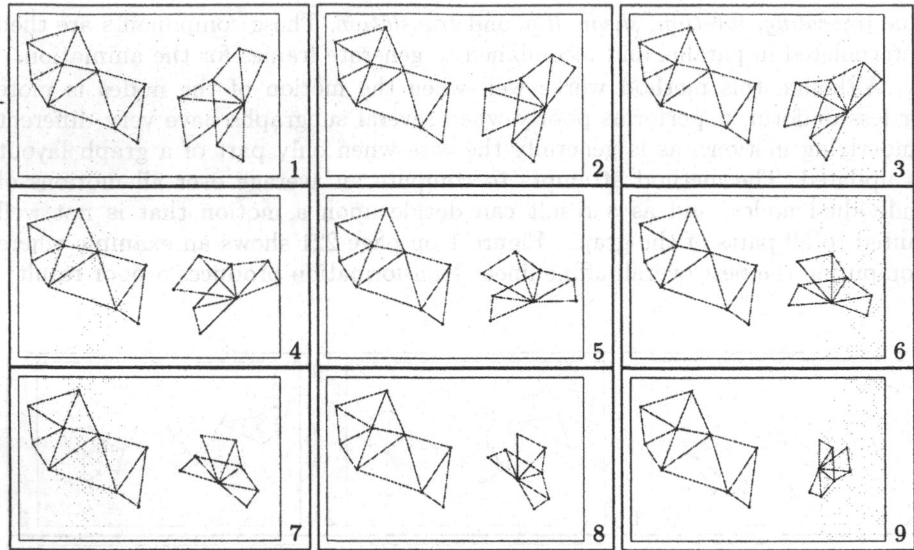

http://www.cs.usyd.edu.au/ carsten/gd01/b.mpg

Fig. 2. Example of a good animation. Different motions are computed for each of the components of the graph. The result is an animation where each subgraph moves in an individual and much more intuitive way.

2 Determining Candidate Transformations

For each node v there is an infinite number of affine linear transformations that can take it from its initial position to its final position. If the transformations applied to individual nodes were chosen arbitrarily and independently, the resulting animation would appear chaotic. In order to effect a smooth animation, it stands to reason that nodes should be grouped together under common transformations whenever possible.

Using this principle, we can restrict the candidate transformations for v to those that v might share with other nodes. In general, three nodes (associated with three non-collinear initial locations and three non-collinear final locations) are required to uniquely determine an affine linear transformation in the two-dimensional plane. Naturally, if four or more nodes were to determine a common affine transformation, this transformation would also be determined by at least one of its subsets of cardinality three. This suggests that the candidate transformations for v could be limited to those generated by triples of nodes that include v. However, the total number of transformations generated would be cubic in the number of nodes. For the purposes of real-time animation, this candidate set would still be far too large to be investigated explicitly.

To avoid paying the cubic cost of generating all candidate transformations of node triples, practical heuristics are needed. In this paper, we use clustering techniques to identify subsets of nodes sharing similar transformations.

3 k-Means

The first clustering method we use is based on the well-known k-means hill-climbing heuristic [12]. The general k-means heuristic for point sets begins with an arbitrary partition $\mathcal{P} = \{P_1, P_2, \ldots, P_k\}$ of the data set into k groups. It then attempts an iterative improvement of the partition, as follows:

- a representative point is computed for each group P_i;
- each element of the data set is assigned to the representative that best suits it, generating a new partition $\mathcal{P}' = \{P'_1, P'_2, \ldots, P'_k\}$.

The process is repeated until an iteration yields no improvement, according to some measure of goodness of a partition.

It is easy to see that this procedure must eventually converge. The great popularity of the k-means method is due to its simplicity and speed. Typically, the number of iterations required is constant, leading to an observed linear-time complexity.

Our adaptation of k-means computes an affine transformation of each group of nodes as the representative of the group, using the linear regression techniques outlined in [7]. In the second step of each iteration, the redistribution of nodes is accomplished by assigning each node to the representative transformation that brings the node closest to its final destination. Improvement can be evaluated according to such measures as the sum of these distances to final destinations, or of squares of these distances. Iteration continues until no further improvement is made.

3.1 Experiments

The following animations show examples of applying the k-means method, where the initial partitions are generated randomly. For all examples, k was set to 10, even though the number of different transformations involved was significantly less. This choice was motivated by two observations. First, as the short-term memory of most users is believed to be able to accommodate roughly 7 items [13], we conclude that computing more than 10 very distinct transformations in one animation would not likely lead to a better preservation of the user's mental map. Second, subclusters derived from a well-associated larger cluster will be associated with very similar transformations, provided that each subcluster contains at least three linearly-independent nodes. This indicates that the method will likely tolerate a choice of k somewhat larger than the minimum.

Figure 3 shows an example where our algorithm successfully identifies three different motions. In addition to an overall rotation of the graph by approximately 180 degrees, two subgraphs also expand in size. The graph is taken from a dataflow analysis application.

Figure 4 refers to the same initial layout as in Figure 3. This time, the animation contains two different motions, a rotation of roughly 180 degrees plus a flip of one connected subgraph. Although the motions are still easy to follow as the animation proceeds, some users may prefer a uniform rotation by the entire graph followed by a flip of a subgraph. As yet, the method does not identify

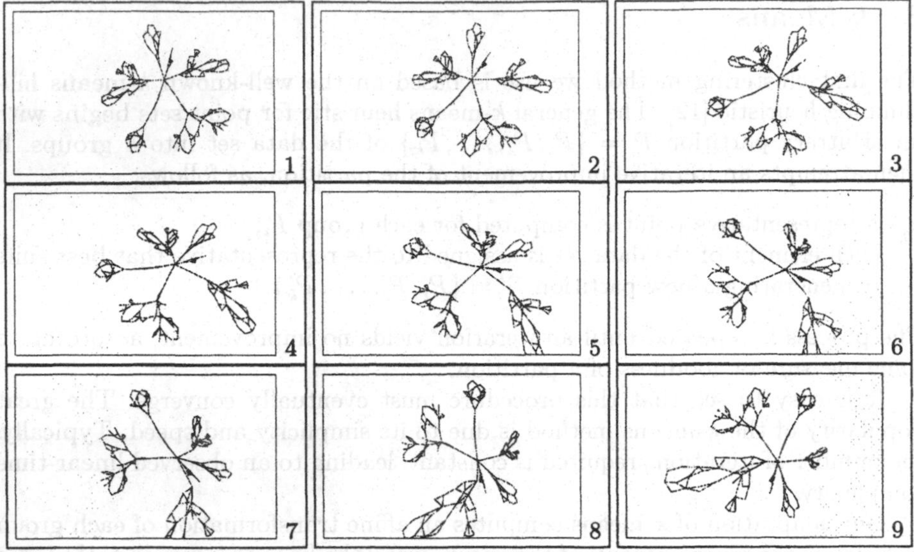

http://www.cs.usyd.edu.au/ carsten/gd01/c.mpg

Fig. 3. Example of an animation using k-means clustering. The algorithm successfully identifies and displays three separate motions.

cases in which animation of subgraphs should be performed in sequence rather than in parallel.

Figure 5 shows an animation on an artificially-generated graph with features chosen in an attempt to confound the method. In particular, the wide variation in inter-node distances gives rise to the possibility of an amplification of error, if a transformation generated by collections of nodes that are close to their final destinations happens to be inappropriately applied to nodes that are far from their final destinations. The animation shows a rotation of the complete graph by roughly 180 degrees, plus a scaling of roughly half the graph. Our algorithm successfully identifies both movements.

The initial layout of Figure 6 is identical to that of Figure 5. However, the positions of the nodes of the final layout are those of Figure 5 with a small amount of random scattering. In this case, no ideal transformation exists, and the introduced noise increases the risk of inappropriate transformation of individual nodes. The animation shows that even in this case, the algorithm successfully discovers the main motions. For some nodes which lie very close together, the algorithm produces small additional motions that are not correct.

3.2 Advantages and Disadvantages

Using k-means to compute the animation in many cases successfully identifies different types of transformations, even in the presence of 'local' noise. Convergence is very fast, and the animations may be viewed in real time.

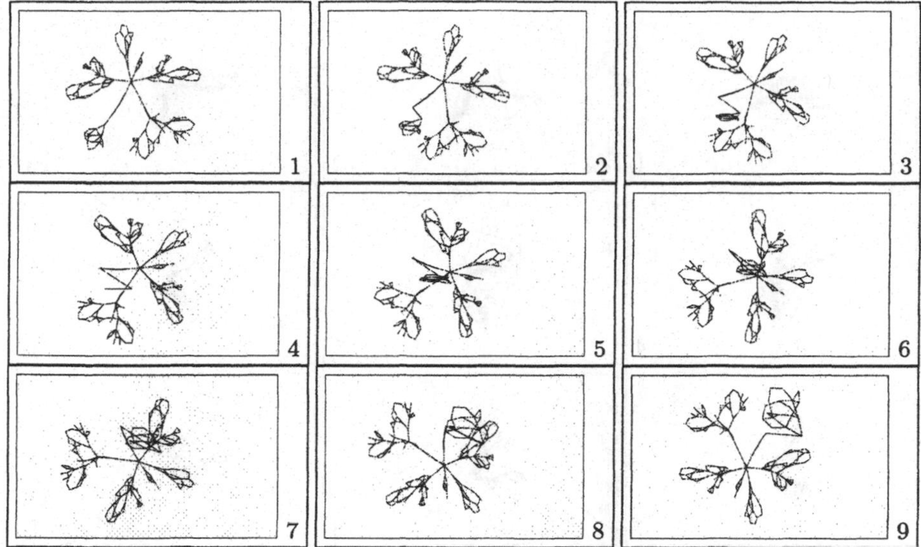

http://www.cs.usyd.edu.au/ carsten/gd01/f.mpg

Fig. 4. Example of an animation using k-means clustering. The algorithm successfully identifies a rotation and a flip.

However, the general k-means method is known to be very sensitive to the choice of partition with which it is initialized [11, page 277]. Its variants, including our method, are not well-equipped to discover or escape from poor clusterings that are nevertheless locally optimal. Much effort has been given to the sensitivity of k-means variants to their initial partitions [1,14,15]. In general settings, failure of the k-means heuristic is more likely when clusters are difficult to distinguish, or when a cluster is split among many groups of the initial partition.

In the context of animation, when clusters are difficult to distinguish, their associated transformations are necessarily similar. Whether two similar clusters are declared to be separate or the same, the resulting animations will not differ greatly.

However, the initial partitioning of a single cluster among many groups is a potentially much more serious problem. This has the effect of a subgraph transformation not being applied in the animation. In our experimentation, we observed a tendency of the method to underutilize rather than overutilize transformations.

4 Distance-Based Clustering

Although popular due to its speed, k-means is not the only clustering strategy that can be applied to graph animation. In various spatial settings, many clustering strategies have been proposed. The criteria they seek to optimize almost

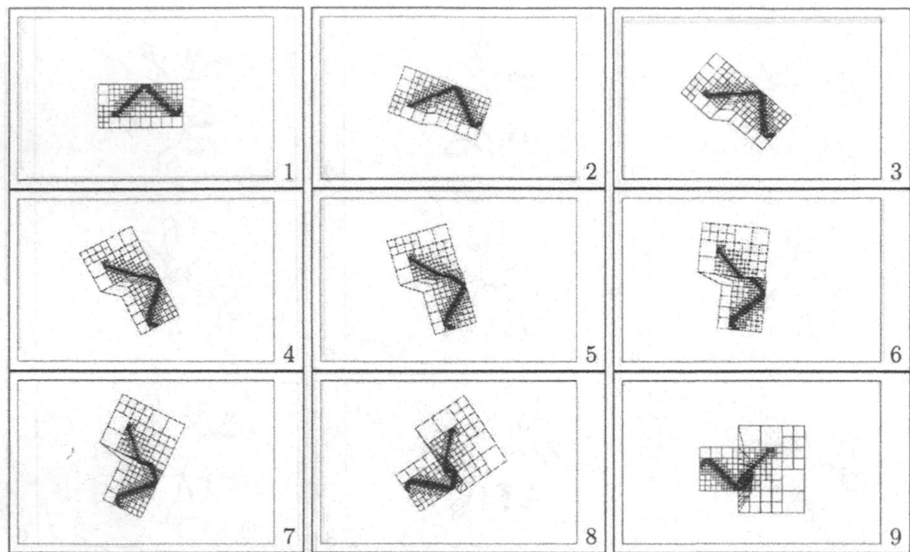

http://www.cs.usyd.edu.au/ carsten/gd01/d.mpg

Fig. 5. Example of an animation using k-means clustering. The algorithm successfully identifies a rotation and a scaling.

invariably make use of some notion of distance, usually based on the Euclidean metric, as it captures the essence of spatial autocorrelation and spatial association [8]. Bottom-up approaches, in which clusters are formed by agglomeration of items that are 'close' together, are in accordance with the view that in spatial application areas, nearby items have more influence upon each other. The same can be argued for the graph layout setting, as one would expect a subgraph transformation to apply to nodes appearing in a contiguous region of the layout.

In the context of animation, the stopping conditions proposed for the k-means animation method can serve as the clustering criterion for more general methods. However, as was mentioned earlier, the problem remains of how to limit the number of triples of nodes examined when generating candidate affine transformations. One solution is to take advantage of the tendency for nodes sharing a common transformation to lie close to one another in the initial and final layouts. By restricting the examination of node triples to those lying close in either the initial or final layouts, a large reduction of the number of triples may be achieved.

4.1 Delaunay Triangulation

A classical data structure from computational geometry, the *Delaunay triangulation*, ideally captures proximity relationships among points in the plane, in a very compact form. The Delaunay triangulation $\mathcal{D}(S)$ of S is a planar graph embedding defined as follows: the nodes of $\mathcal{D}(S)$ consist of the data points of

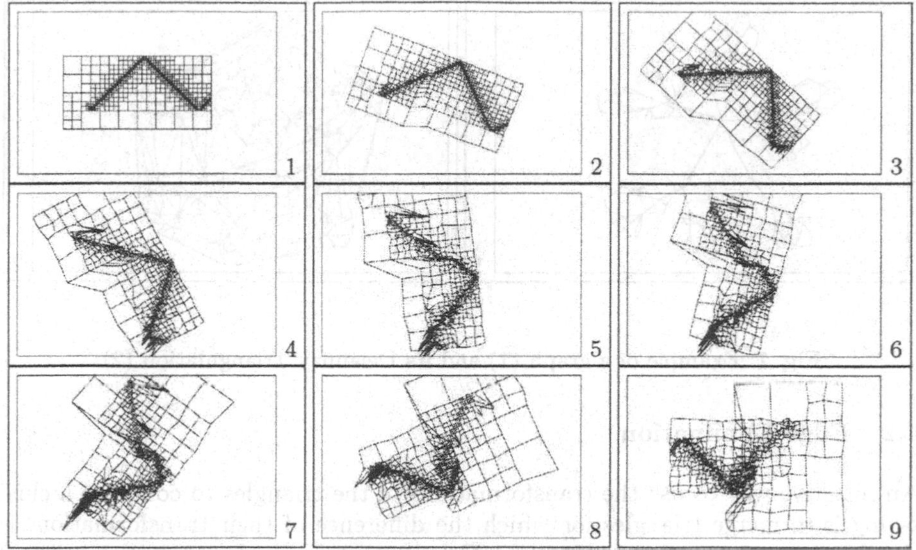

http://www.cs.usyd.edu.au/ carsten/gd01/e.mpg

Fig. 6. Example of an animation using k-means clustering. In addition to a pure rotation and scaling, the final node positions are randomly perturbed by a small offset.

S, and two nodes s_i, s_j are joined by an edge if and only if there exists a circle passing through s_i, s_j having empty interior. The Delaunay triangulation is the graph dual of the well-known Voronoi diagram of S. Some of the interesting features of Delaunay triangulations are listed below; more details can be found in virtually any textbook in computational geometry (for example, [4]).

1. If s_i is the nearest neighbor of s_j from among the data points of S, then (s_i, s_j) is an edge in $\mathcal{D}(S)$.
2. The number of edges in $\mathcal{D}(S)$ is at most $3n-6$, and thus the average number of neighbors of a site s_i in $\mathcal{D}(S)$ is less than 6.
3. The Delaunay triangulation is the most well-proportioned over all triangulations of S, in that the size of the minimum angle over all its triangles is the maximum possible.
4. The triangulation $\mathcal{D}(S)$ can be robustly computed in $O(n \log n)$ time.
5. The minimum spanning tree is a subgraph of the Delaunay triangulation, and in fact, a single-linkage clustering (or dendrogram) can be found in $O(n \log n)$ time from $\mathcal{D}(S)$.

Figure 7 shows a graph (1) and its Delaunay triangulation (2).

For each triangle of the Delaunay triangulation of the initial layout of nodes, we can compute in constant time the affine transformation between the nodes of the triangle to their target positions, by solving a set of six linear equations with six variables. We now can cluster these transformations using clustering techniques suited for the use of generic spatial distance measures.

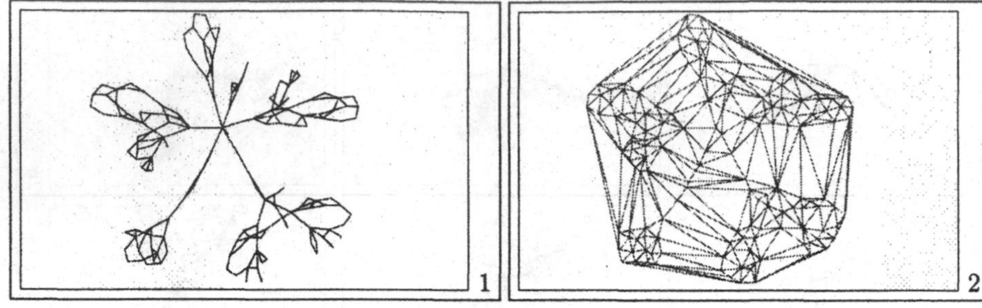

Fig. 7. Example of a graph (1) and its Delaunay Triangulation (2).

4.2 Edge Elimination

An intuitive way to use the transformations of the triangles to compute a clustering is to merge triangles for which the difference of their transformations is smaller than a given threshold value. The elimination of edges results in a small number of large connected regions, the vertices of which can be interpreted as a cluster set. The clusters determined using this method can be animated directly or used to seed other clustering methods, such as k-means, that are sensitive to the quality of their initial partitions.

Figure 8 shows a typical example of the application of the edge elimination method. Figure 8 (1) and (2) show the initial and final drawing of the graph. One part of the graph has been rotated, another part has been scaled and the remaining nodes remained at their initial positions. The graph (3) shows the number of edges (y-axis) separating triangles with differences in their transformations smaller than the threshold (x-axis). Figure 8 (4) shows the clustering obtained by setting the threshold to the point marked by the circle in the graph (3). The degree of darkness of a triangle indicates the number of neighbors with which it has been merged. The distance function is based on the values obtained by decomposing the computed transformation into rotation, scaling, translation, skew, and flip [7]. We can observe that the clusters separate very clearly, and it it easy to produce an appropriate animation from this partition.

Unfortunately, when a significant amount of noise is introduced, in the form of random displacements of nodes, the quality of the result drops sharply. Figure 9 shows the same example with a random displacement of the nodes in the final drawing. Figure 9 (1) and (2) again show the initial and final drawing of the graph, (3) the threshold graph and (4) a clustering obtained by setting the threshold to the point marked in (3). We can see that it is much harder to determine a suitable threshold value in this case. It is however still possible to generate a good seeding for other clustering algorithms by using a small threshold value.

The effect of noise on the quality of the clustering can be explained in terms of the nature of the Delaunay triangulation and of clustering itself. The assumption on which the use of Delaunay clustering was based is that nodes which lie close to each other are more likely to share the same motion than nodes which

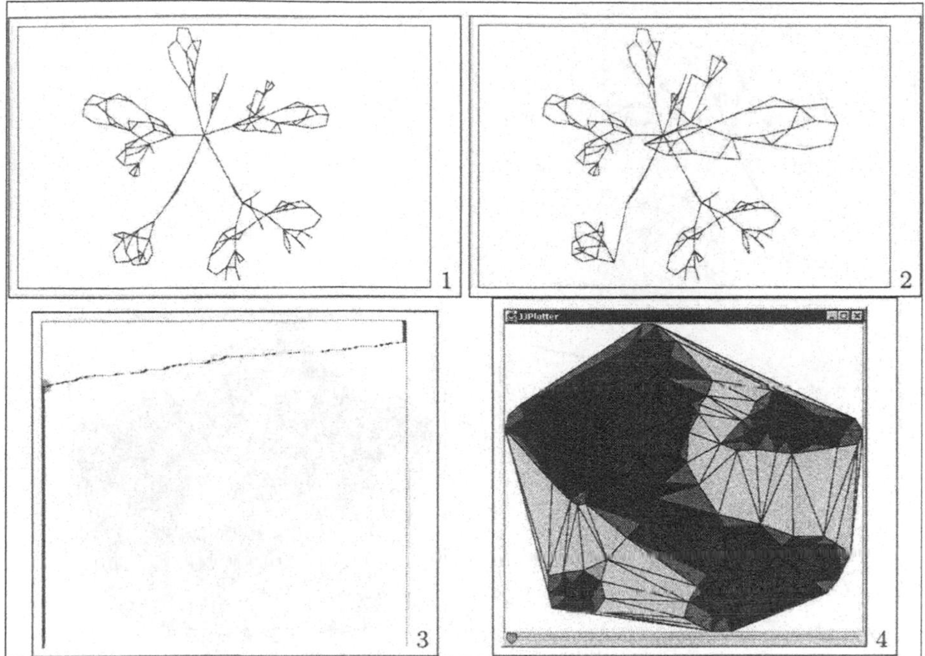

Fig. 8. Example of a clustering obtained by using the edge elimination method. (1) and (2) display the initial and final drawings of the graph. The graph of edges per threshold is displayed in (3), and a clustering obtained by setting the threshold to the point marked in (3) is displayed in (4).

lie further apart. In the Delaunay triangulation, nodes that lie close together form smaller triangles than nodes lying further apart. The displacement of a vertex belonging to a smaller triangle has a greater effect on the transformation generated by the vertices of that triangle, compared to the same displacement of a vertex of a large triangle. Thus displacements tend to reduce the quality of meaningful transformations while having less effect on spurious transformations. This hypothesis agrees with the results of experiments where the graph was designed in such a way that all triangles had approximately the same size. In these cases, useful clusterings were easily generated even in the presence of significant amounts of noise.

5 Conclusion

We have investigated the application of clustering techniques on computing animations of graphs in cases where subgraphs are transformed in different ways. We have seen that in many cases standard techniques such as k-means or edge elimination are able to produce good results. We also discussed the circumstances in which these methods tend to produce poor results. A worthwhile topic for future research would be to apply other clustering techniques to graph animation.

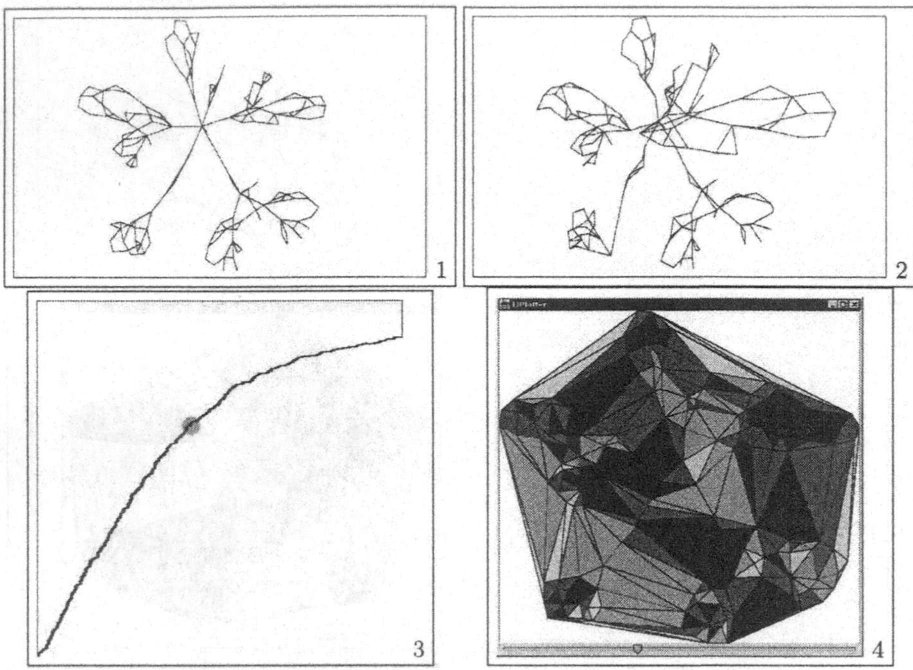

Fig. 9. Example of a clustering obtained by using the edge elimination method. (1) and (2) display the initial and final drawings of the graph. The edges per threshold graph is displayed in (3) and a clustering obtained by setting the threshold to the point marked in (3) is displayed in (4).

References

1. M. S. Aldenderfer and R. K. Blashfield. *Cluster Analysis.* Sage Publications, Beverly Hills, USA, 1984.
2. Giuseppe Di Battista, Peter Eades, Roberto Tamassia, and Ioannis G. Tollis. *Graph drawing: algorithms for the visualization of graphs.* Prentice-Hall Inc., 1999.
3. F. Bertault. A force-directed algorithm that preserves edge-crossing properties. *Information Processing Letters*, 74(1–2):7–13, 2000.
4. Mark de Berg, Marc van Kreveld, and Mark Overmars. *Computational Geometry: Algorithms and Applications*, chapter 9, pages 188–200. Springer Verlag, 2nd edition, 1998.
5. C. Friedrich. The ffGraph library. Technical Report 9520, Universität Passau, Dezember 1995.
6. Carsten Friedrich and Peter Eades. The marey graph animation tool demo. In *Proc. of the 8th Internat. Symposium on Graph Drawing (GD'2000)*, pages 396–406, 2000.
7. Carsten Friedrich and Peter Eades. Graph drawing in motion. *Submitted to Journal of Graph Algorithms and Applications*, 2001.
8. R. P. Haining. *Spatial Data Analysis in the Social and Enviromental Sciences.* Cambridge University Press, UK, 1990.

9. http://www.mpi sb.mpg.de/AGD/. *The AGD-Library User Manual Version 1.1.2.* Max-Planck-Institut für Informatik.
10. Mao Lin Huang and Peter Eades. A fully animated interactive system for clustering and navigating huge graphs. In Sue H. Whitesides, editor, *Proc. of the 6th Internat. Symposium on Graph Drawing (GD'98)*, pages 374–383, 1998.
11. L. Kaufman and P. J. Rousseeuw. *Finding Groups in Data: An Introduction to Cluster Analysis.* John Wiley & Sons, NY, USA, 1990.
12. J. MacQueen. Some methods for classification and analysis of multivariate observations. In L. Le Cam, and J. Neyman, editor, *5th Berkley Symposium on Mathematical Statistics and Probability*, pages 281–297, 1967.
13. G. A. Miller. The magical number seven, plus or minus two: some limits on our capacity for processing information. *The Psychological Review*, pages 63:81–97, 1956.
14. U. Fayyad P. S. Bradley and C. Reina. Scaling clustering algorithms to large databases. In R. Agrawal and P. Stolorz, editor, *Proceedings of the Fourth International Conference on Knowledge Discovery and Data Mining*, pages 9–15, 1998.
15. C. Reina U. Fayyad and P. S. Bradley. Initialization of iterative refinement clustering algorithms. In R. Agrawal and P. Stolorz, editor, *Proceedings of the Fourth International Conference on Knowledge Discovery and Data Mining*, pages 194–198, 1998.

Online Hierarchical Graph Drawing

Stephen C. North and Gordon Woodhull

AT&T Labs - Research
180 Park Ave. Bldg. 103
Florham Park, New Jersey 07932-0971 (U.S.A.)
north,gordon@research.att.com

Abstract. We propose a heuristic for dynamic hierarchical graph drawing. Applications include incremental graph browsing and editing, display of dynamic data structures and networks, and browsing large graphs. The heuristic is an on-line interpretation of the static layout algorithm of Sugiyama, Togawa and Toda. It incorporates topological and geometric information with the objective of making layout animations that are incrementally stable and readable through long editing sequences. We measured the performance of a prototype implementation.

1 Introduction

Graph layout is effective for visualizing relationships between objects. Static layout techniques are well understood, but some applications display graphs that change. Examples include:

- interactive graph editors
- displays of intrinsically dynamic graphs, such as Internet router BGP announcements, or data structures in a running program
- browsers for large graphs based on adjustable subgraphs [10]

The browsing application is motivated by the need for better techniques for visualizing massive graphs [14]. For example, a finite state machine for continuous speech recognition can have more than 5×10^6 transitions. Graphs of Internet structures and biological databases can be even larger. Static layout of large general graphs does not seem feasible: even when the layout computation is tractable, it is difficult for a human to make sense of many thousands of objects unless they are arranged in a regular, predictable structure and it does not seem possible to do this for arbitrary graphs. A helpful alternative could be to show the neighborhood around a movable focus node, or a simplification of the base graph, adjusted interactively.

Informative dynamic graph displays should direct attention to changes while also revealing the graph's global structure. When static layout algorithms based on global optimization are employed, insertion or deletion of even one node or edge can dramatically change the layout. Such instable changes disrupt a user's sense of context, and are uninformative because they do not direct attention to

P. Mutzel, M. Jünger, and S. Leipert (Eds.): GD 2001, LNCS 2265, pp. 232–246, 2002.
© Springer-Verlag Berlin Heidelberg 2002

changes in structure in the underlying graph [9]. An incremental approach is needed.

Hierarchical drawings are often useful in practice. They seem to provide a good match between visual perception and common data analysis tasks such as identifying ancestor-descendant relationships or locating articulation points and bridges. Efficient hierarchical layout algorithms have been devised. Dynamic hierarchical drawing potentially has many of the same benefits, and would be useful in many situations where static layout is being employed. We propose a heuristic, Dynadag, that maintains on-line hierarchical graph drawings.

2 Layout Server Model

Dynadag uses a client-server model with communication based on a shared, managed graph that holds geometric coordinates and other layout attributes. Client and server send changes to each other via *insert*, *modify*, and *delete subgraphs* of the shared graph. The order of these changes is recorded when objects are inserted into subgraphs. The client accumulates changes in the subgraphs and eventually calls the server's *Process* method to obtain a new layout. The server further appends to the *modify subgraph* as it generates a new layout. After the *Process* method finishes, the client may update its display to reflect the new state of the shared graph.

Table 1 lists graph object layout attributes. Coordinates are dimensionless and computed to a client-specified precision (*e.g.* nearest pixel or millimeter). Position attributes are optional in requests. If an insert or modify request does not have a valid position, the server may arbitrarily determine the object's placement. On the other other hand, when the client gives a position (such as when editing a graph interactively, or importing a saved diagram), it is a strong indication to place the object as closely as possible to the request coordinate. Every object also has a flag to request pinning or fixing its position. Thus, some graph objects may be placed manually while others are being managed automatically. Edges can also be given a minimum length and a weight indicating the cost of stretching it. The client further controls the spacing of objects via the separation parameter, which states a minimum horizontal and vertical distance.

The *insert-modify-delete subgraph* mechanism allows the client to make large or small changes to the managed graph. For instance, a client may load an entire external graph into the *insert subgraph* before invoking the server's *Process* method, resulting in a globally optimized layout. Or, a large subgraph may be selected manually and its layout re-computed. At the other extreme, an on-line editor providing direct manipulation may call the *Process* method after every operation.

The change-subgraph interface between clients and servers makes almost no assumptions about how each behaves, and supports multiple layout algorithms. A server is assumed only to make a best effort to process requests and generate a new layout. It is allowed to modify or even ignore requests incompatible with its algorithm. For example, it could align nodes and edges to grid coordinates, or reject non-planar edges or parallel multi-edges. Servers are not responsible for graphical effects such as in-betweening or fading. So animation techniques such as those of of Eades and Friedrich [8] are complementary to our proposal. A significant limitation is that our system does not exploit look-ahead, though doubtless it could produce better off-line animations.

3 Dynadag Heuristic

Most hierarchical graph layout programs use variants of a well-known batch heuristic due to Sugiyama, Tagawa and Toda [16]. This heuristic (STT) draws directed graphs in phases that reduce the search space by solving sub-problems that optimize objectives such as total edge length and crossing number, subject to constraints that edges point downward, nodes not overlap, etc.

The phases of STT are:

1. convert input graph into a directed acyclic graph (DAG) by reversing any cyclic edges
2. assign nodes to discrete levels (ranks), *e.g.* placing root nodes on level 0, their immediate descendants on level 1, etc.
3. convert edges that span multiple levels into chains of model nodes and edges between adjacent levels
4. assign the order of nodes in levels to avoid crossings
5. assign geometric coordinates to nodes and edges, keeping edges (represented as polylines or splines) short and avoiding bends

This ordering of phases prioritizes the aesthetic properties of the resulting layouts. For example, computing a level assignment before determining edge crossings reflects a decision that emphasizing flow is more important than avoiding crossings. Adopting the aesthetic priorities of STT, our dynamic version has the same phases. STT readily lends itself to this modification because each phase depends only on limited information computed by previous phases, and because we can maintain its framework of topological and geometric constraints incrementally as described here.

Dynadag combines the static layout aesthetics of STT and decisions about how to make on-line layouts stable. Measuring how well a dynamic diagram preserves a user's "mental map" is the topic of ongoing research. Without a firm foundation of experimental studies to rely on, we simply assume that basic geometric and topological properties contribute to the a layout's visual stability and readability. Of course, other things being equal, a drawing is more readable when its edges are short and don't have many crossings. These goals often conflict with obvious measures of stability: objects should not move far, and neither the sequence of nodes within a hierarchical level nor the angular order of edges incident on a given node should change much. Because these measures are not comparable, and are handled by different parts of the algorithm, they are not combined into any sort of unified layout quality or stability metrics. We admit that our notion of stability is purely heuristic.

Table 1. Shared graph objects and their attributes

Value	Type	Explanation
$G = (V, E)$	graph object	graph
$u, v, w, \ldots \in V$	node object	node
$e, f, \ldots \in E$	edge object	edge
$\Delta(G)$	coord	min node separation
$L_{i,j}$	node object	jth node in ith level
r_x, r_y	float	precision
$\lambda(v)$	integer	level (rank) assignment
$X(v), Y(v)$	coord	position of node center
$\hat{X}(v), \hat{Y}(v)$	coord	client node position request
$X'(v), Y'(v)$	coord	previous node position
$B(v)$	coord	node shape bounding box
$fixed(v)$	boolean	node movable
$tail(e), head(e)$	node object	endpoints
$C(e)$	coord list	layout spline
$\hat{C}(e)$	coord list	client request spline
$\omega(e)$	float	weight ≥ 0
$\delta(e)$	float	minimum length ≥ 0
$strong(e)$	boolean	strong level constraint

3.1 Main Algorithm

Dynadag maintains an internal *model graph* that satisfies the one-level edge constraint of STT phase three, and holds internal information such as the integer rank assignments of nodes. It also stores the model graph's nodes in a two-dimensional array (or *configuration*) for efficient access. Dynadag's *Process* or main work procedure applies the STT phases incrementally. Each phase examines the insert, modify, and delete subgraphs and updates the model graph, configuration, or objects in the shared graph accordingly. Each phase must perform these computations in a way that is stable with respect to the previous layout, while preserving the layout invariants (*e.g.* hierarchical edges point downward). The objectives and constraints for each phase are shown in table 2.

The main steps of *Process* (algorithm 5) act on the request subgraphs. We will describe each in detail. *Preprocess* conditions the input subgraphs. Some requests trivially fold or cancel, such as an inserting an object and then modifying or deleting it, or modifying the same object multiple times. Likewise, deleting a node implies deleting its incident edges.

3.2 Rerank Nodes

This phase assigns integer levels to the nodes of the graph to maintain the hierarchy, preserve stability, and minimize total edge length, prioritized in that

```
Algorithm Process(inG)
Input: inG: client requests
Output: outG: layout server's updates
(* main procedure to process layout requests *)
outG ←Preprocess(inG)
outG ←RerankNodes(outG)
outG ←ReduceCrossings(outG)
outG ←UpdateGeometry(outG)
return outG
```

```
Algorithm RerankNodes(inG)
(* top level of phase 1- compute new levels λ(v). See table 1 *).
for e ∈ edgeDeletions(G)
        if e is a strong constraint then
                remove constraint arc representing e in CG_y
            else remove ρ(e) and incident arcs from CG_y for
                for v ∈ nodeDeletions(G)
                        remove λ(v), τ(v) and incident arcs in CG_y
                for v ∈ nodeMoveUpdates(G)
                        λ(v) ← mapToRank(RequestCoord(v))
                        if isAStrongMove(v) then
                                for e incident on v
                                        remove constraint arc representing e in CG_y
                                        create edge aux_0 = τ(v), tail(e) with ω(aux_0) = c_rev ω(e)
                                        create edge aux_1 = τ(v), head(e) with ω(aux_1) = ω(e)
                        stabilize λ(v)
```

```
Algorithm ReduceCrossings(M, S)
(* reduce crossings on edges incident to nodes in S *)
pass ←0
best ←crossings(M)
while pass < NPASSES and best > 0
        ntrials ←0
        while pass < NPASSES and ntrials < PATIENCE
                leftward ←pass mod 2 == 0
                downward ←pass mod 4 < 2
                equalPass ←pass mod 8 < 4
                BubbleSortPass(S,leftward,HasMedian(downward),MedianCompare(downward))
                while crossings(M) decreases
                        BubbleSortPass(S,leftward,downward,true,CrossingsCompare)
                current ←crossings(M)
                if current < best then
                        save configuration
                        best ←current
                        ntrials ←0
                    else
                        ntrials ←ntrials +1
        if current > best then
                restore configuration
```

```
Algorithm BubbleSortPass(S, leftward, downward, comparable, compare)
for r in S
        for u in r according to leftward
                if not comparable(u) then continue
                for v after u in r
                        if not (u ∈ S or v ∈ S) then break
                        if not comparable(v) then continue
                        if compare(u, v) then
                                put u after v according to leftward
```

order. The following discussion assumes the hierarchy is drawn top-to-bottom; of course it is simple to orient the hierarchy in other ways by pre- and post-processing. Level assignment employs an integer network simplex solver previously developed for the *dot* layout program [12]. In applying this solver, Dynadag

Table 2. Objectives and constraints

Phase	Objective	Constraints
rerankNodes	$\min \sum_{e=(u,v)\in E} w(e)(\lambda(v)-\lambda(u))$	$\lambda(v) \geq \lambda(u) + \delta(u,v)$
reduceCrossings	crossings	$X(v) = X(u) + 1$
updateGeometry	$\min \sum_{e=(u,v)\in E} w(e)\lvert X(v)-X(u)\rvert$	$X(v) \geq X(u) + \Delta(u,v)$

maintains an auxiliary graph CG_y whose nodes are interpreted as variables and edges as constraints, as shown in tables 3 and 4.

Table 3. Variables in CG_y

Variable	Explanation
$\forall v \in V : \lambda(v)$	level of v or $Y(v)$
$\forall v \in V : \tau(v)$	stable level assignment of v
$\forall e \in E \vert \neg strong(e) : \rho(e)$	lower endpoint of weak edge
$\lambda_{\min}, \lambda_{\max}$	lowest and highest levels

Table 4. Constraints in CG_y

Constraint edge	Weight	Explanation
$\forall v \in V : \lambda(v) - \lambda_{\min} \geq 0$	0	maintain min level
$\forall v \in V : \lambda_{\max} - \lambda(v) \geq 0$	0	maintain max level
$\forall e = (u,v) \in E \vert\ strong(e) : \lambda(v) - \lambda(u) \geq \delta(e)$	$\omega(e)$	strong edge constraint
$\forall e = (u,v) \in E \vert \neg strong(e) : \rho(e) - \lambda(u) \geq 0$	$\omega(e)$	weak edge constraints
$\rho(e) - \lambda(v) \geq \delta(e)$	$c_{\text{rev}}\omega(e)$	

Dynadag treats client edges as either *strong* or *weak* level assignment constraints. A strong edge is always hierarchical: it points downward so its head is on a higher-numbered level than its tail. A *weak* edge is unconstrained and may point downward, upward or sideways across the same level. To favor hierarchical drawing, weak edges usually have a high cost c_{rev} associated with a non-downward orientation, although a client can explicitly set $\omega(e)$ to be small or zero to defeat this bias. Edges are strong unless the client marks them as weak or pins an endpoint. If the algorithm encounters a cycle, it marks the last inserted edge of the cycle as weak; if the cycle is later broken, this edge will point downward.

The network simplex solver we use does not have any intrinsic stability. To compensate, we add explicit variables and constraints that penalize level assignments by their variance from some given assignment (usually the previous layout or a client-suggested coordinate). Adjusting the penalty edge weights changes the tradeoff between minimizing edge length and maintaining geometric stability.

There are a few additional details to obtaining good level assignments. One is that stability constraints are removed on un-pinned nodes when the client inserts the first incident edge: the previous location of a disconnected, unpinned node is assumed unimportant. Also, Dynadag ensures that nodes with slack in their level assignments are brought up near their parents, by adding non-zero weight constraints with reference to a global anchor node.

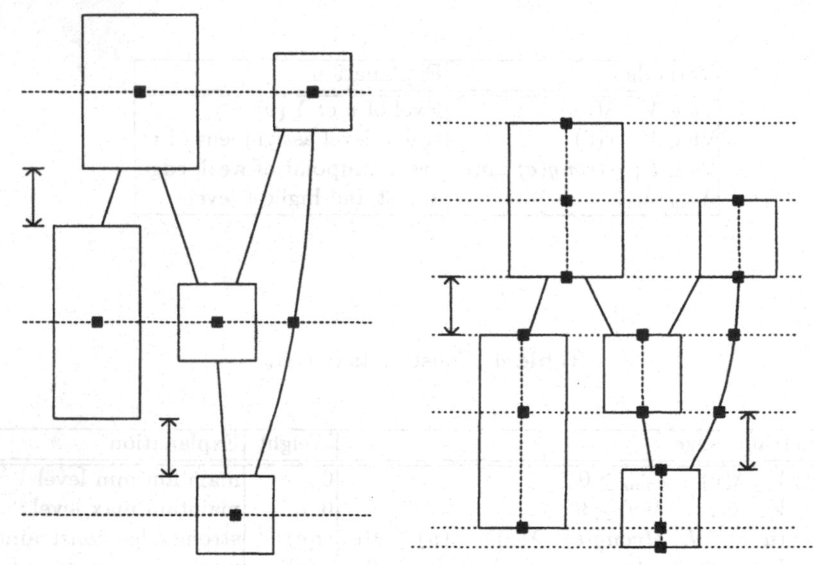

Fig. 1. Multiheight ranking system in CG_y

Dynadag supports two ranking systems. The first, familiar from STT, is intended when nodes all have about the same height. In this case, nodes are aligned on ranks, and ranks are separated enough to fit the nodes. The second system better accommodates large differences in node heights by assigning independent levels to the top and bottom of each node. In this case, ranks simply encode Y coordinates and nodes can potentially span many ranks. The second system is more general than the first, but results in larger model graphs, and can allow edges to cross large nodes unless other heuristics are added to the edge router. When nodes are of the same size, the model graph is approximately twice the input graph's size. Both models are of complexity $O(N^2)$ in the size of the input graph.

At the end of this phase, every node is labeled with its new level assignment, $\lambda(v)$.

3.3 Restore Configuration

This phase updates the configuration $L_{i,j}$ according to the new levels $\lambda(v)$. Long edges will be converted into chains of model nodes, all one level apart. We will write the model node chain of $e = (u, v)$ as $u, \phi_0, \phi_1, \ldots, v$. (Self-arcs and flat edges within the same level are ignored in this phase.) When using the multirank node system, nodes that span ranks are also converted to chains. Thus $L_{i,j}$ gives the order j of all nodes and edges appearing in level i.

Dynadag first moves the pre-existing nodes or node chains to match the new ranks assigned in the previous phase. Then it moves edges by moving the chains to the new ranks, lengthening or clipping them as necessary. If the user has specified coordinates for an edge, Dynadag honors the request by placing the model edge for each rank at the X position determined by intersecting the user path with the rank Y. Otherwise it simply draws the chain in a straight line.

3.4 Minimize Crossings in Configuration

At this point, the configuration fully depicts the requested layout. However, edges may be tangled, and in the multirank node system, may even cross nodes.

Dynadag uses a variant of the *dot* crossing minimization heuristic to eliminate node crossings and avoid edge crossings. First it determines which model graph objects are candidates for adjustment. To the nodes and edges corresponding to objects from the *insert* and *modify* shared subgraphs, it adds edges incident on nodes in these subgraphs. (This neighborhood could be extended to try to improve readability at the expense of stability.)

The crossing minimization heuristic scans the configuration and applies two different sorts: *median sort* and *transposition sort*. As it runs, it records the best configuration found so far; if after some number of passes k the configuration has not improved, it restores the best assignment. Its scans alternate between left-to-right and right-to-left, top-to-bottom and bottom-to-top, to avoid built-in bias.

Median sort rearranges nodes according to the median position of incident nodes in the adjacent rank last visited. Transposition sort exchanges adjacent nodes if this reduces the crossing number. As an optimization, transpose sort employs a sifting matrix [5] to avoid re-counting crossings each scan.

On every scan, either the median sort or the transposition sort may reorder nodes even if the crossing number does not decrease. This allows the heuristic to sometimes escape local minima even when no immediate benefit is evident. We have observed that the transposition sort tends to propagate these attempts up or down edge chains in the graph until it eventually eliminates crossings.

It is important to ignore node crossings on the first scan, because nodes must temporarily move across edges to reduce crossings. Weighting node crossings too heavily prevents the transposition sort from trying these steps. Instead, the heuristic first optimizes the model, ignoring whether model edges belong to real nodes. Then it changes the scoring system to penalize edge-node crossings and especially node-node crossings, and scans the graph with the transposition sort to eliminate most node crossings.

It is not always possible to eliminate edge-node crossings in a strictly hierarchical layout with multirank nodes. If any edge-node crossings are left, Dynadag should specially route these edges non-hierarchically in the last phase, but we have not yet implemented this heuristic.

3.5 Update Geometry

This phase computes the coordinates $X(v)$ for model nodes, re-using the integer network simplex solver from step 2. The linear program's variables and constraints are listed in tables 5 and 6 and are represented in an auxiliary graph CG_x.

Table 5. CG_x variables

Variable	Explanation
λ_{left}	the left boundary of the layout
$\forall v \in G : \chi(v)$	X coordinate of node v
$\forall e \in G : \rho(e)$	left point of e
$\forall v \in G : \tau(v)$	stable anchor of v
$S(L_{i,j})$	width of $L_{i,j}$

Table 6. CG_x constraints

Variable	Explanation
$\forall i : \chi(L_{i,0}) \geq \lambda_{\text{left}}$	maintain left boundary
$\forall i,j : \chi(L_{i+1,j}) \geq \chi(L_{i,j})) + \Delta_x(G) + \frac{S(L_{i,j})+S(L_{i+1,j})}{2}$	separate adjacent nodes
$\forall v \in G : \tau(v) \geq \lambda_{\text{left}}$	maintain left boundary
$\forall e \in G : \chi(head(e)) \geq \rho(e)$	maintain leftmost node of e
$\forall e \in G : \chi(tail(e)) \geq \rho(e)$	maintain leftmost node of e

The objective is:

$$\min \sum_{e=(u,v)\in E} c\omega(e)(\chi(v) - \rho(e) + \chi(u)) - \rho(e)) + \sum_{v\in V}(1-c)(\chi(v) - \tau(v))$$

which has a term for the total weighted edge length (measured by the L_1 norm), and a term for the total distance that nodes move from certain given positions (previous placements or client-requested positions). The constant c trades off stability and edge length minimization.

After node position assignment, Dynadag recomputes edge routes $C(e)$ where needed. New edges must always be routed and existing edges are re-routed when an endpoint has moved or the edge has a model node whose distance is less than $\Delta_x(G)$ from a neighbor in the same level. Edges are drawn one at a time by a 2-D spline fitter [7] whose input is a simple path and a list of barrier segments. It returns a piecewise cubic Bezier curve that is close to the path and does not cross any barrier. For Dynadag to provide these arguments to the spline fitter, it takes the model node path of the edge to be drawn, and computes a constraint polygon that contains the path nodes extended horizontally to $\Delta_x(G)$ from neighboring model nodes on the same ranks, ignoring model nodes of edges that cross the one being routed. (Thus, crossings do not appear artificially "forced" to a certain point.) We also compute the shortest path within the constraint polygon, and provide that, along with the constraint polygon as a list of barrier segments, to the spline fitter.

A final detail is the updating of X coordinates of model nodes ϕ_i of an edge e to reflect the intersections of $C(e)$ with the centerlines of levels that it crosses. In other words, model nodes are moved to match the computed spline.

The *UpdateGeometry* algorithm follows from these details; its listing is omitted to save space.

4 Performance

The asymptotic complexity of the proposed heuristic is dominated by the network simplex algorithm invoked in the first and third phases. Its complexity is $O(IVE)$ per refresh; although I is not provably polynomial, it is often nearly linear in practice. In the second phase, the crossing minimization heuristic is also $O(IVE)$ where I is a small constant that we determine. The edge spline fitter is $O(V^3)$, but often performs quadratically.

We measured the performance of an implementation of the proposed heuristic running on an 1 GHz Intel Pentium PC. The rest graphs were Forrester's World Dynamics graph and the Unix family tree circa 1988, available as `world.dot` and `unix.dot` in the graphviz package from `www.graphviz.org`. To interpret these dynamically, in our experiments we serially inserted nodes (each with its incident edges) until the whole graph was built. Nodes were ordered by breadth-first and depth-first search. We also ran a simple random graph generator. Figure 2 shows frames from an animation. Figure 3 compares running time per update with number of graph objects, for each phase. We made full measurements of running time, static layout quality and stability, and expect to report these in a full version of this paper. We noticed that layouts remain readable throughout long editing sequences. This is fortunate, as we had suspected that they could deteriorate so much as to require frequent instable global reoptimization.

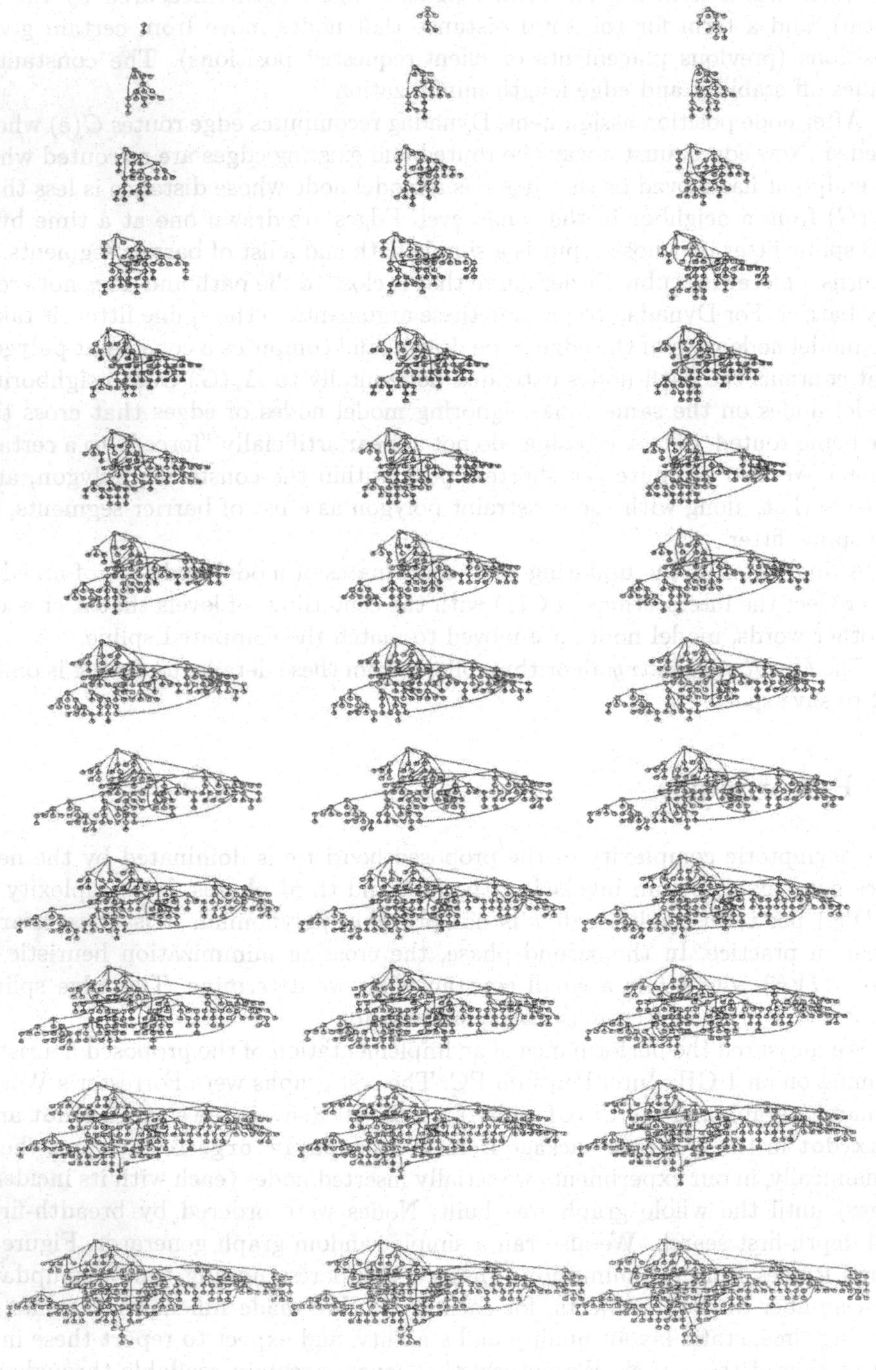

Fig. 2. Sparse random graph, $p(leaf) = 0.90$. Every 5th frame from a sequence of 200 is shown.

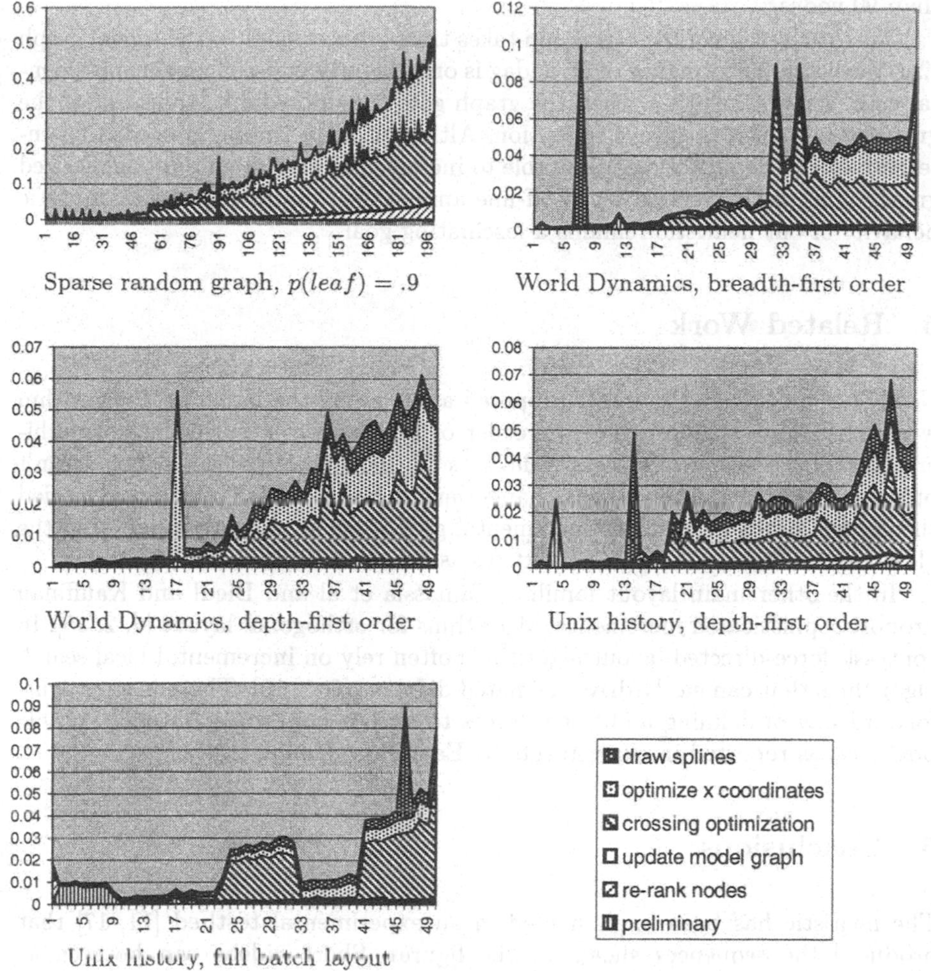

Fig. 3. Time per update in seconds vs. number of nodes for random graphs and two other example graphs. Most updates were less than 0.1 sec. The heuristic was tuned to balance the cost of the different phases.

The performance of our implementation depends upon the density and complexity of the input graph. For sparse graphs with 95% of insertions being leaves, *Process* takes less than a second for each of the first 150 insertions. As the graph gets denser, the greater proportion of long edges increases the cost of crossing optimization and X coordinate assignment. With 90% leaves, Dynadag updates graphs of up to 120 nodes in a second. At 80%, graphs inserted incrementally have many crossings, and Dynadag only runs adequately on graphs with no more than 90 nodes.

Since each stage of the algorithm takes time proportional to the model graph size, the batch performance of Dynadag is only slightly worse than the incremental case. This means that when the graph gets too unreadable, re-layout of the graph or part of it is always an option. Although more tuning is needed for interactive use, the prototype is suitable to incremental display of reasonable sized graphs, and to the generation of off-line animations. Improving the asymptotic behavior of our heuristic remains a fascinating goal.

5 Related Work

Newbery-Paulisch and Bolinger proposed augmenting the batch STT algorithm with constraints that preserve the order of nodes staying within the same hierarchical level between successive layouts [3]. This is a good idea, but doesn't preserve placement when nodes change levels. Eades and Sugiyama identified the general problem of stable incremental graph layout and proposed using the global left-to-right scan order of vertices as the stability criterion [9].

In the other main layout families, Tamassia et al and Biedl and Kauffman propose sophisticated incremental algorithms for orthogonal layout [4, 2, 15]. In contrast, force-directed layout algorithms often rely on incremental local search algorithms that can easily drive animated displays [6, 1, 10]. There is a straightforward way of defining additional forces to anchor nodes near intended stable positions, as reported in experiments by Eades and Huang [13].

6 Conclusions

The heuristic has been implemented in an experimental testbed [11, 17] that produced the sequences shown in the figures. Short videos can be seen at `www.research.att.com/~north/videos/gd2001`.

There are several ways the heuristic and its implementation might be improved:

- implement edge labels as model nodes
- consider flat edges in counting crossings
- improve the *ReduceCrossings* heuristic
- improve sensitivity to the ordering of *moveOldNodes*
- support nested diagrams
- exploit look-ahead in off-line layout
- invent an output-sensitive heuristic

A final remark is that the STT heuristic has proven surprisingly flexible, having accommodated many variants of both static and dynamic layout.

Acknowledgments. John Ellson, Emden Gansner, and John Mocenigo shared many ideas with us and made key contributions to our implementations and user interfaces. We also thank the referees of Graph Drawing 2001 for their suggestions.

References

1. Giuseppe Di Battista, Peter Eades, Roberto Tamassia, and Ioannis G. Tollis. *Graph Drawing: algorithms for the visualization of graphs.* Prentice-Hall, 1999.
2. Therese Bield and Michael Kaufmann. Area-efficient static and incremental graph drawings. In *Proc. 5th European Symposium on Algorithms (ESA'97)*, volume 1284 of *Lecture Notes in Computer Science*, pages 37–52. Springer-Verlag, 1997.
3. K. Bohringer and F. Newbery Paulisch. Using constraints to acheive stability in automatic graph layout algorithms. In *Proceedings of ACM CHI'90*, pages 43–51, 1990.
4. S. S. Bridgeman, J. Fanto, A. Garg, R. Tamassia, and L. Vismara. Interactive Giotto: An algorithm for interactive orthogonal graph drawing. In G. Di Battista, editor, *Graph Drawing '97*, volume 1353 of *Lecture Notes in Computer Science*, pages 303–308, Rome, Italy, 1998. Springer-Verlag.
5. R. Schonfeld C. Matuszewski and P. Molitor. Using sifting for k-layer crossing minimization. In Jan Kratochvíl, editor, *Graph Drawing '99*, volume 1731 of *Lecture Notes in Computer Science*, pages 217–224. Springer-Verlag, 2000.
6. J. Cohen. Drawing graphs to convey proximity: an incremental arrangement method. *ACM Trans. on Computer-Human Interfaces*, 4(11):197–229, 1997.
7. D. Dobkin, E. Gansner, E. Koutsofios, and S. North. Implementing a general-purpose edge router. In G. Di Battista, editor, *Graph Drawing '97*, volume 1353 of *Lecture Notes in Computer Science*, Rome, Italy, 1998. Springer-Verlag.
8. P. Eades and C. Friedrich. The Marey graph animation tool demo. In Joe Marks, editor, *Graph Drawing '00*, volume 1984 of *Lecture Notes in Computer Science*, pages 396–406. Springer-Verlag, 2001.
9. P. Eades, W. Lai, K. Misue, and K. Sugiyama. Layout adjustment and the mental map. *Journal of Visual Languages and Computing*, 6:183–210, 1995.
10. Peter Eades, Robert F. Cohen, and Mao Lin Huang. Online animated graph drawing for web navigation. In G. Di Battista, editor, *Graph Drawing '97*, volume 1353 of *Lecture Notes in Computer Science*, Rome, Italy, 1998. Springer-Verlag.
11. J. Ellson and S. North. TclDG - a Tcl extension for dynamic graphs. In *Proc. 4th USENIX Tcl/Tk Workshop*, pages 37–48, 1996.
12. E. R. Gansner, E. Koutsofios, S. C. North, and K.-P. Vo. A technique for drawing directed graphs. *IEEE Trans. Software Engineering*, 19(3):214–230, 1993.
13. M. Huang and P. Eades. A fully animated interactive system for clustering and navigating huge graphs. In Sue H. Whitesides, editor, *Graph Drawing '98*, volume 1547 of *Lecture Notes in Computer Science*, pages 374–383, Montreal, Canada, 1999. Springer-Verlag.
14. Tamara Munzner. *Interactive visualization of large graphs and networks.* PhD thesis, Stanford University, 2000.
15. A. Papakostas, J. M. Six, and I. G. Tollis. Experimental and theoretical results in interactive orthogonal graph drawing. In S.C. North, editor, *Graph Drawing '96*, volume 1190 of *Lecture Notes in Computer Science*, pages 101–112, 1997.

16. K. Sugiyama, S. Tagawa, and M. Toda. Methods for visual understanding of hierarchical systems. *IEEE Trans. on Systems, Man and Cybernetics*, SMC-11(2):109–125, 1981.
17. G. Woodhull and S. North. Montage - an ActiveX container for dynamic interfaces. In *Proc. 2nd USENIX Windows NT Symposium*, 1998.

Recognizing String Graphs Is Decidable

János Pach and Géza Tóth

Rényi Institute of Mathematics
Hungarian Academy of Sciences
pach@cims.nyu.edu, geza@renyi.hu

Abstract. A graph is called a *string graph* if its vertices can be represented by continuous curves ("strings") in the plane so that two of them cross each other if and only if the corresponding vertices are adjacent. It is shown that there exists a recursive function $f(n)$ with the property that every string graph of n vertices has a representation in which any two curves cross at most $f(n)$ times. We obtain as a corollary that there is an algorithm for deciding whether a given graph is a string graph. This solves an old problem of Benzer (1959), Sinden (1966), and Graham (1971).

1 Introduction

Given a simple graph G, is it possible to represent its vertices by simply connected regions in the plane so that two regions overlap if and only if the corresponding two vertices are adjacent? In other words, is G isomorphic to the *intersection graph* of a set of simply connected regions in the plane? This deceptively simple extension of propositional logic and its generalizations are often referred to in the literature as *topological inference problems* [CGP98a], [CGP98b],[CHK99]. They have proved to be relevant in the area of geographic information systems [E93], [EF91] and in graph drawing [DETT99]. In spite of many efforts [K91a], [K98] (and false claims [SP92], [ES93]), no algorithm was found for their solution. It is known that these problems are at least NP-hard [KM89], [K91b], [MP93].

Since each element of a finite system of regions in the plane can be replaced by a simple continuous arc ("string") lying in its interior so that the intersection pattern of these arcs is the same as that of the original regions, the above problem can be rephrased as follows. Does there exist an algorithm for recognizing *string graphs,* i.e., intersection graphs of planar curves? As far as we know, in this form the question was first asked in 1959 by S. Benzer [B59], who studied the topology of genetic structures. Somewhat later the same question was raised by F. W. Sinden [S66] in Bell Labs, who was interested in electrical networks realizable by printed circuits. Sinden collaborated with R. L. Graham, who communicated the question to the combinatorics community by posing it at the open problem session of a conference in Keszthely, in 1976 [G78]. Soon after G. Ehrlich, S. Even, and R. E. Tarjan [EET76] studied the *"string graph problem"* (see also [K83] and [EPL72] for a special case). The aim of this paper is to answer the

P. Mutzel, M. Jünger, and S. Leipert (Eds.): GD 2001, LNCS 2265, pp. 247–260, 2002.
© Springer-Verlag Berlin Heidelberg 2002

above question in the affirmative: there exists an algorithm for recognizing string graphs.

To formulate our main result precisely, we have to agree on the terminology. Let G be a graph with vertex set $V(G)$ and edge set $E(G)$. A *string representation* of G is an assignment of simple continuous arcs to the elements of $V(G)$ such that no three arcs pass through the same point and two arcs cross each other if and only if the corresponding vertices of G are adjacent. G is a *string graph* if it has a string representation. Every intersection point between two arcs is called a *crossing*. (That is, two arcs may determine many crossings.) For any string graph G, let $\text{ST}(G)$ denote the minimum number of crossings in a string representation of G, and let

$$\text{ST}(n) := \max_{|V(G)|=n} \text{ST}(G),$$

where the maximum is taken over all string graphs G with n vertices.

Theorem 1. *Every string graph with n vertices has a string representation with at most $(2n)^{24n^2+48}$ crossings.*

Using the above notation, we have $\text{ST}(n) \leq (2n)^{24n^2+48}$. On the other hand, it was shown by J. Kratochvíl and J. Matoušek [KM91] that $\text{ST}(n) \geq 2^{cn}$ for a suitable constant c.

Theorem 1 implies that string graphs can be recognized by a finite algorithm. Indeed, by brute force we can try all possible placements of the crossing points along the arcs representing the vertices of the graph and in each case test planarity (which can be done in linear time in the total number of crossings [HT74]).

As was pointed out in [KM91], the representation of string graphs is closely related to the following problem. Let $R \subseteq \binom{E(G)}{2}$ be a set of pairs of edges of G. We say that the pair (G, R) is *weakly realizable* if G can be drawn in the plane so that only pairs of edges belonging to R are *allowed* to cross (but they do not *have* to cross). Such a drawing is called a *weak realization* of (G, R). The minimum number of crossings in a weak realization of (G, R) is denoted by $\text{CR}(G, R)$. Note that the usual *crossing number* of G is equal to $\text{CR}(G, \binom{E(G)}{2})$.) Let

$$\text{CR}(n) := \max_{|V(G)|=n,R} \text{CR}(G, R),$$

where the maximum is taken over all weakly realizable pairs (G, R) on n vertices). It was proved in [KLN91] and [KM91] that the problems of recognizing string graphs and weakly realizing pairs are *polynomially equivalent*. In particular,

$$\text{ST}(n) \leq \text{CR}(n^2) + \binom{n}{2}. \tag{1}$$

Kratochvíl [K98] called it an "astonishing and challenging fact that so far there is no recursive" upper bound known on $\text{ST}(n)$. Our next theorem, which, combined with (1), immediately implies Theorem 1, fills this gap.

Theorem 2. *Let G be a simple graph with m edges, and let (G, R) be a weakly realizable pair. Then (G, R) has a weak realization with at most $(4m)^{12m+24}$ crossings.*

As before, it follows from Theorem 2 that there is a recursive algorithm for deciding whether a pair (G, R) is realizable.

N. Linial has pointed out that the above questions are closely related to estimating the Euclidean distortion of certain metrics induced by weighted planar graphs [LLR95], [R99].

Using completely different (and more elegant) methods, Schaefer and Stefankovič [SS01] have recently established the slightly better bounds $\text{CR}(m) \le m2^m$ and $\text{ST}(n) \le n^2 2^{n^2} + \binom{n}{2}$.

2 Two Simple Properties of Minimal Realizations

In the sequel, let G be a simple graph with n vertices and m edges, let R be a set of pairs of edges, and assume that (G, R) is weakly realizable. Fix a weak realization (drawing) with the *minimum* number of crossings, and assume that this number is at least $(2n)^{12m+24}$. With no loss of generality, all drawings in this paper are assumed to be in *general position*. That is, no edge passes through a vertex different from its endpoints, no three edges have a point in common, and no two edges "touch" each other (i.e., if two edges have a point in common, then they properly cross at this point).

Let A and B be intersection points of two edges $e, f \in E(G)$ and suppose that the portions of e and f between A and B do not have any other point in common. Then the region enclosed by these two arcs is called a *lense*.

Lemma 2.1. *Every lense and its complement contains a vertex of G.*

Proof. By symmetry, it is sufficient to show that every lense contains a vertex. Consider a lense which is minimal by containment, and assume that it is bounded by portions of $e, f \in E(G)$ between A and B. By the minimality, any other edge of G which intersects one side of the lense must also cross the other one. Therefore, replacing the portion of f between A and B by an arc running outside the lense and very close to e, we would reduce the number of crossings in the drawing, contradicting its minimality. □

Deleting from the plane any two arcs, e and f, the plane falls into a number of connected components, called *cells*. At most 4 of them contain and endpoint of e or f. A cell containing no endpoint of e or f is called a *k-cell,* if its boundary consists of k *sides* (subarcs of e and f). Obviously, k must be even, and the sides of a k-cell belong to e and f, alternately. We say that a cell is *empty* if it contains no vertex of G.

Lemma 2.2. *Let e and f be two portions of edges of G that cross each other K times, where $K \ge 16n^3$ and $n \ge 10$. Then e and f determine $M \ge K/(8n^3)$ empty four-cells, $C_1, C_2, \ldots, C_M$, such that C_i and C_{i+1} share a side belonging to f, for every i.*

Proof. The arcs e and f divide the plane into K cells. All but at most 4 of them contain no endpoint of e or f and have an even number of sides.

Define a graph H, whose vertices represent the K cells, and two vertices are joined by an edge if and only if the corresponding cells share a side which belongs to f. Since H is connected and has at most $K-1$ edges, it is a tree. Every leaf of H corresponds to a lense or possibly a cell containing an endpoint of e or f. The number of lenses is at most n, because, by Lemma 2.1, each of them contains a vertex of G. Thus, H has at most n leaves. Consequently, the degree of every vertex of H is at most n.

Delete every vertex of H which corresponds to a cell that (i) either contains an endpoint of e or f, (ii) or contains a vertex of G and has more than 2 sides. The number of deleted vertices is at most $n+4$, each of them has degree at most n, so the resulting forest consists of at most $n(n+4)$ trees. Hence, one of these trees, H', has at least $K' = (K-n-4)/[n(n+4)]$ vertices. Any leaf of H' is either a leaf of H or is connected to a deleted vertex of H. Therefore, H' has at most $2n$ leaves. This implies that H' contains a path with at least $M = (K'-1)/(4n-3) - 1 > K/(8n^3)$ vertices, each of degree 2. The sequence of four-cells corresponding to the vertices of this path meet the requirements of the lemma. □

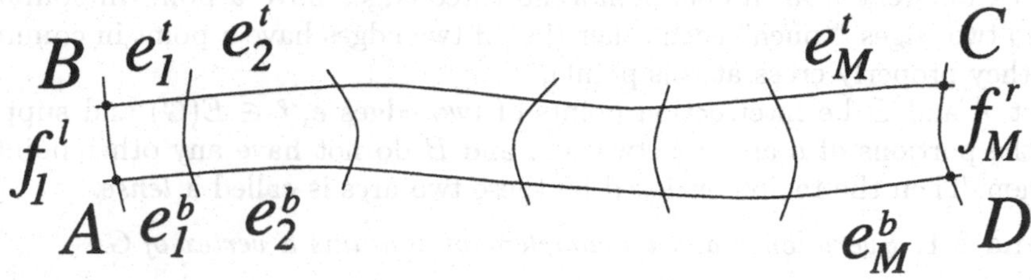

Fig. 1.

A sequence $C_1, C_2, \ldots, C_M$ of four-cells whose existence is guaranteed in Lemma 2.2 is called an *empty* (e,f)*-path* of four-cells. In what follows, we analyze the finer structure of such a path. We assume that e and f are oriented, and we denote the four sides of C_i by e_i^t, f_i^r, e_i^b and f_i^l, in clockwise order. (Here the superscripts stand for *top, right, bottom,* and *left*, repectively, suggesting that on our pictures the boundary pieces belonging to e are "horizontal"; see Figure 1.) For every $1 \le i < M$, we have $f_i^r = f_{i+1}^l$.

Let A denote the common endpoint of f_1^l and e_1^b, that is, $A = f_1^l \cap e_1^b$. Let $B = f_1^l \cap e_1^t$, $C = f_M^r \cap e_M^t$, and $D = f_M^r \cap e_M^b$.

We distinguish two different types of the empty paths, (See Fig. 2) and study them separately in the next two sections.

Type 1: For every i, the two sides of the cell C_i which belong to e have the *same* orientation (i.e., both of them are oriented towards f_i^r or both are oriented towards f_i^l). See Fig. 2 (a), (b).

Type 2: For every i, the two sides of the cell C_i which belong to e have *opposite* orientations. See Fig. 2 (c).

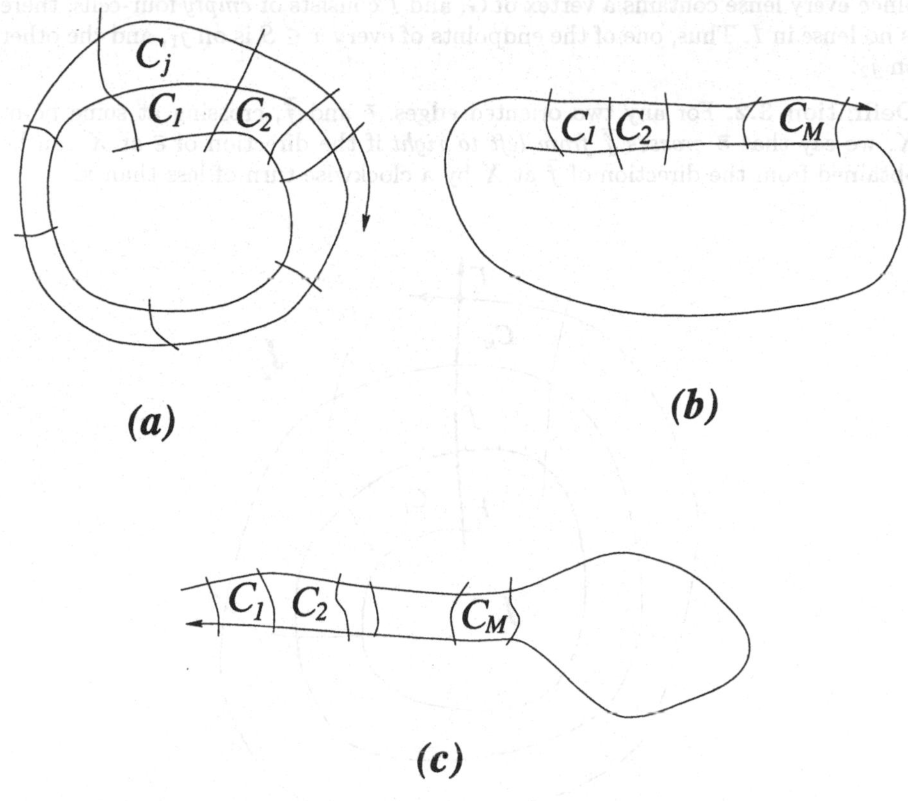

Fig. 2.

3 Empty Paths of Type 1

Lemma 3.1 *Let $C_1, C_2, \ldots, C_M$ be an empty (e, f)-path of four-cells of type 1. Then e_1^t and e_1^b cannot coincide with any side of C_i, $1 < i \leq M/(5m)$.*

Proof. Suppose, in order to obtain a contradiction, that e.g. $e_1^t = e_{j+1}^b$, for some $1 < j < M/(5m)$. This easily implies $e_i^t = e_{j+i}^b$, for every $i \leq M - j$. See Fig 2 (a).

Assume without loss of generality that M can be written in the form $M = M'j+1$, for a suitable $M' \geq 5m$. Let $\hat{f} = f_1^l \cup f_{j+1}^l \cup \cdots \cup f_M^l$, which is a segment of f, and orient it from f_1^l towards f_M^l. Let $F_1 = f_1^l \cap e_1^b$ and $F_2 = f_M^l \cap e_M^t$ be the starting point and the endpoint of $\hat{f}$, resp., and let $I = C_1 \cup C_2 \cup \cdots \cup C_M$. Furthermore, let J_1 denote the region bounded by $j_1 = e_1^b \cup e_2^b \cup \ldots \cup e_j^b \cup f_1^l$

which does not contain I, and let J_2 be the region bounded by $j_2 = e^t_M \cup e^t_{M-1} \cup \ldots \cup e^t_{M-j+1} \cup f^r_M$ which does not contain I.

Let S denote the set of arcs obtained by intersecting the edges of G with I. Since every lense contains a vertex of G, and I consists of *empty* four-cells, there is no lense in I. Thus, one of the endpoints of every $x \in S$ is on j_1, and the other on j_2.

Definition 3.2. For any two oriented edges, $\overline{e}$ and $\overline{f}$, crossing at some point X, we say that $\overline{e}$ *crosses* $\overline{f}$ *from left to right* if the direction of $\overline{e}$ at X can be obtained from the direction of $\overline{f}$ at X by a clockwise turn of less than π.

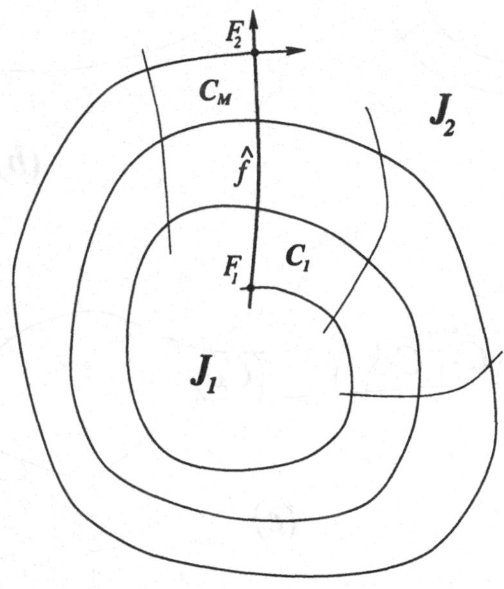

Fig. 3.

Let $\hat{e}$ be the portion of e starting with e^b_1 and ending with e^t_M. Orient it from e^b_1 towards e^t_M. Then $\hat{e}$ crosses $\hat{f}$ from left to right at every crossing. See Fig. 3.

Since I contains no lense, for any orientation of two elements $x, y \in S$, every crossing of these two curves are of the same type, i.e., either x crosses y from left to right at every crossing, or x crosses from right to left at every crossing. In particular, we can pick an orientation of $x \in S$ such that at every crossing it crosses $\hat{f}$ from left to right.

Let $S_1 \subset S$ denote the set of all elements of S that do not cross $\hat{f}$. Let $S_2 \subset S$ (and $S_3 \subset S$) denote the set of all elements starting at a point of j_1 and ending at a point of j_2 (starting at a point of j_2 and ending at a point of j_1, respectively).

We will modify the drawing by re-routing the elements of S_2 so as to reduce the number of crossings. Notice that, since there is no lense in I, the intersection points of x and $\hat{f}$ follow each other in the same order on both x and $\hat{f}$.

Define a binary relation on S_2 as follows. For any $x, y \in S_2$, we say that y *precedes* x (and write $y \prec x$), if x and $\hat{f}$ have two consecutive crossings, X and X', such that y does not intersect the portion of $\hat{f}$ between X and X'.

Claim. The relation $\prec$ is a partial ordering on S_2.

Proof. Suppose that $y \prec x$. The union of the portions of x and $\hat{f}$ between X and X' divides I into two pieces, separating j_1 from j_2. Since y does not cross the portion of $\hat{f}$ between X and X', it must cross the portion of x between X and X' from right to left (see Fig. 4). But then at every other crossing y has to cross x from right to left. This shows that $\prec$ is antisymmetric, because assuming that $x \prec y$, the same argument would show that at each of their crossings y must cross x from left to right, a contradiction.

To show that $\prec$ is transitive, suppose that $z \prec y$ and $y \prec x$. Let Y and Y' be two consecutive crossings between y and $\hat{f}$ such that z does not intersect the portion of $\hat{f}$ between Y and Y'. If the YY' portion of $\hat{f}$ contains the XX' portion, then z does not cross the XX' portion, hence we are done: $z \prec x$. Thus, we can assume that the XX' and YY' portions are disjoint. Suppose without loss of generality that along $\hat{f}$ the order of these four points is X, X', Y, Y'. Let $Y_1, Y_2, \ldots, Y_k$ be all crossings of $\hat{f}$ and y, between the points X' and Y', where $Y_{k-1} = Y, Y_k = Y'$. By our assumptions, every portion $Y_{i-1}Y_i$ of $\hat{f}$ contains at least one crossing with x and at most one crossing with z. Therefore, the XY' portion of $\hat{f}$ contains at least $k+1$ crossings with x and at most $k-1$ crossings with z. Thus, we have $z \prec x$, concluding the proof of the Claim. □

For any $x \in S_2$, define $\text{rank}(x) = |x \cap \hat{f}|$. For any *edge* g of G, let

$$\text{rank}(g) = \{\text{rank}(x) \mid x \in S_2, x \subset g\},$$

and let $\text{rank}(G) = \bigcup_{g \in E(G)} \text{rank}(g)$.

In the proof of the Claim, we showed that if x has two consecutive crossings with $\hat{f}$ with the property that y does not intersect the XX' portion of $\hat{f}$, then y must cross the XX' portion of x. Therefore, if $|\text{rank}(x) - \text{rank}(y)| \geq 2$ for some $x, y \in S_2$, then x and y cross each other. If x and y belong to the same edge of G, then they cannot cross, so in this case we have $|\text{rank}(x) - \text{rank}(y)| \leq 1$. Therefore, for any edge g of G, the set $\text{rank}(g)$ is either empty, or it consists of one integer or two consecutive integers.

Since we have $\text{rank}(\hat{e}) = M' + 1 > 5m$, but $|\text{rank}(G)| \leq 2m$, there is an integer L, $5m > L > 3$ such that $L, L-1, L-2 \notin \text{rank}(G)$.

Now let $S_2 = S^l \cup S^h$ (where l and h stand for 'low' and 'high,' resp.) such that

$$S^l = \{x \in S_2 \mid \text{rank}(x) < L-2\}, \quad S^h = \{x \in S_2 \mid \text{rank}(x) > L\}.$$

Let w be a *minimal* element of S^h with respect to the partial ordering $\prec$. Let $W_1, W_2, \ldots, W_k$ denote the crossings of w and $\hat{f}$, in this order. By the minimality of w, every element of S^h intersects $\hat{f}$ between any W_i and W_{i+1}.

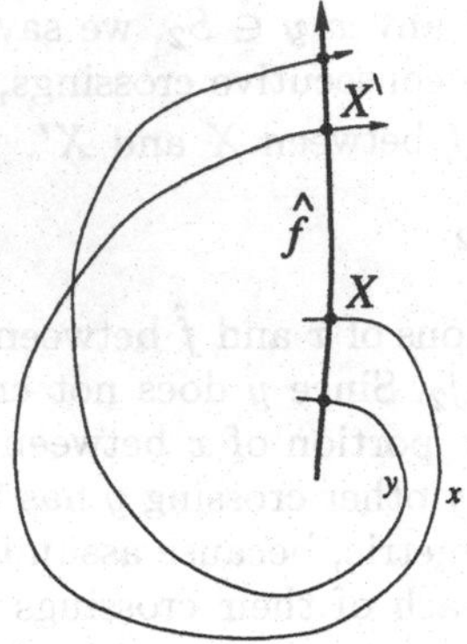

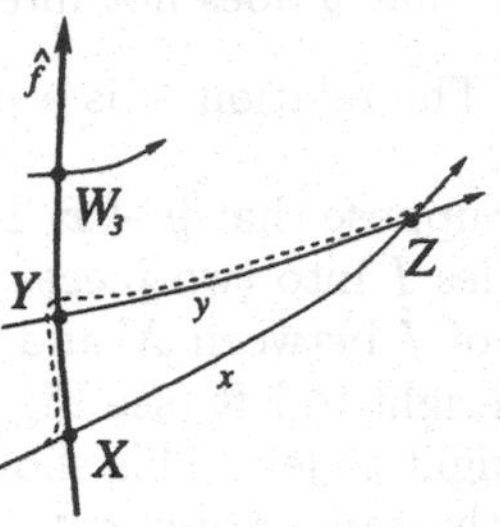

Fig. 4.

By 'shifting' the 'bad' crossings further, we will modify the elements of S^l so that none of them will cross the portion of $\hat{f}$ between F_1 and W_3. Suppose that for some $x \in S^l$, $y \in S^h$, X (resp. Y) is a crossing of x (resp. y) and $\hat{f}$, and no element of S_2 crosses the XY portion of $\hat{f}$, and F_1, X, Y, W_3 follow each other in this order on $\hat{f}$. There are at least $L-3$ crossings of y and $\hat{f}$, which come after Y. Since x and $\hat{f}$ cross at most $L-4$ times, there exists a crossing Z of x and y, which comes after X along x and after Y along y. Let X' be a point on x slightly before X, let Y' be a point of $\hat{f}$ slightly after Y, and let Z' be a point of x slightly after Z. Replace the $X'Z'$ portion of x by a curve running from X' to Y' very close to the XY portion of $\hat{f}$, and from Y' to Z' running very close to the YZ portion of y. This is called an *elementary flip*. See Fig. 4. Any element of S_2 intersects the XZ portion of x and the YZ portion of y the same number of times, therefore, x intersects every other element of S_2 precisely the same number of times as before the the elementary flip.

Do as many elementary flips as possible. When we get stuck, no element of S^l crosses the portion of $\hat{f}$ between F_1 and W_3.

Let P be a point of $\hat{f}$. For any $b \in S_2$, let $B(P)$ and $B'(P)$ be those two crossing points of $\hat{f}$ and b, which follow immediately after P along $\hat{f}$ (if two such point exist). The portions of b and $\hat{f}$ between $B(P)$ and $B'(P)$ divide I into two components. Let $I_b(P)$ denote the one containing j_2. Let $I_2(P) = \bigcup_{b \in S_2} I_b(P)$ and let $I(P) = I \setminus I_2(P)$.

Suppose without loss of generality that $\hat{f}$ is a straight-line segment. Let T be a (very thin) rectangle, whose left side, T^l, coincides with the segment F_1W_3, and whose right side T^r is very close to T^l. For any $y \in S^h$, let $Y_1^l, Y_2^l, \ldots, Y_k^l$ (resp. $Y_1^r, Y_2^r, \ldots, Y_k^r$) be the intersections of y and T^l (resp. T^r). We can assume that the segments $Y_i^l Y_i^r$ are horizontal. Delete $Y_1^r Y_k^l$ from y.

Connect Y_i^r to Y_{i+1}^l by a straight-line segment (inside T), and connect Y_{i+1}^l with Y_{i+1}^r by a curve in $I(Y_{i+1}^l)$ $(1 \le i \le k-1)$, running very close to its boundary. In this way, we obtain a weak realization of (G, R). Any $x \in S^h$ intersects $\hat{f}$ one time less than previously. It is not hard to see that the number of crossings between x and $y \in S$ remained the same or decreased. Therefore, the number of crossings in the above realization is smaller than in the original drawing, a contradiction. □

Lemma 3.1 combined with the next statement shows that it is sufficient to study empty paths of type 2.

Lemma 3.3. *Let $C_1, C_2, \ldots, C_M$ be an empty (e, f)-path of four-cells. Let $\overline{e} = e_1^t \cup e_2^t \cup \cdots \cup e_M^t$, and suppose that $\overline{e}$ and f form an empty $(\overline{e}, f)$-path of type 1, whose length is $L \leq M$.*

Then there is an empty (e, f)-path of type 2, whose length is L.

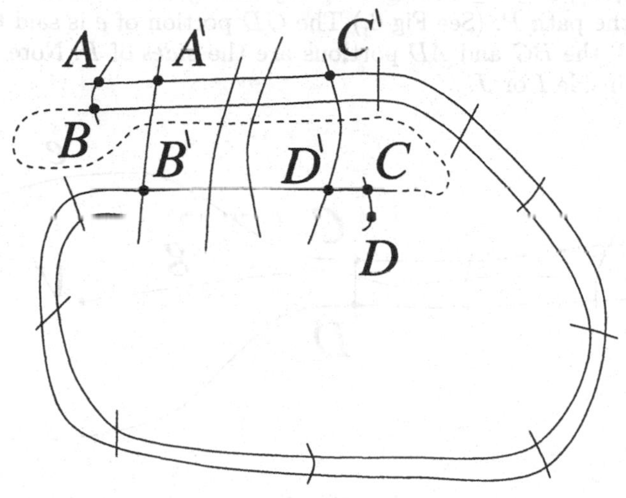

Fig. 5.

Proof. We can assume that $C_1, C_2, \ldots, C_M$ is a path of type 1, otherwise there is nothing to prove. Let A, B, C, and D denote the same as in Figure 1. Then these points follow each other along e in the order A, C, B, D.

Let $\overline{C}_1, \overline{C}_2, \ldots, \overline{C}_L$ be the four-cells of the empty $(\overline{e}, f)$-path and let $\overline{e}_i^t$, $\overline{f}_i^r$, $\overline{e}_i^b$ and $\overline{f}_i^l$ denote the sides of C_i'. Further, let A' be the common endpoint of $\overline{f}_1^l$ and $\overline{e}_1^t$, let $B' = \overline{f}_1^l \cap \overline{e}_1^b$, $C' = \overline{f}_L^r \cap \overline{e}_L^t$, and $D' = \overline{f}_L^r \cap \overline{e}_L^b$. Then A', C', B', and D' follow each other in this order along e, and they all lie between A and C.

Let I denote the region bounded by $\overline{e}_1^b$, $\overline{f}_1^r$, and by the portion of e between $\overline{f}_1^r \cap \overline{e}_1^t$ and B' (cf. Figure 5). Clearly, C and D are in the exterior and in the interior of I, respectively. Since e cannot cross itself, the portion e'' of e between C and B must intersect $\overline{f}_1^r$. It follows that e'' intersects $\overline{f}_L^l, \overline{f}_{L-1}^l, \ldots, \overline{f}_1^l$ in this order. (If e'' enters some $\overline{C}_j$ through the side $\overline{f}_{j+1}^l$, then it must leave $\overline{C}_j$ through the opposite side, $\overline{f}_j^l$, otherwise we would obtain an empty lense contradicting Lemma 2.1. Thus, f and the portion of e between B' and C form an empty (e, f)-path of type 2, whose length is L. □

4 Empty Paths of Type 2 and the Proof of Theorem 2

Let e, f be two portions of edges of G. Suppose that they form an empty (e, f)-path P of type 2, consisting of M four-cells, $C_1, C_2, \ldots, C_M$. We use the notation on Figure 1. Now the arcs $e_1^t, e_2^t, \ldots, e_M^t, e_M^b, \ldots, e_1^b$ follow each other on e in this order. Let I denote the union of the cells C_i $(1 \le i \le M)$, which is a curvilinear quadrilateral, whose vertices are A, B, C, and D, in clockwise order. Furthermore, let J denote the region bounded by f_M^r and the portion of e between C and D.

Suppose that there is a portion of an edge g which intersects $f_1^l, f_2^l, \ldots f_M^l$ in this order and ends at a vertex $v \in V(G)$ lying in J. If g does not intersect any of the sides e_i^t, e_i^b, $1 \le i \le M$, then we say that g (with vertex v) *cuts all the way through* the path P. (See Fig 6.) The CD portion of e is said to be the *top* of the path P, the BC and AD portions are the *sides* of P. Note that there is no piece of e inside I or J.

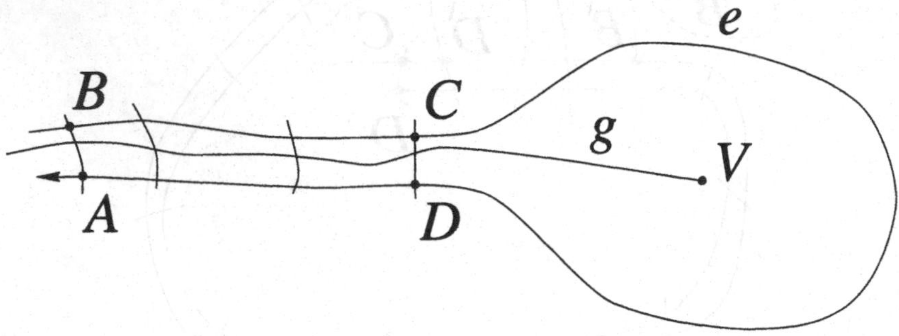

Fig. 6.

Lemma 4.1. *Let e, f be two portions of edges of G that form an empty (e, f)-path P of type 2.*

Then there exists a portion of an edge g which cuts all the way through P.

Proof. We can and will assume without loss of generality that P is *minimal* in the sense that there is no other (e', f)-path P' of type 2, consisting of M four-cells, such that $I' \cup J'$ is strictly contained in $I \cup J$. Obviously, every edge that cuts all the way through P' also cuts all the way through P.

Fix an edge g of G. The boundary of $I \cup J$ may cut g into several portions. Any portion of g which enters $I \cup J$ through the side f_1^l must leave $I \cup J$ at some point, otherwise it meets the requirements in the lemma. Indeed, such a portion can only end at a vertex in J, because I is empty, and it must cross the arcs $f_1^l, f_2^l, \ldots, \ldots, f_M^l, f_M^r$ in this order, because otherwise it would create an empty lense with f.

Suppose there is a portion g' of g entering and leaving $I \cup J$ through the side f_1^l. Using again that all four-cells in P are empty, we obtain that g' must first cross the arcs $f_1^l, f_2^l, \ldots, f_M^l, f_M^r$, in this order, then turn back and cross them another time in the opposite order. Hence, there is a (g', f)-path P' of length M inside $I \cup J$, contradicting the minimality of P.

Therefore, every portion of an edge which enters $I \cup J$ through f_1^l must leave it through the portion of e between B and A. Replacing this portion of e by a curve running very close to f_1^l, but not entering I, we obtain another weak realization of (G, R) with a smaller number of crossings. This contradicts our assumption that the initial realization was minimal. □

Before we turn to the proof of Theorem 2, we have to introduce another notion. For any portion e' of an edge e, an (e', f)-path of empty four-cells is called a *weak* (e, f)*-path.*

Proof. Consider a weak realization of (G, R), in which the number of crossings is minimum, and suppose that this number is at least $(2n)^{12m+24} = (8n^3)^{2m+4}$. Find two edges, e and f, that cross each other $K \geq (8n^3)^{2m+3}$ times.

For every $i = 1, 2, \ldots, 2m+1$, we construct the following objects:

(1) A subsegment $e_i \subseteq e_{i-1}$, where $e_0 = e$;

(2) an (e_i, f)-path P_i of length at least $(8n^3)^{2m-2i+3}$;

(3) a set of *charged pairs*, (g_i, v_i), where each pair consists of an edge g_i of G and one of its endpoints, v_i. Initially, there is no charged pair. In STEP i, we define g_i with endpoint v_i such that (g_i, v_i) will cut all the way through an (e, f)-path $P(g_i, v_i)$ of length $(8n^3)^{2m-2i+3}$. $P(g_i, v_i) \subset P_i$.

(4) $P'(g_i, v_i)$ will be a weak (e, f)-path of length $(8n^3)^{2m-2i+3}$ such that $P(g_i, v_i) \subset P'(g_i, v_i) \subset P_i$, and such that $P'(g_i, v_i)$ is *maximal* with this property.

STEP i: For $i = 1$, let $e_0^+ = e_0^- = e_0 = e$. For $i > 1$, P_{i-1} has two sides, call them e_{i-1}^+ and e_{i-1}^-.

The arc e_{i-1}^- intersects f at least $(8n^3)^{2m-2i+5}$ times, therefore, by the previous lemmas, there is a (e_{i-1}^-, f)-path P^- of length $(8n^3)^{2m-2i+3}$. Since e_{i-1}^+ and e_{i-1}^- are two sides of an (e_{i-1}, f)-path, there is also also a corresponding (e_{i-1}^+, f)-path P^- of length $(8n^3)^{2m-2i+3}$, and either $P^- \subset P^+$ or $P^+ \subset P^-$. Suppose that $P^- \subset P^+$, the other case can be treated analogously. Let $e_i = e_{i-1}^+$, $P_i = P^+$. Let $\hat{f} \subset f$ be a side of one of the four-cells of P_i. Color each point of $\hat{f}$ red, if it is inside $P'(g_j, v_j)$ for some $j < i$, and color the other points blue. The red points form some disjoint arcs of $\hat{f}$, whose endpoints are intersections of $\hat{f}$ and e.

Let F be one of the intersections of e_i^- and $\hat{f}$, and let F' be the intersection of e and $\hat{f}$, closest to F. By our construction, the portion of $\hat{f}$ between F and F' is blue. Suppose without loss of generality that at F e crosses f from left to right.

If the edge e crosses f from right to left at F', then we have an (e, f)-path P of type 2 inside P_i, whose length is $(8n^3)^{2m-2i+3}$. Therefore, there is a pair (g_i, v_i) cutting all the way through P. Since FF' was a blue arc, (g_i, v_i) has not been charged before. Now charge (g_i, v_i) and let $P(g_i, v_i) = P$. Let $P'(g_i, v_i)$ be a weak (e, f)-path of length $(8n^3)^{2m-2i+3}$ such that $P(g_i, v_i) \subset P'(g_i, v_i) \subset P_i$, and such that $P'(g_i, v_i)$ is *maximal* with this property. Go to STEP $i+1$.

On the other hand, if e crosses f from left to right at F', then we have an (e, f)-path P of type 1 inside P_i, whose length $(8n^3)^{2m-2i+3}$. But then the FF'

arcs of e and f together separate the two endpoints of e. (See also the proof of Lemma 3.3.) Let F'' be a point of f very close to F, such that F is between F' and F''. The FF' portions of e and f together separate one of the endpoints, v, from F''. Since e cannot cross itself, there is also an (e, f)-path P of type 2 inside P^- such that (e, v) cuts all the way through it. Clearly, (e, v) was not charged before since the FF' arc of $\hat{f}$ is blue. Now charge $(g_i, v_i) = (e, v)$, and let $P(g_i, v_i) = P$. Let $P'(g_i, v_i)$ be a weak (e, f)-path of length $(8n^3)^{2m-2i+3}$ such that $P(g_i, v_i) \subset P'(g_i, v_i) \subset P_i$, and such that $P'(g_i, v_i)$ is *maximal* with this property. Go to STEP $i + 1$.

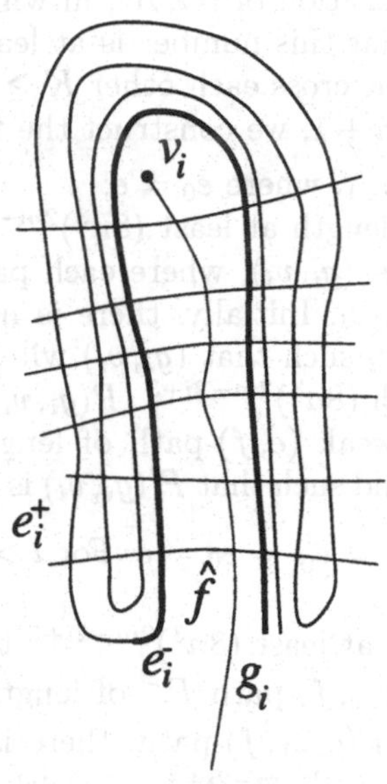

Fig. 7.

The algorithm will terminate after $2m + 1$ steps. At each step we charge a new pair (e, v). This is a contradiction, because the total number of pairs that can be charged is $2m$. Therefore, a minimal weak realization has at most $(2n)^{12m+24} \leq (4m)^{12m+24}$ crossings. This concludes the proof of Theorem 2. □

References

[B59] S. Benzer: On the topology of the genetic fine structure, *Proceedings of the National Academy of Sciences of the United States of America* **45** (1959), 1607–1620.

[CGP98a] Z.-Z. Chen, M. Grigni, and C. H. Papadimitriou, Planar map graphs, in: *STOC '98*, ACM, 1998, 514–523.

[CGP98b] Z.-Z. Chen, M. Grigni, and C. H. Papadimitriou, Planar topological inference, (Japanese) in: *Algorithms and Theory of Computing (Kyoto, 1998) Sūrikaisekikenkyūsho Kōkyūroku* **1041** (1998), 1–8.

[CHK99] Z.-Z. Chen, X. He, and M.-Y. Kao, Nonplanar topological inference and political-map graphs, in: *Proceedings of the Tenth Annual ACM-SIAM Symposium on Discrete Algorithms (Baltimore, MD, 1999)*, ACM, New York, 1999, 195–204.

[DETT99] G. Di Battista, P. Eades, R. Tamassia, and I. G. Tollis, *Graph Drawing*, Prentice Hall, Upper Saddle River, NJ, 1999.

[E93] M. Egenhofer, A model for detailed binary topological relationships, *Geomatica* **47** (1993), 261-273.

[EF91] M. Egenhofer and R. Franzosa, Point-set topological spatial relations, *International Journal of Geographical Information Systems* **5** (1991), 161-174.

[ES93] M. Egenhofer and J. Sharma, Assessing the consistency of complete and incomplete topological information, *Geographical Systems* **1** (1993), 47–68.

[EET76] G. Ehrlich, S. Even, and R. E. Tarjan, Intersection graphs of curves in the plane, *Journal of Combinatorial Theory, Series B* **21** (1976), 8–20.

[EHP00] P. Erdős, A. Hajnal, and J. Pach, A Ramsey-type theorem for bipartite graphs, *Geombinatorics* **10** (2000), 64–68.

[EPL72] S. Even, A. Pnueli, and A. Lempel, Permutation graphs and Transitive graphs, *Journal of Association for Computing Machinery* **19** (1972), 400–411.

[G80] M. C. Golumbic, *Algorithmic Graph Theory and Perfect Graphs*, Academic Press, New York, 1980.

[G78] R. L. Graham: Problem, in: *Combinatorics, Vol. II (A. Hajnal and V. T. Sós, eds.)*, North-Holland Publishing Company, Amsterdam, 1978, 1195.

[HT74] J. Hopcroft and R. E. Tarjan, Efficient planarity testing, *J. ACM* **21** (1974), 549–568.

[K83] J. Kratochvíl, String graphs, in: *Graphs and Other Combinatorial Topics (Prague, 1982)*, Teubner-Texte Math. **59**, Teubner, Leipzig, 1983, 168–172.

[K91a] J. Kratochvíl, String graphs I: The number of critical nonstring graphs is infinite, *Journal of Combinatorial Theory, Series B* **52** (1991), 53–66.

[K91b] J. Kratochvíl, String graphs II: Recognizing string graphs is NP-hard, *Journal of Combinatorial Theory, Series B* **52** (1991), 67–78.

[K98] J. Kratochvíl, Crossing number of abstract topological graphs, in: *Graph drawing (Montreal, QC, 1998), Lecture Notes in Comput. Sci.* **1547**, Springer, Berlin, 1998, 238–245.

[KLN91] J. Kratochvíl, A. Lubiw, and J. Nešetřil, Noncrossing subgraphs in topological layouts, *SIAM J. Discrete Math.* **4** (1991), 223–244.

[KM89] J. Kratochvíl and J. Matoušek, NP-hardness results for intersection graphs, *Comment. Math. Univ. Carolin.* **30** (1989), 761–773.

[KM91] J. Kratochvíl and J. Matoušek, String graphs requiring exponential representations, *Journal of Combinatorial Theory, Series B* **53** (1991), 1–4.

[KM94] J. Kratochvíl and J. Matoušek, Intersection graphs of segments, *Journal of Combinatorial Theory, Series B* **62** (1994), 289–315.

[LLR95] N. Linial, E. London, and Y. Rabinovich, The geometry of graphs and some of its algorithmic applications, *Combinatorica* **15** (1995), 215–245.

[MP93] M. Middendorf and F. Pfeiffer, Weakly transitive orientations, Hasse diagrams and string graphs, in: *Graph Theory and Combinatorics (Marseille-Luminy, 1990), Discrete Math.* **111** (1993), 393–400.

[PS01] J. Pach and J. Solymosi, Crossing patterns of segments, *Journal of Combinatorial Theory, Ser. A*, to appear.

[R99] S. Rao, Small distortion and volume preserving embeddings for planar and Euclidean metrics, in: *Proceedings of the Fifteenth Annual Symposium on Computational Geometry (Miami Beach, FL, 1999)*, ACM, New York, 1999, 300–306.

[SS01] M. Schaefer and D. Stefankovič, Decidability of string graphs, *STOC 01*, to appear.

[S66] F. W. Sinden, Topology of thin film RC circuits, *Bell System Technological Journal* (1966), 1639-1662.

[SP92] T. R. Smith and K. K. Park, Algebraic approach to spatial reasoning, *International Journal of Geographical Information Systems* **6** (1992), 177–192.

On Intersection Graphs of Segments with Prescribed Slopes

Jakub Černý, Daniel Král'*, Helena Nyklová, and Ondřej Pangrác

Department of Applied Mathematics and
Institute for Theoretical Computer Science (ITI)**,
Charles University, Malostranské nám. 25,
118 00 Prague, Czech Republic,
{cerny,kral,nyklova,pangrac}@kam.ms.mff.cuni.cz

Abstract. We study intersection graphs of segments with prescribed slopes in the plane. A sufficient and necessary condition on tuples of slopes in order to define the same class of graphs is presented for both the possibilities that the parallel segments can or cannot overlap. Classes of intersection graphs of segments with four slopes are fully described; in particular, we find an infinite set of quadruples of slopes which define mutually distinct classes of intersection graphs of segments with those slopes.

1 Introduction

Intersection graphs of various types of geometric objects attract attention of researchers and find their applications in many areas of computer science. An intersection graph of a set of geometric objects is the graph whose vertices correspond to the objects of the set and two of them are joined by an edge if and only if the corresponding objects intersect. Intersection graphs of chords in a circle [2,5,6], of arcs of a circle [16], of segments [3,9,13,14,15], of simple curves in the plane [4,8,10], of convex sets in the plane and others [1,11] have been studied intensively. We focus our attention on intersection graphs of segments with prescribed slopes in the plane (see [13]) in this paper. Such classes of graphs have been widely studied. A result of de Fraysseix et al. [7] says that each bipartite planar graph can be represented as an intersection graph of segments with two slopes such that no two segments have an interior point in common.

We study intersection graphs of segments such that the slopes of the segments of one of its realizations are among prescribed slopes $\alpha_1, \ldots, \alpha_k$ (precise definitions of the classes which interest us can be found in Section 2). We study both the classes for which parallel segments can or cannot overlap. The classes of intersection graphs of non–overlapping segments were defined and investigated by Kratochvíl and Matoušek in [13]. They proved that for $k = 3$ this class

* Supported in part by KONTAKT ME337/99

** This research was supported by GAUK 158/99 and GAČR 201/99/0242. Institute for Theoretical Computer Science (ITI) is supported as project LN00A056 by the Ministry of Education of Czech Republic.

P. Mutzel, M. Jünger, and S. Leipert (Eds.): GD 2001, LNCS 2265, pp. 261–271, 2002.
© Springer-Verlag Berlin Heidelberg 2002

of graphs is the same (regardless the choice of the slopes) as the class of intersection graphs of segments with three distinct slopes (the slopes are not prescribed in this case). On the other hand, they also proved in [13] that this does not hold for $k = 4$. We prove a sufficient and necessary condition on the k–tuples of slopes $\alpha_1, \ldots, \alpha_k$ and $\beta_1, \ldots, \beta_k$ in order to define equal classes of intersection graphs of segments with their slopes among these k–tuples (Theorem 1) for both the cases when the segments can and cannot overlap. We further prove that the classes of graphs for distinct choices of slopes are either equal or, neither one of them is a subset of the other (Theorem 2), even if one of them is a class of overlapping segments and the other is a class of non–overlapping segments. Classes of intersection graphs of segments with four distinct slopes are fully described in Theorem 3 and in Corollary 1. Corollary 1 answers the question presented by Jan Kratochvíl during the 1st Graph Drawing conference in Paris in 1993: "Are there only finitely many distinct classes of intersection graphs of segments with prescribed four distinct slopes?" This question can be rephrased in terms introduced in Section 2 as follows: "Are there only finitely many mutually non-equivalent quadruples of slopes?" Corollary 1 provide us infinite number of distinct classes of intersection graphs of segments with prescribed quadruples of slopes.

We give definitions and introduce notation used in the paper together with the previously known results in Section 2. We study when two k–tuples of slopes define the same class of graphs in Section 3. We present an example of a graph which can be realized as an intersection graph of non–overlaping segments with a given k–tuple of slopes and which cannot be realized by possibly overlapping segments with slopes of another given k–tuple in Section 4. We combine the results of Section 3 and Section 4 in Section 5 and we state our main results.

2 Definitions and Basic Properties

A *graph* is a simple undirected graph in the whole paper; if G is a graph, we write $V(G)$ for its vertex set and $E(G)$ for its edge set. Let C be a class of sets indexed by vertices of G. C is an *intersection realization* of a graph G if $c_u \cap c_v \neq \emptyset$ if and only if $uv \in E(G)$ where c_u (c_v) is the set of C indexed by the vertex u (v). A graph is an *intersection graph* with respect to a certain class of objects if it can be realized by objects within this class. We study the following classes of intersection graphs of segments which were defined in [13]:

- SEG is the class of *intersection graphs of segments* in the plane, i.e. those graphs for which there exists a set of segments C such that C is an intersection realization of G (parallel segments may overlap).
- PURE-SEG is the class of *intersection graphs of non–overlapping segments*, i.e. those graphs for which there exists a set of segments C such that the parallel segments of C are disjoint (i.e. do not overlap) and C is an intersection realization of G.
- $k-\mathrm{DIR}(\alpha_1, \ldots, \alpha_k)$ is the class of *intersection graphs of segments with slopes* $\alpha_1, \ldots, \alpha_k$ in the plane, i.e. those graphs for which there exists a set of segments C with slopes among $\alpha_1, \ldots, \alpha_k$ such that C is an intersection real-

ization of G (parallel segments may overlap). We assume w.l.o.g. throughout the paper that $0 = \alpha_1 < \ldots < \alpha_k < \pi$.

- $k -$ PURE-DIR$(\alpha_1, \ldots, \alpha_k)$ is the class of *intersection graphs of non–overlapping segments with slopes* $\alpha_1, \ldots, \alpha_k$ in the plane i.e. those graphs for which there exists a set of segments C with slopes among $\alpha_1, \ldots, \alpha_k$ such that the parallel segments of C are disjoint (i.e. do not overlap) and C is an intersection realization of G. We also assume w.l.o.g. that $0 = \alpha_1 < \ldots < \alpha_k < \pi$ in the paper.
- $k -$ DIR $= \cup_{\alpha_1,\ldots,\alpha_k} k -$ DIR$(\alpha_1, \ldots, \alpha_k)$ is the class of *intersection graphs of segments of at most k slopes.*
- $k -$ PURE-DIR $= \cup_{\alpha_1,\ldots,\alpha_k} k -$ DIR$(\alpha_1, \ldots, \alpha_k)$ is the class of *intersection graphs of non–overlapping segments of at most k slopes.*

Note that SEG $= \cup_{k=1}^{\infty} k-$DIR and PURE-SEG $= \cup_{k=1}^{\infty} k-$PURE-DIR. It is also clear that $k -$ PURE-DIR$(\alpha_1, \ldots, \alpha_k) \subset k -$ DIR$(\alpha_1, \ldots, \alpha_k)$ for an arbitrary choice of $\alpha_1, \ldots, \alpha_k$ and $k -$ PURE-DIR $\subset k -$ DIR (both inclusions are strict).

We use linear transformations of the plane in several proofs in this paper. We describe the transformations by 2×2–matrices of real numbers. If A is such a matrix and if $[x, y]$ is a point in the plane with the coordinates equal to x and y, then the transformation described by A maps the point $[x, y]$ to the point $A[x, y]$ with its coordinates equal to $a_{11}x + a_{12}y$ and $a_{21}x + a_{22}y$.

We say that two k–tuples of slopes $\alpha_1, \ldots, \alpha_k$ and $\beta_1, \ldots, \beta_k$ are *equivalent* if and only if $k -$ PURE-DIR$(\alpha_1, \ldots, \alpha_k) = k -$ PURE-DIR$(\beta_1, \ldots, \beta_k)$. Later (consult Theorem 1), we show that the latter condition is equivalent to $k -$ DIR$(\alpha_1, \ldots, \alpha_k) = k -$ DIR$(\beta_1, \ldots, \beta_k)$. We say that the k–tuple of slopes $\alpha_1, \ldots, \alpha_k$ can be *transformed* to the k–tuple of slopes $\beta_1, \ldots, \beta_k$ if there exists an affine transformation of the plane which transforms one of them to the other, i.e. the following holds: There exists a regular square matrix A of size two such that $\{[t \cos \alpha_i, t \sin \alpha_i], t \in \mathrm{R}, 1 \le i \le k\} = \{A[t \cos \beta_i, t \sin \beta_i], t \in \mathrm{R}, 1 \le i \le k\}$. Note that if $\alpha_1, \ldots, \alpha_k$ can be transformed to $\beta_1, \ldots, \beta_k$, then $\beta_1, \ldots, \beta_k$ can be transformed to $\alpha_1, \ldots, \alpha_k$ (consider the inverse matrix).

The following relations between distinct classes of intersection graphs of non–overlapping segments were proved by Kratochvíl and Matoušek in [13]:

- $k -$ PURE-DIR $\subset (k+1) -$ PURE-DIR for all $k \ge 1$ (the inclusion is strict).
- $2 -$ PURE-DIR $= 2 -$ PURE-DIR(α_1, α_2) for any $0 \le \alpha_1 < \alpha_2 < \pi$.
- $3 -$ PURE-DIR $= 3 -$ PURE-DIR$(\alpha_1, \alpha_2, \alpha_3)$ for any $0 \le \alpha_1 < \alpha_2 < \alpha_3 < \pi$.
- $4 -$ PURE-DIR $\ne 4 -$ PURE-DIR$(0, \pi/4, \pi/2, 3\pi/4)$.

The second and the third result can be restated in our terminology as follows: Any two pairs (triples) of slopes are equivalent.

A finite collection of lines in the plane is called a *line arrangement*; we suppose that no three lines of an arrangement share a common point in the whole paper. Note that two non–homeomorphic line arrangements can be intersection realizations of the same graph. The following lemma (Order Forcing Lemma), originnaly proved in [13], turned to be a really powerful tool:

Lemma 1 (Order Forcing Lemma [12,13]). *Let L be an arrangement of n lines in the plane and let G be the intersection graph of the lines of L. Then there exists a graph G' with the following properties:*

- *G is an induced subgraph of G',*
- *$G' \in$ SEG,*
- *if $G \in k-$ PURE-DIR$(\alpha_1, \dots, \alpha_k)$, then $G' \in k-$ PURE-DIR$(\alpha_1, \dots, \alpha_k)$ (for $k \geq 2$),*
- *for each realization of G' by segments in the plane (with possible overlapping of parallel segments), there exists a region Ω in the realization of G' and a homeomorphism φ from the plane with the arrangement L to Ω such that $\varphi(l_u) = r_u$ where l_u is the line of L corresponding to the vertex u of G and r_u is the segment corresponding to the vertex u of G'.*

We have presented this lemma here in a different (but of course equivalent) form to the one presented in [13]. This form seems to be more useful for our purposes. This lemma assures that there exists a graph G' for each line arrangement L, such that each realization of G' by segments in the plane contains a homeomorphic copy of the line arrangement L. Note that even if the realization of G' may contain overlapping parallel segments, the homeomorphic copy of the line arrangement L consists of non–overlapping segments.

3 Equivalence Results

First, we state an easy lemma stating a sufficient condition for k–tuples of slopes to be equivalent:

Lemma 2. *If two k-tuples of slopes $\alpha_1, \dots, \alpha_k$ and $\beta_1, \dots, \beta_k$ can be transformed one to the other, then they are equivalent.*

Proof. Let $G \in k-$ PURE-DIR$(\alpha_1, \dots, \alpha_k)$ and let C_α be its realization by segments with slopes among $\alpha_1, \dots, \alpha_k$. Let A be the matrix described in the definition of transformation between k–tuples of slopes. Let C_β be the set containing the following segments: If p is a segment of C_α, then the segment $Ap = \{A[x, y], [x, y] \in p\}$ belongs to the set C_β. Note that Ap and Aq intersect (overlap) if and only if p and q intersect (overlap) because A is a regular matrix. Due to the choice of A, the slopes of segments of C_β are among $\beta_1, \dots, \beta_k$ and thus $G \in k-$ PURE-DIR$(\beta_1, \dots, \beta_k)$. This implies that $k-$PURE-DIR$(\alpha_1, \dots, \alpha_k) \subseteq k-$PURE-DIR$(\beta_1, \dots, \beta_k)$. The opposite inclusion can be proved in the same way.

The following lemma provides us some canonical choice of slopes:

Lemma 3. *Let $\alpha_1, \dots, \alpha_k$ be a k-tuple of slopes (where $k \geq 3$). There exists $\pi/2 < \beta_4 < \dots < \beta_k < \pi$ such that $k-$ PURE-DIR$(\alpha_1, \dots, \alpha_k) = k-$ PURE-DIR$(0, \pi/4, \pi/2, \beta_4, \dots, \beta_k)$. In particular if k is eqaul to 3, then $3-$ PURE-DIR$(\alpha_1, \alpha_2, \alpha_3) = 3-$ PURE-DIR$(0, \pi/4, \pi/2)$ for all $\alpha_1, \alpha_2, \alpha_3$.*

Proof. We can assume w.l.o.g. that $0 = \alpha_1 < \ldots < \alpha_k < \pi$ as stated in Section 2. Consider the following matrix A:

$$A = \begin{pmatrix} 1 & -\operatorname{cotg}\alpha_3 \\ 0 & \frac{\sin(\alpha_3-\alpha_2)}{\sin\alpha_2 \sin\alpha_3} \end{pmatrix}$$

The matrix A is regular, since its determinant is non–zero. Note that the following holds:

$$A[\cos\alpha_1, \sin\alpha_1] = A[1,0] = [1,0]$$

$$A[\cos\alpha_2, \sin\alpha_2] = \left[\frac{\sin(\alpha_3-\alpha_2)}{\sin\alpha_3}, \frac{\sin(\alpha_3-\alpha_2)}{\sin\alpha_3}\right]$$

$$A[\cos\alpha_3, \sin\alpha_3] = \left[0, \frac{\sin(\alpha_3-\alpha_2)}{\sin\alpha_2}\right]$$

Thus the matrix A transforms the slope α_1 (α_2, α_3 respectively) to 0 ($\pi/4$, $\pi/2$ respectively). Let β_i (where $4 \le i \le k$) be the slope which the slope α_i is transformed to. It is enough to apply Lemma 2 to the k–tuples $\alpha_1, \ldots, \alpha_k$ and $0, \pi/4, \pi/2, \beta_4, \ldots, \beta_k$ to prove the statement of the lemma. The matrices A and A^{-1} witness that these two k–tuples of slopes can be transformed one to the other.

We haven't used in the proofs of Lemma 2 and Lemma 3 that the parallel segments do not overlap; thus the corresponding lemmas for the classes of overlapping segments also hold:

Lemma 4. *If two k–tuples of slopes $\alpha_1, \ldots, \alpha_k$ and $\beta_1, \ldots, \beta_k$ can be transformed one to the other, then $k - \mathrm{DIR}(\alpha_1, \ldots, \alpha_k) = k - \mathrm{DIR}(\beta_1, \ldots, \beta_k)$.*

Lemma 5. *Let $\alpha_1, \ldots, \alpha_k$ be any given k–tuple of slopes (for $k \ge 3$). Then there exists $\pi/2 < \beta_4 < \ldots < \beta_k < \pi$ such that $k - \mathrm{DIR}(\alpha_1, \ldots, \alpha_k) = k - \mathrm{DIR}(0, \pi/4, \pi/2, \beta_4, \ldots, \beta_k)$. In particular if $k = 3$, then $3 - \mathrm{DIR}(\alpha_1, \alpha_2, \alpha_3) = 3 - \mathrm{DIR}(0, \pi/4, \pi/2)$ for all $\alpha_1, \alpha_2, \alpha_3$.*

4 Non-equivalence Results

Lemma 6. *Let $\pi/2 < \alpha_4 < \ldots < \alpha_k < \pi$ and $\epsilon > 0$ be given. Then, there exists a graph G with three distinguished vertices U, V and W with the following properties:*

- $G \in k - \mathrm{PURE\text{-}DIR}(0, \pi/4, \pi/2, \alpha_4, \ldots, \alpha_k)$,
- $G \notin (k-1) - \mathrm{DIR}$,
- *the segments corresponding to U, V and W have different slopes in any realization of G by (not neceserally non-overlapping) segments with k slopes*

– *let C be any realization of G by (possibly overlapping) segments with k slopes such that the segment corresponding to U (V, W, respectively) has the slope 0 ($\pi/4$, $\pi/2$, respectively). Let $0 < \beta_4 < \ldots < \beta_k < \pi$ be the remaining slopes of the segments of C. It has to hold that $\pi/2 < \beta_4$ and $|\beta_i - \alpha_i| < \epsilon$ for $4 \le i \le k$.*

Proof. We assume w.l.o.g. that ϵ is small enough in order that $\alpha_4 - \epsilon > \pi/2$ and $\alpha_k + \epsilon < \pi$.

We first construct line arrangements $L_4, \ldots, L_k$ and $L'_4, \ldots, L'_k$ and we use later them to construct the desired G. Let i, $4 \le i \le k$, be fixed in the rest of this paragraph. Let us choose integers p and q such that tg $(\alpha_i - \epsilon - \pi/2) < p/q <$ tg $(\alpha_i - \pi/2)$ and let us choose integers p' and q' such that tg $(\alpha_i - \pi/2) <$ $p'/q' <$ tg $(\alpha_i + \epsilon - \pi/2)$. Let L_i and L'_i be the line arrangements drawn in Figure 1; the slope of the bold line is α_i. Let us suppose that we have a line arrangement homeomorphic to L_i. If the slopes of horizontal (diagonal, vertical respectively) lines are precisely 0 ($\pi/4$, $\pi/2$ respectively), then we claim that the slope of the bold line is at least $\pi/2 + \text{arctg}\ (p/q)$: Let x_s be the widths of the cells ($1 \le s \le p$) and let y_t be the heights of the cells ($1 \le t \le q$) of any realization of the line arrangement L_i. The positions of diagonal lines in the cells assure that $x_s > y_t$ for all s and t. These inequalities imply that $\frac{\sum_s x_s}{\sum_t y_t} > \frac{p}{q}$ and thus the slope of the bold line has to be at least $\pi/2 + \text{arctg}\ (p/q)$. Similarly, the slope of the bold line in a line arrangement homeomorphic to L'_i has to be at most $\pi/2 + \text{arctg}\, p'/q'$.

We construct a line arrangement L by placing all the arrangements L_i and L'_i for $4 \le i \le k$ to the plane (see Figure 2); we preserve the horizontal, diagonal and vertical slopes of the lines of all the arrangements and we place the arrangements with the whole lines, i.e. the non–parallel lines of distinct arrangements intersect each other. We do not care at all how the lines of distinct arrangements intersect. The lines of L intersect if and only if they are not parallel. The line arrangement L can be clearly realized by lines with slopes $0, \pi/4, \pi/2, \alpha_4, \ldots, \alpha_k$. Since L contains k lines with distinct slopes such that any pair of them intersects, no arrangement of lines homeomorphic to L can be realized by lines with fewer than k slopes; moreover, the lines in any realization with exactly k slopes which is homeomorphic to L are parallel if and only if they are parallel in L. We choose one of the vertices of the intersection graph corresponding to a horizontal line to be U, one of those corresponding to diagonal lines to be V and one of those corresponding to vertical lines to be W. If the line corresponding to U in a realization is horizontal, i.e. its slope is 0, then the slopes of all the horizontal lines are 0, since all the parallel lines must have the same slope in any realization. The same holds for diagonal and vertical lines. Thus if a homeomorphic copy of L consists of lines of at most k slopes and the lines corresponding to U (V, W respectively) are horizontal (diagonal, vertical respectively), then the slopes of the lines have to differ from $0, \pi/4, \pi/2, \alpha_4, \ldots, \alpha_k$ by at most ϵ; the bold lines of L_i and L'_i have to be parallel as mentioned above in this paragraph.

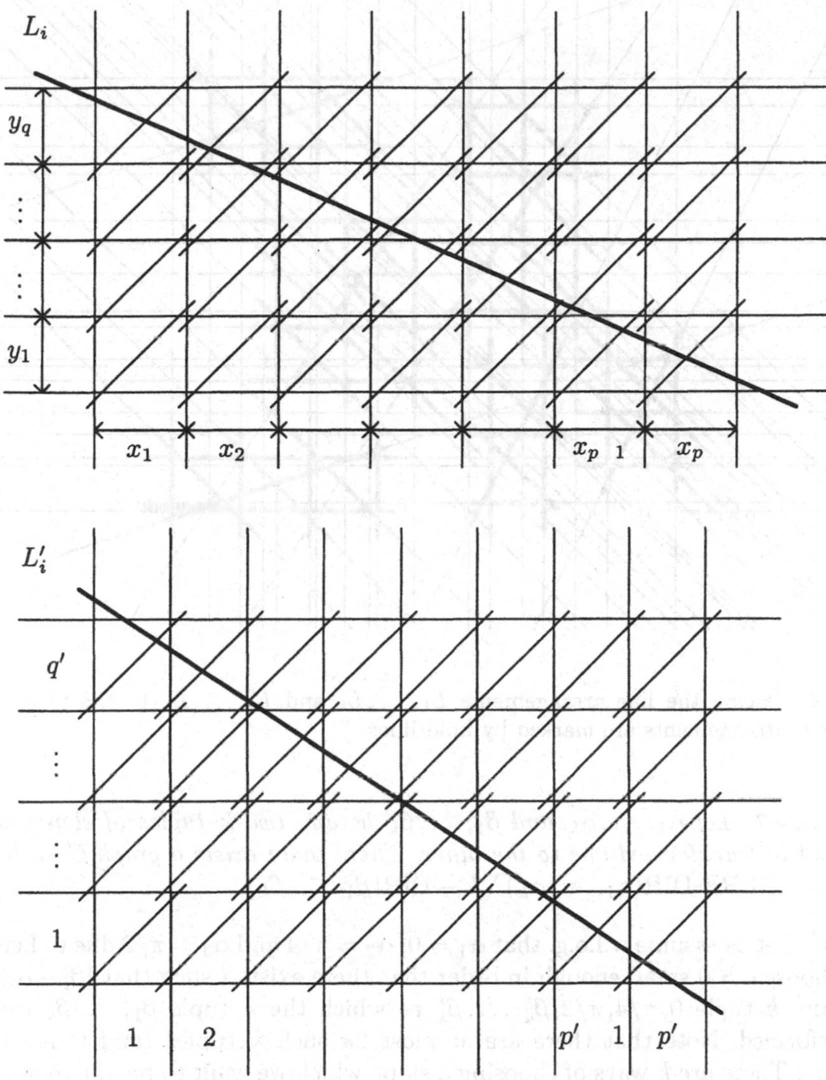

Fig. 1. The line arrangements L_i and L'_i. The slope of the bold line is α_i.

Let G be the graph for the line arrangement L given by Lemma 1. The graph G belongs to $k-$PURE-DIR$(0, \pi/4, \pi/2, \alpha_4, \ldots, \alpha_k)$. Any realization of G by lines of at most k slopes (with or without possible overlapping of parallel segments) contains a homeomorphic copy of the line arrangement L including the distinguished lines/vertices U, V and W.. Thus the graph G is the desired graph from the statement of the lemma.

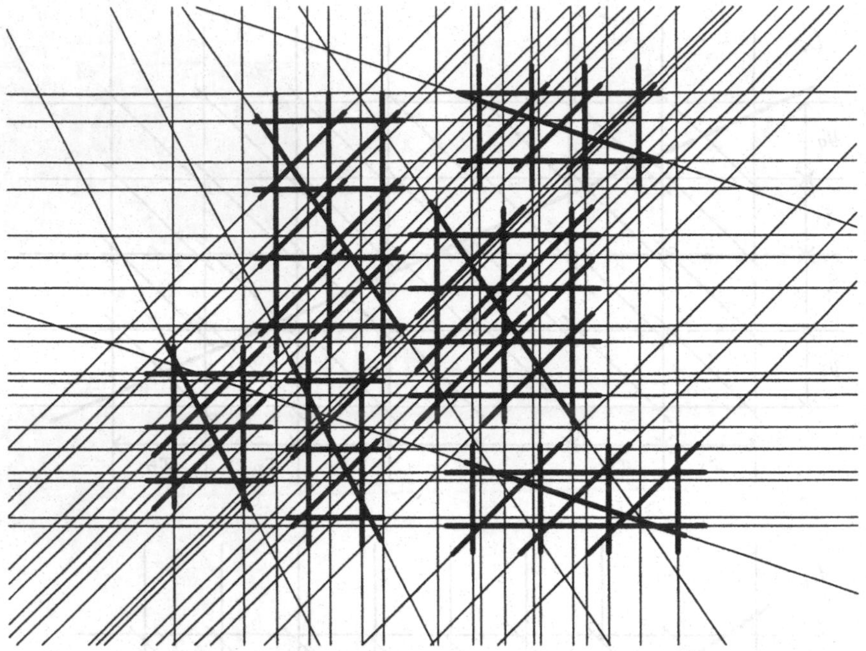

Fig. 2. Placing the line arrangements $L_1, \ldots, L_k$ and $L'_1, \ldots, L'_k$ to the plane. The original arrangements are marked by bold lines.

Lemma 7. *Let $\alpha_1, \ldots, \alpha_k$ and $\beta_1, \ldots, \beta_k$ be any two k–tuples of slopes which cannot be transformed one to the other. Then, there exists a graph G such that $G \in k - \text{PURE-DIR}(\alpha_1, \ldots, \alpha_k) \setminus k - \text{DIR}(\beta_1, \ldots, \beta_k)$.*

Proof. Let us assume w.l.o.g. that $\alpha_1 = 0$, $\alpha_2 = \pi/4$ and $\alpha_3 = \pi/2$ due to Lemma 3. Choose $\epsilon > 0$ small enough in order that there exists i such that $|\beta'_i - \alpha_i| > \epsilon$ for any k–tuple $0, \pi/4, \pi/2, \beta'_4, \ldots, \beta'_k$ to which the k–tuple $\beta_1, \ldots, \beta_k$ can be transformed. Note that there are at most $2k$ such k–tuples (and thus such ϵ exists): There are k ways of choosing a slope which we want to be horizontal and if the horizontal slope is chosen, there two ways of choosing a slope which we want to be the "$\pi/4$" one (only the slopes which neighbour with the "horizontal" one can be chosen).

Let us consider the graph G from Lemma 6 for $\alpha_4, \ldots, \alpha_k$ and ϵ and let U, V and W be its distinguished vertices. Let us suppose that G has an intersection realization by (possibly overlapping) segments with the slopes $\beta_1, \ldots, \beta_k$. Let β_{i_U} (β_{i_V}, βi_W respectively) be the slope of the segment corresponding to the vertex U (V, W respectively). The indices i_U, i_V and i_W of the slopes are mutually distinct due to Lemma 6. Apply the transformation of the plane transforming β_{i_U} to 0, β_{i_V} to $\pi/4$ and β_{i_W} to $\pi/2$ — see the proof of Lemma 2 for details of the construction of this transformation. Let $0, \pi/4, \pi/2, \beta'_4, \ldots, \beta'_k$ be the slopes $\beta_1, \ldots, \beta_k$

after performing this transformation. But the properties of any realization of G by segments which are stated in Lemma 6 contradict our choice of ϵ unless $\alpha_1, \ldots, \alpha_k$ can be transformed to $\beta_1, \ldots, \beta_k$. Thus such realization of G by segments cannot exist and $G \in k-\text{PURE-DIR}(\alpha_1, \ldots, \alpha_k) \setminus k-\text{DIR}(\beta_1, \ldots, \beta_k)$.

5 Main Results

We present our main results in this section. First, we state a necessary and sufficient condition for two k–tuples to be equivalent:

Theorem 1. *The following statements for two k–tuples of slopes $\alpha_1, \ldots, \alpha_k$ and $\beta_1, \ldots, \beta_k$ are equivalent:*

- *The k–tuples $\alpha_1, \ldots, \alpha_k$ and $\beta_1, \ldots, \beta_k$ can be transfomred one to the other.*
- *The k–tuples $\alpha_1, \ldots, \alpha_k$ and $\beta_1, \ldots, \beta_k$ are equivalent, i.e. the following holds: $k-\text{PURE-DIR}(\alpha_1, \ldots, \alpha_k) = k-\text{PURE-DIR}(\beta_1, \ldots, \beta_k)$.*
- *$k-\text{DIR}(\alpha_1, \ldots, \alpha_k) = k-\text{DIR}(\beta_1, \ldots, \beta_k)$.*

Proof. If they can be transformed one to the other, they are equivalent due to Lemma 2; the corresponding claim for overlapping version of the classes is due to Lemma 4. If they cannot be transformed, they are non–equivalent for both the versions of the classes due to Lemma 7 and due to simple fact that $k-\text{PURE-DIR}(\alpha_1, \ldots, \alpha_k) \subset k-\text{DIR}(\alpha_1, \ldots, \alpha_k)$.

We describe the relation between classes of intersection graphs of segments of prescribed slopes in the next theorem:

Theorem 2. *Let $\alpha_1, \ldots, \alpha_k$ and $\beta_1, \ldots, \beta_k$ be two k–tuples. Then the following holds:*

- *If the k–tuples of slopes $\alpha_1, \ldots, \alpha_k$ and $\beta_1, \ldots, \beta_k$ can be transformed one to the other, then $k-\text{PURE-DIR}(\alpha_1, \ldots, \alpha_k) = k-\text{PURE-DIR}(\beta_1, \ldots, \beta_k)$ and $k-\text{DIR}(\alpha_1, \ldots, \alpha_k) = k-\text{DIR}(\beta_1, \ldots, \beta_k)$.*
- *If the k–tuples of slopes $\alpha_1, \ldots, \alpha_k$ and $\beta_1, \ldots, \beta_k$ cannot be transformed one to the other, then $k-\text{PURE-DIR}(\alpha_1, \ldots, \alpha_k) \subset k-\text{DIR}(\alpha_1, \ldots, \alpha_k) \not\subseteq k-\text{DIR}(\beta_1, \ldots, \beta_k)$, $k-\text{PURE-DIR}(\alpha_1, \ldots, \alpha_k) \not\subseteq k-\text{PURE-DIR}(\beta_1, \ldots, \beta_k)$.*

Proof. If the k–tuples of slopes can be transformed one to the other, they are equivalent and thus the corresponding classes of intersection graph of segments are equal as stated in Theorem 1. If they cannot be transformed one to the other, Lemma 7 provides a graph which witness the second statement.

Theorem 3. *Let $0 = \alpha_1 < \alpha_2 < \alpha_3 < \alpha_4 < \pi$ be any quadruple of slopes. There exists exactly one β, $\pi/2 < \beta \leq 3\pi/4$ such that the quadruples $\alpha_1, \alpha_2, \alpha_3, \alpha_4$ and $0, \pi/4, \pi/2, \beta$ are equivalent.*

Proof. Lemma 3 assures existence of $\pi/2 < \beta' < \pi$ such that the quadruples $\alpha_1, \alpha_2, \alpha_3, \alpha_4$ and $0, \pi/4, \pi/2, \beta'$ are equivalent. In case that $\beta' > 3\pi/4$, consider the following matrix:

$$A = \begin{pmatrix} -1 & 0 \\ 0 & -\operatorname{cotg} \beta' \end{pmatrix}$$

The transformation described by A transforms the quadruple $0, \pi/4, \pi/2, \beta'$ to $0, \pi/4, \pi/2, 3\pi/2-\beta'$. Thus choosing $\beta = 3\pi/2-\beta'$ assures the existence of $\pi/2 < \beta \leq 3\pi/4$.

It remains to prove that such β is unique. There are exactly eight matrices (up to multiplication by a constant) which transform the quadruple $0, \pi/4, \pi/2, \beta$ to a quadruple $0, \pi/4, \pi/2, \gamma$ where $\pi/2 < \gamma < \pi$:

$$\begin{pmatrix} 1 & 0 \\ 0 & 1 \end{pmatrix} \begin{pmatrix} 1 & -\operatorname{cotg} \beta \\ \operatorname{cotg} \beta & -\operatorname{cotg} \beta \end{pmatrix} \begin{pmatrix} 0 & 1-\operatorname{cotg} \beta \\ \operatorname{tg} \beta - 1 & 0 \end{pmatrix} \begin{pmatrix} -1 & 1 \\ -1 & \operatorname{cotg} \beta \end{pmatrix}$$

$$\begin{pmatrix} 0 & 1 \\ 1 & 0 \end{pmatrix} \begin{pmatrix} \operatorname{cotg} \beta & -\operatorname{cotg} \beta \\ 1 & -\operatorname{cotg} \beta \end{pmatrix} \begin{pmatrix} \operatorname{tg} \beta - 1 & 0 \\ 0 & 1-\operatorname{cotg} \beta \end{pmatrix} \begin{pmatrix} -1 & \operatorname{cotg} \beta \\ -1 & 1 \end{pmatrix}$$

It is a matter of routine calculation to check that the only quadruples which can be obtained through these transformations are actually only $0, \pi/4, \pi/2, \beta$ and $0, \pi/4, \pi/2, 3\pi/2 - \beta$. This proves the unicity of β.

The immediate previously promised corollary is the following:

Corollary 1. *The classes of graphs* $4-$PURE-DIR$(0, \pi/4, \pi/2, \alpha)$ *for* $\pi/2 < \alpha \leq 3\pi/4$ *are mutually distinct and any class* $4-$PURE-DIR$(\beta_1, \beta_2, \beta_3, \beta_4)$ *is equal to exactly one of them.*

Note that statements of Corollary 1 can be also formulated for $4-$DIR class of graphs (the one with overlapping parallel segments) in the same way.

Acknowledgement. Attention of the authors to intersection graphs of geometric objects were attracted by Jan Kratochvíl who has given lectures on the topic of intersection graphs. The authors are indebted to Jan Kratochvíl for fruitful discussion on the topic, especially on the Order forcing lemma, during his lectures.

References

1. T. Asano: Difficulty of the maximum independent set problem on intersection graphs of geometric objects, Graph theory, combinatorics and applications, vol. 1, Wiley-Intersci. Publ., 1991, pp. 9–18.
2. A. Bouchet: Reducing prime graphs and recognizing circle graphs, Combinatorica 7, 1987, pp. 243–254.
3. N. de Castro, F. J. Cobos, J. C. Dana, A. Marquez, M. Noy: Triangle-free planar graphs as segments intersection graphs, J. Kratochvil (ed.), Graph drawing, 7th international symposium, Štiřín Castle, Czech Republic, proceedings, Springer LNCS 1731, 1999, pp. 341–350.

4. G. Ehrlich, S. Even, R. E. Tarjan: Intersection graphs of curves in the plane, J. Combinatorial Theory Ser. B 21, 1976, no. 1, 8–20.
5. J. C. Fournier: Une caracterization des graphes de cordes, C.R. Acad. Sci. Paris 286A, 1978, pp. 811–813.
6. H. de Fraysseix: A characterization of circle graphs, European Journal of Combinatorics 5, 1984, pp. 223–238.
7. H. de Fraysseix, P. Ossona de Mendez, J. Pach: Representation of planar graphs by segments, Intuitive Geometry 63, 1991, pp. 109–117.
8. M. Goljan, J. Kratochvíl, P. Kučera: String graphs, Academia, Prague 1986.
9. I. B.-A. Hartman, I. Newman, R. Ziv: On grid intersection graphs, Discrete Math. 87, 1991, no. 1, pp. 41–52.
10. V. B. Kalinin: On intersection graphs, Algorithmic constructions and their efficiency (in Russian), Yaroslav. Gos. Univ., 1983, pp. 72–76.
11. S. Klavžar, M. Petkovšek: Intersection graphs of halflines and halfplanes, Discrete Math. 66, 1987, no. 1-2, pp. 133–137.
12. J. Kratochvíl: personal comunication.
13. J. Kratochvíl, J. Matoušek: Intersection Graphs of Segments, Journal of Combinatorial Theory, Series B, Vol. 62, No. 2, 1994, pp. 289–315.
14. J. Kratochvíl, J. Matoušek: NP-hardness results for intersection graphs, Comment. Math. Univ. Carolin. 30, 1989, pp. 761–773.
15. J. Kratochvíl, J. Nešetřil: Independent set and clique problems in intersection defined classes of graphs, Comment. Math. Univ.Carolin. 31, 1990, pp. 85–93.
16. A. C. Tucker: An algorithm for circular-arc graphs, SIAM J. Computing 31.2, 1980, pp. 211–216.

A Short Note on the History of Graph Drawing

Eriola Kruja[1], Joe Marks[2], Ann Blair[1], and Richard Waters[2]

[1] Harvard University, Cambridge, MA 02138
{kruja,amblair}@fas.harvard.edu

[2] MERL — Mitsubishi Electric Research Laboratories, Cambridge, MA 02139
{marks,waters}@merl.com

Abstract. The origins of chart graphics (e.g., bar charts and line charts) are well known [30], with the seminal event being the publication of William Playfair's (1759-1823) *The Commercial and Political Atlas* in London in 1786 [26]. However, the origins of graph drawing are not well known. Although Euler (1707-1783) is credited with originating graph theory in 1736 [12,20], graph drawings were in limited use centuries before Euler's time. Moreover, Euler himself does not appear to have made significant use of graph visualizations. Widespread use of graph drawing did not begin until decades later, when it arose in several distinct contexts. In this short note we present a selection of very early graph drawings; note the apparent absence of graph visualization in Euler's work; and identify some early innovators of modern graph drawing.

1 Early Graph Drawing

Although geometric drawings of various kinds have been used extensively through the ages, drawings that are visual abstractions of mathematical graphs[1] arose relatively recently and have only achieved general currency in the last 150 years.

Fig. 1. Depictions of Morris gameboards from the 13th century. The nodes of these graph drawings are the positions that game counters can occupy. The edges indicate how game counters can move between nodes. Reproduced with permission.

[1] A mathematical graph consists of a set of nodes and a set of edges. An edge connects a pair of nodes.

P. Mutzel, M. Jünger, and S. Leipert (Eds.): GD 2001, LNCS 2265, pp. 272–286, 2002.
© Springer-Verlag Berlin Heidelberg 2002

The earliest forms of graph drawing were probably of Morris or Mill games [35]. Although other early games are based on an underlying notion of a graph, the boards of Mill games depict the graph explicitly (as do some other early games — see [24]). The earliest example of Mill gameboards come from stone carvings in Ancient Egypt. According to Henry Parker ([25], p. 578), they are to be found on roof slabs in a temple began by Rameses I (1400-1366, B.C). Parker describes these carvings in his book, but unfortunately he does not provide any pictures of them. The earliest examples of Mill gameboards to be drawn in a book probably come from the 13th-century "Book of Games," produced under the direction of Alfonso X (1221-1284), King of Castile and Leon [2]. Two of these drawings are shown in Figure 1 [2,35].

The other known examples of ancient graph drawing are family trees that decorated the atria of patrician roman villas. They are described by Pliny the Elder and Seneca, but no examples have survived ([19], p. 111). The earliest surviving examples of genealogical graph drawing are from the Middle Ages [19]. Three examples are shown in Figure 2. The drawing on top, which shows Noah's descendants, is from the 11th century [6].[2] Another religiously inspired genealogy is shown in the bottom left of the figure [8]. In addition to biblical genealogies, family trees of the nobility were popular. The drawing on the bottom right shows a 12th-century genealogy of the Saxon dynasty [34].

Genealogy was not only for religious figures and nobles. Figure 3 contains two sketches of family trees. The top tree is unusual in that it was included in a personal legal document not intended for publication: it was offered as evidence to the Court of Requests during the reign of Elizabeth I by John Stalham to establish his pedigree. He was engaged in a legal dispute over lands and tenements in the parishes of Snelston Alsop and Roston, in the county of Derby, England [28]. The English Public Records Office also contains some pre-Elizabethan examples of sketched family trees in similar legal documents. This usage seems to indicate that educated people understood tree drawings and used them routinely by the 15th century. A later and more extensive family tree is shown at the bottom of Figure 3. It shows the genealogy of the Mannelli family from Florence, Italy, and is one of the earliest and cleanest examples of curved-edge graph drawing [3]. Many more delightful examples of early family trees can be found in the article by Christiane Klapisch-Zuber [19].

Another tree form that appears frequently in medieval literature is one used to depict categories of various kinds. The elegant 14th-century drawing in the left-hand side of Figure 4 shows various cardinal and theological virtues and their more specific subvirtues ([23], p. 30).[3] The less-refined drawing on the right-hand side, also from the 14th century, categorizes various vices ([23], p. 45).

[2] The careful reader will note that the graph in the figure is not a tree, which would seem to indicate some anomalies in Noah's family history!

[3] Murdoch provides source references for all of the figures from his book [23] reproduced here. The drawings shown are from manuscripts held by the British Library, London, and the Bibliothèque Nationale, Paris.

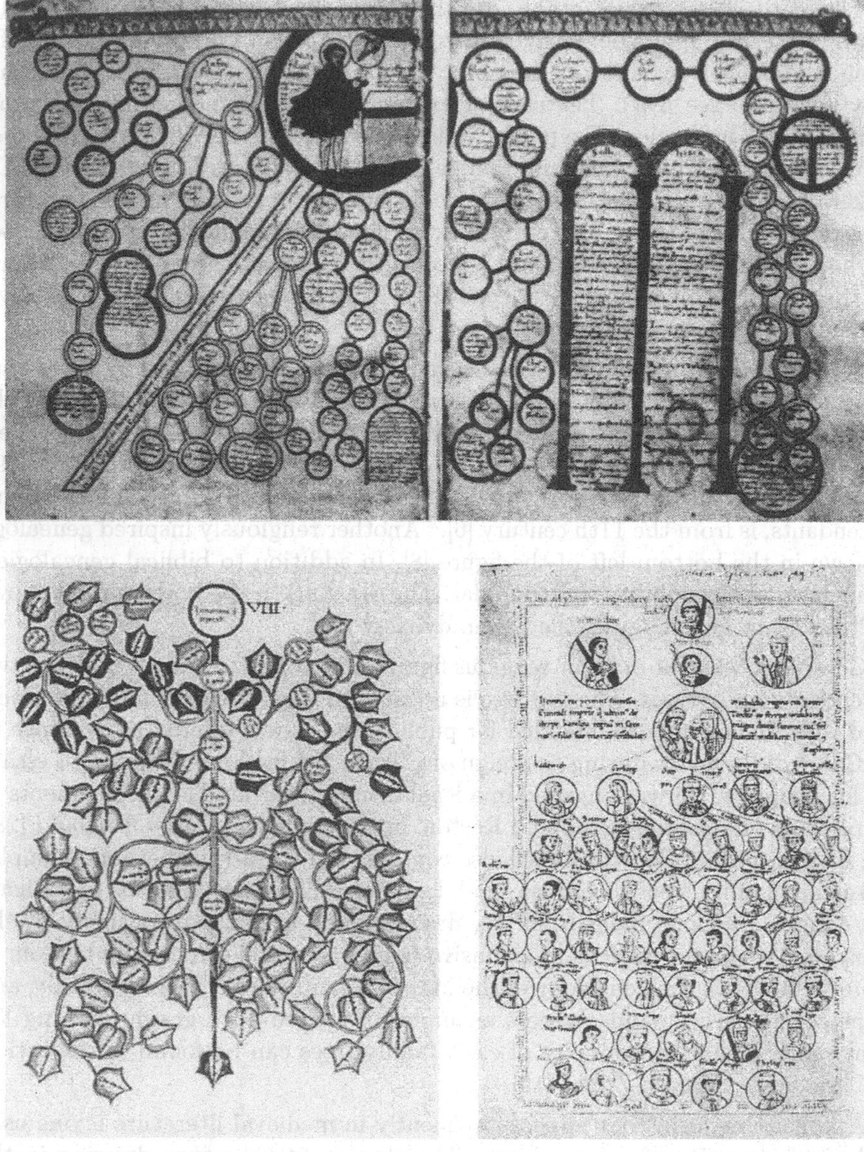

Fig. 2. Family trees that appear in manuscripts from the Middle Ages. Note that the top drawing is spread over two pages in the original manuscript. Reproduced with permission.

Graphs, as opposed to tree drawings, were also used in the Middle Ages to represent and visualize abstract information [23]. For example, certain canonical graph drawings appear regularly in medieval literature. *Squares of opposition* were pedagogical tools used in the teaching of logic, particularly the relations

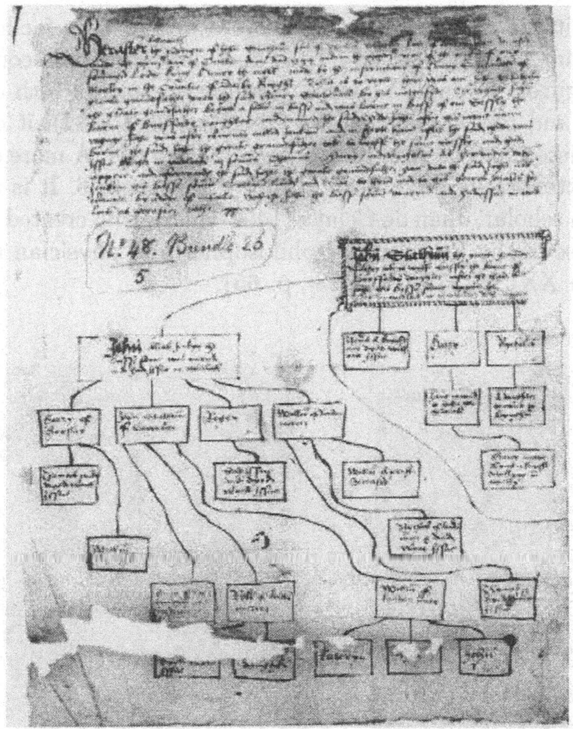

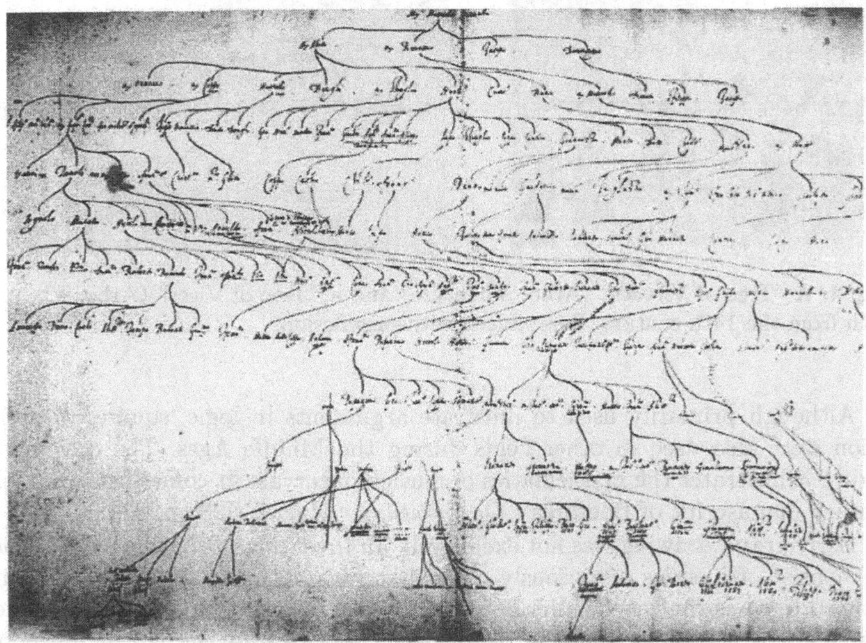

Fig. 3. Family-tree sketches from the 15th and 16th century. Reproduced with permission.

between propositions or syllogisms. They were designed to facilitate the recall of knowledge that students already had, and hence did not contain complete information. Figure 5 contains a simple example due to the 14th-century French mathematician and philosopher, Nicole Oresme (1323-1382): it depicts a basic argument in Aristotelian natural philosophy ([23], p. 67). A more complex graph drawing that extends the basic form is shown in Figure 6. It is due to a 16th-century Spanish scholar, Juan de Celaya (1490-1558), who created the drawing to explain a treatise of the 13th-century philosopher and physician Peter of Spain, later Pope John XXI (1215-1277) ([23], p. 65).

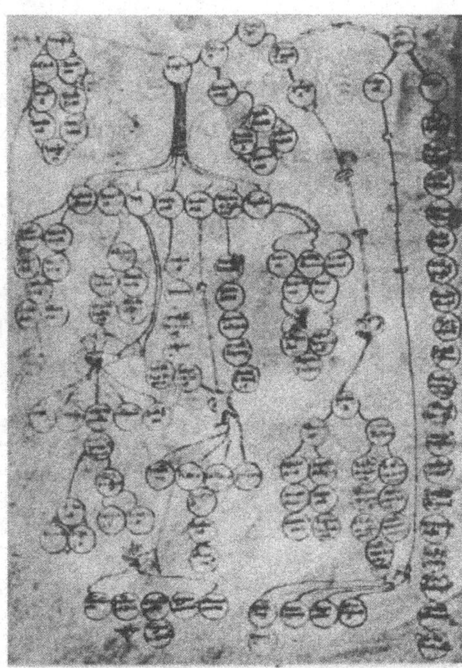

Fig. 4. A "Tree of Virtues" (Arbor Virtutum) and a "Tree of Vices" (Arbor Vitium), both from the 14th century. Reproduced with permission.

Although primarily used to illustrate arguments in logic, squares of opposition were also used in other fields during the Middle Ages. The drawing in Figure 7 illustrates the mathematics of musical intervals. It comes from an 11th-century manuscript of Boethius's *De Instituone Musica* ([23], p. 67).

Early graph drawing was not exclusively an invention of the Old World. *Quipus* consist of a series of variously colored strings attached to a base rope and knotted in ways that encode information idiosyncratically — see Figure 8. They were used by the Incas from the 13th to 16th centuries for accounting purposes and to register important facts and events. Of the few hundred surviving examples, roughly 25% exhibit hierarchical structure [4] and therefore qualify as tree

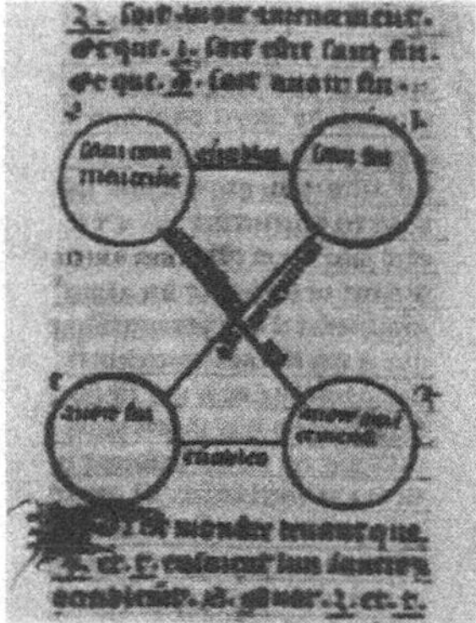

Fig. 5. A simple square of opposition from the 14th century. The nodes represent logical propositions. The edges represent relations between the propositions. Reproduced with permission.

drawings. However, nobody knows exactly what the nodes and edges represented in such graph drawings.

2 Leonhard Euler and Graph Drawing

One might expect the transition from early to modern graph drawing to be marked by Euler's famous paper in 1736 [12] on the path-tracing problem posed by the bridges of Königsberg. In this paper Euler solved the problem by introducing the concept of a graph comprising nodes and edges;[4] it marks the beginning of graph theory as a topic of study. The problem was brought to Euler's attention by his friend Carl Ehler in a letter dated March 9, 1736 [27]. Generalizing from his solution to this particular problem, Euler derived the first results in graph theory [20].

However, 1736 does not mark the beginning of modern graph drawing. Remarkably, in his study of the Königsberg Bridges problem Euler did not use graph visualizations to present or explain his results. Although two sketched maps of Königsberg (see Figure 9) are included in Euler's paper, no graph drawings appear there or anywhere else in his works.

An explanation for Euler's lack of interest in graph visualization can be found in the first paragraph of his paper. Euler began with a reference to Leibniz's (1646-1716) vision of a new kind of geometry without measurement or magnitudes. Leibniz's early discussion of what came to be called the *geometria situs* (and later topology) comes from his letter of September 8, 1679 to Huygens. In this letter Leibniz emphasizes the usefulness of a "new characteristic, completely different from algebra, which will have great advantage to represent to

[4] However, the term "graph" was not coined until 1878 by the English mathematician James Joseph Sylvester (1814-1897) [29].

Fig. 6. A more complex square of opposition from the 16th century. It is a symmetric drawing of K_{12} with labeled nodes and edges. Reproduced with permission.

Fig. 7. Musical intervals drawn in a square of opposition from the 11th century. The nodes (corners) represent numbers and the edges represent named ratios between them (e.g., "octave" and "fifth"). Reproduced with permission.

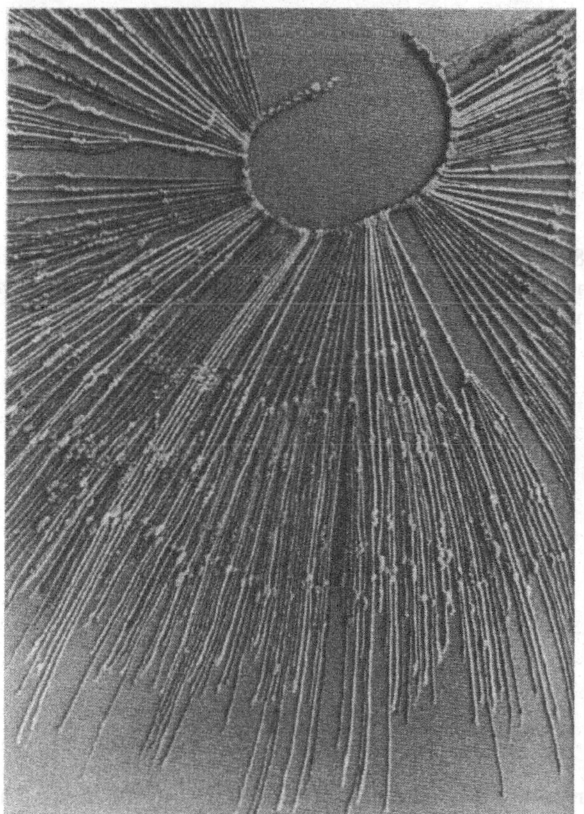

Fig. 8. A quipu in the collection of the Museo National de Anthropologia y Arquelogía, Lima, Peru [4]. Reproduced with permission.

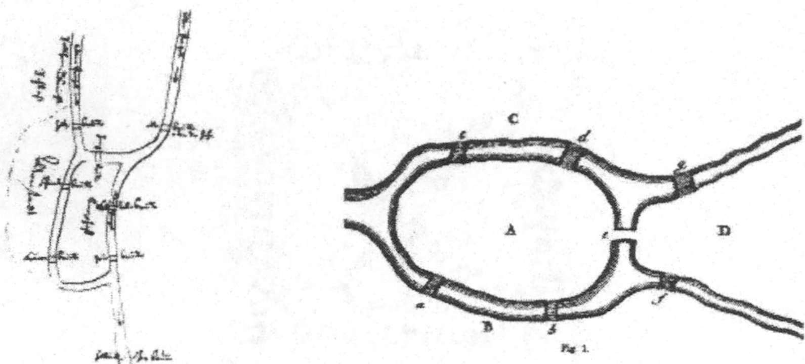

Fig. 9. Ehler's sketched map of Königsberg, 1736 (left), and Euler's more polished version [12]. Euler included one more sketched map (a variant of the first with more bridges included) in his paper, but no abstract graph drawing of the problem. Reproduced with permission.

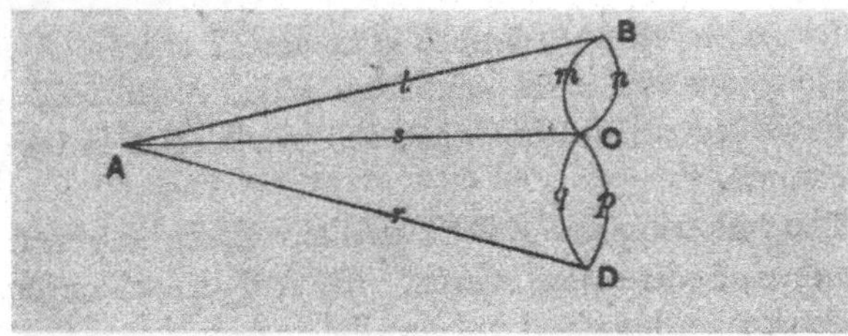

Fig. 10. Ball's 1892 graph-drawing abstraction of the bridges of Königsberg. The nodes represent the land areas and the edges represent the bridges connecting them. Reproduced with permission.

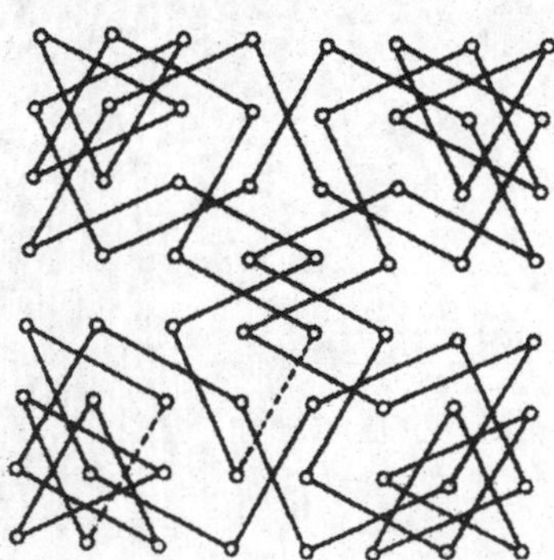

Fig. 11. Vandermonde's 1771 graph drawing of a Knight's Tour. This is actually a drawing of a subgraph of the graph that represents all possible knight moves. In that graph the nodes represent squares on a chessboard and edges represent legal moves. Reproduced with permission.

the mind exactly and naturally, *though without figures*,[5] everything which depends on the imagination" [21]. Euler was probably familiar with a collection of Leibniz's correspondence because of his post at the Imperial Academy of St. Petersburg, with which Leibniz had been closely connected [16]. Given Leibniz's conviction that his geometria situs would involve "neither figures nor models and would not hinder the imagination," Euler may well have been predisposed against the use of graphical visualizations to describe or solve graph-theoretic problems (even though he included geometric figures and function plots in many of his papers). It was probably not until 150 years later that W. W. Rouse Ball (1850-1925) drew the first abstract graph drawing that depicts the Königsberg Bridges problem [32]. Ball's graph drawing (see Figure 10) appeared in his 1892 book on mathematical recreations [5].

Another example of Euler eschewing the use of graph visualization is provided by his work in 1759 on another mathematical puzzle, that of computing a "Knight's Tour" on a chessboard [13]. This problem is to find a sequence of moves that takes a knight to each square of the board exactly once and returns it to its starting square. Twelve years later, in 1771, Vandermonde clarified the problem with a graph drawing [31] (see Figure 11). His inspiration for this drawing is very vivid: "if one supposes that a pin is fixed in the centre of each square, the problem reduces to the determination of a path taken by a thread passed once around each pin and following a rule whose formulation we seek" (translation from [7], p. 24).

3 Early Examples of Modern Graph Drawing

Although the transition from early to modern graph drawing did not coincide exactly with the invention of graph theory, it did follow shortly thereafter. Starting in the late 18th century and early 19th century, graph drawings began to appear more frequently and in more contexts. In mathematics, many papers were then illustrated with graph drawings:

- J.B. Listing (1808-1882) in his 1847 treatise on topology [22] devotes a short section to path tracing in graphs and includes the memorable drawing in Figure 12, which can be drawn in a single stroke.
- Sir William Rowan Hamilton (1805-1865) devised a game based on a non-commutative algebra that he dubbed the "Icosian Calculus." The gameboard consisted of a graph drawing — see Figure 13. Various games and puzzles could be played on this board [17].
- Arthur Cayley's (1821-1895) pioneering work on trees was illustrated with drawings like those in Figure 14 [10].

Around the same time graph drawing also appeared in other fields, such as crystallography and chemistry. René Just Haüy (1743-1822) established the basic principles of crystallography. His abstract drawings of crystals represent a hybrid form of visual abstraction that is part geometric drawing and part three-dimensional graph drawing — see Figure 15 [18].

[5] The italics are ours.

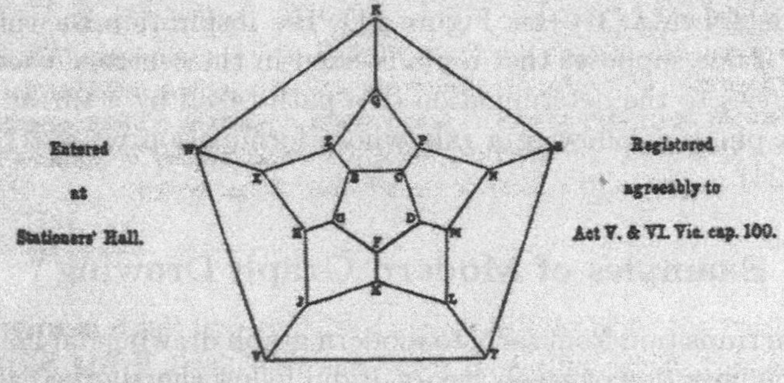
THE ICOSIAN GAME.
Entered at Stationers' Hall.
Registered agreeably to Act V. & VI. Vic. cap. 100.
LONDON:
PUBLISHED AND SOLD WHOLESALE BY JOHN JAQUES AND SON, 102 HATTON GARDEN;
AND TO BE HAD AT MOST OF THE LEADING FANCY REPOSITORIES THROUGHOUT THE KINGDOM.

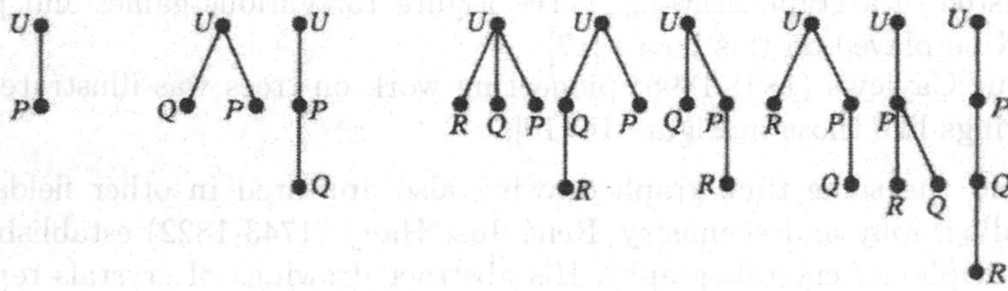

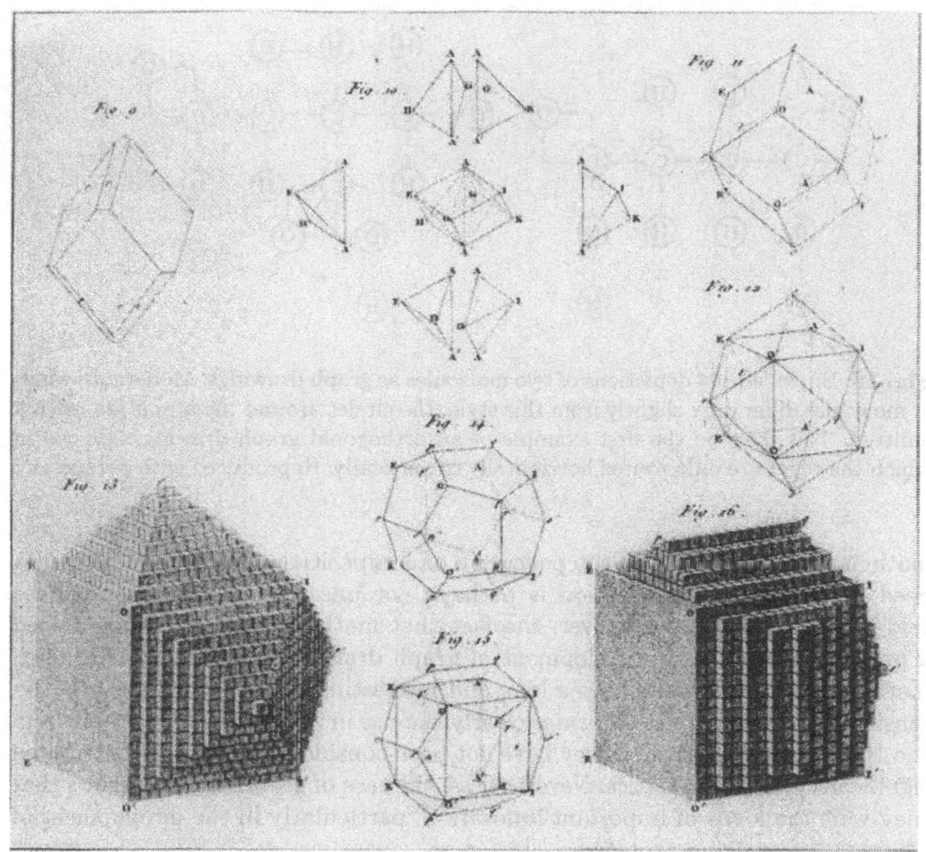

Fig. 15. Drawings from 1784 that depict the geometry of crystal structures but that also foreshadow the use of 3D graph drawing. The graph nodes correspond to corners or apexes of the physical crystal. Edges connect neighboring nodes. Reproduced with permission.

Depicting molecules as graph drawings may seem obvious now, but various other graphical depictions were proposed before Alexander Crum Brown (1838-1922) introduced the familiar drawings shown in Figure 16 [9]. The adoption of Brown's drawing conventions was greatly facilitated by their inclusion in popular lecture notes by E. Frankland [15].

4 Conclusions and Future Work

The main impetus for current research on computer-aided graph drawing is to facilitate the visual analysis of various kinds of complex networked or connected systems, e.g., computer networks, social networks, engineering systems, etc. In contrast, the drawings above show that the driving forces behind the initial development of graph drawing were pedagogy, exposition, record keeping, and

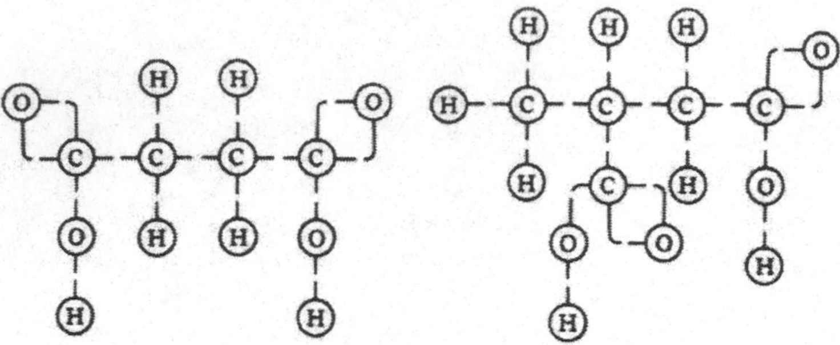

Fig. 16. Brown's 1864 depictions of two molecules as graph drawings. Modern drawings of molecules differ only slightly from this style: the circles around the atoms are usually omitted. This may be the first example of an orthogonal graph drawing, i.e., one in which the edges are only routed horizontally or vertically. Reproduced with permission.

mathematical recreation. That pedagogy and exposition should have created a need for graphical visualization is perhaps not unexpected; the same can be said of record keeping. However, the fact that mathematical recreation played a prominent role in the development of graph drawing is surprising. Although mathematical recreations have a long and interesting history [1,5,11] — the extensive bibliography in [1] indicates clearly the rise in popularity of the topic from the 16th century onward — they have not been considered as having contributed significantly to mathematics. Nevertheless, the case of graph drawing shows that they were the locus of important innovation, particularly in the development of visual aids to solving problems.

We were also surprised to learn that Euler, the originator of modern graph theory, was not a pioneer in graph drawing: his study of graph theory seems to have been completely grounded in symbolic representations. Much as we might like to believe in the fundamental power of drawings to illuminate graph problems, it is clear that some people do not need such visualizations to have insights.

It seems likely that other examples of early modern graph drawing can be found in the literature of other disciplines, e.g., electrical engineering, abstract cartography. One subject for future work is an investigation of how graph drawing emerged in these fields.

Acknowledgments. We gratefully acknowledge the help of the following people: Peter Eades, Renee Hall, Darren Leigh, John Murdoch, Bob O'Hara, Nathaniel Taylor, Brian Tompsett, and the anonymous reviewers.

References

1. W. Ahrens. *Mathematische Unterhaltungen und Spiele.* Druck und Verlag Von B.G. Teubner, Leipzig, 1901.

2. Alfonso X, King of Castile and Leon. *Libros del ajedrez, dados y tablas.* 13th century.
3. S. Ammirato. Alberi genealogici, late 16th century. Biblioteca Riccardiana, Florence, Grandi formati 33, No. 67. Reproduced in [19], Fig. 6.
4. M. Ascher and R. Ascher. *Code of the Quipu: A Study in Media, Mathematics, and Culture.* University of Michigan Press, Ann Arbor, Michigan, 1981.
5. W. W. R. Ball. *Mathematical Recreations and Essays.* The MacMillan Company, New York, 1939. 11th Edition. First published in 1892.
6. Beatus of Liebana. Commentary on the Apocalypse of Saint John, 11th century. Diagram from the Bibliothèque Nationale, Paris, (MS. Lat. 8878, fols. 6v-7r). Reproduced in [33], p. 331.
7. N. L. Biggs and L. Wilson. *Graph Theory 1736-1936.* Clarendon Press, Oxford, 1976. An excellent secondary source on the history of graph theory.
8. Boccaccio. Genealogia deorum, early 15th century. University of Chicago Library, Ms. 100, Tree VIII. Reproduced in [19], Fig. 22.
9. A. C. Brown. On the Theory of Isomeric Compounds. *Transactions of the Royal Society Edinburgh*, 23:707–719, 1864.
10. A. Cayley. On the Theory of the Analytical Forms Called Trees. *Philosophical Magazine*, 4(13):172–176, 1857.
11. M. Édouard Lucas. *Récréations Mathématiques.* Gauthier-Villars, Imprimeur-Libraire, Paris, 1882.
12. L. Euler. Solutio Problematis ad Geometriam Situs Pertinentis. *Commentarii Academiae Scientiarum Imperialis Petropolitanae*, 8:128–140, 1736. Also in Opera Omnia (1) 7, 1923, pp. 1-10. An English translation can be found in [7], pp. 3-8.
13. L. Euler. Solution d'une Question Curieuse qui ne Paroit Soumise a Aucune Analyse. *Mémoires de l'Académie des Sciences de Berlin*, 15:310–337, 1759. Also in Opera Omnia (1) 7, 1923, p. 26-56.
14. B. J. Ford. *Images of Science: A History of Scientific Illustration.* Oxford University Press, New York, 1993.
15. E. Frankland. *Lecture Notes for Chemical Students.* London, 1866. An English translation can be found in [7], pp. 58-60.
16. C. C. Gillispie. *Dictionary of Scientific Biography.* Charles Scribner's Sons, New York, 1973.
17. W. R. Hamilton. The Icosian Game, instruction leaflet, 1859. A copy of this leaflet can be found in [7], pp. 32-35.
18. R. J. Haüy. *Essai d'une théorie sur la structure des crystaux.* 1784. A copy of the drawing can be found in [14], p. 137.
19. C. Klapisch-Zuber. The genesis of the family tree. In W. Kaiser, editor, *I Tatti Studies: Essays in the Renaissance, Volume Four.* Leo S. Olschki, Florence, Italy, 1991.
20. D. König. *Theorie der Endlichen und Unendlichen Graphen.* Akademische Verlagsgesellschaft M.B.H., Leipzig, 1936. Also available in English from Birkhäuser Boston, 1990.
21. W. G. Leibniz. Letter to Christiaan Huygens, September 8, 1679. In I. Gerhardt, editor, *Leibnizens Mathematische Schriften*, volume 2. A. Asher and Co., 1850.
22. J. B. Listing. Vorstudien zur Topologie. *Göttinger Studien*, 1:811–875, 1847. An English translation can be found in [7], pp. 14-16.
23. J. E. Murdoch. *Album of Science – Antiquity and the Middle Ages.* Charles Scribner's Sons, New York, 1984. An excellent secondary source for scientific illustration in ancient and medieval times.

24. H. J. R. Murray. *A History of Board Games Other Than Chess.* Oxford University Press, Oxford, England, 1952.
25. H. Parker. *Ancient Ceylon: An Account of the Aborigines and of Part of the Early Civilisation.* Luzac & Co., London, 1909.
26. W. Playfair. *The Commercial and Political Atlas.* London, 1786.
27. H. Sachs, M. Stiebitz, and R. J. Wilson. An historical note: Euler's Königsberg letters. *Journal of Graph Theory,* 12(1):133–139, 1988.
28. J. Stalham, late 15th century. English Public Records Office, Ref. No. REQ2/26/48. Thanks to Bob O'Hara for researching this and other documents on our behalf at the PRO.
29. J. J. Sylvester. Chemistry and Algebra. *Nature,* 17:284, 1877-8.
30. L. Tilling. Early experimental graphs. *British Journal for the History of Science,* 8:193–213, 1975.
31. A.-T. Vandermonde. Remarques sur les Problèmes de Situation. *Histoire de l'Académie des Sciences (Paris),* 1771. An English translation can be found in [7], pp. 22-26.
32. R. J. Wilson. An Eulerian trail through Königsberg. *Journal of Graph Theory,* 10(3):265–275, 1986.
33. D. Woodward. Medieval mappaemundi. In J. B. Harley and D. Woodward, editors, *The History of Cartography, Volume 1: Cartography in Prehistoric, Ancient, and Medieval Europe and the Mediterranean.* The University of Chicago Press, Chicago & London, 1987.
34. Chronica regia coloniensis, 1150-60. Herzog August Bibliothek, Wolfenbüttel, Cod. Guelf. 74. 3 Aug. 2nd. Reproduced in [19], Fig. 3.
35. `http://www.ahs.uwaterloo.ca/~museum/vexhibit/board/rowgames/mill.html`.

Towards an Aesthetic Invariant for Graph Drawing

Jan Adamec and Jaroslav Nešetřil

Department of Applied Mathematics and Institute of Thoretical Computer sciences (ITI) *
Charles University
Malostranské nám. 25, 11800 Praha, Czech Republic
nesetril@kam.ms.mff.cuni.cz

Abstract. In this paper we do not address the question of visualization, of picture processing of visual information. The information for us is already processed and ,typically, it is of a very simple type such as drawing (however not necessary a graph drawing). What we would like to answer is how to formalize the fact that such a picture (drawing) is *harmonious*. Harmonious we mean in the sense of aesthetic pleasing. We prefer the word harmonious to aesthetic (which is probably more in common usage) as an aesthetic feeling is probably highly individual and we cannot have an ambition to define (or even approach that). We propose an approach which should capture some features of a harmonious picture by means of the notion *Hereditary Fractional Length* (HFL). This approach is based on the analysis of curves [16] which in turn goes back to Steinhaus and Poincaré. The hereditary approach is based on the *dual approach* (it may be viewed as an approach dual to the Piaget's analysis of intelligence), [13]. The Hereditary Fractional Length is preserved by scaling and rotations and it is a very robust parameter which can be computed for a large class of drawings and pictures. This is an important feature as a perception of harmony (and aesthetic pleasure) is a robust feeling. Perhaps this parameter could aid in the hierarchical approach to graph visualization and graph drawing in particular.

1 Introduction

What do have the following pictures same and what different? (The substance and the context are important artistic aspects. However we are interested in formal similarities of these pictures; one picture is a musical score - a sketch by Janáček [24], the other picture one of the Moduli - a sketch by Načeradský and the second author, [21].)

* Supported by a Grant LN00A56 of Czech Ministary of Education

P. Mutzel, M. Jünger, and S. Leipert (Eds.): GD 2001, LNCS 2265, pp. 287–296, 2002.
© Springer-Verlag Berlin Heidelberg 2002

And, very simply, can we distinguish or order in some systematic way the following three figures (graph drawings using a program due to [8])?

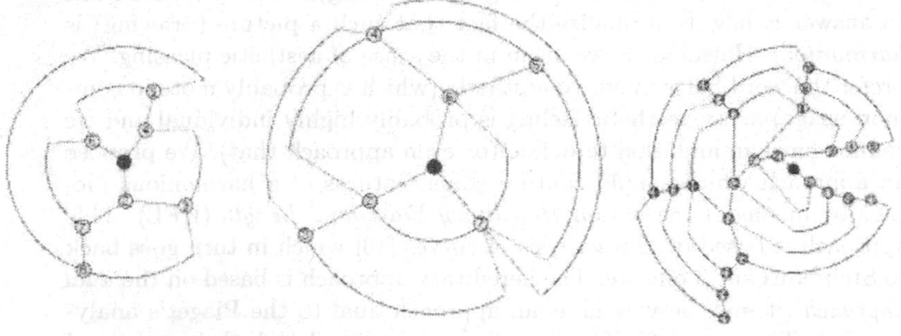

The modern version of these question is *not* how to teach a gifted and collaborating child what is nice and beautiful. Instead we need to teach an individual which is not collaborating at all and who takes every our information deadly seriously and exploits it to the last bit - a computer. People usually do not react this way (and if so, then only in comedies like [9] or [10]; the fact that these great novels have a military setting is then not an accident). In order to "teach" a computer (and even without ambition for teaching, just dealing with it) we need a precision. And precision in the other words calls for some concrete measures or our phenomena, in the other words for *invariants*. The purpose of this note is to suggest such an invariant and to document some experiments which we performed.

2 On Invariants

The traditional principal problem of aesthetics (and art history) - to explain and to predict artistic and aesthetically pleasing took recently an unexpected twist. We do not explain and deal with individual instances, we have to *classify* a vast amount of data and we have to *design* procedures with likely harmonious output. This problem in its manifold variety is interesting already when our objects are well defined compositions composed from simple building blocks such as lines, squares, sticks,.... This in fact is a familiar exercise and training ground of schools of design and architecture and (traditional) art academies. This illustrates difficulty and variety of solutions even of simple situations. This should be not surprising if we realize how many simple lines needed, say Rembrandt or Picasso, to produce full images (for example drawings; 50 lines or even less!).

For our "simple composition from simple building blocks" we would like to create an *invariant* which would help us to categorize and order this compositions. It is difficult to say even on this simple level what it is an invariant, but we can certainly state which properties such an invariant should have:

i. invariant should be an (easy) *computable* aspect of the structure;

ii. invariant should be *consistent* (or invariant, meaning it should not change) under chosen modification of structure;

iii. invariant should be *useful* in that it can be used to catalogue, to order (which structure is "better"), to classify, to distinguish.

We propose here an invariant - called *hereditary fractional length* - to measure an aesthetic quality of a visual data (drawing, scheme, painting, note score, molecular data output and others). This invariant is presented in the next section.

3 Hereditary Fractional Length of a Drawing

Before defining the invariant we want to specify the rules under which the invariant should remain unchanged. Such rules were specified several times and are folklore in the visualization of scientific results. For example the book [4] lists (in section "Aesthetics") the following 11 graphic properties (called *aesthetics* which are commonly adopted:

Crossings, Area, Total Edge Length, Maximal Edge Length, Uniform Edge Length, Total Bends, Maximum Bends, Uniform Bends, Angular Resolution, Aspect Ration and Symmetry.

The names of these criteria are self-explanatory and together they form a very good paradigm for drawing of graphs. However most of them are specific for drawings of graphs (or structures like it) and they do not apply generally (for example to artistic drawings or sketches). Yet we believe that the aesthetic quality of visual algorithms should be tested on *aesthetically charged* objects. This is one of the underlying ideas of our approach.

There is another drawback of the above paradigm. In all of these criteria (with exception of angular resolution, where we want to maximize, and symmetry which is a structural property) we are aiming for a minimization (for example we want to minimize the total length of our drawing). That of course means that we have to optimize these criteria (as they are sometimes mutually pointed against each other) and we have to add to our paradigm preferences among them.

However as optimization problems these criteria are computationally hard (see [4]).

Our approach is different and we believe it could add a new aspect to visualization and analysis of visual data.

Let us first specify objects which are relevant to our method:

A *drawing* D is a finite set of curves in Euclidean plane. (As our drawings are man made we assume that the set is finite.) A curve is a continuous image of unit interval in plane (we mean a "nice" image; this is not place for technicalities).

A *painting* can be any image, including photos. It typically consists of differently colored areas.

An *engraving* is very special kind of object, since it can be included in the previous two classes. However seeing it as a drawing gives a lot more information.

For an infinite line L denote by $i(L, D)$ the number of intersections of the line L and drawing D. We define the *fractional length* $f\ell(D)$ of a drawing D as the expected value of $i(L, D)$ where expectation relates to the random selection of line L. Fractional length could be also called *Combinatorial Entropy* from reasons to be mentioned later.

By virtue of this definition we note the following:

i. The fractional length $f\ell(D)$ is easily evaluated by a random generation of lines;

ii. $f\ell(D)$ is invariant under transposition and rotation;

iii. $f\ell(D)$ is invariant under scaling (i.e. "blowing up").

The role of randomness in the art has been discussed e.g. in [16], [17]. These facts make it possible to evaluate (or very accurately to estimate) fractional length of many drawings schemata, drawings of artists (we systematically tested some of the early works of Picasso [22] and Kandinsky [12] as well some drawings from [21]). We also used the analytical description of tertiary structure of some of the proteins (provided by P. Pančoška, compare [11]). The software developed in [1] uses standard tools of digital image processing, see [2], [23] allows to handle very broad spectrum of examples. Most difficulties are with paintings, because they must be transformed into drawings in order to count number of intersections (several filters are applied on the picture to find its contours, resulting picture is drawing as we defined it before).

Let us state some specific examples: the above Janáček score had fractional length 18.55 which is quite similar to the fractional length 17.87 of the above drawing taken from [21]. The three graph drawings depicted above had fractional lengths (from left to right) 5.70, 5.37, 6.96.

The method is flexible enough to handle complex drawings and pictures. For example classical prints from J. Verne novels when considered as a dense network of individual lines (no blurring). For example the following picture (due to Roux) [26] which serves as the front illustration of [14] has fractional length 136.6.

As expected the $n \times n$ lattice has approximate fractional length n. (As lattices are given analytically we can perform this experiment for very large n).

The following is a rare opportunity to compare a drawing (by Giacometti) and a related photography (which served as a basis of the drawing).

While the drawing has been scanned and the intersection numbers $i(D,T)$ were estimated directly the photo has been first processed by blurring and trace-contours filters. The fractional length of the photo is 14.43, drawing is much more rich with approximate $f\ell(D)$ 26.37.

Here the visual processing enriches the original design so that the drawing reduction has two times higher fractional length than the original taken as a drawing with variable heavy curves (29.64 vs 15.45).

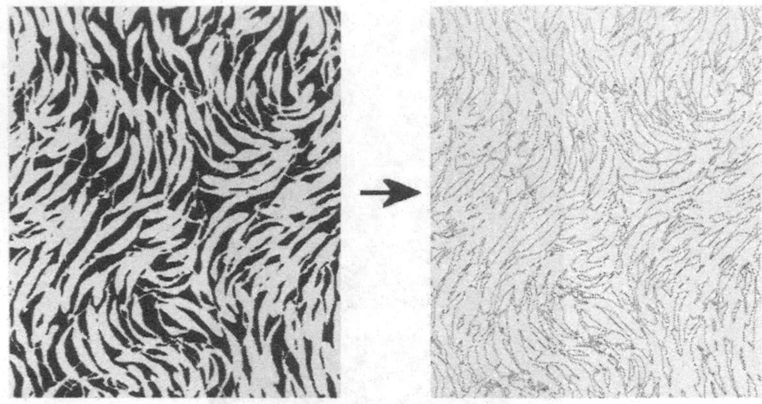

Here are several examples of the same scheme. The fractional lengths are as follows (from left to right): 24.06, 8.73, 17.86

What do these numbers mean? What is a significance of the fractional length of a drawing? The definition of the fractional length is motivated by the research done by prof. Michel Mendes France in a series papers devoted to the analysis of curves, [16]. He defines the temperature $T(D)$ of a curve D by the following formula

$$T = (\log \frac{E(i(L,D))}{E(i(L,D)) - 1})^{-1}$$

where expectation relates to the random selection of the line L.

He related this parameter to *entropy, dimension* and other parameters which he defined in an analogy to *fractal theory, statistical physics* and *geometric probability.* These definitions rest on classical theorems due to Steinhaus [25] (which in turn goes to Poincaré). One can show that the $\log fl(D)$ is a good approximation of the entropy of a curve. Thus the name *Combinatorial Entropy* for the Fractional length.

By viewing a drawing as a set of curves and thinking of *eulerian trail* [14](in each component of the drawing) as a new curve of double length (as we traverse every segment of a drawing twice) we can define the the temperature by the same formula for a more general class of pictures (drawings). For large number of intersections (of a drawing D with a line L) the temperature $T(D)$ is approximately equal to the average value of $i(D,L) - 1 = f\ell(D) - 1$ while the entropy $H(D)$ is approximated by the fractional length $f\ell(D)$. It follows (and this is Steinhaus' theorem) that the fractional length $f\ell(D)$ of a drawing D is approximated by the ratio

$$2\ell(D)/c(D)$$

where $\ell(D)$ denotes the total length of the drawing D and $c(D)$ denotes the length of circumference of the (convex closure) of D. The last formula motivates and justifies the name fractional length of our invariant: a drawing D has its fractional length equal to r iff the same (fractional) number of crossings is obtained when we traverse the circumference of D by $r/2$ times. As will be seen from the sequel we could not use the definition of either temperature or entropy as we demand that pure invariant has strong hereditary properties (see below). These properties (proportionality when considering a part of the picture) do not hold for neither temperature nor entropy.

Remark. We cannot resist the temptations to mention that similar ratios occur in art history. For example [3] is basing his esthetic theory on a ratio $M = C/O$ where O denotes the "order" or "how many tricks are involved " and C denotes the "complexity" which is proportional to "a preliminary effort necessary to perceive the object " and O stands for "harmony, symmetry or order". Clearly this is meant symbolically and, as demonstrated in [3], the ratio has to be interpreted in individual cases. Our approach may be seen (on this symbolic level) as a statistical verification of these ideas. Motivated by the romantic symbolism of Marcel Duchamp this *romantic algebra* was used many more times with much less rigour, [7]. There is no romanticism in our formulas, these are exact results.

Based on these interpretations $f\ell(D)$ measures the information content and the amount of work which the artist (explicitly) put into his drawing. On the other hand side, by comparing various drawings of the same object (or theme) the lower $f\ell(D)$ indicates the elegance and simplicity of the output. However as such the fractional length captures only the global properties of the drawing (expressed by the total length of the drawing and the circumference). Still, via the average number of intersections it captures, implicitly, many properties of the drawing. In [13] we proposed a technique (based on the algebraical structure *cogroup*) to generate harmonious objects. In our setting this can be formulated as follows:

Hereditary Fractional Length Thesis

A *harmonious (or aesthetically pleasing) drawing or design has a fractional length in each of its (meaningful) parts proportional to the global fractional length. Here a meaningful part is a part which reflects the properties of* D. *A lower bound is obviously* $\mathrm{logi}(D, L)$.

Clearly HFLT depends on the selection of (meaningful) parts - fields and as such it is only a program. HFL can be visualized as a matrix which has a hereditary structure (fields) where each field has a corresponding fractional length. There are at most $f\ell(D)$ fields and the depth of the hierarchical structure is at most $\log f\ell(D)$ (i.e. very efficient). How this is done is shown on the following examples.

We performed experiments with HFLT based on the regular subdivision of grids containing the circumference of our pictures and find some supporting evidence. Two particular results are included in this report. The Janáček score depicted above as one of the first illustrations of this paper gives results summarized by the following table (the middle of a rectangle indicates the fractional length of that rectangle).

					1,45	1,28	1,03								
		—2,36—		—1,60—		—2,11—									
		1,97	1,66		1,47	1,75	1,02								
	—2,39—				—4,51—				—1,68—				—2,57—		
		1,44	1,63		1,39	2,50	1,44						1,20	2,63	1,61
		—1,83—		—2,17—		—4,07—		—1,32—		—1,81—		—1,25—		—2,74—	
		1,06	1,24	1,00	2,16	2,53	2,55	1,34	1,03	1,97	1,11	1,23	1,00	1,24	1,05
			—8,02—								—9,61—				
		1,54	2,00	1,03	1,89	2,37	1,81	1,38	1,11	2,62	1,89	2,72	1,59	2,43	1,53
—1,25—		—2,38—		—2,72—		—3,92—		—2,28—		—4,40—		—3,77—		—3,52—	
1,00	1,36		1,64	1,01	2,28	2,21	2,08	1,63	1,83	2,60	2,40	2,62	1,93	2,52	1,43
	—3,03—				—7,28—				—6,69—				—7,29—		
1,32	1,64	1,00	2,11	1,36	3,24	2,44	1,98	1,39	1,76	3,29	2,11	1,68	2,35	2,36	1,35
—2,27—		—2,24—		—4,40—		—3,94—		—2,13—		—4,56—		—4,06—		—3,45—	
1,55	1,83	1,00	1,50	1,65	3,31	2,24	1,83	1,85	1,03	2,22	2,08	1,95	2,96	2,57	1,45
							—18,55—								
1,20	1,83	2,70	1,38	1,67	1,21	1,32	1,94	3,29	1,74	1,43	1,01	1,73	2,43	2,69	1,48
—3,31—		—3,26—		—2,23—		—3,46—		—4,48—		—2,39—		—3,07—		—3,37—	
1,89	2,78	2,08	1,45	1,04	1,75	2,41	2,26	3,04	1,62	2,01	1,70	1,18	2,04	1,99	1,07
	—6,52—				—5,84—				—7,22—				—5,36—		
1,16	2,16	2,62	2,60	1,07	2,49	2,11	2,02	2,36	1,34	1,97	2,75	1,39	2,14	1,01	
—2,18—		—4,35—		—3,38—		—3,57—		—3,79—		—4,41—		—3,70—		—1,21—	
1,00	1,49	2,60	2,12	1,06	2,91	1,78	2,22	3,16	1,71	2,12	2,90	1,89	2,42	1,14	
			—9,89—								—11,16—				
	1,35	3,14	1,23	1,26	2,52	1,00	2,22	2,65	1,45	1,73	2,67	1,96	2,37	1,01	
—1,86—		—4,31—		—2,67—		—2,08—		—3,08—		—4,33—		—3,95—		—1,01—	
1,05	1,94	3,24	1,68	1,17	1,20	1,01	1,59	1,64	1,25	2,19	2,80	2,44	2,00		
	—5,94—				—2,78—				—5,57—				—4,58—		
1,00	2,35	2,49	1,11						1,14	2,37	2,27	2,14	2,57	1,40	
—2,84—		—2,54—						—1,18—		—3,42—		—3,40—		—1,40—	
1,00	2,04	1,50	1,00						1,01	1,17	1,52	1,36	1,04		

For the following drawing (of our institute in Prague)

we found the following distribution of fractional lengths:

														1,72	2,10
						1,00		1,76		1,71		1,34		3,40	
							1,00	1,02	1,76	1,38	1,68	1,23	1,23	1,50	2,77
					3,25				6,86				7,04		
						1,00	1,56	2,14	2,73	2,65	2,05	2,22	2,78	1,98	3,22
						3,26		4,91		4,96		4,98		5,33	
						2,30	2,75	3,13	3,22	2,77	3,41	2,94	2,60	3,21	3,23
			7,54								15,20				
	1,00			1,02	1,30	3,42	2,65	1,33	3,50	2,54	1,73	1,60	2,58	2,79	2,84
1,00				2,28		5,08		4,75		4,08		4,04		5,20	
				1,00	2,38	2,63	2,26	2,97	2,73	2,15	2,72	2,11	2,72	3,14	2,53
	1,00				7,67				8,33				9,27		
				1,31	2,97	3,00	1,86	1,80	2,97	2,95	1,72	1,58	2,74	3,34	2,75
				3,12		4,96		4,18		4,28		4,56		5,71	
				1,11	2,07	2,95	2,64	1,75	3,06	3,12	1,76	2,14	3,43	3,73	2,26
							27,62								
				1,12	2,54	2,64	2,80	2,29	2,48	2,32	2,89	2,27	1,86	3,86	2,23
		1,24		3,97		5,62		4,22		4,73		4,42		5,82	
			1,24	2,46	3,10	3,09	3,39	2,04	2,83	2,38	2,67	2,47	2,96	3,92	2,56
	3,64				10,64				8,40				9,62		
		1,21	2,73	2,80	3,61	3,77	2,46	2,05	2,49	2,14	2,55	2,11	1,77	3,66	2,86
2,00		3,84		5,88		5,98		4,28		4,43		3,52		6,05	
1,11	2,12	2,31	2,98	2,82	3,44	3,44	2,93	2,24	2,80	2,16	2,70	2,22	1,97	3,63	2,86
			14,65								16,44				
1,00	2,21	1,88	3,14	2,89	3,49	3,39	2,86	2,63	2,22	2,52	1,79	2,34	3,82	3,06	2,64
2,19		4,06		5,45		5,57		4,69		4,48		5,23		5,58	
		1,22	3,06	2,81	2,45	2,38	2,95	2,62	2,65	2,86	2,41	2,94	2,66	3,58	2,59
	3,97				9,59				8,51				8,25		
			1,40	2,62	3,17	2,99	2,77	3,56	2,81	2,74	3,42	2,49	1,34	2,21	2,89
		1,40		3,44		4,87		4,49		3,74		2,33		3,49	
					1,22	2,24	1,98	1,33	1,12	1,00				1,00	1,62

4 Conclusion

Hereditary Fractional Length (or *Hereditary Combinatorial Entropy*) attempts to characterize harmonious and aesthetically pleasing pictures by a balanced distribution of Fractional Lengths (of Combinatorial Entropies) of their *meaningful* parts. One could also say that this measures not the simplicity (or *complexity*) of a picture but rather a relative *balance* and *order* of otherwise complex picture. This is perhaps in accord with our artistic experience and it has also an intuitive similarity with Birkhoff thesis mentioned above. It is pleasing to note that graphical objects as diverse as a musical score (i.e. Janáček sketch above) and an artistic drawing share a high degree of similarity thus perhaps pointing their similar (formal) aesthetic appeal.

Clearly many more examples have to be drawn but it seems that HFLT thesis is a good approximation to the notion of a harmonious drawing. But in any case an invariant is only one feature of otherwise very complicated phenomenon of visualization. But we believe that this, perhaps first, invariant for harmonious (i.e. aesthetic) drawing will inspire some further work in this direction.

References

1. J. Adamec: Kreslení grafů, diploma thesis, Charles University, Prague (2001).
2. G. A. Baxes: Digital Image Processing. Principles and Applications, Wiley, 1994.
3. G. D. Birkhoff: A Mathematical Theory of Aesthetics and its Application tp Poetry and Music, The Rice Institute Pamphlet, vol. XIX, 3 (1932), 342p.
4. G. DiBattista, P. Eades, R. Tamassia, I. G. Tollis: Graph Drawing. Algorithms for the Visualization of Graphs. Prentice Hall 1999.
5. H. Damisch: The Origins of Perspective, MIT Press 1994, original French edition Flamarion 1987.

6. H. Damisch: Le travail de l'art: vers une topologie de la couleur? In: [21].
7. T. de Duve: Kant after Duchamp, MIT Press 1998.
8. H. de Fraysseix: A drawing software (personal communication).
9. J. Hašek: Osudy dobrého vojáka Švejka, 1920 (in English: *The Good Soldier Schweik*).
10. J. Heller: Catch-22, 1961
11. V. Janota, J. Nešetřil, P. Pančoška: Spectra Graphs and Proteins. Towards Understanding of Protein Folding. In: Contemporary Trends in Discrete Mathematics, AMS, DIMACS Series 49, 1999, pp. 237 - 255.
12. W. Kandinsky: Point and Line to Plane, Dover Publications 1979.
13. J. Kabele, J. Nešetřil: Remarks on Radically Different Aesthetic - a computational compromise (to appear).
14. J. Matoušek, J. Nešetřil: Invitation to Discrete Mathematics, Oxford Univ. Press 1998.
15. K. Mehlhorn, S. Naher: LEDA - A platform for combinatorial and geometric computing, Cambridge Univ. Press, 1999.
16. M. Mendès France: The Planck Constant of a Curve. In: Fractal Geometry and Analysis (J. Bélair, S. Dubuc, eds.) Kluwer Acad. Publ. 1991, pp. 325 - 366.
17. M. Mendès France, J. Nešetřil: Fragments of a Dialogue, KAM Series 95 - 303, Charles University Prague (a Czech translation in Atelier 1997).
18. J. Nešetřil: The Art of Drawing. In: Graph Drawing (ed. J. Kratochvíl), Springer Verlag,1999.
19. J. Nešetřil: Mathematics and Art, From The Logical Point of View, 2, 2/93 (1994), 50 - 72.
20. J. Nešetřil: Aesthetics for Computers or How to Measure a Harmony (to appear in Visual Mind (ed. M. Emmer), MIT Press).
21. J. Načeradský, J. Nešetřil: Antropogeometrie I, II (Czech and English), Rabas Gallery, Rakovník, 1998 (ISBN 80-85868-25-3).
22. P. Picasso: Picasso - Der Zeichner 1893-1929, Diogenes, 1982.
23. W. K. Pratt: Digital Image Processing, Wiley, 1978.
24. M. Štědroň: Leoš Janáček and Music of 20. Century, Nauma, Brno, 1998 (in Czech).
25. H. Steinhaus: Length, shape and area, Colloq. Math. 3 (1954), 1 - 13.
26. J. Verne: Sans Dessus Dessous, J. Helzel (Paris), 1889. J. Nešetřil: Mathematics and Art, From The Logical Point of View, 2, 2/93 (1994), 50 - 72.

Orthogonal Drawings with Few Layers*

Therese Biedl[1,**], John R. Johansen[1], Thomas Shermer[2,**], and David R. Wood[3,***]

[1] Department of Computer Science, University of Waterloo, Canada,
{biedl,jrjohans}@uwaterloo.ca

[2] School of Computing Science, Simon Fraser University, Canada,
shermer@cs.sfu.ca

[3] Basser Department of Computer Science, The University of Sydney, Australia,
davidw@cs.usyd.edu.au

Abstract. In this paper, we study 3-dimensional orthogonal graph drawings. Motivated by the fact that only a limited number of layers is possible in VLSI technology, and also noting that a small number of layers is easier to parse for humans, we study drawings where one dimension is restricted to be very small. We give algorithms to obtain point-drawings with 3 layers and 4 bends per edge, and algorithms to obtain box-drawings with 2 layers and 2 bends per edge. Several other related results are included as well. Our constructions have optimal volume, which we prove by providing lower bounds.

1 Introduction

Motivated by experimental evidence suggesting that displaying a graph in three dimensions is better than in two [22,23], there is a growing body of research in 3-dimensional graph drawing. Orthogonal drawings, in which edges are drawn as axis-parallel polylines, is a popular layout style with applications in VLSI circuit layout. Since present-day VLSI technology limits circuits to a few layers, consideration of the number of layers for orthogonal drawings are important. In this paper we present bounds on the volume and the number of bends in orthogonal graph drawings with only a few (2 or 3) layers. As well as VLSI concerns, we are motivated by the effective visualisation of 3-D orthogonal drawings in which one wishes to minimise the depth of a 3-D drawing displayed on a screen.

The (*3-dimensional*) *orthogonal grid* is the cubic lattice, consisting of *grid points* with integer coordinates, together with the axis-parallel *grid lines* determined by these points. We use the word *box* to mean an axis-parallel box with integral boundaries. At each grid point in a box B that is extremal in some direction $d \in \{\pm X, \pm Y \pm Z\}$, we say there is *port* on B in direction d. One grid point

* The authors would like to thank Stephen Wismath for helpful discussions.

** Supported by NSERC.

*** Supported by the ARC. Completed while visiting the School of Computer Science, McGill University, Canada.

P. Mutzel, M. Jünger, and S. Leipert (Eds.): GD 2001, LNCS 2265, pp. 297–311, 2002.
© Springer-Verlag Berlin Heidelberg 2002

can thus define up to six incidents ports. For each dimension $I \in \{X, Y, Z\}$, an I*-line* is a line parallel to the I-axis, an I*-segment* is a line-segment within an I-line, and an I*-plane* is a plane perpendicular to the I-axis.

Let $G = (V, E)$ be an undirected graph without loops, $n = |V|$ and $m = |E|$. Let Δ be the maximum degree of G; a graph with maximum degree at most Δ is called a Δ*-graph.* An *orthogonal (box-)drawing* of G represents vertices by pairwise non-intersecting boxes and edges by pairwise disjoint grid paths connecting the endpoints of the edge. An orthogonal drawing with a particular shape of box representing every vertex, e.g., point or cube, is called an orthogonal *shape*-drawing. An orthogonal point-drawing can only exist for 6-graphs.

From now on, we use the term *drawing* to mean a 3-dimensional orthogonal drawing. Furthermore, the graph-theoretic terms 'vertex' and 'edge' also refer to their representation in a drawing. The *size* of a vertex v is denoted by $X(v) \times Y(v) \times Z(v)$, where for each $I \in \{X, Y, Z\}$, $I(v)$ is the number of I-planes intersecting v. The number of ports of v is called its *surface*, denoted by *surface*(v). The number of grid points in a box is called its *volume.*

Various criteria have been proposed in the literature to evaluate the aesthetic quality of a particular drawing. The primary criterion considered in this paper is that one dimension of the bounding box should be very small (2 or 3 units). For convenience we choose this to be the Z-dimension, and refer to the Z-planes of such a drawing as *layers.* We also consider the following secondary criteria.

First, the volume of a drawing should be small, where the *volume* of a drawing is that of the smallest axis-aligned box, called the *bounding box*, which encloses the drawing. Minimising the number of bends is also an important aesthetic criterion for orthogonal drawings. A drawing with no more than b bends per edge is called a b*-bend drawing.* Minimising either the volume or the total number of bends in a drawing is NP-hard [11].

For box-drawings the size and shape of a vertex with respect to its degree are also considered an important measures of aesthetic quality. A vertex v is α*-degree-restricted* if $\text{surface}(v) \leq \alpha \cdot \deg(v) + o(\deg(v))$. If for some constant α, every vertex v is α-degree-restricted, then the drawing is said to be *degree-restricted*; we use the term α-degree-restricted if we want to specify constant α. A drawing is said to be *strictly* α-degree-restricted if $\text{surface}(v) \leq \alpha \cdot \deg(v)$ for all vertices v, that is, no smaller-order terms are allowed.

The aspect ratio of a vertex v is normally defined to be the ratio between its largest and smallest side; that is, $\max\{X(v), Y(v), Z(v)\} / \min\{X(v), Y(v), Z(v)\}$. Since we are primarily concerned with drawings in a constant number of layers we define the *aspect ratio* of a vertex v to be $\max\{X(v), Y(v)\} / \min\{X(v), Y(v)\}$. We say that a drawing has *bounded aspect ratios* if there exists a constant r such that all vertices have aspect ratio at most r.

1.1 Box-Drawings

Algorithms to produce orthogonal box-drawings have been studied in [1,4,7,8,14, 19,28,29]. Lower bounds for the volume of orthogonal box-drawings have been presented in [1,7,8,14]. Table 1 summarises the known bounds on the volume and maximum number of bends per edge with various aesthetic criteria. We include the number of layers in each construction as well.

Table 1. Bounds on the volume, the number of layers and the maximum number of bends in box-drawings (assuming $m \in \Omega(n)$).

Lower Bound volume	reference	Upper Bound volume	layers	bends	graphs	reference
bounded aspect ratio / degree-restricted						
$\Omega(m^{3/2})$	[8]	$O(nm\sqrt{\Delta})$	$O(\sqrt{\Delta})$	2	simple	[4]
$\Omega(m^{3/2})$	[8]	$O(m^2)$	$O(\sqrt{m})$	5	multigraphs	[8]
$\Omega(m^{3/2})$	[8]	$O(m^{3/2})$	$O(\sqrt{m})$	6	multigraphs	[8]
bounded aspect ratio / not necessarily degree-restricted						
$\Omega(m^2)$	Thm. 7	$O(m^2)$	2	2	multigraphs	Thm. 3
no bounds on aspect ratio / degree-restricted						
$\Omega(m^{3/2})$	[8]	$O(n^3\Delta)$	$O(\Delta)$	2	simple	[4]
$\Omega(m^2)$	Thm. 7	$O(m^2)$	2	3	multigraphs	Thm. 5
$\Omega(m^{3/2})$	[8]	$O(m^{3/2})$	$O(\sqrt{m})$	6	multigraphs	[8]
no bounds on aspect ratio / not necessarily degree-restricted						
$\Omega(mn)$	[1]	$O(n^3)$	$O(n)$	1	simple	[7]
$\Omega(m\sqrt{n})$	[8]	$O(mn^{3/2})$	$O(\min\{n, m/\sqrt{n}\})$	1	simple	[29]
$\Omega(mn)$	Thm. 7	$O(mn)$	2	2	multigraphs	Thm. 2
$\Omega(m\sqrt{n})$	[8]	$O(n^{5/2})$	$O(\sqrt{n})$	3	simple	[7]
$\Omega(m\sqrt{n})$	[8]	$O(m\sqrt{n})$	$O(\sqrt{n})$	4	simple	[8]

Orthogonal box-drawings with a constant number of layers have not been studied previously. In this paper we prove the following results.

- Every graph has a 2-bend box-drawing in an $m \times n \times 2$ grid. This volume bound matches the best known upper bound for 2-bend box-drawings [29], but uses only 2 layers, as opposed to $O(m/n)$ layers in [29].
- Every graph has a degree-restricted 3-bend drawing in a $(m+n) \times (m + \frac{3}{2}n + 1) \times 2$ grid. This volume bound shows that 3 bends per edge suffice for degree-restricted drawings with $O(m^2)$ volume; previously 5 bends per edge were needed [8]. Additionally, we use only 2 layers, as opposed to $O(\sqrt{m})$ layers in [8].
- Every graph has a 2-bend drawing with bounded aspect ratios in a $(\frac{3}{4}m + \frac{1}{2}n) \times (\frac{3}{4}m + \frac{1}{2}n) \times 2$ grid. This volume bound shows that 3 bends per edge suffice for bounded aspect ratio drawings with $O(m^2)$ volume; previously 5 bends per edge were needed [8]. Additionally, we use only 2 layers, as opposed to $O(\sqrt{m})$ layers in [8].

We prove the following lower bounds on the volume of drawings with k layers (assuming that no vertices are "above" each other; see Section 5).

- There are graphs that need $\Omega(mn/k^5)$ volume in any drawing with at most k layers.
- There are graphs that need $\Omega(m^2/k^5r)$ volume in any drawing with at most k layers and aspect ratios at most r.
- There are graphs that need $\Omega(m^2/k^5)$ volume in any drawing with at most k layers which is strictly α-degree-restricted for some $\alpha \in o(n/k^3)$. Typically, α is a small constant, so the assumption $\alpha \in o(n/k^3)$ is reasonable.
- No drawing with a constant number of layers can be both degree-restricted and have bounded aspect ratios.

1.2 Point-Drawings

Algorithms for producing point-drawings have been presented in [3,9,10,11,13, 15,17,19,26,25,27]. A lower bound of $\Omega(n^{3/2})$ for the volume of point-drawings was established by Kolmogorov and Barzdin [15]. Lower bounds for the number of bends in point-drawings were established by Wood [30].

Table 2. Upper Bounds for 3-Dimensional Orthogonal Point-Drawing

Graphs	Max. (Avg.) Bends	Bounding Box	Volume	Reference
multigraph	7	$O(\sqrt{n}) \times O(\sqrt{n}) \times O(\sqrt{n})$	$\Theta(n^{3/2})$	[12,13]
multigraph	6	$O(\sqrt{n}) \times O(\sqrt{n}) \times O(n)$	$O(n^2)$	[13]
multigraph	5	$O(n) \times O(n) \times O(1)$	$O(n^2)$	[9]
multigraph	4	$O(n) \times O(n) \times O(1)$	$O(n^2)$	Thm. 1
multigraph $\Delta \leq 4$	3	$O(n) \times O(n) \times O(1)$	$O(n^2)$	[13]
multigraph	5	$O(\sqrt{n}) \times O(n) \times O(n)$	$O(n^{5/2})$	[13]
simple	4 ($2\frac{2}{7}$)	$O(n) \times O(n) \times O(n)$	$2.13n^3$	[25,27]
multigraph	3	$O(n) \times O(n) \times O(n)$	$8n^3$	[12,13]
multigraph	3	$O(n) \times O(n) \times O(n)$	$4.63n^3$	[19]
multigraph	3	$O(n) \times O(n) \times O(n)$	$n^3 + o(n^3)$	[26]
simple $\Delta \leq 5$	2	$O(n) \times O(n) \times O(n)$	n^3	[25,27]

Point-drawings with a constant number of layers were first studied by Eades *et al.*[13] for 4-graphs. Their algorithm produces a 3-bend point-drawing in a $2n \times (n+2) \times 3$ grid. A corresponding lower bound is $\Omega(n^2/k)$ volume for point-drawings with at most k layers [3], hence $\Omega(n^2)$ volume is necessary for drawings with a constant number of layers. Closson *et al.* [9] were the first to give drawings with a constant number of layers for any 6-graph; their algorithm produces a 5-bend point-drawing of a 6-graph in a $7n \times 5n \times 5$ grid. At the expense of allowing one more bend per edge, the the authors present a fully dynamic algorithm that supports the on-line insertion and deletion of vertices and edges in $O(1)$ time.

In this paper, we describe an algorithm to produce a 4-bend point-drawing of a 6-graph in a $3n \times 2n \times 3$ grid. Thus, we establish that with only 4 bends per edge, $O(n^2)$ volume can be obtained (the previous best volume bound for 4-bend point-drawings was $O(n^3)$ [13]). Also, our volume is $24n^2$, improving on the bound of $175n^2$ for the algorithm in [9]. If the graph has maximum degree 5, we can obtain drawings with only two layers.

2 Toolkit

In this section we give a number of introductory results which will be employed by our algorithms to follow. They can be considered to be an orthogonal graph drawer's toolkit.

A *cycle cover* of a directed graph is a spanning subgraph consisting of directed cycles. The following result, which can be considered as three applications of the classical result of Petersen that "every regular graph of even degree has a 2-factor" [21], has an algorithmic proof by Eades *et al.* [13].

Lemma 1 ([13]). *If G is an n-vertex 6-graph then there exists a directed graph G' (possibly with loops) such that:*

1. *G is a subgraph of the underlying undirected graph of G'.*
2. *Each vertex of G' has in-degree* 3 *and out-degree* 3.
3. *G' can be partitioned into three arc-disjoint cycle covers.*

G' and the cycle covers can be computed in $O(n)$ time. □

We will need the following lemma which slightly strengthens a previous result [8,13].

Lemma 2. *The edges of a graph G can be coloured red and blue so that the number of monochromatic edges incident to each vertex v is at most* $\frac{1}{2}\deg(v)+1$.

Proof. Pair the odd degree vertices in G, and add an edge between the paired vertices. All vertices now have even degree. In particular, the degree of a vertex v is now $2\lceil\frac{1}{2}\deg(v)\rceil$. Alternately colour the edges red and blue by following an Eulerian tour of G starting at an inserted edge (if any). Thus, there are at most $\lceil\frac{1}{2}\deg(v)\rceil+1$ monochromatic edges incident to v. In fact, all vertices v, except the starting vertex in the Eulerian tour, have at most $\lceil\frac{1}{2}\deg(v)\rceil$ monochromatic incident edges. If there were no inserted edges then every vertex has even degree in the original G, and the number of monochromatic edges incident to v is at most $\frac{1}{2}\deg(v)+1$. If there were some inserted edges then, since we started the Eulerian tour at an inserted edge, the number of monochromatic edges incident to a vertex v is at most $\lceil\frac{1}{2}\deg(v)\rceil \leq \frac{1}{2}\deg(v)+\frac{1}{2}$. □

3 Point Drawings

In this section we describe an algorithm for producing point-drawings in a constant number of layers. Our algorithm is based on the decomposition of a 6-graph into three cycle covers, and the classification of edges according to the relative positions of the endpoints in an arbitrary ordering of the vertices along a 2-dimensional diagonal. This approach was first introduced in the 3-BENDS algorithm of Eades *et al.* [13]. The difference between the 3-BENDS algorithm and the algorithm which follows is that we use a 2-dimensional diagonal layout of the vertices, whereas the 3-BENDS algorithm uses a 3-dimensional diagonal layout. A 2-dimensional diagonal vertex layout is also used by Closson *et al.* [9].

Theorem 1. *Every 6-graph $G = (V, E)$ has a 4-bend point-drawing in a $3n \times 2n \times 3$ grid.*

Proof. Consider the following algorithm.

1. Compute G' and a cycle cover decomposition of G'; see Lemma 1. Label the cycle covers and the arcs in G' *red*, *green* and *blue*.
2. Let $V = (v_1, \ldots, v_n)$ be an arbitrary linear ordering of the vertices.
3. For each directed cycle C in the cycle decomposition, and for each arc $\overrightarrow{v_iv_j} \in C$ with $\overrightarrow{v_jv_k}$ the next arc in C, classify $\overrightarrow{v_iv_j}$ as follows, depending on the relative values of i, j and k. If $i < j < k$ then $\overrightarrow{v_iv_j}$ is *normal increasing.* If $i > j > k$ then $\overrightarrow{v_iv_j}$ is *normal decreasing.* If $i < j > k$ then $\overrightarrow{v_iv_j}$ is *increasing to a local maximum.* If $i > j < k$ then $\overrightarrow{v_iv_j}$ is *decreasing to a local minimum.*
4. Position vertices considering the red cycle cover as follows: For each vertex v_j, $1 \leq j \leq n$, suppose $\overrightarrow{v_iv_j}$ is the red arc entering v_j. If $\overrightarrow{v_iv_j}$ is normal increasing or decreasing to a local minimum then set $Y_j = 2j$. Otherwise $\overrightarrow{v_iv_j}$ is normal decreasing or increasing to a local maximum; set $Y_j = 2j - 1$. Position v_j at $(3j, Y_j, 0)$.
5. For each arc $\overrightarrow{v_iv_j}$ in the red cycle cover of G' (with v_iv_j in G), route v_iv_j using a Y-port at v_i and a Y-port at v_j. Whether a Y^+-port or a Y^- port is used depends on the classification of $\overrightarrow{v_iv_j}$. More precisely:
   a) If $\overrightarrow{v_iv_j}$ is *normal increasing*, as in Fig. 1(a), route v_iv_j with the 4-bend edge: $(3i, Y_i, 0) \to (3i, Y_j - 1, 0) \to (3i, Y_j - 1, 1) \to (3j, Y_j - 1, 1) \to (3j, Y_j - 1, 0) \to (3j, Y_j, 0)$.
   b) If $\overrightarrow{v_iv_j}$ is *normal decreasing*, as in Fig. 1(b), route v_iv_j with the 4-bend edge: $(3i, Y_i, 0) \to (3i, Y_j + 1, 0) \to (3i, Y_j + 1, 1) \to (3j, Y_j + 1, 1) \to (3j, Y_j + 1, 0) \to (3j, Y_j, 0)$.
   c) If $\overrightarrow{v_iv_j}$ is *increasing to a local maximum*, as in Fig. 2(a), route v_iv_j with the 4-bend edge: $(3i, Y_i, 0) \to (3i, Y_j + 1, 0) \to (3i, Y_j + 1, 1) \to (3j, Y_j + 1, 1) \to (3j, Y_j + 1, 0) \to (3j, Y_j, 0)$.
   d) If $\overrightarrow{v_iv_j}$ is *decreasing to a local minimum*, as in Fig. 2(b), route v_iv_j with the 4-bend edge: $(3i, Y_i, 0) \to (3i, Y_j - 1, 0) \to (3i, Y_j - 1, 1) \to (3j, Y_j - 1, 1) \to (3j, Y_j - 1, 0) \to (3j, Y_j, 0)$.

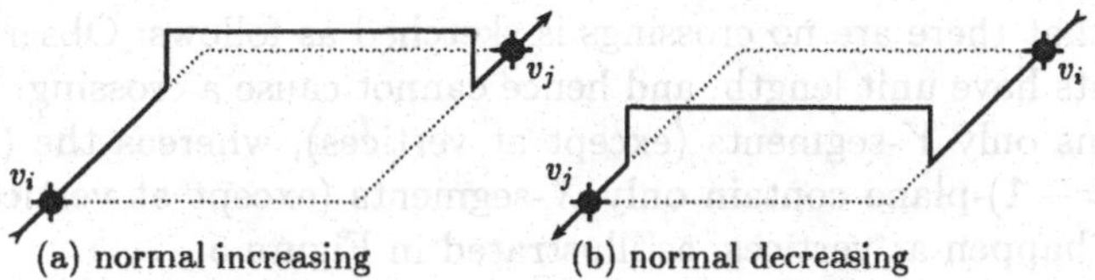

Fig. 1. Normal edge routes in the red cycle cover.

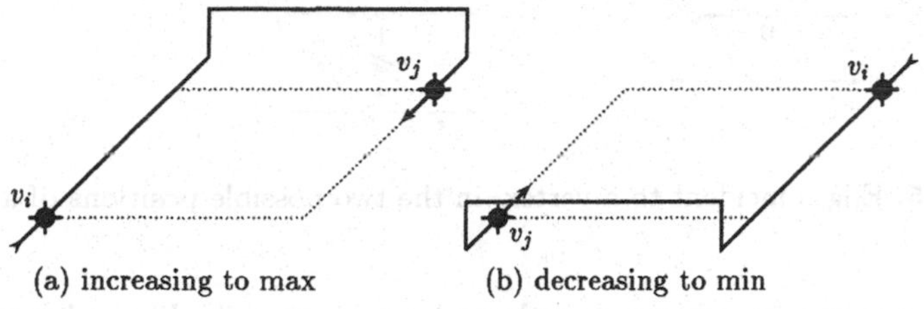

Fig. 2. Local min/max edge routes in the red cycle cover.

6. For each arc $\overrightarrow{v_iv_j}$ in the green cycle cover of G' (with v_iv_j in G), route v_iv_j using the Z^+-port at v_i and a X^--port at v_j. More precisely, route v_iv_j with the 4-bend edge. $(3i, Y_i, 0) \to (3i, Y_i, 1) \to (3j-1, Y_i, 1) \to (3j-1, Y_i, 0) \to (3j-1, Y_j, 0) \to (3j, Y_j, 0)$. See Fig. 3.

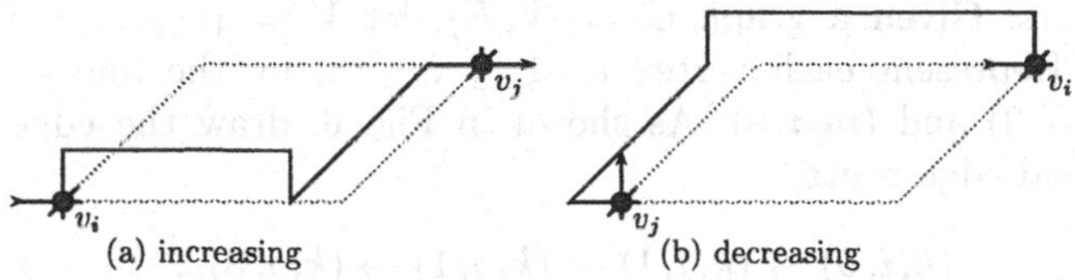

Fig. 3. Edge routes in the green cycle cover.

7. For each arc $\overrightarrow{v_iv_j}$ in the blue cycle cover of G' (with v_iv_j in G) route $\overrightarrow{v_iv_j}$ using the Z^--port at v_i and the X^+-port at v_j. More precisely, route v_iv_j with the 4-bend edge: $(3i, Y_i, 0) \to (3i, Y_i, -1) \to (3j+1, Y_i, -1) \to (3j+1, Y_i, 0) \to (3j+1, Y_j, 0) \to (3j, Y_j, 0)$. See Fig. 4.

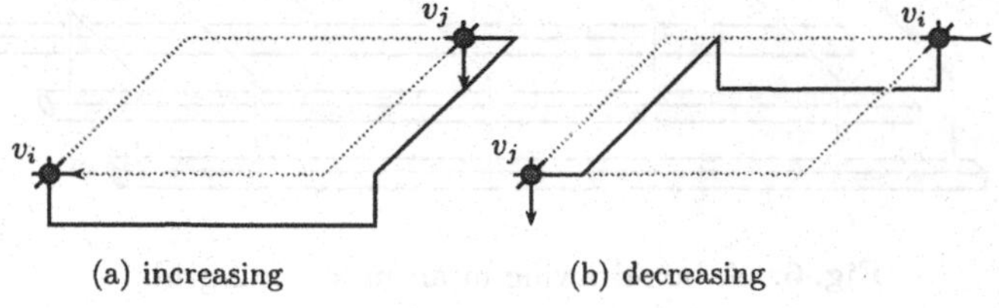

Fig. 4. Edge routes in the blue cycle cover.

A proof that there are no crossings is sketched as follows: Observe first that all Z-segments have unit length, and hence cannot cause a crossing. The (Z=0)-plane contains only Y-segments (except at vertices), whereas the (Z=1)-plane and the (Z= − 1)-plane contain only X-segments (except at vertices). Finally, no crossings happen at vertices, as illustrated in Figure 5. □

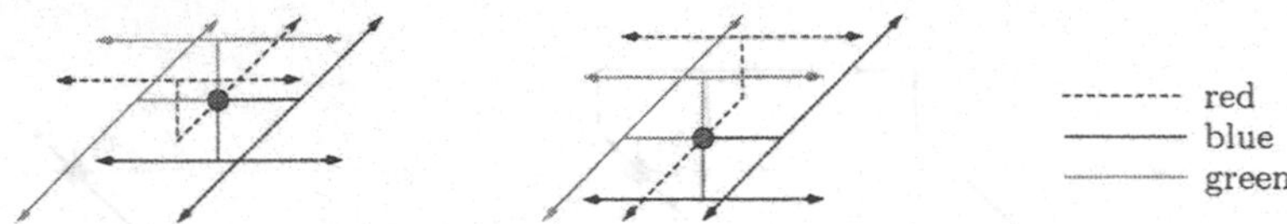

Fig. 5. Edges incident to a vertex, in the two possible positions of a vertex.

For 5-graphs, one can save one layer by rerouting the blue edge that uses the bottom port to another free port. Details are omitted.

4 Box Drawing Algorithms

Now we turn to graphs with arbitrarily high degrees.

A simple algorithm: The following simple algorithm produces a box-drawing with two layers. Given a graph $G = (V, E)$, let $V = (v_1, \ldots, v_n)$ and $E = (e_1, \ldots, e_m)$. Represent each vertex v_i, $1 \leq i \leq n$, by the line-segment with endpoints $(1, i, 0)$ and $(m, i, 0)$. As shown in Fig. 6, draw the edge $e_k = v_i v_j$ with the 2-bend edge route

$$(k, i, 0) \rightarrow (k, i, 1) \rightarrow (k, j, 1) \rightarrow (k, j, 0) \ .$$

Clearly there are no edge crossings. We thus have the following result.

Theorem 2. *Every graph has a 2-bend box-drawing in an $m \times n \times 2$ grid.* □

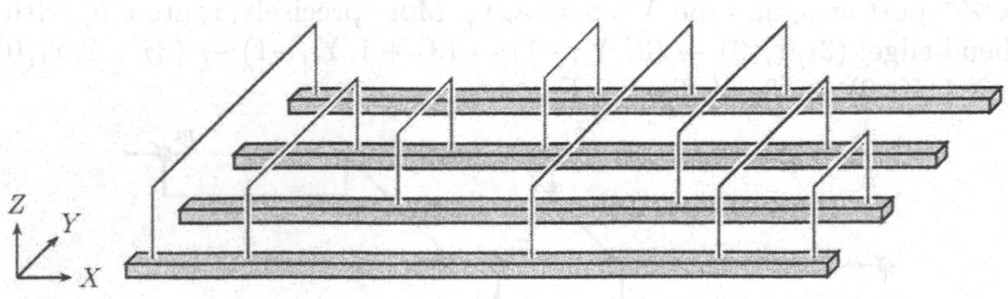

Fig. 6. A box-drawing in an $m \times n \times 2$ grid.

Lifting 2-dimensional drawings: Another method for producing box-drawings with two layers is to start with a 2-dimensional drawing with crossings. Vertices are then represented by boxes of height two, and edges are constructed by routing the X-segment of a 2-dimensional edge in the (Z=0)-plane and the Y-segment in the (Z=1)-plane connected by unit-length Z-segments at each bend. The resulting 3-dimensional edge has twice as many bends as the original 2-dimensional edge. This method was called *lifting half-edges* in [4]. We apply this method to the algorithms of Biedl and Kaufmann [6], Papakostas and Tollis [20], and Wood [28].

These algorithms all produce degree-restricted 2-dimensional drawings, hence in the resulting 3-dimensional drawings we have $X(v)+Y(v) \in O(\deg(v))$ for all vertices v. While this is not necessarily a degree-restricted drawing, we will see in Lemma 3 that $X(v) + Y(v) \in \Omega(\deg(v))$ is required for all drawings with a constant number of layers. Two of the algorithms also produce bounded aspect ratios, which is transferred while lifting them to the third dimension.

Theorem 3. *Every graph has:*

(a) a 2-bend drawing in a $\frac{m+n}{2} \times \frac{m+n}{2} \times 2$ *grid [6] (see also [4, Theorem 3]),*
(b) a 2-bend drawing in a $(m-1) \times (\frac{m}{2}+2) \times 2$ *grid [20],*
(c) a 2-bend drawing in a $(\frac{3}{4}m+\frac{1}{2}n) \times (\frac{3}{4}m+\frac{1}{2}n) \times 2$ *grid such that each vertex* v *has aspect ratio at most* $2+O(\frac{1}{deg(v)})$ *[6],*
(d) a 2-bend drawing in a $(\frac{3}{4}m+\frac{5}{8}n) \times (\frac{3}{4}m+\frac{5}{8}n) \times 2$ *grid such that every vertex has aspect ratio one [28].* □

4.1 Degree-Restricted Box-Drawings

We obtain results for degree-restricted box-drawings in two ways. One possible drawing is obtained by lifting the 2-dimensional drawing by Biedl and Kant [5]. This yields drawings with 4 bends per edge. Next, we describe an algorithm that uses only 3 bends per edge.

Lifting the drawing by Biedl/Kant: In [5], the first author and Kant gave an algorithm for 2-dimensional point-drawings, which can be extended to give 2-dimensional box-drawings of graphs with arbitrarily high degrees. The resulting grid size is $(m-n+1) \times (m-n/2+n_2/2)$, where n_2 is the number of vertices of degree 2. This drawing also has the property that vertex v is drawn as a $1 \times \lceil \deg(v)/2 \rceil$ line segment.

We can obtain a 3-dimensional drawing by applying the lifting half-edges technique. Every vertex now becomes a $1 \times \lceil \deg(v)/2 \rceil \times 2$ box; hence the drawing is 2-degree restricted. Since every edge in the 2-dimensional drawing has at most 2 bends, every edge in the resulting 3-dimensional drawing has at most 4 bends.

Theorem 4. *Every graph has a 4-bend box-drawing in a* $(m-n+1) \times m \times 2$*-grid such that every vertex is 2-degree restricted.* □

A drawing with 3 bends per edge: In this section we describe an algorithm for producing degree-restricted box-drawings with only 3 bends per edge.

Theorem 5. *Every graph has a 3-bend box-drawing in an* $(m+n) \times (m+\frac{3}{2}n+1) \times 2$ *grid such that every vertex is 3-degree restricted.*

Proof. Let $(v_1, \ldots, v_n)$ be a an arbitrary linear ordering of the vertices of G. As described in Lemma 2, colour the edges red and blue such that at most $\frac{1}{2}\deg(v)+1$ edges of the same colour are incident to v. Define $W(v) = \lfloor \frac{1}{2}\deg(v) \rfloor + 1$.

Represent v_i as a $W(v_i) \times 1 \times 2$-box, and place it such that its leftmost points have X-coordinate $\sum_{j<i} W(v_j)$ and Y-coordinate $Y_i = \sum_{j<i} W(v_j)$. All vertex boxes share the (Z=0)-plane and the (Z=1)-plane; see also Fig. 7.

Assign unique Y^- ports at v_i with $Z = 0$ (respectively, $Z = 1$) coordinates to the red successor (predecessor) edges of v_i. For each vertex v_i, denote by X_i the largest X-coordinate of the box of v_i. We route a red edge $e = v_i v_j$ ($i < j$) as follows: Assume that e was assigned the port with X-coordinate $X_i - \alpha$ at v_i and $X_j - \beta$ at v_j. Then we draw $v_i v_j$ with the 3-bend edge route: $(X_i - \alpha, Y_i, 0) \to (X_i - \alpha, Y_i - \alpha - 1, 0) \to (X_j - \beta, Y_i - \alpha - 1, 0) \to (X_j - \beta, Y_i - \alpha - 1, 1) \to (X_j - \beta, Y_j, 1)$, as illustrated in Fig. 7.

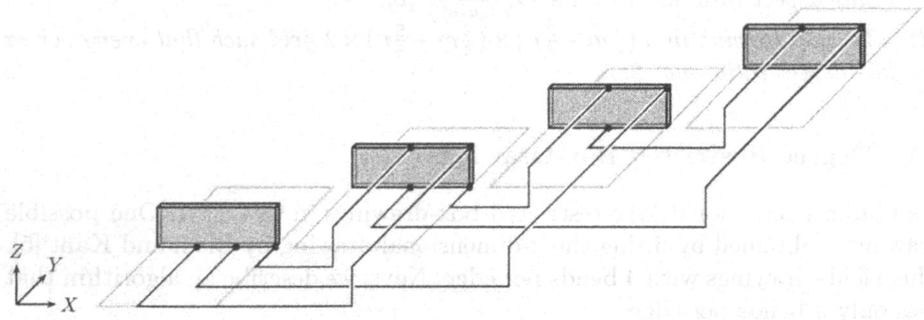

Fig. 7. A selection of red edge routes in a degree-restricted box-drawing.

Blue edges are routed similarly using Y^+ ports. For space reasons we omit the proof that edges do not cross.

The width of the drawing is $\sum_{i=1}^{n} W(v_i) \leq \sum_{i=1}^{n}(\frac{1}{2}\deg(v_i)+1) \leq m+n$, and the depth is at most $\sum_{i=1}^{n} W(v_i)+W(v_n) \leq m+n+\frac{1}{2}\deg(v_n)+1 \leq m+\frac{3}{2}n+1$. □

Note that we can obtain a smaller volume by using a vertex ordering and edge colouring that guarantees that the number of red predecessors and successors is not too unbalanced. If we use the *median placement heuristic* (see [2]) to obtain such an ordering, one can show that the width reduces to $\frac{3m}{4} + \frac{9n}{8}$, and the depth becomes $\frac{3m}{4} + \frac{13n}{8} + 1$.

5 Box Drawing Lower Bounds

In this section, we give lower bounds. These lower bounds hold under the assumption that no vertices are "above each other", defined precisely as follows: The *airspace* of a vertex v are all those points that have a common X-coordinate and a common Y-coordinate with v; see also Fig. 8. We say that a drawing has *no vertices above each other* if the airspaces of any two vertices are distinct.

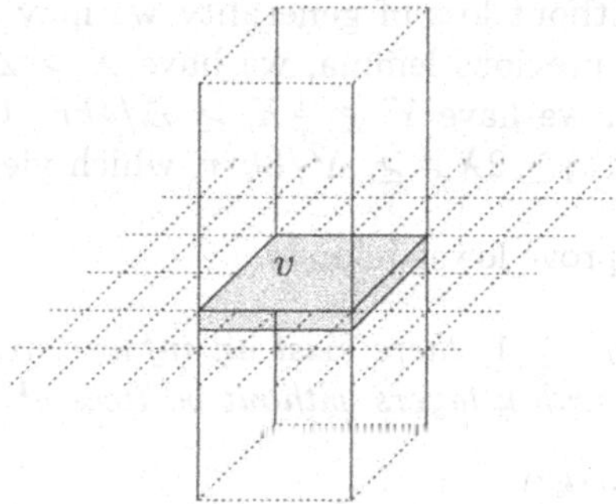

Fig. 8. The airspace of a vertex v.

Note that all our algorithms produce drawings without vertices above each other. Also, in VLSI design, vertices would normally not be put above each other to avoid interference. Likewise, for visualisation purposes vertices above each other might easily obscure each other, which should be avoided. Hence, assuming that no vertices are above each other is not an unreasonable assumption.

We start with a lemma discussing the dimensions of each vertex.

Lemma 3. *Assume that Γ is a drawing with k layers, with no vertices above each other. Then $X(v) + Y(v) \geq \deg(v)/2k$ for every vertex v.*

Proof. Let v be an arbitrary vertex. Let P_1 be the set of X-ports and Y-ports of v, and let P_2 be those points on the boundary of the airspace of v that are not ports of v. Note that $|P_1 \cup P_2| \leq k \cdot 2(X(v) + Y(v))$, since there are k layers.

We claim that each incident edge e of v must use one element of P_1 or P_2, without counting elements in P_1 or P_2 repeatedly. This holds if the port of e at v is an X-port or a Y-port, because then we assign to e the port in P_1. If the port of e at v is a Z-port, then e must somewhere enter the airspace of v. (Note that e must be outside v's airspace at the other endpoint, since no two vertices have intersecting airspaces.) We assign this point of entry into the airspace of v to e. No two such edges can use the same point because edge routes are disjoint.

Thus, we must have $k \cdot 2(X(v) + Y(v)) \geq \deg(v)$, which yields the claim. □

As a consequence of this lemma, no drawing with $O(1)$ layers can be both degree-restricted and have bounded aspect ratios at the same time.

Theorem 6. *Let Γ be a drawing in k layers without vertices above each other. If Γ is strictly α-degree-restricted and has aspect ratios at most r, then*

$$\alpha \geq \Delta/8k^2r,$$

where Δ is the maximum degree of the graph. In particular, not all of α, k, r can be constant unless Δ is a constant.

Proof. Let v be a vertex of maximum degree, and assume v is represented by an $X \times Y \times Z$-box. Without loss of generality we may assume $X \geq Y$. Since $X + Y \geq \Delta/2k$ by the previous lemma, we have $X \geq \Delta/4k$. Since the aspect ratio of v is at most r, we have $Y \geq \frac{1}{r}X \geq \Delta/4kr$. The surface of v hence satisfies $\alpha \cdot \Delta \geq \text{surface}(v) \geq 2XY \geq \Delta^2/8k^2r$, which yields the claim. □

Now we proceed to prove lower bounds.

Theorem 7. *For every $k \geq 1$, there exist an infinite number of graphs G such that for any drawing Γ with k layers without vertices above each other*

- *Γ has volume $\Omega(mn/k^5)$.*
- *if Γ has aspect ratios at most r, then Γ has volume $\Omega(m^2/k^5r)$.*
- *if Γ is strictly α-degree-restricted, where $\alpha \in o(n/k^3)$, then Γ has volume $\Omega(m^2/k^5)$.*

Proof. We use as graphs the so-called Ramanujan-graphs; see [16] for their definition and [8,1] for some of their properties. For our proof, all we need to know is that for a fixed k, there exists an infinite number of Ramanujan-graphs such that for any two vertex sets V_1, V_2 with $|V_1|, |V_2| \geq n/4k$ there are at least $C \cdot m/k^2$ edges between V_1 and V_2, for some constant C independent of k. Let G_k be such a graph; we know that G_k is d-regular for some constant d, so $m = dn/2$.

Consider an arbitrary drawing Γ of G_k without vertices above each other, and assume it is contained in an $X \times Y \times k$-grid. Similar as in lower bound proofs in [7,8], we show a lower bound by distinguishing whether many vertices are intersected by one grid-line or not. For space reasons we omit rounding details and assume that n is divisible by $4k$.

Case 1. One grid line intersects at least $n/2k$ vertices:
Assume that there exists a grid line, say an X-line, that intersects at least $n/2k$ vertices. Let $v_1, \ldots, v_t$ be the vertices intersected by the X-line, listed in order of occurrence along the line. Let X_0 be a not necessarily integer X-coordinate such that the $(X{=}X_0)$-plane intersects none of these t vertices and separates the first $n/4k$ of them from the remaining ones, of which there are at least $n/4k$. We will refer to the first set as V_+ and the second set as V_-.

By assumption at least $C \cdot m/k^2$ edges connect V_+ and V_-. These edges cross the $(X{=}X_0)$-plane, which thus must contain at least $C \cdot m/k^2$ points having integer Y- and Z-coordinates. Hence $Yk \geq C \cdot m/k^2$. The three different claims are now proved as follows:

- The X-line intersects the vertices $v_1, \dots, v_t$, $t \geq n/2k$, hence $X \geq t \geq n/2k$ and the volume of Γ is $XYk \geq Cmn/2k^3 \in \Omega(mn/k^3)$.
- If we know a bound r on the aspect ratio, then $Y(v_i) \leq rX(v_i)$, and therefore by Lemma 3, $X(v_i) \geq d/2k(1+r)$. Therefore $X \geq X(v_1) + \cdots + X(v_t) \geq dt/2k(1+r) \geq dn/4k^2(1+r) = m/2k^2(1+r)$ which yields $XYk \geq Cm^2/3k^4(1+r) \in \Omega(m^2/k^4r)$.
- Assume that Γ is strictly α-degree-restricted; we may assume $\alpha \leq Cn/4k^3$ by $\alpha \in o(n/k^3)$. Let Y_0 be the Y-coordinate of the X-line, and define $Y_+ = Y_0 + \alpha d/2$ and $Y_- = Y_0 - \alpha d/2$. Note that $v_1, \dots, v_t$ are contained in the range $Y_- \leq Y \leq Y_+$, since a vertex with surface at most αd extends at most $\alpha d/2$ in Y-direction. Define P to be the grid points (see also Figure 9)
$$P = \{(X,Y,Z) : X{<}X_0, Y{=}Y_-\} \cup \{(X,Y,Z) : X{<}X_0, Y{=}Y_+\}$$
$$\cup \{(X,Y,Z) : X{=}X_0, Y_- \leq Y \leq Y_+\}.$$

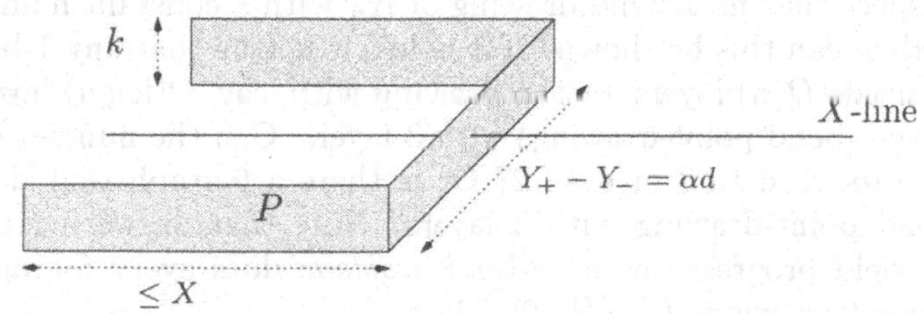

Fig. 9. The set P separates V_- from V_+.

The points in P separate the vertices in V_- from the vertices V_+. Hence, the Cm/k^2 edges between V_- and V_+ must use a grid point in P, so $|P| \geq Cm/k^2$. Note that $|P| \leq 2Xk + k\alpha d \leq 2Xk + Cdn/4k^2 = 2Xk + Cm/2k^2$ by $\alpha \leq Cn/4k^3$. Therefore $X \geq Cm/2k^3$, and $XYk \geq C^2m^2/2k^5 \in \Omega(m^2/k^5)$.

Case 2: No grid line intersects many vertices:

Now assume that no grid line intersects at least $n/2k$ vertices. Since there are at most k Z-planes, there must exist a $(Z{=}Z_0)$-plane that intersects at least n/k vertices.

As an $(X{=}X_0)$-plane is swept from smaller to larger values of X_0, the Y-line determined by the intersection of this $(X{=}X_0)$-plane with the $(Z{=}Z_0)$-plane sweeps the $(Z{=}Z_0)$-plane. At any time, this Y-line intersects at most $n/2k$ vertices by assumption. We can therefore find a (not necessarily integral) value X_0 such that there are at least $n/4k$ vertices to the left of the $(X{=}X_0)$-plane and not intersected by it, and there are at least $n/4k$ vertices to the right of the $(X{=}X_0)$-plane and not intersected by it. (See [7] for details of finding X_0.)

By assumption at least Cm/k^2 edges connect the vertices to the left and to the right of the $(X{=}X_0)$-plane. These edges cross the $(X{=}X_0)$-plane at a grid point, hence $Yk \geq Cm/k^2$. Similarly one shows $Xk \geq Cm/k^2$, therefore $XYk \geq C^2m^2/k^5 \in \Omega(m^2/k^5)$. This proves all claims. □

6 Conclusion

In this paper, we have studied 3-dimensional orthogonal graph drawings with few layers. We gave algorithms both for point-drawings (using 3 layers) and for box-drawings (using 2 layers). Note that one cannot hope for fewer layers, unless one allows edges to overlap each other, or crossings to occur. Our constructions are optimal with respect to the volume, as they match (within a constant) the lower bounds, some of which were provided in this paper as well. Some open problems that deserve attention are outlined in the following:

- What results can be shown for dynamic drawings with few layers? The algorithm in [9] can be extended to a dynamic setting by adding one more bend per edge. Is this possible for our algorithms as well? Note that in a dynamic setting we cannot rely on the cycle-decomposition of Lemma 1, as updating a cycle-decomposition appears to be impossible in constant time.
- We suspect that no 1-bend drawing of K_n with a constant number of layers exist. How can this be shown? If it holds, is it true that any 1-bend drawing of K_n needs $\Omega(n)$ layers, or is a drawing with, say, $O(\log n)$ layers possible?
- We gave 4-bend point-drawings with 3 layers. Can the number of bends per edge be reduced to 3 or even 2? Or is there a 6-graph that does not have a 2-bend point-drawing with 3 layers? Note that answering this question would yield progress on the *2-bend problem*: does every 6-graph have a 2-bend point-drawing? (See [13,25,30]).

References

1. T. Biedl. 1-bend 3-D orthogonal box-drawings: Two open problems solved. *J. Graph Algorithms Appl.*, 5(3):1–15, 2001.
2. T. Biedl, T. Chan, Y. Ganjali, M. Hajiaghayi, and D. R. Wood. Balanced vertex-orderings of graphs. Technical Report CS-AAG-2001-01, Basser Department of Computer Science, The University of Sydney, Australia, 2001.
3. T. C. Biedl. Heuristics for 3D-orthogonal graph drawings. In *Proc. 4th Twente Workshop on Graphs and Combinatorial Optimization*, pages 41–44, June 1995.
4. T. C. Biedl. Three approaches to 3D-orthogonal box-drawings. In Whitesides [24], pages 30–43.
5. T. Biedl and G. Kant A Better Heuristic for Orthogonal Graph Drawings. *Comput. Geom. Theory Appl.* 9:158–180, 1998.
6. T. C. Biedl and M. Kaufmann. Area-efficient static and incremental graph drawings. In R. Burkhard and G. Woeginger, editors, *Proc. Algorithms: 5th Annual European Symp. (ESA'97)*, volume 1284 of *Lecture Notes in Comput. Sci.*, pages 37–52. Springer, 1997.
7. T. C. Biedl, T. Shermer, S. Whitesides, and S. Wismath. Bounds for orthogonal 3-D graph drawing. *J. Graph Algorithms Appl.*, 3(4):63–79, 1999.
8. T. Biedl, T. Thiele, and D. R. Wood. Three-dimensional orthogonal graph drawing with optimal volume. In Marks [18], pages 284–295.
9. M. Closson, S. Gartshore, J. Johansen, and S. K. Wismath. Fully dynamic 3-dimensional orthogonal graph drawing. *J. Graph Algorithms Appl.*, 5(2):1–34, 2000.

10. G. Di Battista, M. Patrignani, and F. Vargiu. A split&push approach to 3D orthogonal drawing. *J. Graph Algorithms Appl.*, 4(3):105–133, 2000.
11. P. Eades, C. Stirk, and S. Whitesides. The techniques of Komolgorov and Bardzin for three dimensional orthogonal graph drawings. *Inform. Proc. Lett.*, 60(2):97–103, 1996.
12. P. Eades, A. Symvonis, and S. Whitesides. Two algorithms for three dimensional orthogonal graph drawing. In S. North, editor, *Proc. Graph Drawing: Symp. on Graph Drawing (GD'96)*, volume 1190 of *Lecture Notes in Comput. Sci.*, pages 139–154. Springer, 1997.
13. P. Eades, A. Symvonis, and S. Whitesides. Three dimensional orthogonal graph drawing algorithms. *Discrete Applied Math.*, 103:55–87, 2000.
14. K. Hagihara, N. Tokura, and N. Suzuki. Graph embedding on a three-dimensional model. *Systems-Comput.-Controls*, 14(6):58–66, 1983.
15. A. N. Kolmogorov and Ya. M. Barzdin. On the realization of nets in 3-dimensional space. *Problems in Cybernetics*, 8:261–268, March 1967.
16. A. Lubotzky, R. Phillips, and P. Sarnak, Ramanujan graphs. *Combinatorica* **8** (1988), 261–277.
17. B. Y. S. Lynn, A. Symvonis, and D. R. Wood. Refinement of three-dimensional orthogonal graph drawings. In Marks [18], pages 308–320.
18. J. Marks, editor. *Proc. Graph Drawing: 8th International Symp. (GD'00)*, volume 1984 of *Lecture Notes in Comput. Sci.* Springer, 2001.
19. A. Papakostas and I. G. Tollis. Algorithms for incremental orthogonal graph drawing in three dimensions. *J. Graph Algorithms Appl.*, 3(4):81–115, 1999.
20. A. Papakostas and I. G. Tollis. Efficient orthogonal drawings of high degree graphs. *Algorithmica*, 26:100–125, 2000.
21. J. Peterson. Die Theorie der regulären Graphen. *Acta. Math.*, 15:193–220, 1891.
22. C. Ware and G. Franck. Viewing a graph in a virtual reality display is three times as good as a 2D diagram. In A. L. Ambler and T. D. Kimura, editors, *Proc. IEEE Symp. Visual Languages (VL'94)*, pages 182–183. IEEE, 1994.
23. C. Ware and G. Franck. Evaluating stereo and motion cues for visualizing information nets in three dimensions. *ACM Trans. Graphics*, 15(2):121–140, 1996.
24. S. Whitesides, editor. *Proc. Graph Drawing: 6th International Symp. (GD'98)*, volume 1547 of *Lecture Notes in Comput. Sci.* Springer, 1998.
25. D. R. Wood. The DLM algorithm for three-dimensional orthogonal graph drawing in the general position model. Submitted; see Technical Report CS-AAG-2001-04, Basser Department of Computer Science, The University of Sydney, 2001.
26. D. R. Wood. Minimising the number of bends and volume in three-dimensional orthogonal graph drawings with a diagonal vertex layout. Submitted; see Technical Report CS-AAG-2001-03, Basser Department of Computer Science, The University of Sydney, 2001.
27. D. R. Wood. An algorithm for three-dimensional orthogonal graph drawing. In Whitesides [24], pages 332–346.
28. D. R. Wood. Multi-dimensional orthogonal graph drawing with small boxes. In J. Kratochvil, editor, *Proc. Graph Drawing: 7th International Symp. (GD'99)*, volume 1731 of *Lecture Notes in Comput. Sci.*, pages 311–222. Springer, 1999.
29. D. R. Wood. Bounded degree book embeddings and three-dimensional orthogonal graph drawing. Submitted, 2001.
30. D. R. Wood. Lower bounds for the number of bends in three-dimensional orthogonal graph drawings. In Marks [18], pages 259–271.

Bounded Degree Book Embeddings and Three-Dimensional Orthogonal Graph Drawing

David R. Wood*

Basser Department of Computer Science
The University of Sydney
Sydney NSW 2006
Australia

Abstract. A *book embedding* of a graph consists of a linear ordering of the vertices along a line in 3-space (the *spine*), and an assignment of edges to half-planes with the spine as boundary (the *pages*), so that edges assigned to the same page can be drawn on that page without crossings. Given a graph $G = (V, E)$, let $f : V \to \mathbb{N}$ be a function such that $1 \leq f(v) \leq \deg(v)$. We present a Las Vegas algorithm which produces a book embedding of G with $O(\sqrt{|E|} \cdot \max_v \lceil \deg(v)/f(v) \rceil)$ pages, such that at most $f(v)$ edges incident to a vertex v are on a single page. This algorithm generalises existing results for book embeddings. We apply this algorithm to produce 3-D orthogonal drawings with one bend per edge and $O(|V|^{3/2}|E|)$ volume, and *single-row* drawings with two bends per edge and the same volume. In the produced drawings each edge is entirely contained in some Z-plane; such drawings are without so-called *cross-cuts*, and are particularly appropriate for applications in multilayer VLSI. Using a different approach, we achieve two bends per edge with $O(|V||E|)$ volume but with cross-cuts. These results establish improved bounds for the volume of 3-D orthogonal graph drawings.

1 Introduction

This paper presents a Las Vegas algorithm for producing book embeddings of a graph with bounds on the number of edges incident to a vertex on a single page. This algorithm is used as the basis of algorithms for producing three-dimensional orthogonal graph drawings with one and two bends per edge. We focus on drawings appropriate for applications in multilayer VLSI. Throughout this paper, $G = (V, E)$ is a undirected simple connected graph. We denote the number of vertices of G by $n = |V|$, the number of edges of G by $m = |E|$, and the maximum degree of G by $\Delta(G)$, or Δ if the graph in question is clear.

1.1 Book Embeddings

A *book* consists of a line in 3-space, called the *spine*, and a number of *pages*, each a half-plane with the spine as boundary. A *book embedding* (π, ρ) of a

* Supported by the Australian Research Council. Completed while visiting the School of Computer Science, McGill University, Canada. davidw@cs.usyd.edu.au.

P. Mutzel, M. Jünger, and S. Leipert (Eds.): GD 2001, LNCS 2265, pp. 312–327, 2002.
© Springer-Verlag Berlin Heidelberg 2002

graph consists of a linear ordering π of the vertices, called the *spine ordering*, along the spine of a book and an assignment ρ of edges to pages so that edges assigned to the same page can be drawn on that page without crossings. That is, for any two edges vw and xy, if $v <_\pi x <_\pi w <_\pi y$ then $\rho(vw) \neq \rho(xy)$. The *book thickness* of a graph G is the minimum number of pages in a book embedding of G. For graphs with genus γ, Malitz [13] proved that the book thickness is $O(\sqrt{\gamma})$. Since $\gamma \leq m$, the book thickness is $O(\sqrt{m})$, a result proved independently by the same author [14].

Note that a book embedding may route all of the edges incident to a vertex on a single page. In this paper we study book embeddings where the number of edges incident to a vertex on a single page is bounded. We define the *pagedegree* of a vertex v to be the maximum number of edges incident to v on a single page. A *bounding function* of a graph $G = (V, E)$ is a function $f : V \rightarrow \mathbb{N}$ such that $1 \leq f(v) \leq \deg(v)$ for all vertices $v \in V$. For some bounding function f of G, a *degree-f* book embedding of G is one in which the pagedegree of every vertex v is at most $f(v)$. If for all vertices $v \in V$, $f(v) = c$ for some constant c, a degree-f book embedding is simply called a *degree-c* book embedding.

In this paper we establish that, for an arbitrary bounding function f of a graph G, there is a degree-f book embedding with $O(\sqrt{m\, Q_f(G)})$ pages, where $Q_f(G) = \max_{v \in V} \lceil \deg(v)/f(v) \rceil$. We describe a Las Vegas algorithm which determines this book embedding in $O(m \log^2 n \log\log m)$ time with high probability. This result, and its proof, generalises the above-mentioned $O(\sqrt{m})$ bound on the book thickness due to Malitz [14].

1.2 Three-Dimensional Orthogonal Graph Drawing

The *three-dimensional orthogonal grid* is the cubic lattice, consisting of *grid-points* with integer coordinates, together with the axis-parallel *grid-lines* determined by these points. We use the word *box* to mean a three-dimensional axis-parallel box with integral boundaries. Boxes are possibly degenerate, in the sense that they may be rectangles, line-segments or even a single grid-point. The number of grid-points along the edge of a box parallel to the X-axis (respectively, Y-axis and Z-axis) is called the *width* (*depth* and *height*) of the box. Note that this is one more than the actual length. A $W \times D \times H$ box has width W, depth D and height H. For each dimension $I \in \{X, Y, Z\}$, an *I-line* is a line parallel to the I-axis, an *I-segment* is a line-segment within an I-line, and an *I-plane* is a plane perpendicular to the I-axis.

A *three-dimensional orthogonal drawing* of a graph represents the vertices by pairwise non-intersecting boxes in the three-dimensional orthogonal grid. An edge vw is represented by a sequence of contiguous segments of grid-lines possibly bent at grid-points, between the surfaces of the boxes of v and w. The intermediate grid-points along the path representing an edge do not intersect the box of any vertex or any other edge route. From now on, we use the term *drawing* to mean a three-dimensional orthogonal drawing, and the graph-theoretic terms 'vertex' and 'edge' will also refer to their representation in a drawing.

The *volume* of a drawing is the number of grid-points in the smallest axis-aligned box, called the *bounding box*, which encloses the drawing. The volume and the maximum number of bends per edge are the most commonly proposed measures for determining the aesthetic quality of a drawing. A drawing with at most b bends per edge is called a *b-bend drawing.* For a graph G, denote by $\text{vol}(G, b)$ the minimum volume, taken over all b-bend drawings of G. Let $\text{vol}(n, m, b)$ be the maximum, taken over all graphs G with n vertices and m edges, of $\text{vol}(G, b)$. Thus, $\text{vol}(n, m, b)$ is a volume bound within which we can draw all graphs with n vertices and m edges, and with at most b bends per edge. This paper establishes improved upper bounds on $\text{vol}(n, m, 1)$ and $\text{vol}(n, m, 2)$.

A drawing with height k and with all vertices having height k is said to be in the *k-PCB* (*P*rinted *C*ircuit *B*oard) model, as defined by Aggarwal *et al.* [2]. Such drawings, which we call *multilayer drawings*, are an appropriate model for multilayer VLSI circuits. In multilayer VLSI, vertical edge segments between different Z-planes, called *cross-cuts*, lead to a deterioration in performance with an increase in the likelihood of faulty chips [2]. Therefore drawings without cross-cuts are particularly desirable. In this paper we observe that for drawings with a fixed maximum number of bends per edge, permitting cross-cuts allows for drawings with less volume.

We consider three types of multilayer drawings, which are defined by the relative positions of the vertices in a Z-plane. A multilayer drawing has a *two-dimensional general position* vertex layout if no two vertices are intersected by a single X- or Y-plane. For example, the vertices may have a two-dimensional *diagonal* layout (see Fig. 3). We say a multilayer drawing has a *linear* vertex layout if every vertex is intersected by a single X- or Y-plane (see Fig. 4). Multilayer drawings with a linear vertex layout are particularly appropriate in *single-row* VLSI routing problems (see [12] for example).Finally, in a *grid* vertex layout the vertices are positioned in a square grid (see [4,7]).

1-Bend Drawings: Biedl *et al.* [7] construct 1-bend multilayer drawings of the complete graph K_n with $O(n^3)$ volume; thus $\text{vol}(K_n, 1) \in O(n^3)$ and $\text{vol}(n, m, 1) \in O(n^3)$. Vertices are Z-lines of length n positioned in a 2-dimensional diagonal vertex layout, and each edge is routed in some Z-plane. Biedl *et al.* [7] suggest a relationship between the assignment of Z-planes to edges in orthogonal drawings and the assignment of pages to edges in book embeddings. Implicit in their 1-bend drawing of K_n is a degree-1 book embedding of K_n with n pages.

Drawings with one bend per edge were also studied by Wood [18], who shows that given a book embedding of a graph G with P pages, there is a 1-bend drawing of G with $O(nmP)$ volume. Applying the book embedding algorithm of Malitz [13], it follows that every graph with genus γ has a 1-bend drawing with $O(nm\sqrt{\gamma})$ volume, and since $\gamma \leq m$, it follows that $\text{vol}(n, m, 1) \in O(nm^{3/2})$.

The first contribution of this paper is an algorithm for producing 1-bend multilayer drawings without cross-cuts and with $O(mn^{3/2})$ volume; thus $\text{vol}(n, m, 1) \in O(mn^{3/2})$. The algorithm is based on our result for bounded degree book embeddings. Compared with the above-mentioned result of Wood [18], this represents an improvement by a factor of $\Theta\left(\sqrt{\frac{m}{n}}\right)$ for the volume of 1-bend drawings. For graphs with $m \in O(n^{3/2})$ this is the best known upper bound for the volume of 1-bend drawings. (For graphs with $m \in \Omega(n^{3/2})$, using the 1-bend drawing of K_n [7] produces a drawing with less volume.)

A lower bound for the volume of 1-bend drawings was established by Biedl [3], who shows that the Δ-regular n-vertex *Ramanujan* graph $G_{n,\Delta}$ has $\Omega(n^2\Delta)$ volume in any 1-bend drawing. Hence $\text{vol}(n, m, 1) \in \Omega(nm)$. Since $K_n = G_{n,n-1}$, the 1-bend drawing of K_n [7] has optimal volume; that is, $\text{vol}(K_n, 1) \in \Theta(n^3)$.

2-Bend Drawings: A 2-bend multilayer drawing of K_n in the linear layout model, with no cross-cuts and with $O(n^3)$ volume is presented by Biedl *et al.* [7]. The second contribution of this paper is an algorithm for producing 2-bend multilayer drawings with no cross-cuts and with $O(mn^{3/2})$ volume. This algorithm is based on our result for bounded degree book embeddings and employs a linear vertex layout. Again, for graphs with $m \in O(n^{3/2})$ this is an improved bound compared to the above result.

The algorithm of Biedl [5] produces 2-bend drawings with $O(n^2\Delta)$ volume in which vertices are *degree-restricted*; that is, the surface area of each vertex is proportional to its degree. This feature is more appropriate for applications in visualisation rather than VLSI. Strictly speaking, this algorithm does not produce multilayer drawings. However, a trivial modification does produce multilayer drawings while maintaining the volume bound, at the expense of losing the degree-restriction property. Regardless, this algorithm is not particularly appropriate for multilayer VLSI since the produced drawings have long cross-cuts.

The third contribution of this paper is an algorithm which produces 2-bend drawings with (short) cross-cuts and with $O(nm)$ volume. Hence $\text{vol}(n, m, 2) \in O(nm)$, which is an improvement on the above bound in [5] for all graphs. The crucial step in this algorithm is the application of an *equitable* edge-colouring result of Hakimi and Kariv [10].

3-Bend and 4-Bend Drawings: A 3-bend multilayer drawing of K_n with no cross-cuts and with $O(n^{5/2})$ volume is presented by Biedl *et al.* [7]; thus $\text{vol}(n, m, 3) \in O(n^{5/2})$. By placing the vertices in a $O(n^{1/2}) \times O(n^{1/2})$ grid and routing each edge in a distinct Z-plane, a simple algorithm by Biedl *et al.* [4] produces a 3-bend multilayer drawings without cross-cuts and with $O(nm)$ volume. A more complicated algorithm by the same authors produces 4-bend multilayer drawings with cross-cuts and with $O(mn^{1/2})$ volume. They also prove that the Ramanujan graph $G_{n,\Delta}$ requires $\Omega(\Delta n^{3/2})$ volume in any drawing. Thus $\text{vol}(n, m, b) \in \Theta(mn^{1/2})$ for all $b \geq 4$. Table 1 summarised the known upper bounds on the volume of drawings in the multilayer VLSI model.

The remainder of the paper is organised as follows. Our algorithm for producing degree-f book embeddings is presented in Section 2. In Section 3 and Section 4 we describe our algorithms for producing 1-bend and 2-bend drawings, respectively. In Section 5 we conclude with some open problems.

Table 1. Upper bounds on the volume of multilayer drawings.

# bends	bounding box	vol(n, m, b)	layout	cross-cuts	reference
1	$O(n) \times O(n) \times O(n)$	$O(n^3)$	diagonal	no	[7]
	$O(n) \times O(n) \times O(m^{3/2})$	$O(nm^{3/2})$	diagonal	no	[18]
	$O(n) \times O(n) \times O(mn^{-1/2})$	$O(n^{3/2}m)$	diagonal	no	Thm. 2
2	$O(n) \times O(n) \times O(n)$	$O(n^3)$	linear	no	[7]
	$O(n) \times O(n) \times O(mn^{-1/2})$	$O(n^{3/2}m)$	linear	no	Thm. 3
	$O(n) \times O(n) \times O(\Delta)$	$O(n^2\Delta)$	gen. pos.	yes	[5]
	$O(n) \times O(n) \times O(mn^{-1})$	$O(nm)$	diagonal	yes	Thm. 4
3	$O(n^{1/2}) \times O(n^{1/2}) \times O(n^{3/2})$	$O(n^{5/2})$	grid	no	[7]
	$O(n^{1/2}) \times O(n^{1/2}) \times O(m)$	$O(nm)$	grid	no	[4]
4	$O(n^{1/2}) \times O(n^{1/2}) \times O(mn^{-1/2})$	$O(n^{1/2}m)$	grid	yes	[4]

2 Bounded Degree Book Embeddings

This section describes a generalisation of the Las Vegas algorithm of Malitz [14] for producing degree-f book embeddings. The following definitions are from [14]. A *2-coloured bipartite graph* is a bipartite graph $G = (V_L \cup V_R, E)$ whose vertices have been coloured LEFT and RIGHT such that adjacent vertices are coloured differently. Note that a bipartite graph with k connected components has 2^k vertex 2-colourings. For some edge $e \in E$, $L(e)$ refers to the end-vertex of e in V_L, and $R(e)$ refers to the end-vertex of e in V_R. A *canonical ordering* of a 2-coloured bipartite graph $G = (V_L \cup V_R, E)$ is a linear ordering of the vertices of G such that all LEFT vertices precede all RIGHT vertices.

Let π be a canonical ordering of a 2-coloured bipartite graph $G = (V_L \cup V_R, E)$. Two edges vw and xy are said to *cross* if $v <_\pi x <_\pi w <_\pi y$. Two edges are *disjoint* if they have no common endpoint and they do not cross. Two edges *intersect* if they have a common endpoint or they cross. For (traditional) book embeddings the number of pairwise crossing edges provides a lower bound on the number of pages, whereas for degree-1 book embeddings the number of pairwise intersecting edges plays the same role. G is *completely intersecting* with respect to π if E can be labelled $e_1, e_2, \ldots, e_k$ such that

$$L(e_1) \leq_\pi L(e_2) \leq_\pi \cdots \leq_\pi L(e_k) \text{ and } R(e_1) \leq_\pi R(e_2) \leq_\pi \cdots \leq_\pi R(e_k) .$$

Intuitively, G is completely intersecting with respect to π, if in a degree-1 book embedding with π as the spine ordering, every edge must be placed on a unique page, as shown in Fig. 1.

Lemma 1. *If a 2-coloured bipartite graph G is completely intersecting with respect to some canonical ordering then G is a forest.*

Proof. Let π be a canonical ordering of G. Suppose to the contrary that G is not a forest and G is completely intersecting with respect to π. Then G contains

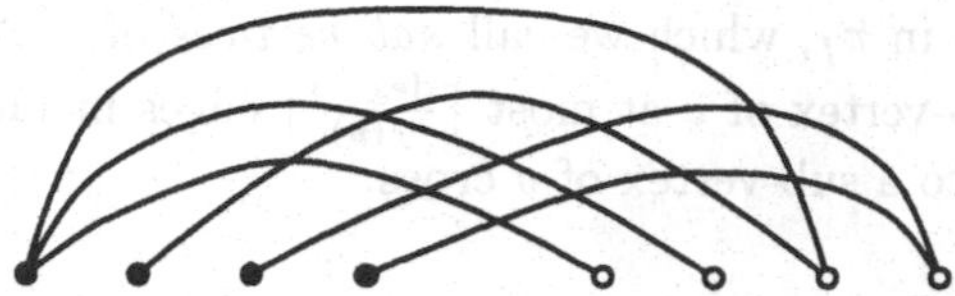

Fig. 1. A completely intersecting canonical ordering of a graph.

a cycle $(v_1, w_1, v_2, w_2, \ldots, v_k, w_k, v_{k+1})$ with $v_1 = v_{k+1}$ for some $k \geq 2$. Without loss of generality we can assume that v_1 is the leftmost vertex. We proceed by induction on i with the following induction hypothesis: "For every $i \geq 1$, $v_i <_\pi v_{i+1}$ and $w_i <_\pi w_{i+1}$."

To prove the basis of the induction, observe that if $w_2 <_\pi w_1$ then v_1w_1 does not intersect v_2w_2, and hence $w_1 <_\pi w_2$. By our initial assumption, $v_1 <_\pi v_2$. $w_1 <_\pi \cdots <_\pi w_i$. If $v_{i+1} <_\pi v_i$ then $v_{i+1}w_i$ does not intersect v_iw_{i-1}. Thus $v_i <_\pi v_{i+1}$. If $w_{i+1} <_\pi w_i$ then v_iw_i does not intersect $v_{i+1}w_{i+1}$. Thus $w_i <_\pi w_{i+1}$. Therefore the inductive hypothesis holds, which is a contradiction as it implies that $v_1 <_\pi v_{k+1}$ and $v_1 = v_{k+1}$. □

The next lemma for completely intersecting sets of edges, is the analogue of Lemma 2.2 in [14] for completely crossing sets of edges. Generalising a result of Tarjan [16], it says that book thickness can be determined efficiently if the spine ordering is a canonical ordering of a bipartite graph.

Lemma 2. *Let π be a canonical ordering of a 2-coloured bipartite graph $G = (V_L \cup V_R, E)$ with m edges and n vertices. If at most k edges are completely intersecting with respect to π, then a k-page degree-1 book embedding of G with spine ordering π can be determined in $O(m \log\log n)$ time.*

Proof. Define a poset $(E, \preceq)$ as follows. For all $e_1, e_2 \in E$ let

$$e_1 \preceq e_2 \stackrel{\text{def}}{=} e_1 = e_2 \text{ or } (L(e_2) <_\pi L(e_1) \text{ and } R(e_1) <_\pi R(e_2)) .$$

It is a simple exercise to check that $\leq$ is reflexive, transitive and antisymmetric, and thus is a partial order. Two edges are incomparable under $\preceq$ if and only if they intersect. Thus an antichain is a completely intersecting set of edges, and a chain is a set of pairwise disjoint edges. By Dilworth's Theorem [9] there is a decomposition of E into k chains where k is the size of the largest antichain. That is, there is a k-page degree-1 book embedding of G with spine ordering π. The time complexity can be achieved using a dual form of the algorithm by Heath and Rosenberg [11, Theorem 2.3]. □

An equivalent result to Lemma 2 is given by Malucelli and Nicoloso [15]. To enable this lemma to be extended to degree-f book embeddings, consider the following construction. Let π be a linear ordering of the vertices of a graph $G = (V, E)$, and suppose f is a bounding function of G. We define a graph $G_{\pi,f}$ and a linear ordering π_f of $G_{\pi,f}$ as follows. Replace each vertex $v \in V$ by $f(v)$

consecutive vertices in π_f, which we call *sub-vertices* of v. As shown in Fig. 2, connect to each sub-vertex of v at most $\lceil\frac{\deg(v)}{f(v)}\rceil$ edges incident to v so that no two edges incident to a sub-vertex of v cross.

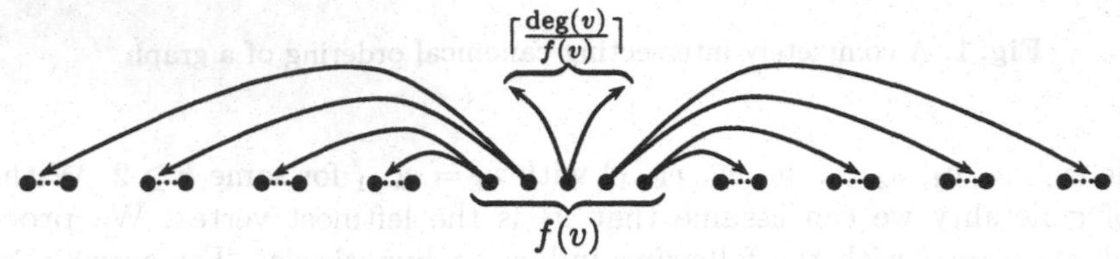

Fig. 2. Constructing π_f

Lemma 3. *Let f be a bounding function, and let π be a canonical ordering of a 2-coloured bipartite graph $G = (V_L \cup V_R, E)$ with m edges and n vertices. If at most k edges of $G_{\pi,f}$ are completely intersecting with respect to π_f, then a k-page degree-f book embedding of G with spine ordering π can be determined in $O(m \log\log(\sum_v f(v)))$ time.*

Proof. Apply Lemma 2 to $G_{\pi,f}$ with spine ordering π_f, to obtain a degree-1 book embedding (π_f, ρ) of $G_{\pi,f}$ with at most k pages. In (π_f, ρ), the pagedegree of a sub-vertex is at most one. Thus, in the book embedding (π, ρ) of G, the pagedegree of v is at most $f(v)$; that is, (π, ρ) is a degree-f book embedding of G. The time bound follows from Lemma 2 and that $G_{\pi,f}$ has $\sum_v f(v)$ vertices. □

To prove the main theorem of this section, we will consider a random linear ordering of V. Recall that $Q_f(G) = \max_v\lceil\frac{\deg(v)}{f(v)}\rceil$.

Lemma 4. *Let f be a bounding function and let π be a random canonical ordering of a 2-coloured forest $T = (V_L \cup V_R, E)$ with $n = |V_L \cup V_R|$ vertices. The probability that $T_{\pi,f}$ is completely intersecting with respect to π_f is at most*

$$\frac{2^n (Q_f(T))^{|E|}}{|E|!} .$$

Proof. The probability that $T_{\pi,f}$ is completely intersecting with respect to π_f is the number of canonical orderings π of T for which $T_{\pi,f}$ is completely intersecting with respect to π_f, divided by the number of canonical orderings of T. If $T_{\pi,f}$ is completely intersecting with respect to π_f then all edges incident to a vertex v must be incident to the same sub-vertex of v in π_f, and thus, T is completely intersecting with respect to π. (Note that this implies that $\Delta(T) \leq Q_f(T)$.) Thus, the desired probability is at most the number of canonical orderings π of T in which T is completely intersecting, divided by the number of canonical orderings of T.

We first bound the number of canonical orderings of T for which T is completely intersecting. Initially suppose T is connected; that is, $n = |E| + 1$. For some fixed ordering $(v_1, v_2, \ldots, v_l)$ of V_L, an ordering of V_R which makes T completely intersecting must be of the form

$$\{R(e) : v_1 \in e\}, \{R(e) : v_2 \in e\}, \ldots, \{R(e) : v_l \in e\} .$$

Similarly, if $(w_1, w_2, \ldots, w_r)$ is a fixed ordering of V_R, then an ordering of V_L which makes T completely intersecting must be of the form

$$\{L(e) : w_1 \in e\}, \{L(e) : w_2 \in e\}, \ldots, \{L(e) : w_l \in e\} .$$

The vertices within each set $\{R(e) : v_i \in e\}$ and $\{L(e) : w_i \in e\}$ possibly can be permuted. Thus the number of canonical orderings of T which are completely intersecting is at most $\prod_x \deg_T(x)!$.

We claim that $\prod_x \deg_T(x)! \leq \Delta(T)^{|E|}$. To prove this claim, we proceed by induction on $|E|$. The basis of the induction with $|E| = 1$ is trivial. Suppose for all connected trees $T' = (V', E')$ with $|E'| < |E|$ that $\prod_{x \in V'} \deg_{T'}(x)! \leq \Delta(T')^{|E'|}$. Let v be a leaf of T incident to the edge vw. Let $T' = (V', E') = T \setminus \{vw\}$. Since $\deg_{T'}(w) = \deg_T(w) - 1$, and by the inductive hypothesis applied to T',

$$\prod_{x \in V} \deg(x)! = \deg(w) \prod_{x \in V'} \deg_{T'}(x) \leq \deg(w) \cdot \Delta(T')^{|E|-1} \leq \Delta(T)^{|E|} . \quad (1)$$

Thus the claim is proved.

Now suppose T is disconnected. Then T has $n - |E|$ connected components. Suppose the connected components have edge sets $E_1, E_2, \ldots, E_{n-|E|}$. For T to be completely intersecting, the LEFT vertices in each connected component must be consecutive in the ordering, and similarly for the RIGHT vertices. Within V_L, the components can be ordered $(n - |E|)!$ different ways. For a fixed ordering of the connected components of V_L, for T to be completely intersecting, the components of V_R must be ordered the same way. By (1), the number of canonical orderings which are completely intersecting is at most

$$(n - |E|)! \prod_{i=1}^{n-|E|} \Delta(T)^{|E_i|} \leq (n - |E|)!\, \Delta(T)^{|E|} .$$

The number of canonical orderings of T is $|V_L|! \cdot |V_R|!$. Thus, the probability that a random canonical ordering of T is completely intersecting is at most

$$\frac{(n - |E|)!\, \Delta(T)^{|E|}}{|V_L|! \cdot |V_R|!} \leq \frac{(n - |E|)!\, \Delta(T)^{|E|}}{\left(\frac{n}{2}!\right)^2} \leq \frac{2^n\, (n - |E|)!\, \Delta(T)^{|E|}}{n!} \leq \frac{2^n\, \Delta(T)^{|E|}}{|E|!} ,$$

where the final three inequalities follow from well-known and easily proved facts concerning factorials. The result holds, since $\Delta(T) \leq Q_f(T)$. □

We now prove the main result of this section. It's proof is a generalisation of Theorem 2.3 in [14], which is based on ideas from Theorem 4.7 in Chung *et al.* [8].

Theorem 1. *Let f be a bounding function of a graph $G = (V, E)$ with m edges. There exists a degree-f book embedding of G with $O(\sqrt{m\, Q_f(G)})$ pages.*

Proof. Let $n' = |V|$, and denote $Q_f(G)$ by Q. Since G is connected, $m \geq n' - 1$. If $m = n' - 1$ then G is a tree. By considering a pre-order traversal of G, it is easily seen that G has a book embedding (π, ρ) with one page [8]. The graph $G_{\pi,f}$ is a forest with maximum degree Q, and thus has a edge-colouring χ with Q colours. A book embedding (π, χ) of G is a degree-f book embedding of G with $Q \leq \sqrt{\Delta Q} \leq \sqrt{m Q}$ pages. Thus the result is proved for trees.

Now assume $m \geq n'$. Let $n = 2^{\lceil \log n' \rceil}$, and add $n - n'$ isolated vertices to G. (Unless stated otherwise all logarithms are base 2.) G now has n vertices, with n a power of 2. Clearly, $n \leq 2n'$, and $n \leq 2m$.

Let π be a random linear ordering of V. For each j, $1 \leq j \leq \log n$, divide the linear ordering π into 2^j sections each with the same number of vertices, and label the sections from left to right L, R, L, R, etc. The edges whose endpoints are in adjacent L-R sections (but not adjacent R-L sections) are called *j-level* edges. Note that every edge of G appears in a unique level, and edges in adjacent L-R sections in some j-level are canonically ordered by π.

For each j, $1 \leq j \leq \log n$, let A_k^j be the event that there exists a k-edge 2-coloured subgraph T of G such that:

- T consists solely of j-level edges,
- T is canonically ordered with respect to π, and
- $T_{\pi,f}$ is completely intersecting with respect to π_f.

By Lemma 1, such a subgraph T is a forest. The probability that A_k^j occurs

$$\mathbf{P}\left\{A_k^j\right\} < \underbrace{\binom{m}{k} 2^k}_{(1)} \cdot \underbrace{2^{j-1}}_{(2)} \cdot \underbrace{\binom{\frac{n}{2^j}}{l}\binom{\frac{n}{2^j}}{r}\frac{l!\,r!\,(n-l-r)!}{n!}}_{(3)} \cdot \underbrace{\frac{2^{l+r}Q^k}{k!}}_{(4)} ,$$

where:

(1) is an upper bound on the number of k-edge 2-coloured forests T with no isolated vertices;
(2) is the number of pairs of adjacent L-R sections in the j-level;
(3) is an upper bound on the probability that π canonically orders T in the fixed pair of adjacent j-level sections, where T has l LEFT vertices and r RIGHT vertices; and
(4) is the probability that T is completely intersecting, by Lemma 4 and since $Q_f(T) \leq Q$.

Since $\binom{a}{b} \leq \frac{a^b}{b!}$,

$$\mathbf{P}\left\{A_k^j\right\} < \frac{(2m)^k}{k!} \cdot 2^{j-1} \cdot \left(\frac{n}{2^j}\right)^{l+r} \frac{(n-l-r)!}{n!} \cdot \frac{2^{l+r}Q^k}{k!} .$$

Applying Stirling's Formula, where $\mathbf{e}$ is the base of the natural logarithm,

$$\mathbf{P}\left\{A_k^j\right\} < (2m)^k \cdot 2^{j-1} \cdot \left(\frac{n}{2^j}\right)^{l+r} \sqrt{\frac{n-l-r}{n}} \left(\frac{n-l-r}{\mathbf{e}}\right)^{n-l-r} \left(\frac{\mathbf{e}}{n}\right)^n \cdot \frac{2^{l+r} Q^k \mathbf{e}^{2k}}{k^{2k+1}}.$$

Now, $n-l-r<n$. By elementary properties of a forest, $k+1 \le l+r \le 2k$. Since $l+r \le 2\frac{n}{2^j}$, we have $k \le \frac{n}{2^{j-1}}$, and hence $2^{j-1} \le \frac{n}{k} \le \frac{2m}{k}$. Thus,

$$\begin{aligned}\mathbf{P}\left\{A_k^j\right\} &< (2m)^{k+1} \cdot \left(\frac{1}{2^j}\right)^{k+1} n^{(l+r)+(n-l-r)-n} \cdot \mathbf{e}^{-(n-l-r)+n+2k} \cdot \frac{2^{2k} Q^k}{k^{2(k+1)}} \\ &< \left(\frac{8\mathbf{e}^4 m Q}{2^j k^2}\right)^{k+1}.\end{aligned}$$

Define $k_j = 4\mathbf{e}^2 \sqrt{\frac{mQ}{2^j}}$. Since $m \ge \frac{n}{2}$ and $Q \ge 1$,

$$\mathbf{P}\left\{A_{k_j}^j\right\} < \left(\frac{1}{2}\right)^{1+4\mathbf{e}^2\sqrt{mQ/2^j}} < \frac{1}{2}\left(\frac{1}{2}\right)^{2\sqrt{2}\,\mathbf{e}^2\sqrt{n/2^j}}.$$

Consider the event that $A_{k_j}^j$ occurs for some j, $1 \le j \le \log n$.

$$\mathbf{P}\left\{\bigcup_{j=1}^{\log n} A_{k_j}^j\right\} < \frac{1}{2} \sum_{j=1}^{\log n} \left(\frac{1}{2}\right)^{2\sqrt{2}\,\mathbf{e}^2\sqrt{n/2^j}}.$$

By induction on N, the following can be proved.

$$\forall a > 1, \forall b \ge \frac{\sqrt{2} - \log_a(a-1)}{\sqrt{2}-1}, \sum_{j=1}^{N} \left(\frac{1}{a}\right)^{b\sqrt{2^{N-j}}} < \left(\frac{1}{a}\right)^{b-1}.$$

Applying this fact with $N = \log n$, $a = 2$ and $b = 2\sqrt{2}\,\mathbf{e}^2$,

$$\mathbf{P}\left\{\bigcup_{j=1}^{\log n} A_{k_j}^j\right\} < \frac{1}{2}\left(\frac{1}{2}\right)^{2\sqrt{2}\,\mathbf{e}^2-1} = \left(\frac{1}{2}\right)^{2\sqrt{2}\,\mathbf{e}^2}.$$

Thus,

$$\mathbf{P}\left\{\bigcap_{j=1}^{\log n} \overline{A_{k_j}^j}\right\} = \mathbf{P}\left\{\overline{\bigcup_{j=1}^{\log n} A_{k_j}^j}\right\} = 1 - \mathbf{P}\left\{\bigcup_{j=1}^{\log n} A_{k_j}^j\right\} > 1 - \left(\frac{1}{2}\right)^{2\sqrt{2}\,\mathbf{e}^2} > 0.99999.$$

This says that for the random linear ordering π, with (very high) positive probability, $A_{k_j}^j$ does not occur for all j, $1 \le j \le \log n$. Therefore, there exists a linear ordering π' of V such that $A_{k_j}^j$ does not occur for all j. That is, in each pair of adjacent L-R sections in the j-level, there is no completely intersecting subgraph in π'_f with at least k_j edges. For each pair of adjacent L-R sections in level j, apply Lemma 3 to the subgraph of $G_{\pi',f}$ consisting of j-level edges with endpoints in that pair of sections (using the canonical ordering π'_f). By using the same set of pages for j-level edges, we obtain a degree-f book embedding of G with spine ordering π', and with the number of pages at most

$$\sum_{j=1}^{\log n} k_j = 4\mathrm{e}^2\sqrt{mQ}\sum_{j=1}^{\log n}\sqrt{\frac{1}{2^j}} < \frac{4\mathrm{e}^2\sqrt{mQ}}{\sqrt{2}-1} < 72\sqrt{mQ} \ . \qquad \square$$

Corollary 1. *Let f be a bounding function of graph $G = (V, E)$ with n vertices and m edges. There is a Las Vegas algorithm which will compute, with high probability, a degree-f book embedding of G with $O(\sqrt{m\,Q_f(G)})$ pages in $O(m\log^2 n \log\log m)$ time.*

Proof. Consider the following Las Vegas algorithm to compute the book embedding whose existence is proved in Theorem 1.

1. Choose a random linear ordering π of V.
2. Partition the edges into j-levels with respect to π.
3. Embed each set of j-level edges in its own set of pages (using Lemma 3 applied to $G_{\pi,f}$ as described above).
4. If the total number of pages is at most $\frac{4\mathrm{e}^2\sqrt{mQ}}{\sqrt{2}-1}$ then halt. Otherwise repeat from Step 1.

The time taken for each iteration within in each j-level is $O(m\log\log(\sum_v f(v)))$ by Lemma 3. Since $f(v) \leq \deg(v)$, $\sum_v f(v) \in O(m)$, and the time taken for each iteration is $O(m\log n\log\log m)$. For each iteration of the above algorithm, we say the algorithm *fails* if the randomly chosen linear ordering π does not admit a degree-f book embedding with at most $(4\mathrm{e}^2\sqrt{mQ})/(\sqrt{2}-1)$ pages. The probability of failure is at most $2^{-2\sqrt{2}\mathrm{e}^2}$. If we repeat the above algorithm at most $\log n$ times, the probability of failure every time is at most $2^{-2\sqrt{2}\mathrm{e}^2\log n} = n^{-2\sqrt{2}\mathrm{e}^2} \to 0$ as $n \to \infty$. Thus, with probability tending to 1 as $n \to \infty$, the above algorithm will determine a degree-f book embedding of G with at most $72\sqrt{m\,Q_f(G)}$ pages in $O(m\log^2 n\log\log m)$ time. $\square$

Note that Theorem 1 with the bounding function $f(v) = \deg(v)$ is the same result proved by Malitz [14], and the above proof is based on Malitz's idea of defining j-levels and applying Dilworth's Theorem to a partial ordering of the edges in each level. However, our proof differs in two respects. First, we do not assume that $j \leq k$, as is the case in [14, page 76] (also see [13, page 92]). Furthermore, we do not use a book embedding of the complete graph $K_{\sqrt{n}}$ for levels $j = \frac{1}{2}\log n + 1, \frac{1}{2}\log n + 2, \ldots, \log n$.

3 1-Bend Drawings

The following simple result highlights the relationship between degree-1 book embeddings and 1-bend drawings.

Lemma 5. *If a graph $G = (V, E)$ has a degree-1 book embedding (π, ρ) with P pages then it has a 1-bend drawing with $O(n^2 P)$ volume.*

Proof. Let $(v_1, v_2, \ldots, v_n)$ be the numbering of vertices corresponding to the linear ordering π. Represent vertex v_i by a Z-line at (i, i), and draw an edge $v_i v_j$ $(i < j)$ with the route $(i, i, \rho(vw)) \to (j, i, \rho(vw)) \to (j, j, \rho(vw))$. Two edges can only intersect if they have the same Z-coordinate; that is they are on the same page. Such an intersection would imply a crossing in the book embedding. Hence no two edges routes intersect; see Fig. 3(a). □

Given a book embedding (π, ρ) of a graph $G = (V, E)$, consider the following elementary method for producing a degree-1 book embedding. First, apply Vizing's Theorem [17] to obtain an edge-colouring χ of G with $\Delta(G) + 1$ colours. (In fact a greedy edge-colouring with $O(()\Delta)$ colours will suffice.) Let $\rho'(vw) = \rho(vw) \cdot \chi(vw)$ for all edges $vw \in E$. Then (π, ρ') is a degree-1 book embedding with $O(\Delta P)$ pages. By Lemma 5, there is a 1-bend drawing with $O(n^2 \Delta P)$ volume. This bound is reduced to $O(nmP)$ by Wood [18]. The results in [18] discussed in the introduction follow from the bounds on P in [13,14].

The following algorithm for producing 1-bend drawings exploits a degree-f book embedding where $f(v)$ is proportional to the ratio of the degree of v and the average degree of the graph. In the produced drawing, the width and depth of a vertex v equals $f(v)$. Edges in the same page are routed in a single Z-plane.

Theorem 2. *Let $G = (V, E)$ be a graph with $n = |V|$ vertices and $m = |E|$ edges. There is a 1-bend drawing of G with $O(mn^{3/2})$ volume.*

Proof. Let f be the bounding function defined by $f(v) = \lceil \frac{n}{2m} \deg(v) \rceil$ for all vertices $v \in V$. Then

$$Q_f(G) = \max_{v \in V} \frac{\deg(v)}{\lceil \frac{n}{2m} \deg(v) \rceil} \leq \frac{2m}{n} .$$

That is, $Q_f(G)$ is at most the average degree of G. By Theorem 1, there is degree-f book embedding (π, ρ) of G with $P = O(\sqrt{m \frac{m}{n}}) = O(mn^{-1/2})$ pages.

For each vertex v, let $S_v = \sum_{w <_\pi v} f(w)$. Represent v by the $f(v) \times f(v) \times P$ box with minimum corner at $(S_v, S_v, 0)$. Clearly, vertices do not intersect. The ith *successor* of v on a page p is the edge vx_i in the list $vx_1, vx_2, \ldots, vx_k$ of edges incident to v on page p such that $v <_\pi x_1 <_\pi x_2 <_\pi \cdots <_\pi x_k$. The ith *predecessor* of v on page p is the edge vx_i in the list $vx_1, vx_2, \ldots, vx_k$ of edges incident to v on page p such that $x_k <_\pi x_{k-1} <_\pi \cdots <_\pi x_1 <_\pi v$. For every edge vw $(v <_\pi w)$, if vw is the ith successor of v on page $\rho(vw)$ and the jth predecessor of w on page $\rho(vw)$, then draw vw with the 1-bend edge route $(S_v + f(v) - 1,\ S_v + f(v) - i,\ \rho(vw)) \to (S_w + j - 1,\ S_v + f(v) - i,\ \rho(vw)) \to (S_w + j - 1,\ S_w,\ \rho(vw))$, as illustrated in Fig. 3(b).

Two edges can only intersect if they have the same Z-coordinate; that is they are on the same page. Such an intersection would imply a crossing in the book embedding. Hence no two edges intersect. The width and depth of the bounding box is $\sum_v \left(\lceil \frac{n}{2m} \deg(v) \rceil + 1 \right) \leq 2n + \frac{n}{2m}(2m) = 3n$. The height of the bounding box is $P = O(mn^{-1/2})$. Thus the volume is $O(mn^{3/2})$. □

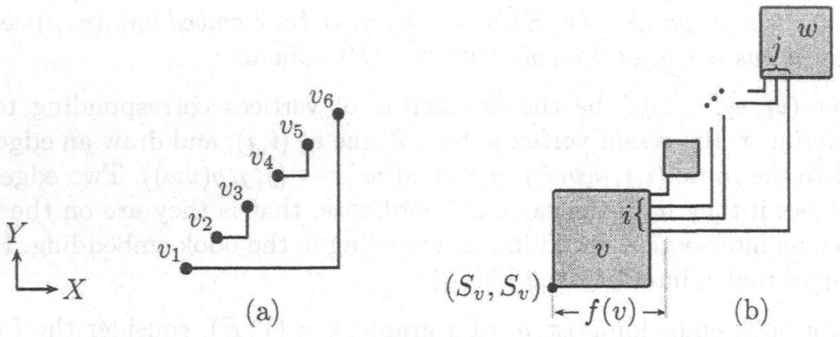

Fig. 3. 1-bend edge routes within a Z-plane.

4 2-Bend Drawings

We now present our algorithms for producing 2-bend drawings, the first with a linear vertex layout, and the second with a diagonal vertex layout.

Theorem 3. *Every graph $G = (V, E)$ with n vertices and m edges has a 2-bend linear drawing with $O(mn^{3/2})$ volume.*

Proof. Let f be the same bounding function defined in Theorem 2. By Theorem 1 there is degree-f book embedding (π, ρ) of G with $P = O(mn^{-1/2})$ pages. For each vertex $v \in V$, let $S_v = \sum_{w <_\pi v} f(w)$. Represent v by the $f(v) \times 1 \times P$ box with minimum corner at $(S_v, 0, 0)$. Clearly, vertices do not intersect.

Denote by $\omega(vw)$, the *pagewidth* of an edge vw in (π, ρ); that is, the maximum number of edges cut by a line-segment contained in $\rho(vw)$, perpendicular to the spine ordering, and with endpoints on vw and the spine.

For every vertex $v \in V$ and page p, the ith *neighbour* of v on page p is the edge vx_i in the list $vx_1, vx_2, \ldots, vx_k$ of edges incident to v on page p such that $x_1 <_\pi x_2 <_\pi \cdots <_\pi x_k$. For every edge vw ($v <_\pi w$), if vw is the ith neighbour of v on page $\rho(vw)$, and vw is the jth neighbour of w on page $\rho(vw)$, then draw vw with the 2-bend edge route $(S_v + i - 1,\ 0,\ \rho(vw)) \rightarrow (S_v + i - 1,\ \omega(vw),\ \rho(vw)) \rightarrow (S_w + j - 1,\ \omega(vw),\ \rho(vw)) \rightarrow (S_w + j - 1,\ 0,\ \rho(vw))$, as illustrated in Fig. 4.

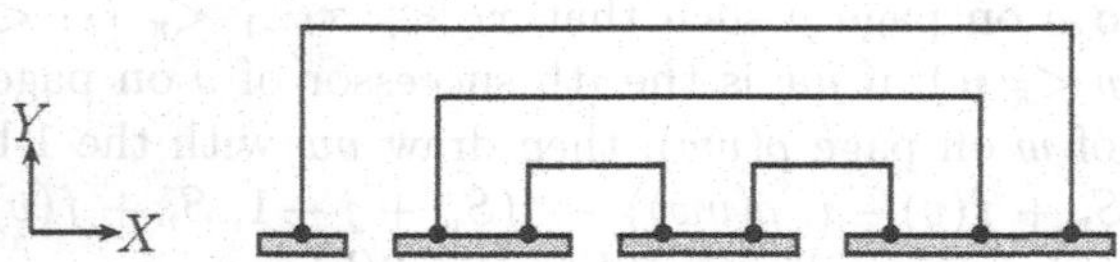

Fig. 4. A Z-plane within a 2-bend linear drawing

Two edges can only intersect if they have the same Z-coordinate; that is, they are on the same page of the book embedding. Clearly Y-segments do not intersect. An X-segment intersecting a Y-segment would imply a crossing in the book embedding. Two X-segments on the same page either have no X-coordinate in common, or they are nested, and therefore have different pagewidth, and thus have different Y-coordinates. Hence no two edges routes intersect.

The maximum pagewidth and hence the depth of the bounding box is at most the width of the bounding box. As in Theorem 2, the width and hence the depth of the bounding box is $O(n)$. The height of the bounding box is $P = O(mn^{-1/2})$, and thus the volume is $O(mn^{3/2})$. □

The next drawing algorithm, which does not use a book embedding, exploits a (non-proper) edge-colouring in which the colours are evenly distributed about the edges incident to each vertex. The colour of an edge determines its 'height' in the drawing.

Theorem 4. *Every graph $G = (V, E)$ with n vertices and m edges has a 2-bend multilayer drawing with cross-cuts and with $O(nm)$ volume.*

Proof. Hakimi and Kariv [10, Theorem 3] prove that for every $k \in \mathbb{N}$, every graph has a (non-proper) edge k colouring such that the number of monochromatic edges incident to a vertex v is at most $\lceil \frac{1}{k}(\deg(v) + 1)\rceil$. Apply this result with $k = \lceil \frac{2m}{n} \rceil$, to obtain an edge $\lceil \frac{2m}{n} \rceil$-colouring χ of G, such that the number of monochromatic edges incident to a vertex v is at most $\lceil (\deg(v) + 1)/\lceil \frac{2m}{n} \rceil \rceil \leq \lceil \frac{n}{2m}(\deg(v)+1)\rceil$. Let π be an arbitrary linear ordering of V. For each vertex v, let $S_v = \lceil \frac{n}{2m}(\deg(v)+1)\rceil$ and $T_v = \sum_{w<_\pi v} S_w$. Represent v by the $S_v \times S_v \times 2\lceil \frac{2m}{n} \rceil$ box with minimum corner at $(T_v, T_v, 0)$. Clearly, vertices do not intersect.

Define the ith *successor* and ith *predecessor* of a vertex v with colour c as in Theorem 2 but with "colour" replacing "page". For every edge vw $(v <_\pi w)$, if vw is the ith successor of v coloured $\chi(vw)$, and vw is the jth predecessor of w coloured $\chi(vw)$, then draw vw with the 2-bend edge route $(T_v + S_v - 1, T_v + S_v - i, 2\,\chi(vw)) \to (T_w + j - 1, T_v + S_v - i, 2\,\chi(vw)) \to (T_w + j - 1, T_v + S_v - i, 2\,\chi(vw) - 1) \to (T_w + j - 1, T_w, 2\,\chi(vw) - 1)$.

The ith successor of a vertex v coloured c is a unique edge, and similarly for predecessors. Thus two X-segments have different Z-coordinates or different Y-coordinates, and any two Y-segments have different Z-coordinates or different X-coordinates. Hence such edge-segments do not intersect. Since X-segments have even Z-coordinates, and Y-segments have odd Z-coordinates, no X-segment intersects a Y-segment. The unit-length Z-segment in each edge does not intersect any other edges as this would imply that one of the adjacent X- or Y-segments would have been involved in an intersection. Thus no two edges intersect. The width and depth of the bounding box is $\sum_v \left(\lceil \frac{n}{2m}(\deg(v) + 1)\rceil + 1\right) \leq 2n + \frac{n}{2m}(2m+n) \leq 2n + \frac{n}{2m}(4m) = 4n$. The height of the bounding box is $O(\frac{m}{n})$. Therefore the volume is $O(nm)$. □

5 Conclusion

We have presented a Las Vegas algorithm for producing book embeddings with bounds on the pagedegree of each vertex. This algorithm is used to produce

1-bend and 2-bend drawings without cross-cuts. Using an approach based on equitable edge-colourings we described an algorithm for producing 2-bend drawings with cross-cuts and with the best known volume upper bound. For all of our algorithms constant-factor improvements are easily possible. For example, in the 1-bend algorithm half the pages can be routed in the space with Y-coordinate greater than X-coordinate. In the 2-bend algorithm with a diagonal layout, 2-D general position vertex layouts [6] can be used to reduce the width and depth.

We finish with some open problems. First, what are values of $\mathrm{vol}(n,m,1)$, $\mathrm{vol}(n,m,2)$ and $\mathrm{vol}(n,m,3)$? The best known bounds are $\Omega(nm) \ni \mathrm{vol}(n,m,1) \in O(\min\{mn^{3/2}, n^3\})$, and for $b \in \{2,3\}$, $\Omega(n^{1/2}m) \ni \mathrm{vol}(n,m,b) \in O(nm)$. For $b \geq 4$, $\mathrm{vol}(n,m,b) \in \Theta(mn^{1/2})$. The algorithm in [4] produces multilayer drawings with $O(mn^{1/2})$ volume but with cross-cuts. Does every graph have a multilayer drawing with no cross-cuts and with $O(mn^{1/2})$ volume? Some progress towards a positive answer to this question is presented in [19]. If the answer is negative, is there a lower bound of $\Omega(nm)$ for the volume of multilayer drawings with no cross-cuts regardless of the number of bends?

Acknowledgements. Thanks to Therese Biedl for stimulating discussions on this topic, and to everyone in the School of Computer Science at McGill University, especially Sue Whitesides, for their generous hospitality.

References

1. *Proc. EuroConference on Combinatorics, Graph Theory and Applications (COMB'01)*, Electronic Notes in Discrete Mathematics, to appear.
2. A. Aggarwal, M. Klawe, and P. Shor. Multilayer grid embeddings for VLSI. *Algorithmica*, 6(1):129–151, 1991.
3. T. Biedl. 1-bend 3-D orthogonal box-drawings: Two open problems solved. *J. Graph Algorithms Appl.*, 5(3):1–15, 2001.
4. T. Biedl, T. Thiele, and D. R. Wood. Three-dimensional orthogonal graph drawing with optimal volume. In J. Marks, editor, *Proc. Graph Drawing: 8th International Symp. (GD'00)*, volume 1984 of *Lecture Notes in Comput. Sci.*, pages 284–295. Springer, 2001.
5. T. C. Biedl. Three approaches to 3D-orthogonal box-drawings. In S. Whitesides, editor, *Proc. Graph Drawing: 6th International Symp. (GD'98)*, volume 1547 of *Lecture Notes in Comput. Sci.*, pages 30–43. Springer, 1998.
6. T. C. Biedl and M. Kaufmann. Area-efficient static and incremental graph drawings. In R. Burkhard and G. Woeginger, editors, *Proc. Algorithms: 5th Annual European Symp. (ESA'97)*, volume 1284 of *Lecture Notes in Comput. Sci.*, pages 37–52. Springer, 1997.
7. T. C. Biedl, T. Shermer, S. Whitesides, and S. Wismath. Bounds for orthogonal 3-D graph drawing. *J. Graph Algorithms Appl.*, 3(4):63–79, 1999.
8. F. R. K. Chung, F. T. Leighton, and A. L. Rosenberg. Embedding graphs in books: a layout problem with applications to VLSI design. *SIAM J. Algebraic Discrete Methods*, 8(1):33–58, 1987.
9. R. P. Dilworth. A decomposition theorem for partially ordered sets. *Ann. of Math. (2)*, 51:161–166, 1950.

10. S. L. Hakimi and O. Kariv. A generalization of edge-coloring in graphs. *J. Graph Theory*, 10(2):139–154, 1986.
11. L. S. Heath and A. L. Rosenberg. Laying out graphs using queues. *SIAM J. Comput.*, 21(5):927–958, 1992.
12. E. S. Kuh, T. Kashiwabara, and T. Fujisawa. On optimum single-row routing. *IEEE Trans. Circuits and Systems*, 26(6):361–368, 1979.
13. S. M. Malitz. Genus g graphs have pagenumber $O(\sqrt{g})$. *J. Algorithms*, 17(1):85–109, 1994.
14. S. M. Malitz. Graphs with E edges have pagenumber $O(\sqrt{E})$. *J. Algorithms*, 17(1):71–84, 1994.
15. F. Malucelli and S. Nicoloso. Optimal partition of a bipartite graph into non-crossing matchings. In [1].
16. R. Tarjan. Sorting using networks of queues and stacks. *J. Assoc. Comput. Mach.*, 19:341–346, 1972.
17. V. G. Vizing. On an estimate of the chromatic class of a p-graph. *Diskret. Analiz No.*, 3:25–30, 1964.
18. D. R. Wood. Three-dimensional orthogonal graph drawings with one bend per edge. Submitted. See *Three-Dimensional Orthogonal Graph Drawing.* Ph.D. thesis, School of Computer Science and Software Engineering, Monash University, Australia, 2000.
19. D. R. Wood. Geometric thickness in a grid of linear area. In [1].

Straight-Line Drawings on Restricted Integer Grids in Two and Three Dimensions (Extended Abstract)*

Stefan Felsner[1], Giuseppe Liotta[2], and Stephen Wismath[3]

[1] Freie Universität Berlin, Fachbereich Mathematik und Informatik, Takustr. 9, 14195 Berlin, Germany. felsner@inf.fu-berlin.de.

[2] Dipartimento di Ingegneria Elettronica e dell'Informazione, Università degli Studi di Perugia, Via Duranti, 06100 Perugia, Italy. liotta@diei.ing.unipg.it.

[3] Dept. of Mathematics and Computer Science, U. of Lethbridge, Alberta, T1K-3M4 Canada. wismath@cs.uleth.ca.

Abstract. This paper investigates the following question: Given an integer grid ϕ, where ϕ is a proper subset of the integer plane or a proper subset of the integer 3d space, which graphs admit straight-line crossing-free drawings with vertices located at the grid points of ϕ? We characterize the trees that can be drawn on a two dimensional $c \cdot n \times k$ grid, where k and c are given integer constants, and on a two dimensional grid consisting of k parallel horizontal lines of infinite length. Motivated by the results on the plane we investigate restrictions of the integer grid in 3 dimensions and show that every outerplanar graph with n vertices can be drawn crossing-free with straight lines in linear volume on a grid called a prism. This prism consists of $3n$ integer grid points and is universal – it supports all outerplanar graphs of n vertices. This is the first algorithm that computes crossing-free straight line 3d drawings in linear volume for a non-trivial family of planar graphs. We also show that there exist planar graphs that cannot be drawn on the prism and that the extension to a $n \times 2 \times 2$ integer grid, called a box, does not admit the entire class of planar graphs.

1 Introduction

This paper deals with crossing-free straight-line drawings of planar graphs in 2 and 3 dimensions. Given a graph G, we constrain the vertices in a drawing of G to be located at integer grid points and aim at computing drawings whose area/volume is small. A rich body of literature has been published on such straight-line drawings in 2d. Typically, these papers focus on lower bounds on

* Research supported in part by the CNR Project "Geometria Computazionale Robusta con Applicazioni alla Grafica ed al CAD", the project "Algorithms for Large Data Sets: Science and Engineering" of the Italian Ministry of University and Scientific and Technological Research (MURST 40%); and by the Natural Sciences and Engineering Council of Canada.

P. Mutzel, M. Jünger, and S. Leipert (Eds.): GD 2001, LNCS 2265, pp. 328–342, 2002.
© Springer-Verlag Berlin Heidelberg 2002

the area required by drawings of specific classes of graphs and on the design of algorithms that possibly match these lower bounds. A very limited list of mile-stone papers in this field includes the works by de Fraysseix, Pach, and Pollack [7,8] and by Schnyder [20] who independently showed that every n-vertex triangulated planar graph has a crossing-free straight-line drawing such that the vertices are at grid points, the size of the grid is $O(n) \times O(n)$, and that this is worst case optimal; the work by Kant [16,17], Chrobak and Kant [3], Schnyder and Trotter [21], Felsner [12] and Chrobak, Goodrich, and Tamassia [4] who studied convex grid drawings of triconnected planar graphs in an integer grid of quadratic area; and the many papers proving that linear or almost-linear area bounds can be achieved for classes of trees, including the result by Garg, Goodrich and Tamassia [13] and the result by Chan [2]. Summarizing tables and more references can be found in the book by Di Battista, Eades, Tamassia, and Tollis [9].

While the problem of computing small-sized crossing-free straight-line drawings in the plane has a long tradition, the 3d counterpart has received less attention. Chrobak, Goodrich, and Tamassia [4] gave an algorithm for constructing 3d convex drawings of triconnected planar graphs with $O(n)$ volume and non-integer coordinates. Cohen, Eades, Lin and Ruskey [6] showed that every graph admits a straight-line crossing-free 3d drawing on an integer grid of $O(n^3)$ volume, and proved that this is asymptotically optimum. Calamoneri and Sterbini [1] showed that all 2-, 3-, and 4-colourable graphs can be drawn in a 3d grid of $O(n^2)$ volume with $O(n)$ aspect ratio and proved a lower bound of $\Omega(n^{1.5})$ on the volume of such graphs. For r-colourable graphs, Pach, Thiele and Tóth [18] showed a bound of $\theta(n^2)$ on the volume. Garg, Tamassia, and Vocca [14] showed that all 4-colorable graphs (and hence all planar graphs) can be drawn in $O(n^{1.5})$ volume and with $O(1)$ aspect ratio but using a grid model where the coordinates of the vertices may not be integral.

In this paper we study the problem of computing drawings of graphs on integer 2d or 3d grids that have small area/volume. The area/volume of a drawing Γ is measured as the number of grid points contained in or on a *bounding box* of Γ, *i.e.* the smallest axis-aligned box enclosing Γ. Note that along each side of the bounding box the number of grid points is one more than the actual length of the side.

We approach the drawing problem with the following point of view: Instead of "squeezing" a drawing onto a small portion of a grid of unbounded dimensions, we assume that a grid of *specified* dimensions is given and we consider which graphs have drawings that fit that restricted grid. For example, it is well-known that there are families of graphs that require $\Omega(n^2)$ area to be drawn in the plane, the canonical example being a sequence of $n/3$ nested triangles (see [8, 5,20]). Such graphs can be drawn on the surface of a 3 dimensional triangular prism of linear volume and using integer coordinates. Thus a natural question is whether there exist specific restrictions of the 3d integer grid of linear volume that can support straight-line crossing-free drawings of meaningful families of graphs. For planar graphs the best known results for 3 dimensional crossing-free straight-line drawings on an integer grid are by Calamoneri and Sterbini [1] who show $O(n^2)$ volume for general planar graphs and by Eades, Lin and Ruskey [6] who show $O(n \log n)$ volume for trees. By following the above described approach we have been able to design the first algorithms that draw significant families of planar graphs in an integer 3d grid requiring only $O(n)$ volume.

The main contributions of this paper are combinatorial characterizations and negative results on the drawability of graphs on 2d and 3d restricted integer grids and new drawing algorithms for some classes of graphs. An overview of the results is as follows:

- We characterize those trees that can be drawn on an integer restricted 2d grid consisting of k consecutive infinite horizontal grid lines (for a given positive integer k) and where edges can connect either collinear vertices or vertices that are one unit apart in their y-coordinates; we also present a linear time recognition and drawing algorithm for this class of trees.
- We study those trees that can be drawn on an integer restricted 2d grid of dimensions $c \cdot n \times k$, where c and k are two given integer positive constants and n is the number of vertices of the tree. In this case we relax one of our drawing constraints to allow adjacent vertices to be more than one unit apart in their y-coordinates, and show that this family of drawable trees coincides with those studied within the drawing convention of the previous item. A consequence of our characterization is that for any given k and c there always exist some trees that are not drawable on the $c \cdot n \times k$ grid.
- Motivated by the results on restricted integer 2d grids we explore the capability of restricted integer 3d grids for supporting linear volume drawings of graphs. In particular, we focus on two types of 3d integer grids to be defined subsequently, both having linear volume, called the *prism* and the *box*. We show that all outerplanar graphs can be drawn in linear volume on a prism. Note that this is the first result on 3d straight-line drawings of a significant class of planar graphs that achieves linear volume with integer coordinates.
- We further explore the class of graphs that can be drawn on a prism by asking whether the prism is a *universal* integer 3d grid for all planar graphs. We answer this question in the negative by exhibiting examples of planar graphs that cannot be drawn on a prism. We also investigate the relationship between prism-drawable and Hamiltonian graphs.
- We extend our study to box-drawability and present a characterization of the box-drawable graphs. While the box would appear to be a much more powerful grid than the prism, we prove that not all planar graphs are box-drawable.

2 Preliminaries

We assume familiarity with basic graph drawing, and computational geometry terminology; see for example [19,9]. Since in the remainder of the paper we shall study crossing-free straight-line drawings of planar graphs, from now on we shall simply talk about "graphs" to mean "planar graphs" and about "drawings" to mean "crossing-free straight line drawings". We use the terms "vertex" and "edge" for both the graph and its drawing.

We will draw graphs such that vertices are located at integer grid points. The dimensions of a grid are specified as the number of different grid points along each side of a bounding box of the grid. In 2 dimensions, a $p \times q$ grid consists of p grid points along the x-axis and of q grid points along the y-axis. In 3 dimensions, a $p \times q \times r$ grid consists of p grid points along the x-axis, q grid points along the y-axis, and r grid points along the z-axis; p, q and r are referred to as the x-, y-, and z-*dimension* of the grid, respectively.

We shall deal with the following grids and drawings.

- A 2d 1-*track* (or simply a *track*) is a $\infty \times 1$ grid; a 1-*track* drawing of a graph G is a drawing of G where the vertices are at distinct grid points of the track.
- A 2d *strip* is a $\infty \times 2$ grid; note that a strip consists of two tracks. A *strip drawing* of a graph G is a drawing of G with the vertices located at distinct grid points of the strip and the edges either connect vertices on the same track or connect vertices on different tracks.
- Let k be a given positive integer value. A 2d k-*track* grid is a $\infty \times k$ grid consisting of k consecutive parallel tracks. A k-*track drawing* of a graph G is a drawing of G where the vertices are at distinct grid points of the k-track and edges are only permitted between vertices that are either on the same track or that are one unit apart in their y-coordinates.
- Let k and c be two given positive integer values. In a $c \cdot n \times k$-*grid drawing* of a graph G the vertices are located at distinct grid points and the edges can connect any pair of vertices on that grid.
- We will also study two different types of $n \times 2 \times 2$ grids. A *box* is a $n \times 2 \times 2$ grid where each side of the bounding box is also a grid line. Therefore, a box has four tracks which lie on two parallel planes and are one grid unit apart from each other. A *prism* is a $n \times 2 \times 2$ grid obtained by removing a track from a box. Figure 1 shows an example of a box of size $4 \times 2 \times 2$ and an example of a prism.

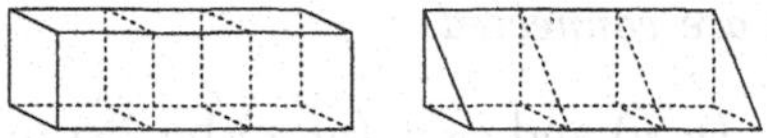

Fig. 1. A Box and a Prism

Note that k-track drawings differ from the so-called k-level drawings (see, e.g. [15]) as in a k-track drawing (consecutive) vertices on the same track are permitted to be joined by an edge and the given graph is undirected.

Let ϕ be one of the grids defined above. We say that a graph G is ϕ *drawable* if G admits a ϕ drawing Γ where each vertex is mapped to a distinct grid point of ϕ.

Property 1. A graph is 1-track drawable if and only if it is a simple path

While in a k-track drawing no edge can connect vertices that are on non-consecutive tracks, in a $c \cdot n \times k$-grid this is allowed. As the following property shows, this difference has immediate consequences on the families of k-track drawable and $c \cdot n \times k$-grid drawable graphs, and the graph K_4 provides an example.

Property 2. Let c, k be two positive integers. There exist graphs with n vertices that are $c \cdot n \times k$-grid drawable but are not k-track drawable.

3 Two-Dimensional Restricted Grids

In this section we characterize the family of k-track drawable trees and the family of $c \cdot n \times k$-grid drawable trees. In contrast to Property 2, we show that these two families of trees are actually the same. We also give linear time recognition and drawing algorithms for these trees. The approach is as follows. We first study strip-drawable trees, then we extend the result to the k-track grid, and finally we show that the result also holds on a $c \cdot n \times k$ grid.

3.1 Strip-Drawable Trees

By Property 1 we have that all paths are strip-drawable, since they are in fact 1-track drawable. A tree is defined as *2-strict* if it contains a vertex of degree greater than or equal to three. An immediate consequence of Property 1 is the following.

Property 3. A 2-strict tree is not 1-track drawable.

An edge is defined as a *core edge* if its removal results in two 2-strict components. For an edge $e = (u, v)$, we refer to the two subtrees resulting from its removal as T_u and T_v.

Lemma 1. *Core edges are connected.*

Proof. (sketch) Let $e_1 = (u, v)$ and $e_2 = (w, x)$ be 2 core edges and consider any edge on the tree between them. Each such edge receives one 2-strict component from T_u and one from T_x and thus must be core.

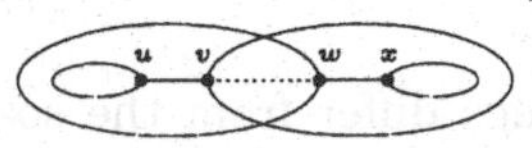

Fig. 2. Core edges are connected

Lemma 2. *A tree is strip drawable if and only if its core edges form a path.*

Proof. (sketch) ($\Rightarrow$) (by contradiction) By the previous lemma, if the core edges do not form a path, then there is a vertex v with at least 3 incident core edges $(v, a), (v, b), (v, c)$ – see Figure 3. If the subtrees T_a, T_b, T_c are drawable then by Property 3 their associated drawings $\Gamma_a, \Gamma_b, \Gamma_c$ each require 2 tracks. There is no location for v that permits a crossing-free connection to all 3 subdrawings.

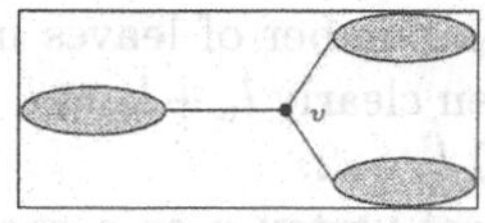

Fig. 3. T is not strip-drawable if the core edges are not a path

($\Leftarrow$) If the core edges form a path of at least 2 vertices, then draw them consecutively on track t_1. Consider an arbitrary non-core edge $e = (u, v)$ with u on track t_1. Since e is non-core, T_v must *not* be 2-strict and is thus 1-track

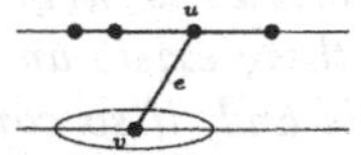

Fig. 4. Drawing a tree on a strip

drawable. Therefore v can be placed on track t_2 with the drawing of T_v also on the same track as in Figure 4.

There is one degenerate case to consider. If there are no core edges (*i.e.* a path of length 0), then either the tree has no vertex of degree 3 and is in fact 1-track drawable, or there exists at most one vertex v with neighbours $w_1, w_2, \ldots w_k$ and each T_{w_i} is *not* 2-strict. Each of the subtrees can thus be drawn on track t_2 and v on track t_1 as in Figure 5.

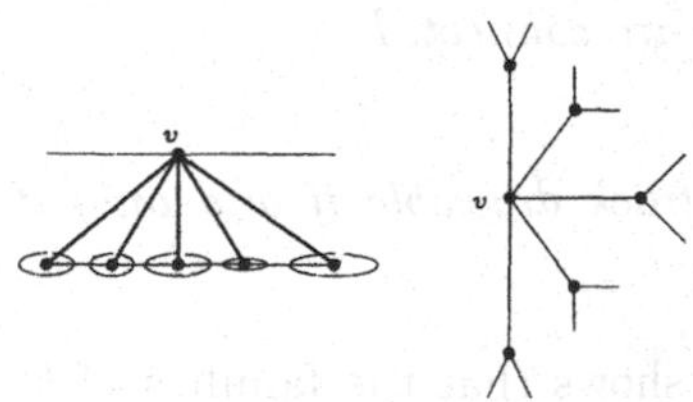

Fig. 5. A degenerate core path

Lemma 3. *Let T be a tree with n vertices. There exists an $O(n)$-time algorithm that recognizes whether T is strip-drawable and, if so, computes a strip drawing of T.*

Proof. (sketch) Note that a tree is 2-strict iff it has more than 2 leaves; thus counting leaves is the crucial operation. First the core edges must be established and then the path condition on the core edges checked.

With each edge $e = (u, v)$ we associate 2 counters: l_u will be the number of leaves in T_u, and l_v will be the number of leaves in T_v. Let l be the number of leaves in the entire tree T. Then clearly $l_u + l_v = l$. By the previous observation, e is a core edge iff both l_u and $l_v > 2$.

Choose an arbitrary non-leaf vertex r as a root. Each vertex v reports the number of leaves in the subtree below it to its parent u – thus establishing l_u for the edge (u, v) and hence l_v. If v has no children then it is a leaf and reports 1. A simple recursive function can be used to implement this counting step in linear time.

Finally, checking that the core edges form a path is also easily accomplished in linear time as is the production of a strip drawing.

Theorem 1. *A tree T with n vertices is strip drawable if and only if its core edges form a path. Furthermore, there exists an $O(n)$-time algorithm that determines whether T is strip drawable and, if so computes a strip drawing of T.*

3.2 k-Track and $c \cdot n \times k$-Grid Drawable Trees

The results of Theorem 1 can be extended to k tracks by generalizing some of the concepts of the previous section. A tree is *k-strict* if it contains a vertex adjacent to at least three subtrees that are $(k-1)$-strict. An edge is a *k-core edge* if its removal results in two k-strict components. The proofs of some of the following lemmas are similar to the case when $k = 2$ and are omitted in this extended abstract.

Property 4. A k-strict tree is not $(k-1)$-track drawable.

Lemma 4. *k-core edges are connected.*

Lemma 5. *A tree is k-track drawable if and only if the k-core edges form a path.*

The following lemma shows that the families of k-track drawable trees and $c \cdot n \times k$-grid drawable trees coincide.

Lemma 6. *Let c, k be two positive integer. A tree T is $c \cdot n \times k$-grid drawable if and only if it is k-track drawable.*

Proof. (sketch) Since a k-track drawing is a restricted form of a $c \cdot n \times k$ grid drawing where c is any given positive integer constant, it suffices to show that if T has a $c \cdot n \times k$ grid drawing, then the k-core edges form a path which can be shown by contradiction – the proof follows the form of Lemma 2.

Theorem 2. *Let T be a tree with n vertices and let c and k be two positive integer constants. The following three statements are equivalent:*

1. *T is $c \cdot n \times k$-grid drawable.*
2. *T is k-track drawable.*
3. *the k-core edges of T form a path.*

Furthermore, there exists an $O(n)$-time algorithm that determines whether T satisfies the above conditions and computes a k-track and a $c \cdot n \times k$-grid drawing of T.

One consequence of the previous theorem is the existence of non-drawable trees – ternary trees for example provide the critical strict components.

Corollary 1. *The complete ternary tree of height k is not drawable on a $c \cdot n \times (k-1)$ grid, for any positive integer c.*

4 Three-Dimensional Drawings of Outerplanar Graphs

In this section we show that all outerplanar graphs are prism-drawable by providing a linear time algorithm that computes this drawing. This is the first known three-dimensional straight-line drawing algorithm for the class of outerplanar graphs that achieves $O(n)$ volume on an integer grid.

A high level description of our drawing algorithm, called `Algorithm Prism Draw`, is as follows. Let G be an outerplanar graph with a specified *outerplanar embedding*, i.e. a circular ordering of the edges incident around each vertex such that all vertices of G belong to the external face. `Algorithm Prism Draw` computes a prism drawing of G by executing two main steps. Firstly a 2d drawing of G is computed on a grid that consists of $O(n)$ horizontal tracks and such that adjacent vertices are at grid points whose y-coordinates differ by at most one by visiting G in a breadth-first fashion. Secondly, the drawing is "wrapped" on the faces of a prism by folding it along the tracks.

`Algorithm Prism Draw`

input: An outerplanar graph G with a given outerplanar embedding.
output: A prism drawing of G.

`Step 1. The 2d Drawing Phase:` A 2d grid drawing Γ of G where vertices are assigned to distinct tracks is computed as follows.

- Add a dummy vertex d on the external face and an edge connecting d to an arbitrary vertex v.

- mark d; i:=0; currx:=0
- draw v on track t_0 by setting X(v) := currx; Y(v) :=i
- currx := currx +1
- mark v
- while there are unmarked vertices of G do
  - visit the vertices on track t_i from left to right and for each encountered vertex u do
    - ∗ let w be a marked neighbour of u in G
    - ∗ visit the neighbours of u in counterclockwise order starting from w, and for each encountered vertex r such that r is unmarked do
      - · draw r on track t_{i+1} by setting X(r) := currx; Y(r) := i+1
      - · currx := currx +1
      - · mark r
  - i := i+1

`Step 2: The 3d Wrapping Phase`: A prism drawing Γ' is obtained by folding Γ along its tracks as follows.

- for each vertex v of Γ define its coordinates $X'(v)$, $Y'(v)$ and $Z'(v)$ in Γ' by setting:
  - $X'(v) := X(v)$
  - if $Y(v) = 0, 1$ *mod* 3 then $Y'(v) := 0$, else $Y'(v) := 1$
  - if $Y(v) = 0, 2$ *mod* 3 then $Z'(v) := 0$, else $Z'(v) := 1$

`End of Algorithm Prism Draw`

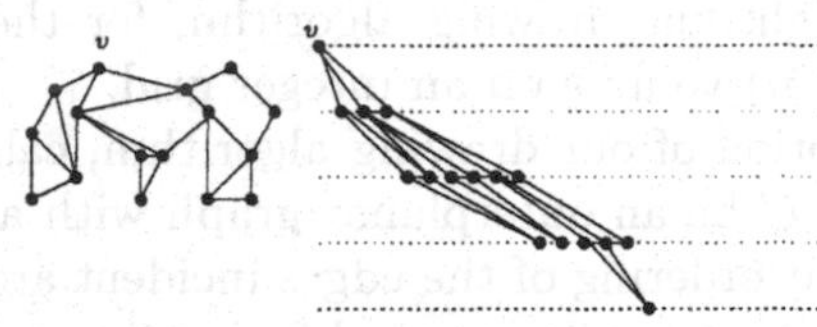

Fig. 6. An outerplanar graph drawn by `Step 1` of `Algorithm Prism Draw`.

Figure 6 shows an example of the output of Step 1 of the algorithm. The correctness of `Algorithm Prism Draw` is established via the following observations:

- No two vertices of Γ are assigned the same X-coordinate.
- Every vertex assigned to track t_{i+1} has a neighbour on track t_i, for $i \geq 0$.
- No pair of edges between vertices on adjacent tracks intersect.

Theorem 3. *Every outerplanar graph G with n vertices admits a crossing-free straight line grid drawing in 3 dimensions in optimal $O(n)$ volume. Furthermore, there exists an algorithm that computes such a drawing of G in $O(n)$ time and with the vertices of G drawn on the grid points of a prism.*

5 Prism-Drawable Graphs

Since by Theorem 3 all outerplanar graphs can be drawn on a prism, it is natural to investigate the class of graphs that are prism-drawable. For example, note that the family of planar graphs consisting of a sequence of nested triangles and that are known to require $\Omega(n^2)$ area in the plane, can be drawn on the prism (and thus have $O(n)$ volume). It is also clear that since any drawing on the prism can be augmented by edges to form a convex polytope, by the theorem of Steinitz only planar graphs are prism-drawable. In this section, we give a characterization for the prism-drawable graphs and show that not all planar graphs are in this class. Figure 7 shows a graph, and its prism drawing.

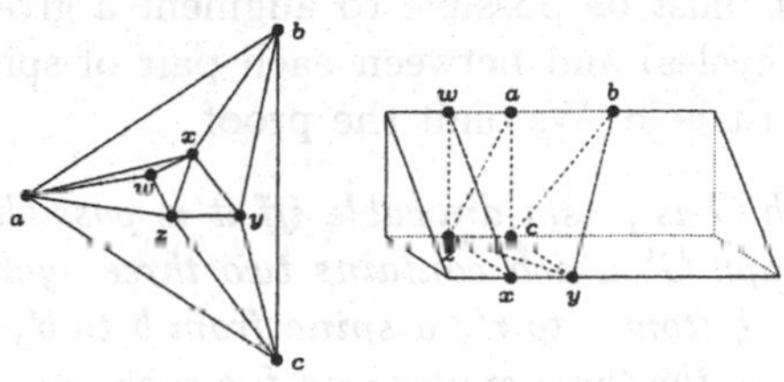

Fig. 7. A prism-drawable graph G and its drawing

5.1 Characterization of Prism-Drawable Graphs

An essential prerequisite of our characterization of prism-drawable graphs, is the study of the strip-drawable graphs since a prism effectively consists of three strips. We define a *spine* in a graph as a sequence of adjacent vertices with no chord. The characterization of strip-drawable graphs proposed in this section notes that in a strip drawing, there must exist 2 potential spines and that edges incident to vertices on both spines must not cross. This formulation will be generalized in subsequent sections to larger grid sets.

Theorem 4. *A graph G is strip-drawable iff it is possible to augment G with edges to produce a graph G' which contains two pairs of adjacent vertices r, b and r', b' and there exists a spine from r to r', a spine from b to b' with all vertices of G on the two spines and if there exists an edge (r_i, b_j) then there are no edges of the form (r_k, b_l) with ($k < i$ and $l > j$) or ($k > i$ and $l < j$).*

Proof. (sketch) Refer to Figure 8. Given a strip drawing, it is clear that edges can be added along the 2 tracks to form the 2 spines, and to add an edge between the leftmost pair on the two tracks and between the rightmost pair on the 2 tracks. Since no pair of edges intersect, the non-crossing conditions are maintained.

Given an augmented graph with the required properties, a valid strip drawing is obtained by drawing each spine on a separate track, and the edge conditions

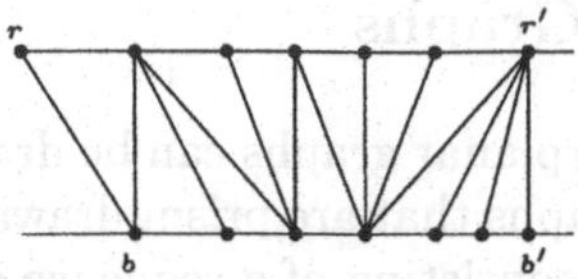

Fig. 8. Strip characterization

ensure there are no crossings. Vertices on each spine can be placed at consecutive integral X-coordinates thus ensuring that the strip is at most of length n.

The characterization of prism drawable graphs generalizes Theorem 4 to 3 dimensions, namely it must be possible to augment a given graph to obtain 3 spines with 2 lids (3 cycles) and between each pair of spines the non-crossing condition on edges must hold. We omit the proof.

Theorem 5. *A graph G is prism-drawable iff it is possible to augment G with edges to produce a graph G' which contains two three cycles r, b, g and r', b', g' and there exists a spine from r to r', a spine from b to b', a spine from g to g' with all vertices of G on the three spines and for each pair of spines $x \rightarrow x'$ and $y \rightarrow y'$ $(x, y = r, b, g,\ x \neq y)$, if (x_i, y_j) is an edge, then there are no edges of the form (x_k, y_l) with $(k < i$ and $l > j)$ or $(k > i$ and $l < j)$.*

5.2 Prism-Drawability and Planarity

Define a graph G as *strictly prism-drawable* if it is prism-drawable and all prism drawings of G have at least one edge on each facet of the prism. Note that K_4 is strictly prism-drawable. Our goal is to show the existence of series-parallel graphs that are not prism-drawable, and the graph P of Figure 9 is a more useful strictly prism-drawable graph for this purpose.

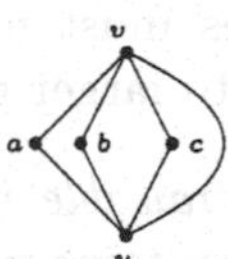

Fig. 9. A strictly prism-drawable graph P

Lemma 7. *The graph P in Figure 9 is strictly prism-drawable.*

Lemma 8. *Let G be a 1-connected graph that has a cut vertex whose removal separates the graph into h strictly prism-drawable components $(h \geq 3)$; then G is not prism-drawable.*

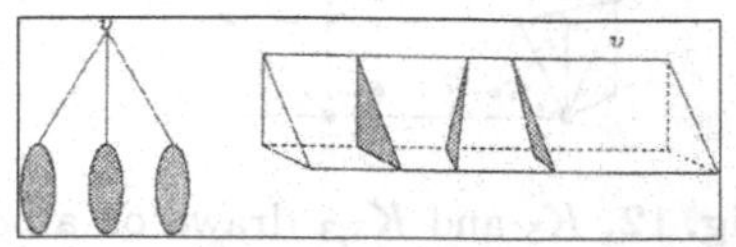

Fig. 10. A non-prism-drawable graph

Proof. (sketch) Figure 10 outlines the necessary construction. Consider a prism drawing Γ_i of G_i. Γ_i has a 3-cycle which defines a plane that intersects all three facets of the prism, because G_i is strictly prism-drawable. Thus, each Γ_i slices the prism into $h+1$ slices. Now there is no location for v that permits it to be connected crossing-free to all Γ_i without crossing at least one 3-cycle.

It is not critical that v be a cut vertex in the previous proof – any vertex connected to 3 or more strictly prism-drawable subgraphs will suffice and thus the previous lemma generalizes; however we omit the stronger result in this extended abstract.

Theorem 6. *There exist series-parallel graphs that are not prism-drawable.*

Proof. (sketch) The graph in Figure 11 has the required properties.

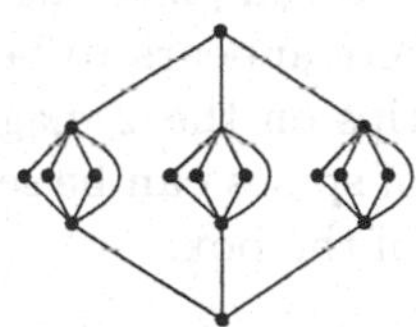

Fig. 11. A series-parallel graph that is not prism-drawable

Theorem 7. *Not all maximal planar graphs are prism-drawable. Also, the family of maximal planar prism drawable graphs is a proper subset of the family of Hamiltonian planar graphs.*

Proof. (sketch) It is not hard to show that all prism drawings of maximal planar graphs are Hamiltonian. Since there exist maximal planar graphs which are not Hamiltonian it immediately follows that not all maximal planar graphs are prism-drawable.

It is also possible to use the spine characterization more directly to produce planar graphs that are not prism-drawable – even Hamiltonian.

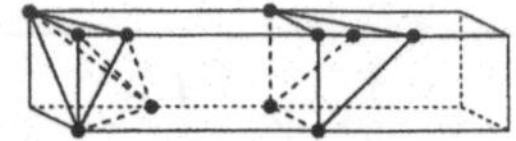

Fig. 12. K_5 and $K_{3,3}$ drawn on a box

6 Box-Drawable Graphs

The family of box-drawable graphs is clearly a superset of the class of prism-drawable graphs. Furthermore, there exist non-planar graphs that are box-drawable; for example K_5 and $K_{3,3}$ are box-drawable as shown in figure 12. Note however that K_6 is not box-drawable and we will show that there exist planar graphs (and even series-parallel graphs) that are not box-drawable. We refer to the 6 possible strips on which edges can be drawn as the *facets* of the box although two such facets appear inside the bounding box.

Theorem 8. *A graph G is box-drawable iff it is possible to augment G with edges to produce a graph G′ which contains two four cycles* r, b, g, o *and* r', b', g', o' *and there exists a spine from* r *to* r'*, a spine from* b *to* b'*, a spine from* g *to* g'*, and a spine from* o *to* o' *with all vertices of G on the four spines and for each pair of spines* $x \to x'$ *and* $y \to y'$*, if* (x_i, y_j) *is an edge, then there are no edges of the form* (x_k, y_l) *with* $(k < i$ *and* $l > j)$ *or* $(k > i$ *and* $l < j)$*.*

Proof. (sketch) As in the previous characterizations for the strip and prism, a box drawing can be augmented to complete the spines and the 6 facets must form valid strips. In this case there appears to be an extra condition necessary to ensure that the 2 pairs of strips on the 2 diagonally opposite spines do not intersect, however these 2 pairs of spines can be separated to ensure no diagonal crossings appear in the interior of the box.

Examples of graphs that are not box-drawable can be constructed using Theorem 8.

Theorem 9. *Not all planar graphs are box-drawable.*

7 Conclusions, Extensions, and Open Problems

In this paper we showed that all outerplanar graphs can be drawn in linear volume on a prism. We gave efficient characterizations of the trees that are drawable in 2 dimensions on an $n \times k$ grid. Classes of planar graphs that are not prism-drawable nor box-drawable were also provided. There remain several interesting problems and directions for further research.

1. Can all outerplanar graphs be drawn in $o(n^2)$ area on a 2d integer grid? Does there exist a 2d universal grid set of $o(n^2)$ area that supports all outerplanar graphs?

2. Characterize the graphs drawable on an $n \times k$ grid.
3. Can all planar graphs be drawn in $O(n)$ volume on a 3d integer grid? Does there exist a 3d universal grid set of $O(n)$ volume that supports all planar graphs?
4. **Aspect Ratio**: Our results about linear volume come at the expense of aspect ratio. Is it possible to achieve both linear volume and $o(n)$ aspect ratio for outerplanar graphs? We conjecture that it is in fact not possible in 2d to simultaneously attain $O(n)$ area and $O(1)$ aspect ratio for some classes of planar graphs.

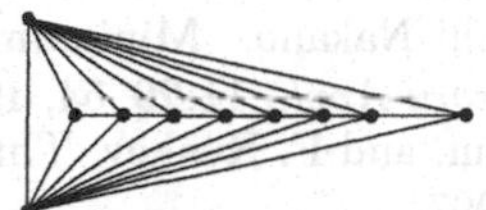

Fig. 13. A graph S_n with poor aspect ratio?

Conjecture 1. There is no fixed constant k for which the family of graphs S_n (in Figure 13) can be drawn in a 2d integer grid of size $k\sqrt{n} \times \sqrt{n}$.

Note that the graph S_n can be drawn on a $n \times 3$ grid (and hence in $O(n)$ area but with *linear* aspect ratio).

5. **Grid Drawings and Pathwidth**: The notion of the *pathwidth* of a graph has been well-studied in the graph theory literature – for definitions, see for example the book by Diestel [10]. A connection between pathwidth and layered graphs was established in [11]. It is not difficult to show that for trees there is a strong connection between pathwidth and grid drawings as summarized in the following propositions.

Proposition 1. *For a tree T, pathwidth(T)=*$\min_k$ *(T is drawable on a* $n \times k$ *grid).*

Proposition 2. *If G is a planar graph then pathwidth(G)* $\leq \min_k$ *(G is drawable on k-tracks).*

We have not been able to determine whether there are planar graphs with *pathwidth*$(G) < \min_k$ (G is drawable on a k-strip). If not, a linear time algorithm for recognizing graphs of pathwidth $\leq k$ would provide us with a linear time algorithm to recognize k-track drawable graphs. We believe that 2-strip drawable graphs can be recognized in polynomial time, however the complexity of the recognition of k-strip drawable graphs for $k \geq 3$ remains open.

Acknowledgments. We thank Helmut Alt, Pino di Battista, Hazel Everett, Ashim Garg, and Sylvain Lazard for useful discussions related to the results contained in this paper.

References

1. T. Calamoneri and A. Sterbini. Drawing 2-, 3-, and 4-colorable graphs in $o(n^2)$ volume. In S. North, editor, *Graph Drawing (Proc. GD '96)*, volume 1190 of *Lecture Notes Comput. Sci.*, pages 53–62. Springer-Verlag, 1997.
2. T.M. Chan. A near-linear area bound for drawing binary trees. In *Proc. 10th Annu. ACM-SIAM Sympos. on Discrete Algorithms.*, pages 161–168, 1999.
3. M. Chrobak and G. Kant. Convex grid drawings of 3-connected planar graphs. *Internat. J. Comput. Geom. Appl.*, 7(3):211–223, 1997.
4. Marek Chrobak, Michael T. Goodrich, and Roberto Tamassia. Convex drawings of graphs in two and three dimensions. In *Proc. 12th Annu. ACM Sympos. Comput. Geom.*, pages 319–328, 1996.
5. Marek Chrobak and Shin ichi Nakano. Minimum-width grid drawings of plane graphs. *Comput. Geom. Theory Appl.*, 11:29–54, 1998.
6. R. F. Cohen, P. Eades, T. Lin, and F. Ruskey. Three-dimensional graph drawing. *Algorithmica*, 17:199–208, 1997.
7. H. de Fraysseix, J. Pach, and R. Pollack. Small sets supporting Fary embeddings of planar graphs. In *Proc. 20th ACM Sympos. Theory Comput.*, pages 426–433, 1988.
8. H. de Fraysseix, J. Pach, and R. Pollack. How to draw a planar graph on a grid. *Combinatorica*, 10(1):41–51, 1990.
9. G. Di Battista, P. Eades, R. Tamassia, and I. G. Tollis. *Graph Drawing.* Prentice Hall, Upper Saddle River, NJ, 1999.
10. Reinhard Diestel. *Graph theory.* Graduate Texts in Mathematics. 173. Springer, 2000. Transl. from the German. 2nd ed.
11. V. Dujmovic, M. Fellows, M. Hallett, M. Kitching, G. Liotta, C. McCartin, N. Nishimura, P. Ragde, F. Rosamond, M. Suderman, S. Whitesides, D. R. Wood. On the Parameterized Complexity of Layered Graph Drawing. *ESA*, 1–12, 2001.
12. Stefan Felsner. Convex drawings of planar graphs and the order dimension of 3-polytopes. *Order – accepted to appear.*
13. A. Garg, M. T. Goodrich, and R. Tamassia. Planar upward tree drawings with optimal area. *Internat. J. Comput. Geom. Appl.*, 6:333–356, 1996.
14. A. Garg, R. Tamassia, and P. Vocca. Drawing with colors. In *Proc. 4th Annu. European Sympos. Algorithms*, volume 1136 of *Lecture Notes Comput. Sci.*, pages 12–26. Springer-Verlag, 1996.
15. Michael Juenger and Sebastian Leipert. Level planar embedding in linear time. In J. Kratochvil, editor, *Graph Drawing (Proc. GD '99)*, volume 1731 of *Lecture Notes Comput. Sci.*, pages 72–81. Springer-Verlag, 1999.
16. G. Kant. A new method for planar graph drawings on a grid. In *Proc. 33rd Annu. IEEE Sympos. Found. Comput. Sci.*, pages 101–110, 1992.
17. G. Kant. Drawing planar graphs using the canonical ordering. *Algorithmica*, 16:4–32, 1996.
18. János Pach, Torsten Thiele, and Géza Tóth. Three-dimensional grid drawings of graphs. In G. Di Battista, editor, *Graph Drawing (Proc. GD '97)*, volume 1353 of *Lecture Notes Comput. Sci.*, pages 47–51. Springer-Verlag, 1997.
19. F. P. Preparata and M. I. Shamos. *Computational Geometry: An Introduction.* Springer-Verlag, 3rd edition, October 1990.
20. W. Schnyder. Embedding planar graphs on the grid. In *Proc. 1st ACM-SIAM Sympos. Discrete Algorithms*, pages 138–148, 1990.
21. W. Schnyder and W. T. Trotter. Convex embeddings of 3-connected plane graphs. *Abstracts of the AMS*, 13(5):502, 1992.

Low-Distortion Embeddings of Trees*

Robert Babilon, Jiří Matoušek, Jana Maxová, and Pavel Valtr

Department of Applied Mathematics and
Institute for Theoretical Computer Science (ITI)
Charles University, Malostranské nám. 25,
118 00 Prague, Czech Republic,
{babilon,matousek,jana,valtr}@kam.ms.mff.cuni.cz

Abstract. We prove that every tree $T = (V, E)$ on n vertices can be embedded in the plane with distortion $O(\sqrt{n})$; that is, we construct a mapping $f: V \to \mathbf{R}^2$ such that $\rho(u,v) \le \|f(u) - f(v)\| \le O(\sqrt{n}) \cdot \rho(u,v)$ for every $u, v \in V$, where $\rho(u,v)$ denotes the length of the path from u to v in T (the edges have unit lengths). The embedding is described by a simple and easily computable formula. This is asymptotically optimal in the worst case. We also prove several related results.

1 Introduction

Embeddings of finite metric spaces into Euclidean spaces or other normed spaces that approximately preserve the metric received considerable attention in recent years. Numerous significant results have been obtained in the 1980s, mainly in connection with the local theory of Banach spaces. Later on, surprising algorithmic applications and further theoretical results were found in theoretical computer science; see, for example, [2], [1], [4], [6], and references therein.

Here we consider only embeddings into the spaces $\mathbf{R}^d$ with the Euclidean metric. The quality of an embedding is measured by the *distortion*. Let (V, ρ) be a finite metric space. We say that a mapping $f: V \to \mathbf{R}^d$ has distortion at most D if there exists a real number $\alpha > 0$ such that for all $u, v \in V$,

$$\alpha \cdot \rho(u,v) \le \|f(u) - f(v)\| \le D \cdot \alpha \cdot \rho(u,v).$$

This definition permits scaling of all distances in the same ratio r, in addition to the distortion of the individual distances by factors between 1 and D. Since the image in $\mathbf{R}^d$ can always be re-scaled as needed, we can choose the factor α at our convenience.

We study embeddings with the dimension d fixed, and mainly the case $d = 2$, i.e. embeddings into the plane. A straightforward volume argument shows that the distortion required to embed into $\mathbf{R}^d$ the n-point metric space with all distances equal to 1 is at least $\Omega(n^{1/d})$. In [5], it was proved that general metric spaces may require even significantly bigger distortions; namely, for every

* The research was supported by project LN00A056 of the Ministry of Education of the Czech Republic and by Charles University grants No. 158/99 and 159/99.

P. Mutzel, M. Jünger, and S. Leipert (Eds.): GD 2001, LNCS 2265, pp. 343–351, 2002.
© Springer-Verlag Berlin Heidelberg 2002

fixed d, there are n-point metric spaces that need $\Omega(n^{1/\lfloor(d+1)/2\rfloor})$ distortion for embedding into $\mathbf{R}^d$.

Better upper bounds can be obtained for special classes of metric spaces. A metric ρ on a finite set V is called a *tree metric* if there is a tree $T = (V, E)$ (in the graph-theoretic sense) and a weight function $w: E \to (0, \infty)$ such that $\rho(u, v)$ is the length of the path connecting the vertices u and v in T, where the length of an edge $e \in E$ is $w(e)$. It was conjectured in [5] that, for fixed d, all n-point tree metrics can be embedded into $\mathbf{R}^d$ with distortion $O(n^{1/d})$. If true, this is best possible, as the example of a star with $n-1$ leaves and unit-length edges shows (a volume argument applies). Gupta [3] proved the somewhat weaker upper bound $O(n^{1/(d-1)})$.

Here we make a step towards establishing the conjecture. We deal with the planar case, where the gap between the lower bound of $\sqrt{n}$ and Gupta's $O(n)$ upper bound is the largest (in fact, the $O(n)$ upper bound holds even for general metric spaces and embeddings into $\mathbf{R}^1$ [5]). So far we can only handle the case of unit-length edges.

Theorem 1. *Every n-vertex tree with unit-length edges, considered as a metric space, can be embedded into* $\mathbf{R}^2$ *with distortion* $O(\sqrt{n})$. *The embedding is described by a simple explicit formula and can be computed efficiently.*

Gupta's result actually states that if the considered tree has at most ℓ *leaves*, then an embedding into $\mathbf{R}^d$ with distortion $O(\ell^{1/(d-1)})$ is possible. We show that here the dependence on ℓ *cannot* be improved. Let $F_{\ell,m}$ (the fan with ℓ leaves and path length m) denote the tree consisting of ℓ paths of length m each glued together at a common vertex (root):

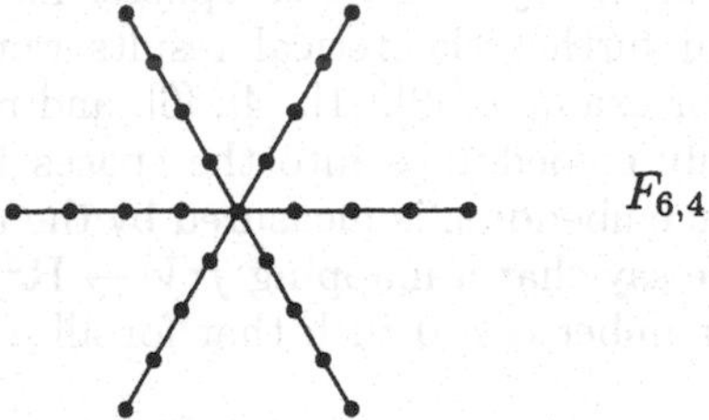

Proposition 1. *For every fixed* $d \geq 2$ *and every* $m, \ell \geq 2$ *every embedding of* $F_{\ell,m}$ *into* $\mathbf{R}^d$ *requires distortion* $\Omega(\ell^{1/(d-1)})$ *if* $\ell \leq m^{\frac{d(d-1)}{d+1}}$ *and distortion* $\Omega(\ell^{1/d} m^{1/(d+1)})$ *if* $\ell \geq m^{\frac{d(d-1)}{d+1}}$.

Theorem 1 provides embeddings which are optimal in the worst case, i.e. there are trees for which the distortion cannot be asymptotically improved. But optimal or near-optimal embeddings of special trees seem to present interesting challenges, and sometimes low-distortion embeddings are aesthetically pleasing and offer a good way of drawing the particular trees. We present one interesting example concerning the fan $F_{\sqrt{n},\sqrt{n}}$. Here the "obvious" embedding as in the above picture, as well as the embedding from Theorem 1, yield distortions

$O(\sqrt{n})$. In Section 4, we describe a somewhat surprising better embedding, with distortion only $O(n^{5/12})$. This is already optimal according to Proposition 1.

Several interesting problems remain open. The obvious ones are to extend Theorem 1 to higher dimensions and/or to trees with weighted edges. Another, perhaps more difficult, question is to extend the class of the considered metric spaces. Most significantly, we do not know an example of an n-vertex planar graph (with weighted edges) whose embedding into $\mathbf{R}^2$ would require distortion larger than about $\sqrt{n}$. If it were possible to show that all planar-graph metrics can be embedded into the plane with $o(n)$ distortion, it would be a neat metric condition separating planar graphs from non-planar ones, since suitable n-vertex subdivisions of any fixed non-planar graph require $\Omega(n)$ distortion [5].

2 Proof of Theorem 1

Notation. Let T be a tree (the edges have unit lengths) and let ρ be the shortest-path metric on $V = V(T)$. One vertex is chosen as a root. The *height* $h(v)$ of a vertex $v \in V$ is its distance to the root. Let π_v denote the path from a vertex v to the root.

For every vertex we fix a linear (left-to-right) ordering of its children. This defines a partial ordering $\preceq$ on V: we have $u \prec v$ iff $u \notin \pi_v$, $v \notin \pi_u$, and π_v goes right of π_u at the vertex where π_u and π_v branch. We define

$$\operatorname{sgn}_v(u) = \begin{cases} 0 & \text{if } u \in \pi_v \text{ or } v \in \pi_u \\ +1 & \text{if } u \prec v \\ -1 & \text{if } v \prec u. \end{cases}$$

Further we define $a_v(u)$ as the distance of v to the nearest common ancestor of u and v, and $\ell(v) = |\{u \in V(T) : u \prec v\}|$.

Construction. Our construction resembles Gupta's construction [3] to some extent, but we needed several new ideas to obtain significantly better distortion.

First we describe the idea of the contruction.

The embedded tree is placed in some angle; the root of the tree is placed in the vertex of the angle. In the first step we put vertices of height h on the horizontal line $y = 2\sqrt{n}h$. Our angle is divided into subangles which correspond to subtrees from the root and whose sizes are proportional to the number of vertices in the subtrees. The vertices with height 1 are placed on the first horizontal line in the middle of the corresponding subangles. We continues in the same way for every subtree, with each subangle translated so that its vertex lies at the root of the subtree.

In the second (and last) step we raise the y-coordinates of the vertices by appropriate amounts between 0 and $\sqrt{n}$.

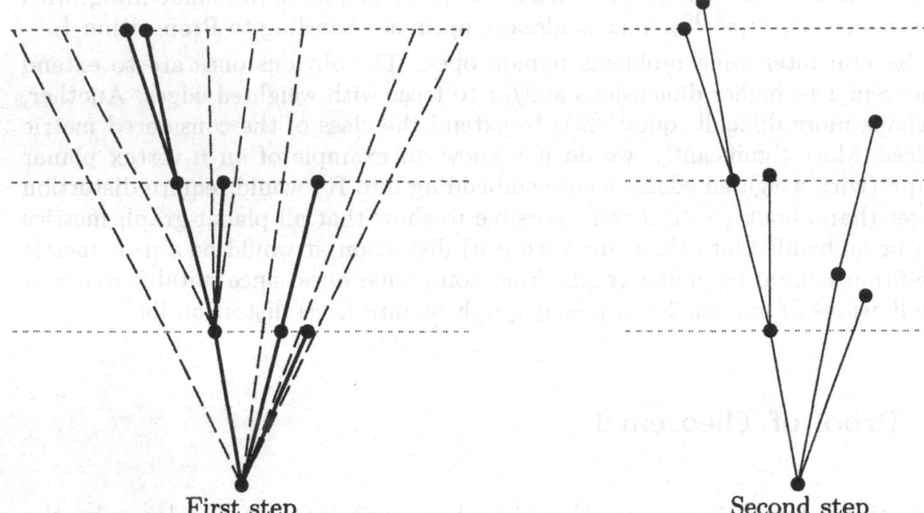

Now we describe the embedding more formally; actually we use a slightly modified embedding for which the formulas are quite simple.

First we define an auxiliary function $k(v)$:

$$k(v) = \sum_{w \in V} \operatorname{sgn}_v(w) a_v(w).$$

The embedding $f: V \to \mathbf{R}^2$ is now defined by $f(v) = (x(v), y(v))$, where

$$\begin{aligned} x(v) &= n^{-1/2} k(v) \\ y(v) &= 2\sqrt{n}\, h(v) + (\ell(v) \bmod \sqrt{n}). \end{aligned}$$

Distortion. We now estimate the distortion of the embedding defined above. First we show that the maximum expansion of the tree distance of any two vertices is at most $O(\sqrt{n})$. By the triangle inequality, it suffices to consider two vertices connected by an edge. So let u be the father of v in T; it suffices to verify that $\|f(u) - f(v)\| = O(\sqrt{n})$. We clearly have $|y(u) - y(v)| \le 3\sqrt{n}$, and so it suffices to prove that $|x(u) - x(v)| = O(\sqrt{n})$, which will follow from $|k(u) - k(v)| \le n$. We have

$$|k(v) - k(u)| \le \sum_{w \in V(T)} \left| \operatorname{sgn}_u(w) a_u(w) - \operatorname{sgn}_v(w) a_v(w) \right|.$$

We always have $|a_u(w) - a_v(w)| \le 1$, as $\rho(u, v) = 1$. Moreover, it cannot happen that $\operatorname{sgn}_u(w) = -\operatorname{sgn}_v(w)$, and so the contribution of each w to the above sum is at most 1. Therefore $|k(u) - k(v)| \le n$ as claimed.

Next, we are going to prove that $\|f(u) - f(v)\| = \Omega(\rho(u, v))$ for every $u, v \in V$; this will finish the proof of Theorem 1.

First we consider the situation when $h(u) = h(v)$; this is the main part of the proof, and the case where u and v have different levels will be an easy extension of this. So let $h(u) = h(v)$ and $a = a_v(u)$, and assume that $u \prec v$.

Lemma 1. $\ell(v) - \ell(u) \geq a$

This is because all the vertices on π_u not lying on π_v are counted in $\ell(v)$ but not in $\ell(u)$. □

Corollary 1. *If* $\ell(v) - \ell(u) < \frac{1}{2}\sqrt{n}$ *then* $|y(u) - y(v)| \geq \frac{1}{2}\rho(u, v)$.

Indeed, we have $|y(u) - y(v)| = |(\ell(u) \bmod \sqrt{n}) - (\ell(v) \bmod \sqrt{n})| \geq |\ell(u) - \ell(v)| \geq a = \frac{1}{2}\rho(u, v)$. □

Thus, the difference in the y-coordinate takes care of u and v whenever $\ell(v) - \ell(u) \leq \frac{1}{2}\sqrt{n}$. Next, we need to show that if this is not the case then the x-coordinate takes care of u and v.

Lemma 2. $k(v) - k(u) \leq [\ell(v) - \ell(u)] \cdot a$.

Corollary 2. *If* $\ell(v) - \ell(u) \geq \frac{1}{2}\sqrt{n}$ *then* $x(v) - x(u) = n^{-1/2}[\ell(v) - \ell(u)] \cdot a \geq \frac{1}{2}a \geq \frac{1}{4}\rho(u, v)$. □

Combining this corollary with Corollary 1 yields $\|f(u) - f(v)\| = \Omega(\rho(u, v))$ for all $u, v \in V$ with $h(u) = h(v)$.

Proof of Lemma 2. We have $k(v) - k(u) = \sum_{w \in V} t_{u,v}(w)$, where $t_{u,v}(w) = \mathrm{sgn}_v(w)a_v(w) - \mathrm{sgn}_u(w)a_u(w)$. First we check that $t_{u,v} \geq 0$ for all w. We always have $\mathrm{sgn}_v(w) \geq \mathrm{sgn}_u(w)$, and so the only case which might cause trouble is $\mathrm{sgn}_v(w) = \mathrm{sgn}_u(w) = 1$. In this case $w \prec u \prec v$, and it is easy to check that then $a_v(w) \geq a_u(w)$, which shows $t_{u,v} \geq 0$ in all cases.

Next, we verify that if w contributes 1 to $\ell(v) - \ell(u)$, which means $w \prec v$ and $w \not\prec u$, then $t_{u,v}(w) \geq a$. We distinguish two cases for such w: either $u \prec w \prec v$ or w and u lie on a common path to the root, i.e. $w \in \pi_u$ or $u \in \pi_w$.

In the first case, $u \prec w \prec v$, $\mathrm{sgn}_u(w) = -1$ and $\mathrm{sgn}_v(w) = 1$, and so $t_{u,v}(w) = a_u(w) + a_v(w)$. It is easy to see that if u and v are vertices with $a_u(v) = a_v(u) = a$ then $\max(a_u(w), a_v(w)) \geq a$ for all $w \in V$ and so $t_{u,v}(w) \geq a$ in this case.

In the second case, we have $\mathrm{sgn}_v(w) = 1$ and $\mathrm{sgn}_u(w) = 0$, and so $t_{u,v}(w) = a_v(w) = a$, since the nearest common ancestor of u and v is the same as the nearest common ancestor of w and v. The proof of Lemma 2 is complete. □

Claim. If two vertices u, v are on different levels, then

$$\|f(v) - f(u)\| \geq \frac{1}{12\sqrt{2}}\rho(u, v).$$

Proof. Without loss of generality, suppose that $h(u) < h(v)$ and $u \prec v$. Set $a := a_u(v)$, and let w be the vertex on the path from u to v with $h(w) = h(u)$, $\rho(u, w) = 2a$.

If $\rho(u,v) > 3a$ then $\|f(v) - f(u)\| \geq y(v) - y(u) \geq (\rho(u,v) - 2a)\sqrt{n} \geq \frac{\sqrt{n}}{3}\rho(u,v) \geq \frac{1}{3}\rho(u,v)$.

If $\rho(u,v) \leq 3a$ and $a < 4\sqrt{n}$ then $\|f(v) - f(u)\| \geq \sqrt{n} > a/4y(v) - y(u) \geq \frac{1}{12}\rho(u,v)$.

Finally, suppose that $\rho(u,v) \leq 3a$ and $a \geq 4\sqrt{n}$. Since the y-axis and the extension of any edge span an angle smaller than 45°, the path from $f(w)$ to $f(v)$ lies in the cone bounded by the two halflines emanating from $f(w)$ diagonally upwards and spaning the angle 45° with the y-axis. The point $f(u)$ lies at distance at least $\frac{1}{\sqrt{2}}(x(w) - x(u) + y(w) - y(u)) > \frac{1}{\sqrt{2}}(x(w) - x(u) - \sqrt{n}) \geq \frac{1}{\sqrt{2}}(a/2 - \sqrt{n}) \geq \frac{1}{\sqrt{2}}\frac{a}{4} \geq \frac{1}{12\sqrt{2}}\rho(u,v)$ from this cone (in the second inequality we used Corollary 2). Thus, $\|f(v) - f(u)\| > \frac{1}{12\sqrt{2}}\rho(u,v)$ in this case. □

3 Proof of Proposition 1

Let $F_{\ell,m}$ be defined as in Section 1; that is, $F_{\ell,m}$ denotes the tree consisting of ℓ paths of length m glued at the root r. We show that every embedding of $F_{\ell,m}$ into $\mathbf{R}^d$ requires distortion $\Omega(\ell^{1/(d-1)})$ if $\ell \leq m^{\frac{d(d-1)}{d+1}}$ and distortion $\Omega(\ell^{1/d}m^{1/(d+1)})$ if $\ell \geq m^{\frac{d(d-1)}{d+1}}$.

Let $f\colon V(F_{\ell,m}) \to \mathbf{R}^d$ be any non-contracting embedding with distortion D (i.e. for all $u,v \in V$, $\rho(u,v) \leq \|f(u) - f(v)\| \leq D \cdot \rho(u,v)$). We choose a real number R such that there are $\ell/2$ of the images of endvertices of $F_{\ell,m}$ contained inside the ball B of radius R around $f(r)$ or on its boundary and there are $\ell/2$ of them otside the ball or on its boundary. Let S denote the sphere of radius R around $f(r)$ (i.e. the boundary of B). Let $P_1, \ldots, P_{\ell/2}$ denote the $\ell/2$ paths for which the images of their endvertices are not contained in B.

Each $f(P_i)$ intersects S at least once (by $f(P_i)$ we mean $f(V(P_i))$ together with straight lines between any $f(u)$, $f(v)$ such that $\{u,v\} \in E(P_i)$). Let m_i denote the number of vertices on $f(P_i)$ inside B before the first intersection with S. We put $\mu = \min\{m_i \mid i = 1, \ldots, \ell/2\}$. Since $\ell/2$ of the endvertices are contained in B, we immediately have, by a volume argument, $R = \Omega(m\ell^{1/d})$. From the choice of μ it follows that $R \leq D\mu$.

Let P be a spherical shell of width D around B; that is

$$P = B(f(r), R + D) \setminus \operatorname{int} B(f(r), R).$$

At least $\ell/2$ vertices (one from each path P_i, $i = 1, \ldots, \ell/2$) are contained in P. We consider a ball of radius μ around each of these vertices. These balls are disjoint (because the tree-distance between any two of the vertices is at least 2μ) and they are contained in μ-neighbourhood of P (see the picture). Therefore we have, for a suitable constant $c > 0$,

$$R^{d-1}(D + 2\mu) \geq c\ell\mu^d.$$

Thus there are two possibilities: either $R^{d-1}D = \Omega(\ell\mu^d)$ or $R^{d-1}\mu = \Omega(\ell\mu^d)$.

If $R^{d-1}D = \Omega(\ell \cdot \mu^d)$, then $D^{d+1} = \Omega(\ell \cdot R)$, since $R \leq D\mu$. And since $R = \Omega(m\ell^{1/d})$ we have $D^{d+1} = \Omega(\ell \cdot m \cdot \ell^{1/d})$. Thus $D = \Omega(\ell^{1/d} \cdot m^{1/(d+1)})$. If

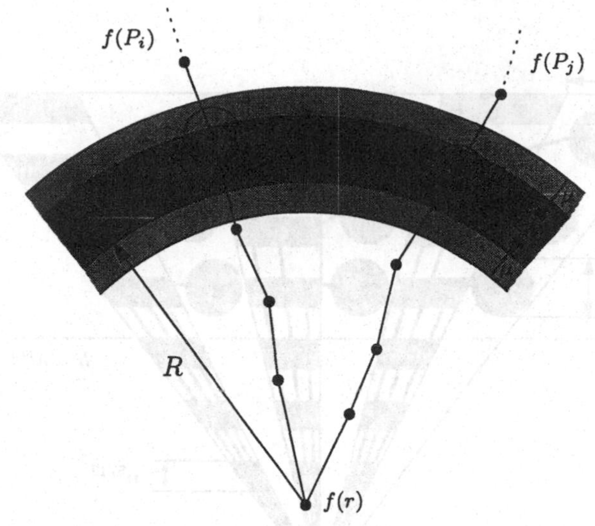

Fig. 1. Illustration to the proof of Proposition 1.

$R^{d-1}\mu = \Omega(\ell \cdot \mu^d)$, then since $R \leq D\mu$, we immediately have $D = \Omega(\ell^{1/(d-1)})$. Thus we know that $D \geq \min\{\Omega(\ell^{1/d} \cdot m^{1/(d+1)}), \Omega(\ell^{1/(d-1)})\}$. It means that $D = \Omega(\ell^{1/d} \cdot m^{1/(d+1)})$ if $\ell \geq m^{\frac{d(d-1)}{d+1}}$ and $D = \Omega(\ell^{1/(d-1)})$ if $\ell \leq m^{\frac{d(d-1)}{d+1}}$.

4 A Low-Distortion Embedding of a Subdivided Star

In this section we describe an asymptotically optimal embedding of $F_{\sqrt{n},\sqrt{n}}$ into $\mathbf{R}^2$, with distortion $O(n^{5/12})$. The other $F_{\ell,m}$ can be embedded similarly, with distortion as in Proposition 1.

The embedding is sketched in Fig 2. The whole tree is embedded into the shaded trapezoids (there are $n^{1/4} \cdot 2n^{1/3}$ of them) and into the shaded discs (there are $n^{1/4} \cdot n^{1/4}$ of them). The root of $F_{\sqrt{n},\sqrt{n}}$ is embedded onto the bottom vertex of the whole triangle.

In the following, we describe the embedding in more detail. It will be done in two steps. In the first step, we embed vertices up to the level $h = 2n^{1/3}$ into the dashed trapezoids on the corresponding level. In the second step, we embed all the other vertices on the diameters of the discss and change the positions of some vertices on levels between $n^{1/3}$ and $2n^{1/3}$.

Step 1. We start with the vertices on level $h = 2n^{1/3}$. There are $n^{1/2}$ vertices on this level. We divide them into $n^{1/4}$ groups, each of size $n^{1/4}$. Each of these groups is packed into one of the dashed trapezoids (on the last level). This is possible since the area of each trapezoid on this level is $\Omega(n^{11/12})$, which is

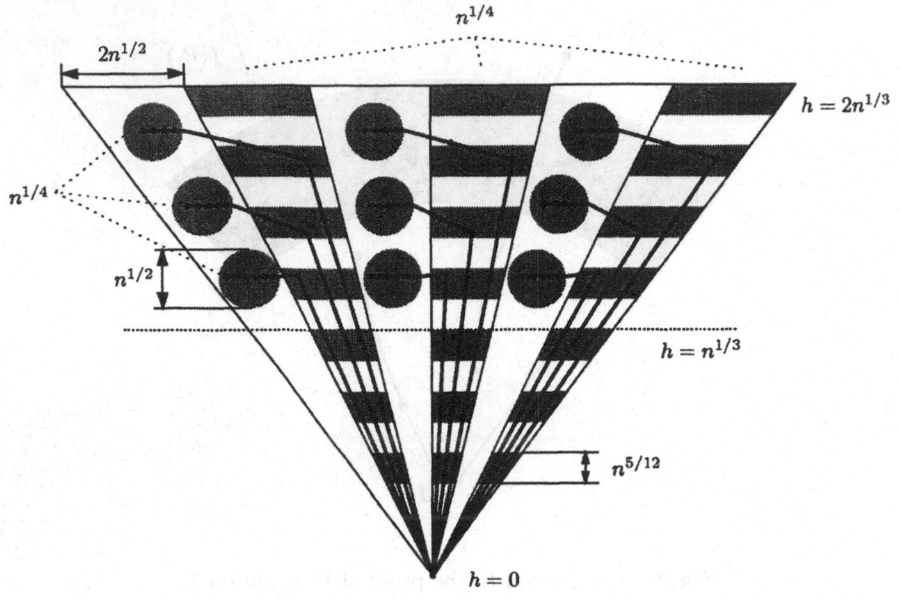

Fig. 2. The embedding of $F_{\sqrt{n},\sqrt{n}}$.

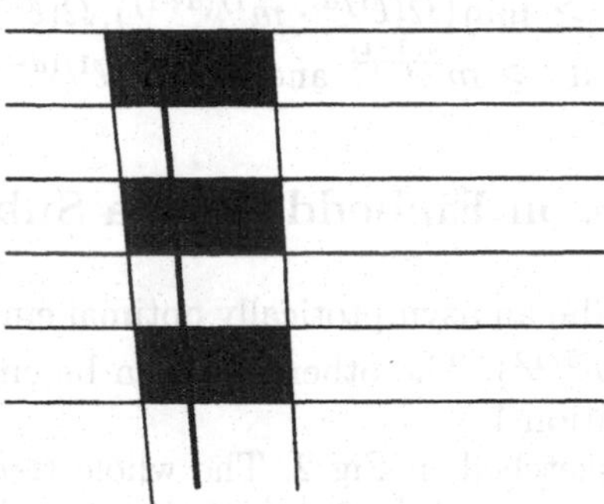

Fig. 3. The placement of vertices in lower levels.

exactly the required area (we have $n^{1/4}$ vertices, any two of them must be at distance at least $\Omega(n^{1/3})$). When all the vertices on the level $h = 2n^{1/3}$ are embedded, we embed the vertices on lower levels. We connect every vertex x on the last level by a straight line to the root. On this line we embed all the vertices on the path from x to the root. We do this in such a way that the distance Δ (measured on the connecting line) from the vertex to the bottom of the dashed trapezoid is the same for all the vertices on the path from x to the root (see Fig. 3). It is easy to check that the distance between the vertices on the last level guarantees that the vertices in the lower levels are also far enough from each other. Indeed, if we consider two vertices x_1 and y_1 at distance d in level h_1 (i.e. their tree-distance is $2h_1$) then for any x_2 on the path from x_1 to the root and any y_2 on the path from y_1 to the root, both in the level h_2, their distance is at least dh_2/h_1 (while their tree-distance is $2h_2$).

Thus, the construction described so far gives an embedding of $F_{\sqrt{n},2n^{1/3}}$ with distortion $O(n^{5/12})$. In the next step we modify this embedding into an embedding of $F_{\sqrt{n},\sqrt{n}}$.

Step 2. In this step we embed all the vertices on the remaining levels. We also change the positions of some vertices on the levels between $n^{1/3}$ and $2n^{1/3}$. Left to the $n^{1/4}$ columns of trapezoids there are $n^{1/4}$ columns of discs (and there are $n^{1/4}$ discs in each column). In each of the discs one path ends. A more precise description follows.

We start from the bottom and continue upwards. For each disc we consider the level nearest to its equator (below the equator). We take the leftmost vertex in the shaded trapezoid on this level and change the embedding of its successors: all of its successors are embedded on the (possibly extended) equator of the considered disc. The distance between any two neighbours on this path is 1 in the embedding. We continue with the next higher level (with a smaller number of paths). In the last level, there is only one path to end. The resulting embedding is indicated in Fig 2.

It is straightforward to check that the resulting embedding of $F_{\sqrt{n},\sqrt{n}}$ yields distortion $O(n^{5/12})$ as claimed. □

Acknowledgment. We would like to thank Helena Nyklová, Petra Smolíková, Ondřej Pangrác, Robert Šámal, and Tomáš Chudlarský for useful discussions on the problems considered in this paper.

References

1. Y. Bartal. Probabilistic approximation of metric spaces and its algorithmic applications. In *Proc. 37th Ann. IEEE Sympos. on Foundations of Computer Science*, pages 184–193, 1996.
2. U. Feige. Approximating the bandwidth via volume respecting embeddings. *J. Comput. Syst. Sci*, 60:510–539, 2000.
3. A. Gupta. Embedding tree metrics into low dimensional Euclidean spaces. *Discrete Comput. Geom.*, 24:105–116, 2000.
4. N. Linial, E. London, and Yu. Rabinovich. The geometry of graphs and some its algorithmic applications. *Combinatorica*, 15:215–245, 1995.
5. J. Matoušek. Bi-Lipschitz embeddings into low-dimensional Euclidean spaces. *Comment. Math. Univ. Carolinae*, 31:589–600, 1990.
6. S. Rao. Small distortion and volume respecting embeddings for planar and Euclidean metrics. In *Proc. 15th Annual ACM Symposium on Comput. Geometry*, 1999.

Insight into Data through Visualization

Eduard Gröller

Institute of Computer Graphics and Algorithms
Vienna University of Technology
A-1040 Vienna, Austria
groeller@cg.tuwien.ac.at
http://www.cg.tuwien.ac.at/home/

Abstract. Computer graphics, scientific visualization, information visualization, and graph drawing are areas which deal with visual information layout. They all use the remarkable properties of the human visual perception to rapidly absorb and analyse visual information. The paper discusses important visualization aspects and gives examples of how visualization techniques facilitate insight into data characteristics. The connection between visualization and graph drawing is shortly discussed.

1 Introduction

Computer graphics and visualization have come a long way in developing techniques and tools to support the human user in data analysis and investigation. As stated by R. Hamming: "The purpose of computing is insight, not numbers". The human visual system is well adapted to rapidly absorb and interpret large quantities of visual information. Scientific visualization deals with the representation, manipulation, and rendering of scientific data (measured, simulated or modelled data). Volume visualization and flow visualization are two important subbranches of scientific visualization. Several examples, e.g., medical visualization, shall illustrate how visual representations facilitate the understanding of underlying data characteristics. Scientific visualizations are often characterized by an inherent spatial context. This is not the case with information visualization. Information visualization on the other hand deals with interactive visualizations of abstract, often higher-dimensional, unstructured, and large data sets (e.g., data base content). Issues in information visualization are to select appropriate visual metaphors and to provide tools for efficient user interaction. With scientific visualization the user goal is often an analysis of the data, whereas with information visualization exploration and browsing through the abstract data is of major concern. Information visualization is also used to handle hierarchical or relational data. Such data is best represented in graph structures. These applications act as bridge between the visualization and graph drawing field. This paper shall help in intensifying collaborative contacts between these two areas. The primary goal is to show how visualization uses the considerable capabilities of the human visual perception for fast information acquisition and processing.

P. Mutzel, M. Jünger, and S. Leipert (Eds.): GD 2001, LNCS 2265, pp. 352–366, 2002.
© Springer-Verlag Berlin Heidelberg 2002

2 Computer Graphics

In general computer graphics [FvDFH96], [HB96] deals with computer supported methods and techniques to process images, animation sequences, and visual information. Computer graphics comprises generative computer graphics, image processing, and pattern recognition. Generative computer graphics takes an abstract, textual description to generate, manipulate, and render images. With image processing images are modified to facilitate further analysis by a human user or by a computer. Modifications include error elimination, contrast enhancement, detection of edges and homogeneous areas. Pattern recognition uses images to extract descriptive information which characterizes the content of the image. In the following we concentrate on generative computer graphics including interactivity. The three areas of computer graphics are, however, overlapping. Techniques of generative computer graphics are often used in displaying image-processing results. On the other hand rendering of computer-generated images may involve post-processing steps which thematically rather belong to image processing. The interactive input to graphical systems sometimes involves pattern-recognition algorithms. Certain subareas, e.g., color systems, equally belong to all three areas. Also most graphical devices are used in all three areas. Several important application fields of generative computer graphics include data visualization, presentation graphics, computer aided design (CAD), architecture, medical applications, geographic information systems (GIS), computer animation and movies, marketing, virtual reality, computer games, edutainment.

2.1 Human Visual Perception

Already in ancient times visual representations, e.g., rock paintings, petroglyphs, Egyptian tomb paintings, Minoan frescoes, expressed culture and traditions of primitive peoples and early civilizations. One reason for the popularity of pictorial presentations lies in the fact, that about 70% of all the information a person absorbs is acquired through the visual system. Human visual perception is a rather complicated multi-stage process. Important structures of the eye include the iris, lens, pupil, cornea, retina, vitreous humor, optic disk and optic nerve. The retina contains photoreceptors, i.e., rods and cones. Rods are the sensors on the retina which are responsible to detect greyscales and low levels of light. Cones are responsible for color perception and exist in three types. There are cones for detecting "red"-, "green"-, and "blue"-light respectively. Therefore the human being is called a trichromat. There are about 20 times more rods than cones. Rods predominantly occur near the periphery of the retina whereas cones are concentrated near the center of the eye. Also the three types of color-sensitive cones occur in greatly varying numbers, i.e., there are many more "green"-light sensitive cones than "blue"-light sensitive ones. The retina further contains components which process the electrical information coming from the photoreceptors. Processing includes detection of intensities, edges, motion and color-signal transformations. The optic nerves from each eye transport the visual information to the brain, where it is combined and analysed in several stages. Combining the two perspective views from the left and right eye respectively enables stereoscopic viewing. This facilitates the detection of spatial relations among viewed objects and distance estimations. Other effects that influence visual perception are: focus, accomodation, adaption, visual acuity, contrast sensitivity, simultaneous contrast, Mach-banding, color constancy, depth, shape, and texture detection. Human visual perception rapidly scans, recognizes and recalls images. It thus provides a high-bandwidth data-channel to the human brain. Visual perception, however, is a quite intricate process. Therefore it is essential to be aware of the major perceptual components in order to be able to produce expressive and effective visual representations.

2.2 Computer-Graphics Workflow

Image generation in 3D computer graphics consists of a pipeline of consecutive steps (figure 1). The major stages in the computer-graphics workflow are modelling, rendering, and display. In the **modelling** stage the user specifies the scene data which shall be displayed. This includes object geometry, surface characteristics, lighting environment, motion and camera parameters. Objects are usually constructed through a modelling software which provides the resulting 3D data either as internal data structure or sequentially on file. The **rendering** stage performs the conversion of the 3D data into a 2D image in several steps. Transformations, e.g., translations, rotations, scalings, involve coordinate changes to modify position, size, and shape of objects. Mapping projects the 3D scene onto a 2D image plane. Usually perspective or parallel projection is used. With clipping parts of the scene which are outside the viewing frustum and therefore invisible according to the current camera parameters are eliminated. In case of occluding

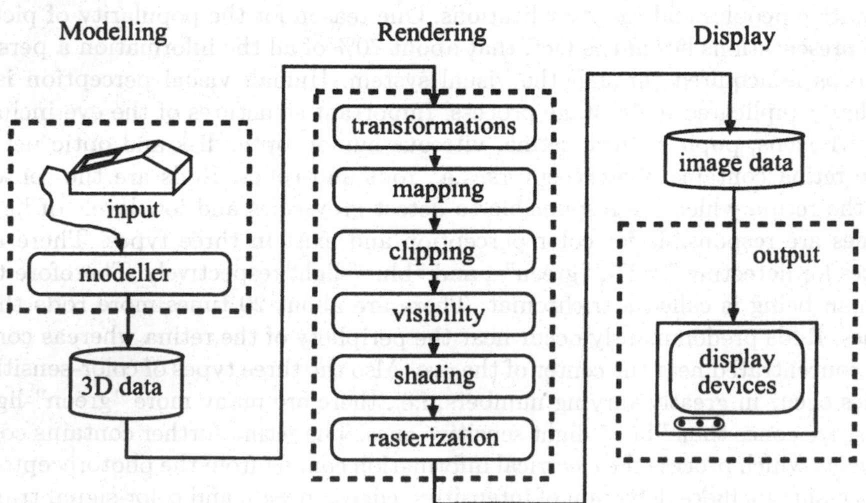

Fig. 1. Computer-graphics workflow

objects a visibility algorithm determines the visible parts of the scene, i.e., objects closest to the view point. Shading takes surface characteristics and lighting conditions to determine the intensity or color of visible object regions. A small angle of light incidence (light source is directly above an object) produces bigger brightness values. As nowadays most display units are raster-oriented (e.g., screen displays) the geometry data, e.g., triangles, must be converted into raster data. This is done through raster-conversion algorithms. In the **display** stage the 2D data of the rendering step are either directly displayed or are stored in one of the many available image file formats.

2.3 Applications of Computer Graphics

In the following we shortly and exemplarily discuss two of the many applications of computer graphics: computer-aided design (CAD) and computer graphics in education and training. **Computer-aided design (CAD)** deals with the use of computers to support and automatize construction and drawing processes. Computer graphics is a small but important part of CAD tasks. CAD is applied in the design of architectural models (figure 2), automobiles, aircrafts, computers, textiles (figure 3), and many other products. CAD programs typically offer the designer a multi-window environment, which depicts enlarged parts or different views of objects. Special rendering styles, e.g., wireframe rendering, allow an interactive and incremental construction process. CAD programs facilitate the efficient placement of generic shapes. Also connections between objects, annotations, and measurements are automatically included in design results. Variations or alternatives of a specific design are easily producible. A CAD model also enables a computer simulation of important object characteristics like performance, stress or flow behavior. Simulation of realistic lighting conditions and

Fig. 2. Architectural modelling [Tra95]

Fig. 3. Textile modelling [GRS95]

surface rendering create a realistic image of the final product. In architectural applications CAD supports the design of floor plans, wiring, electrical outlets, complex lighting simulations, and walk-throughs (animations). CAD is a subarea of fields like CAE (computer-aided engineering) or CAP (computer-aided planing). **Education and training** has been an early driving force for computer graphics. Flight simulators are a typical example. They are special systems which provide persuasive visual feedback. The user is immersed in a training situation which is as realistic as possible. Other simulators are concerned with, e.g., air-traffic control, ships, heavy-equipment, car traffic. Models of physical systems or equipment allow to explore the functionality and behavior at reduced cost and risk. Understanding of processes is eased through computer-graphics views (e.g, cut-away views) which are not possible in reality.

2.4 Extending the Computer-Graphics Workflow

The traditional computer-graphics workflow (figure 1) is based on the desktop metaphor: data input and output is primarily done through keyboard, mouse and display screen. There are several possibilities to extend this traditional workflow. **Virtual reality** and **augmented reality** are two rather new areas, which fully or partially immerse a user into a computer generated virtual environment. These approaches use novel input and output techniques like (see-through) head-mounted-displays, large projection units, 3D input devices. Figure 4 illustrates Studierstube [SFH+01], an augmented reality system where several users can collaboratively interact with virtual information. Essential components of virtual or augmented reality include immersion, real-time user interaction and feedback, and graphical (realistic) representation of the virtual world.

Typically the computer-graphics workflow simulates realistic image generation which means that elaborate lighting calculations (ray tracing, radiosity) are performed to emulate images of real-world scenes as closely as possible. **Non-photorealistic rendering** on the other hand simulates different drawing styles: contour and silhouette rendering, pen and ink drawings, water color simulation, replication of brush strokes previously used by artists. Figure 5 shows a medical illustration [CMH+01] which uses non-photorealistic rendering.

Perspective or parallel projection are first approximations of how the human eye perceives a 3D scene. With computer graphics more general (artistic) mapping strategies may be utilized. The 3D information may not only be mapped to a 2D image plane but onto a more general (curved) image surface. There may not be only one fixed view point but several view points within a single image. Again artists have already used such approaches. Figure 6 to the left shows a perspective rendering of a 3D object. In Figure 6 to the right the object is mapped by using an extended camera [LG96], i.e., the image surface is a bent cylinder, view points are changing from top to bottom of the image by moving along the cylinder axis.

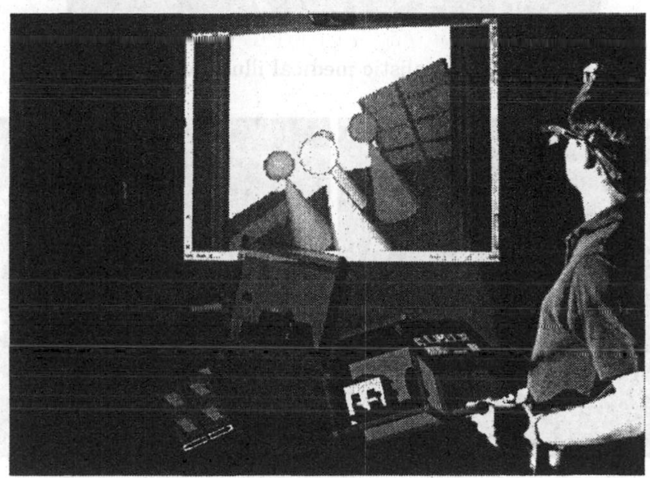

Fig. 4. Studierstube augmented reality system [SFH+01]

3 Scientific Visualization

Scientific visualization [SM00] deals with the representation, manipulation, interaction, and rendering of scientific data. It allows to form a mental vision, image, or picture of something not visible or present to the sight, or of an abstraction. Visualization has been long used before the advent of the computer era. Typical examples include: geographic maps, nautical charts, weather charts, x-ray images, experimental flow visualization. The primary goal of visualization is to allow better insight into the data. Depending on the data to visualize and the visualization goal an appropriate technique must be selected among the large

Fig. 5. Non-photorealistic medical illustration [CMH⁺01]

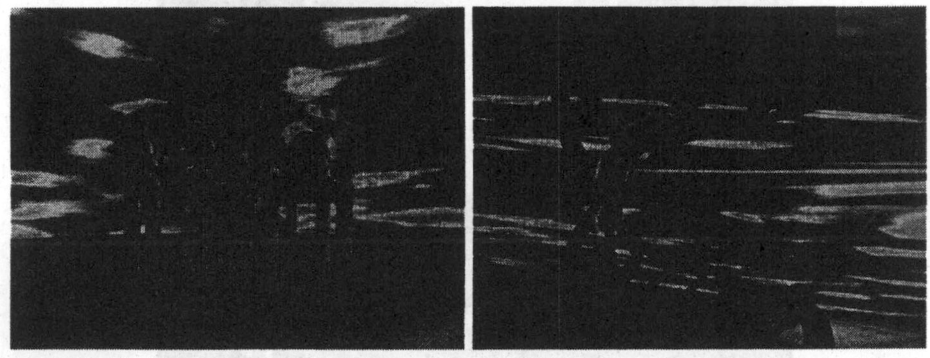

Fig. 6. Perspective projection (left), extended camera (right) [LG96]

variety of available methods. The great diversity of inhomogeneous data sources include medical data, flow data, abstract data, geographic data, historical data (archeology), microscopic data (molecular physics), macroscopic data (astronomy). Data sources differ according to domain (continuous, discrete, grid), data range, dimensionality, data type (scalar, vector, tensor), structuring (sequential, relational, hierarchical, networked). Depending on the level of knowledge about the data scientific visualization tries to accomplish one of three high-level goals. If not much is known about the data, visualization offers exploration tools to support the user in finding hypotheses about the data. In case there are already hypotheses visualization is used to analyze the data with the goal of confirmation or refutation of these hypotheses. In case all relevant data characteristics are already known visualization is an efficient tool for communicating results, e.g., to non experts. Specific visualization goals are: identification (which data values are given in a certain area of the domain), localization (which parts of the domain do have a certain data value), correlation analysis, data comparison (spatial, temporal), distribution analysis (extrema, outliers, frequencies, clusters), categorization and classification.

3.1 Visualization Pipeline

A visualization process encompasses several consecutive steps which are combined into the visualization pipeline (figure 7). **Data acquisition** produces data due to measurements (e.g., computed tomography in medicine), simulations (e.g., finite element methods for flow simulation), or modelling (e.g., differential equations to model a physical phenomenon). **Data enhancement** processes the raw data from the acquisition step. Data enhancement includes filtering (noise removal), resampling (to a different grid), data completion, calculating derived data (e.g., gradient information), data interpolation and reconstruction. **Visualization mapping** assigns visual properties to the data. This is the most crucial step in the visualization pipeline. Depending on the visualization mapping the same data might produce totally different images, and different parts of the data will be emphasized to a varying degree. This is also a crucial difference to photo-realistic computer graphics where real-world scenes are the target to approximate. In many cases with visualization there is no intuitive or obvious geometric representation of a specific data. Therefore only the desired investigation goal guides the selection of the most appropriate visual representation. As an example, medical data on a 3D grid (voxel data) can be visually represented by iso-surfaces. Another rather different representation would be to assign colors and opacities to each of the grid points. High dimensional objects can be represented through a great variety of geometric objects (icons, glyphs) whose

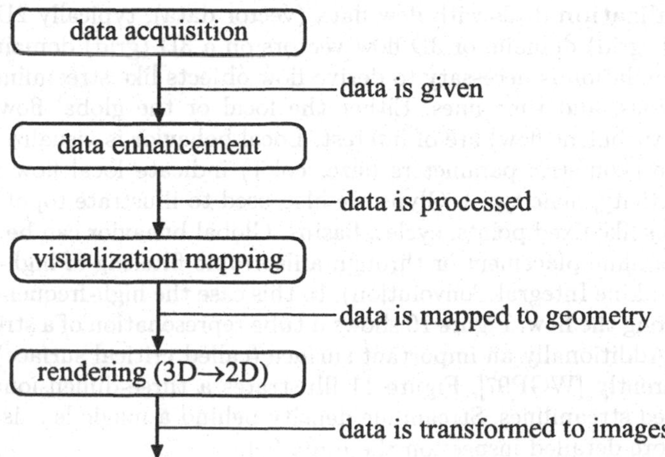

Fig. 7. Visualization pipeline

properties (size, color, shape) encode the underlying information. **Rendering** is the last step in the visualization pipeline and uses computer- graphics techniques (projection, visibility, shading, compositing calculations) to produce images or animation sequences. Interactivity is very important in analyzing the data and can range from interactive manipulation of rendering parameters (e.g., camera modification) to changing visualization-mapping parameters (e.g., changing an iso-value). In the most general case, which usually is quite difficult to achieve, also the data acquisition process is interactively affected (computational steering).

3.2 Volume Visualization, Flow Visualization

Volume visualization and flow visualization are the major subbranches of scientific visualization. **Volume visualization** deals with displaying volume data, i.e., scalar data on a (regular) 3D grid. There are basically two types of visualization mappings for volume data. The first class of techniques derive iso-surfaces, e.g., skin and bone surface, which are rendered with traditional surface-based computer-graphics techniques. A typical representative is the marching-cubes algorithm. The second class of techniques do not produce any intermediate geometric structure but render the volume data directly. So-called transfer functions assign colors and opacities to individual voxels. These visual values are accumulated, e.g., along viewing rays, to produce semitransparent renderings of the volume data. With these techniques also the interior of volumetric structures are made visible. Typical representatives are volume ray casting, splatting, shear-warp-factorization, and Fourier volume rendering. Figure 8 shows the semi-transparent volume-rendering of a CT head. Skin and blood vessels are depicted semi-transparently whereas the skull is rendered opaquely [KG01]. In figure 9 the CT data of a human colon is virtually stretched and unfolded. This allows a fast overview on the entire organ surface and facilitates polyp detection [BWKG01].

Flow visualization deals with flow data (vector data), typically 2D flow vectors on a 2D (grid) domain, or 3D flow vectors on a 3D (grid) domain. Often a numerical simulation is necessary to derive flow objects like streamlines, streaklines, path lines, and time lines. Either the local or the global flow behavior (laminar vs. turbulent flow) are of interest. Local behavior is visualized through glyphs whose geometric parameters (size, color) indicate local flow properties (velocity, vorticity, helicity,...). Glyps are also used to illustrate topological flow characteristics like fixed points, cycles, basins. Global behavior can be visualized through streamline placement or through anisotropic filtering of high-frequency textures (e.g., Line Integral Convolution). In this case the high-frequency texture is smeared along the flow. Figure 10 shows a tube represenation of a streamline in a flow field. Additionally an important surface (called critical surface) is shown semi-transparently [WGP97]. Figure 11 illustrates a three-dimensional flow by a set of dashed streamlines. Streamline density behind a magic lens is increased to allow a more detailed inspection there [FG98].

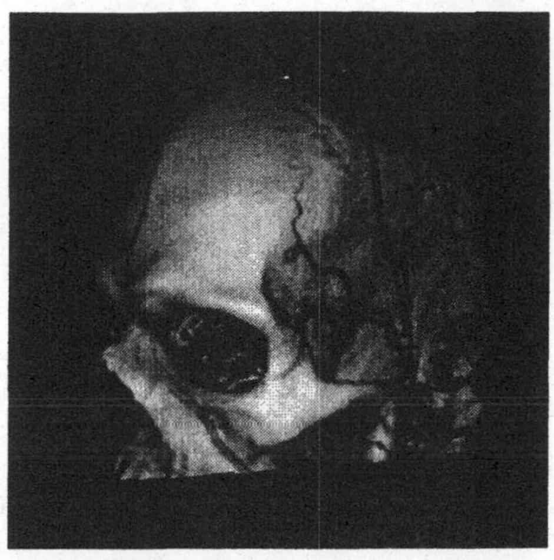

Fig. 8. Semi-transparent volume-rendering of CT head [KG01]

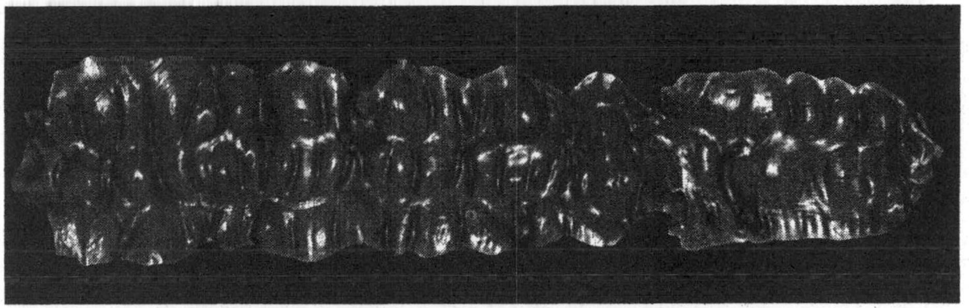

Fig. 9. Virtual colon unfolding for polyp detection [BWKG01]

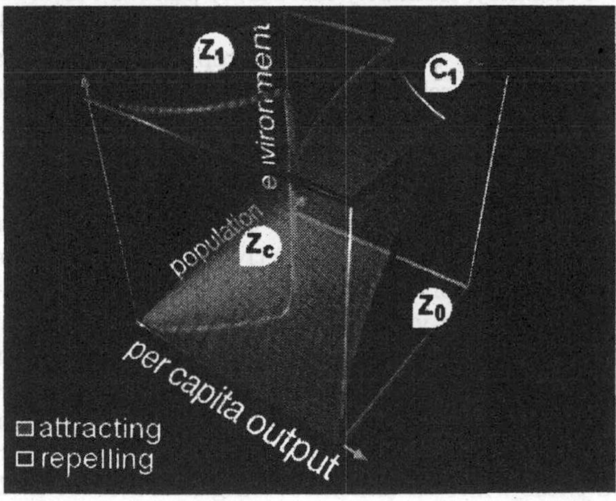

Fig. 10. Flow visualization with streamline and critical surface [WGP97]

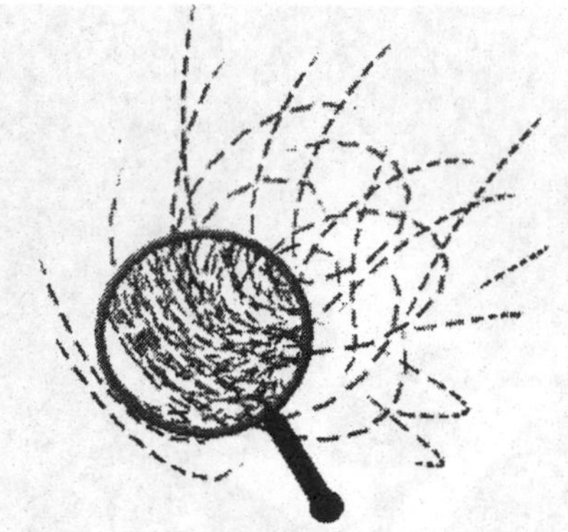

Fig. 11. Magic lens for 3D flow visualization [FG98]

4 Information Visualization

Information visualization is "the use of computer-supported, interactive, visual representations of **abstract** data to amplify cognition" [CMS99]. As opposed to scientific visualization information visualization deals with abstract often high-dimensional data with mostly no inherent spatial structure. Major issues are to specify an appropriate easy to understand visual metaphor and to allow flexible user interaction. The primary goal of information visualization is the explorative investigation of large data bases. After gaining a broad overview of the data, zoom and filter operations are applied, and details are depicted on demand. Abstract data include linear data (tables, program source files, chronological lists), hierarchies (tree structures), networks (general graph structures, webs), attribute metadata (with type, size, age, in n-dimensional space), and information retrieval results (word co-occurency, similarity measures) [And01]. If the number of data elements to display is limited, then two or three dimensional graphical primitives are useful. A file system, for example, can be represented in 2D as a hyperbolic projection or in 3D as a hierarchical arrangement of cones (cone trees, cam trees). If the data base is very large focus-and-context approaches apply. The most interesting (and usually small) part of the data is represented in detail (focus). For orientation purposes the remaining, much larger part of the data is displayed as context information (concise representation on a higher level of abstraction). In the field of document visualization perspective wall and document lens realize such a focus-and-context approach, where the user can interactively determine the focus area. Often fisheye views and distorted views realize the focus-and-context principle. In figure 12 another realization of focus-and-context is shown. SDOF (semantic depth of field) [KMH01] represents objects in the focus sharply and objects in the context blurry. In figure 12 to the left the white knight and those black chessmen threatening it are in focus. In figure 12 to the right the white knight and those white chessmen covering it are in focus. The blurry representation of the remainder allows a preattentive concentration on the objects in focus.

Fig. 12. SDOF: Semantic depth of field [KMH01]

Looking for dependencies in high-dimensional data with many attributes poses the problem of having to represent many dimensions simultaneously. With parallel coordinates individual dimensions are represented as vertical parallel lines. An n-dimensional data point is represented by connecting the individual coordinate components on these parallel lines. Figure 13 shows an extension of this concept: instead of parallel lines parallel planes are used. A six-dimensional curve is shown by projecting it into three planes. Corresponding projected points in adjacent planes are connected with each other [WLG97]. Another possibility

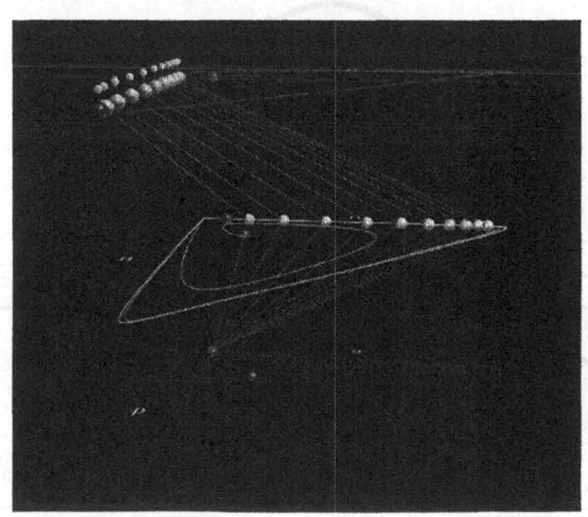

Fig. 13. Extended parallel coordinates

is the mapping of high-dimensional attributes to geometric properties of a two or three dimensional glyph. Individual data dimensions are encoded in length, size, and shape or in the arrangement of geometric objects. The appearance of the glyphs and their position facilitate the perception of dependencies and correlations. Graph drawing algorithms are also a good choice in depicting these interconnections. In general the typically very large and high-dimensional information spaces require efficient search strategies (browsing, dynamic queries, visual data mining). Often different views of the same data are linked together, i.e., manipulating data in one view simultaneously effects all the other views also.

5 Visualization and Graph Drawing

We have now discussed concepts of computer graphics, scientific visualization and information visualization. Comprising these areas broadly under the term visualization we now shortly discuss the connection to graph drawing. Figure 14 puts visualization and graph drawing into context and lists some topics where both areas can profit from each other. Visualization predominantly deals with how to draw objects, whereas graph drawing is often concerned with layout problems, i.e., where to draw objects. Visualization deals with many different types of data and graph drawing is focussed on abstract graph data. Typically visualization handles (very) large data sizes (giga and tera bytes). Due to the high complexity of many layout algorithms graph drawing is concerned with relatively small data sizes. Visualization is an applied field in the sense that the performance of techniques is evaluated through user studies and runtime analyses. Graph drawing is more theoretical by studying theoretical complexity bounds. From the visualization side information visualization is the area closest

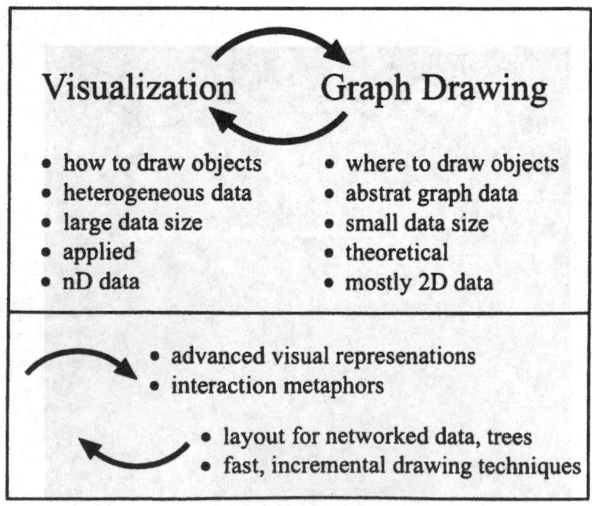

Fig. 14. Comparison: visualization vs. graph drawing

to graph drawing. It is often dealing with networked data that can be represented as graphs, so graph drawing approaches are applicable. Visualization can provide graph drawing with advanced visual representations (shading, visibility, complex node and edge shapes, annotations). Information visualization has also come up with a set of interesting interaction metaphors which might also be useful for graph drawing applications. These are, e.g., focus-and-context, distortion techniques, hierarchical representations (e.g., level-of-detail with smooth transitions), interactive manipulation (insertion or deletion of edges and nodes). Graph drawing on the other hand has useful contributions for visualization whenever correlated and networked data must be displayed as graphs or trees. The layout expertise of the graph drawing community is not yet fully used in the visualization area. In visualization data sets are often incrementally increased or decreased (e.g., subset selection). Therefore it would also be very interesting to get incremental layout strategies from the graph drawing field. Graph drawers know where to put information, visualizers know how to put the information (visual representation). More interaction between these two groups will be fruitful to both areas.

References

[And01] K. Andrews. Information visualisation. http://www2.iicm.edu/ivis, 2001.

[BWKG01] A. Vilanova Bartroli, R. Wegenkittl, A. König, and E. Gröller. Virtual colon unfolding. In *IEEE Visualization 2001 Proceedings*. IEEE Computer Society, October 2001.

[CMH+01] B. Csebfalvi, L. Mroz, H. Hauser, A. König, and E. Gröller. Fast visualization of object contours by non-photorealistic volume rendering. *Computer Graphics Forum*, 20(3):C–452–C–460, September 2001.

[CMS99] St. Card, J. Mackinlay, and B. Shneiderman. *Readings in Information Visualization*. Morgan Kaufmann, 1999.

[FG98] A. Fuhrmann and E. Gröller. Real-time techniques for 3D flow visualization. In *IEEE Visualization 1998 Proceedings*, pages 305–312. IEEE Computer Society, October 1998.

[FvDFH96] J. Foley, A. van Dam, S. Feiner, and J. Hughes. *Computer Graphics: Principles and Practice, Second Edition in C*. Addison-Wesley, Reading, MA, 1996.

[GRS95] E. Gröller, R. T. Rau, and W. Strasser. Modeling and visualizatin of knitwear. *IEEE Transactions on Visualization and Computer Graphics*, 1(4):302–310, 1995.

[HB96] D. Hearn and M. P. Baker. *Computer Graphics, C Version*. Prentice Hall, 1996.

[KG01] A. König and E. Gröller. Mastering transfer function specification by using VolumePro technology. In *Spring Conference on Computer Graphics 2001*, pages 279–286, April 2001.

[KMH01] R. Kosara, S. Miksch, and H. Hauser. Semantic depth of field. In *IEEE Symposium on Information Visualization Proceedings*. IEEE Computer Society, October 2001.

[LG96] H. Löffelmann and E. Gröller. Ray tracing with extended cameras. *Journal of Visualization and Computer Animation*, 7(4):211–228, October 1996.

[SFH+01] D. Schmalstieg, A. Fuhrmann, G. Hesina, Zs. Szalavari, L. M. Encarnação, M. Gervautz, and W. Purgathofer. The Studierstube augmented reality project. In *SIGGRAPH 2001 Course Notes (no. 27, Augmented Reality: The Interface is Everywhere)*. ACM Siggraph, 2001.

[SM00] H. Schumann and W. Müller. *Visualisierung - Grundlagen und allgemeine Methoden*. Springer, 2000.

[Tra95] Ch. Traxler. Representation and realistic rendering of natural scenes with directed cyclic graphs. http://www.cg.tuwien.ac.at/research/rendering/csg-graphs/index.html, 1995.

[WGP97] R. Wegenkittl, E. Gröller, and W. Purgathofer. Visualizing the dynamical behavior of wonderland. *IEEE Computer Graphics & Applications*, 17(6):71–79, December 1997.

[WLG97] R. Wegenkittl, H. Löffelmann, and E. Gröller. Visualizing the behavior of higher dimensional dynamical systems. In *IEEE Visualization '97 Proceedings*, pages 119–125. IEEE Computer Society, October 1997.

Floor-Planning via Orderly Spanning Trees*

Chien-Chih Liao[1], Hsueh-I. Lu[2], and Hsu-Chun Yen[3]

[1] Department of Electrical Engineering, National Taiwan University
Taipei 106, Taiwan, Republic of China
henry@cobra.ee.ntu.edu.tw

[2] Institute of Information Science, Academia Sinica
Taipei 115, Taiwan, Republic of China
hil@iis.sinica.edu.tw
http://www.iis.sinica.edu.tw/~hil

[3] Department of Electrical Engineering, National Taiwan University
Taipei 106, Taiwan, Republic of China
yen@cc.ee.ntu.edu.tw
http://www.ee.ntu.edu.tw/~yen

Abstract. *Floor-planning* is a fundamental step in VLSI chip design. Based upon the concept of *orderly spanning trees*, we present a simple $O(n)$-time algorithm to construct a floor-plan for any n-node plane triangulation. In comparison with previous floor-planning algorithms in the literature, our solution is not only simpler in the algorithm itself, but also produces floor-plans which require fewer module types. An equally important aspect of our new algorithm lies in its ability to fit the floor-plan area in a rectangle of size $(n-1) \times \left\lfloor \frac{2n+1}{3} \right\rfloor$.

1 Introduction

In VLSI chip design, *floor-planning* [20,16] refers to the process of, given a graph whose nodes (respectively, edges) representing functional entities (respectively, interconnections), partitioning a rectangular chip area into a set of non-overlapping rectilinear polygonal modules (each of which describes a functional entity) in such a way that the modules of adjacent nodes share a common boundary. For example, Figure 1(b) is a floor-plan of the graph in Figure 1(a).

Early stage of the *floor-planning* research focused on using *rectangular modules* as the underlying building blocks. A floor-plan using only rectangles to represent nodes is called a *rectangular dual*. It was shown in [13,14,15] that a plane triangulation G admits a rectangular dual if and only if G has four exterior nodes, and G has no *separating triangles*. (A separating triangle, which is also known as complex triangle [21,20], is a cycle of three edges enclosing some nodes in its interior.) As for floor-planning general plane graphs, Yeap and Sarrafzadeh [21] showed that rectilinear modules with at most two concave corners are sufficient and necessary.

* Research supported in part by NSC Grant 90-2213-E-002-100.

P. Mutzel, M. Jünger, and S. Leipert (Eds.): GD 2001, LNCS 2265, pp. 367–377, 2002.
© Springer-Verlag Berlin Heidelberg 2002

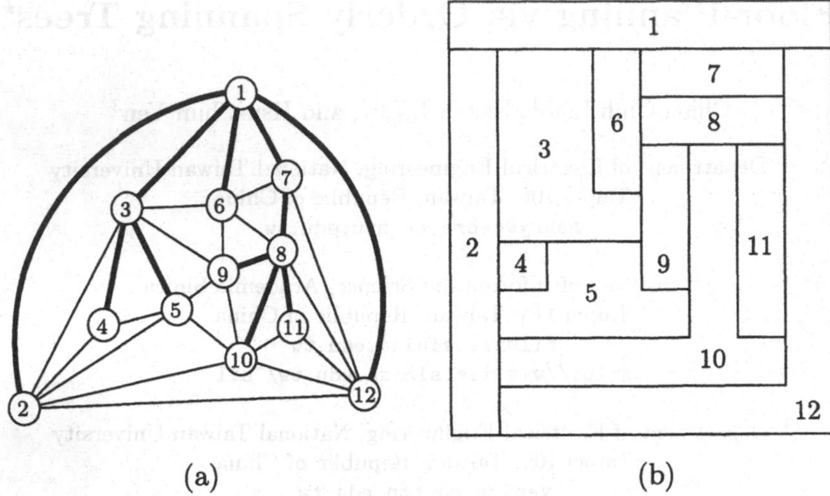

Fig. 1. (a) A plane triangulation G, where an orderly spanning tree T of G rooted at node 1 is drawn in dark. The node labels show the counterclockwise preodering of the nodes in T. (b) A floor-plan of G.

In a subsequent study of floor-planning, He [9] measured the complexity of a module in terms of the number of its constituent rectangles, as opposed to the amount of concave corners. A module that is a union of k or fewer disjoint rectangles is called a *k-rectangular module*. Since any rectilinear module with at most two concave corners can be constructed by three rectangular modules, the result of Yeap and Sarrafzadeh [21] implies the feasibility of floor-planning plane graphs using 3-rectangular modules. He [9] presented a linear-time algorithm to construct a floor-plan of a plane triangulation using only 2-rectangular modules. He's floor-planning algorithms consists of three phases: The first phase utilizes the *canonical ordering* [6,11,12] to assign nodes on separating triangles. The second phase involves the so-called *vertex expansion* operation to break all separating triangles. The third phase adapts rectangular-dual algorithms [1,2,8, 12] to finalize the drawing of the floor-plan. Figure 2 depicts the shapes of the 2-rectangular modules required by He's algorithm. For convenience, these four shapes are referred to as *I-module*, *L-module*, *T-module*, and *Z-module* throughout the rest of this paper.

In this paper, we provide a "simpler" linear-time algorithm that computes "compact" floor-plans for plane triangulations. The "compactness" of the output floor-plans is an important advantage of our algorithm over previous results [9, 21]. Specifically, the output of our algorithm for an n-node plane triangulation has area no more than $(n-1) \times \lfloor \frac{2n+1}{3} \rfloor$. Previous work [9,21], however, reveals no such area information. What "simplicity" means is two-fold:

- First, as opposed to the multiple-phase approach of [9,21], our algorithm is based upon a recent development of *orderly spanning trees* [3], which provides

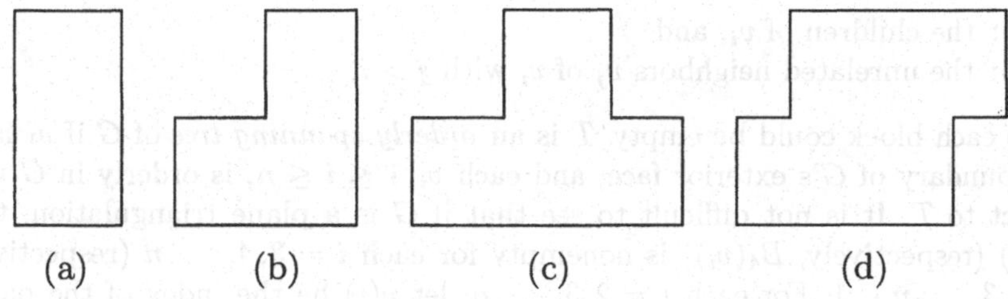

Fig. 2. Four types of modules required by He's floor-planning algorithm [9]: (a) I-module, (b) L-module, (c) T-module, and (d) Z-module. Our algorithm does not need Z-modules.

an extension of *canonical ordering* [6,11,12] to plane graphs not required to be triconnected and an extension for *realizer* [19,18] to plane graphs not required to be triangulated. Our approach bypasses the somewhat complicated rectangular-dual phase. Aside from the two applications of orderly spanning trees reported in [3] (namely, succinct encodings for planar graphs with efficient query support [10,17,4] and 2-visibility drawings for planar graphs [7]), our investigation here finds another interesting application of orderly spanning trees.

- Second, the floor-plan design of our algorithm is "simpler" (in comparison with [9]) in its own right, in the sense that I-modules, L-modules, and T-modules suffice. (Recall that Z-modules are needed by He's algorithm [9].)

The remainder of this paper is organized as follows. Section 2 gives the preliminaries. Section 3 presents our linear-time floor-planning algorithm as well as its correctness proof. Section 4 concludes the paper.

2 Preliminaries

A *plane graph* is a planar graph equipped with a fixed planar embedding. The embedding of a plane graph divides the plane into a number of connected regions, each of which is called a *face*. The unbounded face of G is called the *exterior face*, whereas the remaining faces are *interior faces*. G is a *plane triangulation* if G has at least three nodes and the boundary of each face, including the exterior face, of G is a triangle. Let T be a rooted spanning tree of a plane graph G. Two nodes are *unrelated* in T if they are distinct and neither of them is an ancestor of the other in T. An edge of G is *unrelated* with respect to T if its endpoints are unrelated in T. Let $v_1, v_2, \ldots, v_n$ be the counterclockwise preordering of the nodes in T. A node v_i is *orderly* in G with respect to T if the neighbors of v_i in G form the following four blocks in counterclockwise order around v_i:

$B_1(v_i)$: the parent of v_i,
$B_2(v_i)$: the unrelated neighbors v_j of v_i with $j < i$,

$B_3(v_i)$: the children of v_i, and
$B_4(v_i)$: the unrelated neighbors v_j of v_i with $j > i$,

where each block could be empty. T is an *orderly spanning tree* of G if v_1 is on the boundary of G's exterior face, and each $v_i, 1 \leq i \leq n$, is orderly in G with respect to T. It is not difficult to see that if G is a plane triangulation, then $B_2(v_i)$ (respectively, $B_4(v_i)$) is nonempty for each $i = 3, 4, \ldots, n$ (respectively, $i = 2, 3, \ldots, n-1$. For each $i = 2, 3, \ldots, n$, let $p(i)$ be the index of the parent of v_i in T. Let $w(i)$ denote the number of leaves in the subtree of T rooted at v_i. Let $\ell(i)$ and $r(i)$ be the functions such that $v_{\ell(i)}$ (respectively, $v_{r(i)}$) is the last (respectively, first) neighbor of v_i in $B_2(v_i)$ (respectively, $B_4(v_i)$) in counterclockwise order around v_i. For example, in the example shown in Figure 1(a), one can easily verify that node 3 is indeed orderly with respect to T, where $B_1(3) = \{1\}$, $B_2(3) = \{2\}$, $B_3(3) = \{4, 5\}$, $B_4(3) = \{6, 9\}$, $p(3) = 1$, $w(3) = 2$, $\ell(3) = 2$, and $r(3) = 9$. When G is a plane triangulation, it is known [3] that for each edge (v_i, v_j) of $G - T$ with $i < j$, at least one of $i = \ell(i)$ and $j = r(i)$ holds. To be more specific, if $i = 2$ and $j = n$, then both $3 = \ell(n)$ and $n = r(3)$ hold; otherwise, precisely one of $i = \ell(i)$ and $j = r(i)$ holds.

The concept of orderly spanning tree for planar graphs [3] extends that of *canonical ordering* [6,11,12] for plane graphs not required to be triconnected and that of *realizer* [19,18,5] for plane graphs not required to be triangulated. Specifically, when G is a plane triangulation, (i) if T is an orderly spanning tree of G, then the counterclockwise preordering of the nodes of T is always a canonical ordering of G, and (ii) if (T_1, T_2, T_n) is a realizer of G, where T_i is rooted at v_i for each $i = 1, 2, n$, then each T_i plus both external edges of G incident to v_i is an orderly spanning tree of G. Our floor-planning algorithm is based upon the following lemma.

Lemma 1 (see [3]). *Given an n-node plane triangulation G, an orderly spanning tree T of G with at most $\lfloor \frac{2n+1}{3} \rfloor$ leaves is obtainable in $O(n)$ time.*

A *floor-plan* F of G is a partition of a rectangle into n non-overlapping rectangular modules $r_1, r_2, \ldots, r_n$ such that v_i and v_j are adjacent in G if and only if the boundaries of r_i and r_j share at least one non-degenerated line segment. The *size* of F is the area of the rectangle being partitioned by F with the convention that the corners of all modules are placed on integral grid points. For example, the size of the floor-plan shown in Figure 1(b) is 8×7.

3 Our Floor-Planning Algorithm

This section proves the following main theorem of the paper.

Theorem 1. *Given an n-node plane triangulation G with $n \geq 3$, a floor-plan F of G can be constructed in $O(n)$ time such that*

1. *F consists of I-modules, L-modules, and T-modules only, and*
2. *the size of F is bounded by $(n-1) \times \lfloor \frac{2n+1}{3} \rfloor$.*

Let T be an orderly spanning tree of G, where $v_1, v_2, \ldots, v_n$ is the counterclockwise preordering of T. Our floor-planning algorithm is described as follows. Pictures of intermediate steps are shown to illustrate how our algorithm obtains the floor-plan in Figure 1(b) for the plane graph G with respect to the orderly spanning tree T shown in Figure 1(a).

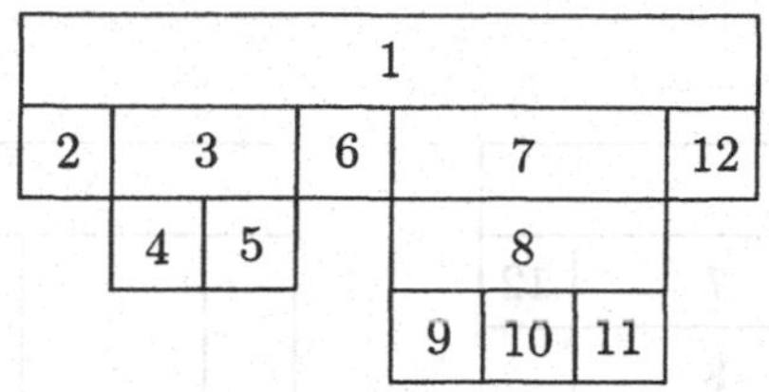

Fig. 3. Step 1: visibility drawing of T.

Algorithm FLOORPLAN(G, T)

Step 1. Produce a (vertical) visibility drawing of T as follows: For each $i = 1, 2, \ldots, n$, if v_i is a leaf of T, then draw v_i as a unit square; otherwise, draw v_i as a $1 \times w(i)$ rectangle. Place each node beneath its parent such that the children of each node is placed in the same order as in T.

Comment. For example, Figure 3 shows the resulting visibility drawing for the T shown in Figure 1(a).

Step 2. Turn the above visibility drawing of T into a 2-visibility drawing of G by stretching the nodes downward in the least necessary amount such that v_i and v_j are horizontally visible to each other if and only if v_i and v_j is an unrelated edge of G with respect to T.

Comment. For example, Figure 4 shows how to obtain the resulting 2-visibility drawing for the plane triangulation shown in Figure 1(a). As a matter of fact, this is how Chiang et al. [3] obtained their 2-visibility drawing of G with respect to T with size at most $(n-1) \times w(v_1)$. Note that the resulting drawing still satisfies the property that the bottom boundary of each internal node of T is completely occupied by the top boundaries of its children.

Step 3. First, grow a horizontal branch for v_n from boundary of v_n visible to v_2 such that the left boundary of the horizontal branch touches v_2. Second, for each $i = 3, 4, \ldots, n-1$, grow horizontal branches for v_i from the boundaries of v_i visible to $v_{\ell(i)}$ and $v_{r(i)}$ such that the left (respectively, right) boundary of the horizontal branches touches $v_{\ell(i)}$ (respectively, $v_{r(i)}$). Furthermore, when extending the boundary of v_i, we also extend the boundaries of the descendants of v_i to maintain the property that the bottom boundary of

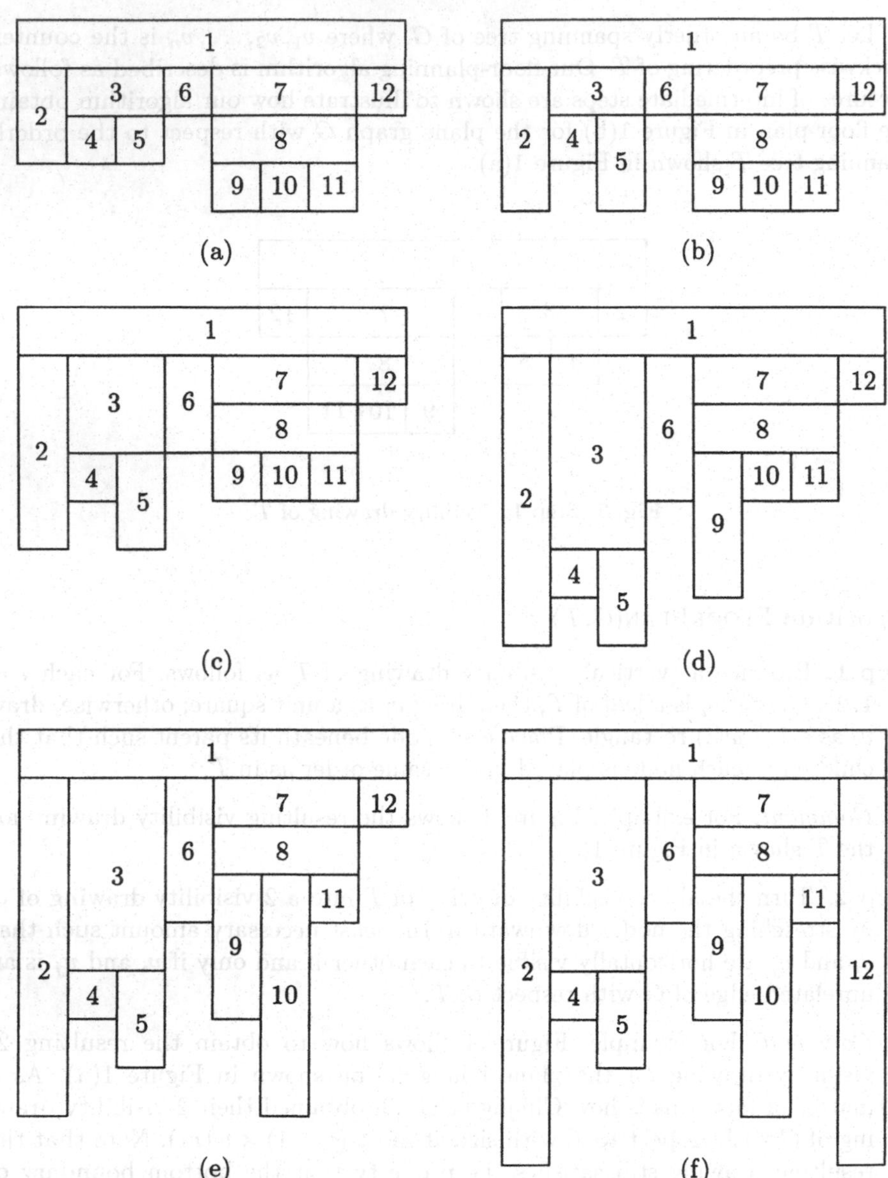

Fig. 4. Step 2: ensuring the horizontal visibility between v_i and each node in $B_2(v_i)$ for (a) nodes 3 and 4, (b) node 5, (c) nodes 6–8, (d) node 9, (e) node 10, and (f) nodes 11 and 12.

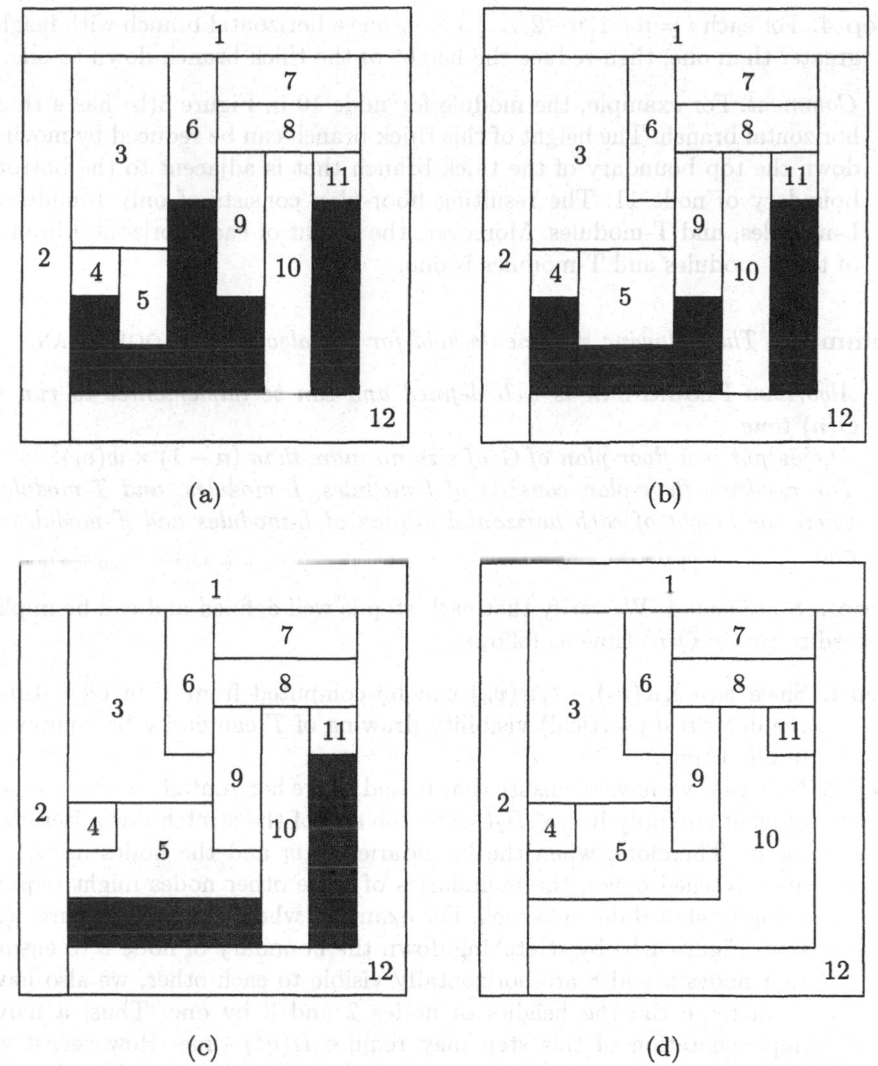

Fig. 5. Step 3: growing the horizontal branches for (a) node 12, (b) node 3, (c) nodes 4 and 5, and (d) nodes 6–11.

each internal node of T is completely occupied by the top boundaries of its children.

Comment. For example, Figure 5 illustrates how to obtain the resulting drawing for the plane triangulation shown in Figure 1(a). Note that when the horizontal branch of node 3 is extended to the right by one unit to touch the left boundary of node 9, the right boundary of node 5 is also extended to the right by the same amount.

Step 4. For each $i = n-1, n-2, \ldots, 3$, if v_i has a horizontal branch with height greater than one, then reduce the height of the thick branch down to one.

Comment. For example, the module for node 10 in Figure 5(b) has a thick horizontal branch. The height of this thick branch can be reduced by moving down the top boundary of the thick branch that is adjacent to the bottom boundary of node 11. The resulting floor-plan consists of only I-modules, L-modules, and T-modules. Moreover, the height of each horizontal branch of the L-modules and T-modules is one.

Lemma 2. *The following statements hold for our algorithm* FLOOR-PLAN.

1. *Algorithm* FLOORPLAN *is well defined and can be implemented to run in $O(n)$ time.*
2. *The output is a floor-plan of G of size no more than $(n-1) \times w(v_1)$.*
3. *The resulting floor-plan consists of I-modules, L-modules, and T-modules, where the height of each horizontal branch of L-modules and T-modules is one.*

Proof. Statement 1. We verify that each step is well defined and can be implemented to run in $O(n)$ time as follows.

Step 1. Since $w(v_1), w(v_2), \cdots, w(v_n)$ can be computed from T in $O(n)$ time, the described (vertical) visibility drawing of T can easily be computed in $O(n)$ time.

Step 2. Note that we have to ensure that v_i and v_j are horizontally visible to each other if and only if $v_j \in B_2(v_i)$ at the end of the stretch-down iteration for v_i. Therefore, when the boundaries of v_i and the nodes in $B_2(v_i)$ are stretched down, the boundaries of some other nodes might require being stretched down as well. For example, when we obtain Figure 4(c) from Figure 4(b) by stretching down the boundary of node 6 to ensure that nodes 6 and 8 are horizontally visible to each other, we also have to increase the the heights of nodes 2 and 3 by one. Thus, a naive implementation of this step may require $\Omega(n^2)$ time. However, if we mark the line segments to be stretched down that is to the left of v_j and do not actually stretch down the nodes until the completion of the for-loop, then this step can be implemented to run in $O(n)$ time.

Step 3. A naive implementation of this step may require $\Omega(n^2)$ time, since growing the horizontal branches for a node may cause boundary change for its descendants. However, the time complexity can be reduced to $O(n)$ by adapting the "lazy strategy" similar to the one used in the previous step. Since each unrelated edge (v_i, v_j) of $G - T$ with $i < j$ and $(v_i, v_j) \neq (v_2, v_n)$ satisfies exactly one equality of $i = \ell(j)$ and $j = r(i)$, this step is well defined. For the same reason, the resulting drawing is a partition of a rectangle into n rectilinear regions. (That is, there is no gap among modules in the rectangle.) To prove that the resulting drawing is indeed a floor-plan of G, it suffices to show that growing a horizontal branch of v_i is to reach the boundary of v_j does not result in new adjacency among these rectilinear modules. Suppose v_k is a node whose bottom boundary touches the top bottom of the horizontal branch of v_i. Assume for a contradiction that v_k is not adjacent to v_i in G. Since the resulting drawing of the previous step is a 2-visibility drawing of G, there must be a node $v_{k'}$ lies between v_i and v_k preventing their horizontal visibility to each other. It follows that there is a face of G containing at least four nodes $v_i, v_j, v_k, v_{k'}$, contradicting the fact that G is triangulated.

Step 4. By the fact that T is an orderly spanning tree of G and G is a plane triangulation, one can see that if v_i grows a horizontal branch to reach v_j, then there must be a unique node v_k whose bottom boundary touches the top boundary of that horizontal branch of v_i. It is also not difficult to verify that both (v_i, v_k) and (v_j, v_k) are unrelated edges G with respect to T. Thus, in the resulting drawing of the previous step, the left and right boundaries of v_k have to touch v_i and v_j. Therefore, the height of that horizontal branch of v_i can be reduced to one by moving downward the bottom boundary of v_k, which is also the top boundary of that horizontal branch, without changing the adjacency of v_k to other nodes in the floor-plan. Clearly, each height-reducing operation takes $O(1)$ time, so this step runs in $O(n)$ time. Since the for-loop of this step proceeds from $i = n-1$ down to 3, each horizontal branch has height exactly one at the end of this step.

Statement 2. As shown by Chiang et al. [3], the resulting drawing of Step 2 is a 2-visibility of G of size no more than $(n-1) \times w(v_1)$. Since Steps 3 and 4 do not affect the adjacency among the rectilinear modules, the statement is proved.

Statement 3. By the definition of Step 3, one can easily verify that the resulting floor-plan consists of I-modules, L-modules, and T-modules. By the height-reducing operation performed on the horizontal branches in Step 4, the statement is proved. □

We are ready to prove the main theorem as follows.

Proof. [for Theorem 1] Straightforward by Lemmas 1 and 2. □

4 Conclusion

A linear-time algorithm for producing compact floor plans for plane triangulations has been designed, Our algorithm is based upon a newly developed technique of orderly spanning trees with bounded number of leaves [3]. In comparison with previous work on floor-planning plane triangulations [9], our algorithm is simpler in the algorithm itself as well as in the resulting floor-plan in the sense that the Z-modules required by [9] is not needed in our design. Another important feature of algorithm algorithm is the upper bound $(n-1) \times \lfloor \frac{2n+1}{3} \rfloor$ on the area of the output floor-plan. Previous work [9,21] does not provide any area bounds on their outputs. Investigating whether the $(n-1) \times \lfloor \frac{2n+1}{3} \rfloor$ area is worst-case optimal is an interesting future research direction. Another interesting possibility is modifying our floor-plan algorithm into an algorithm to compute rectangular-dual with bounded area for plane triangulation without separating triangles.

References

1. J. Bhasker and S. Sahni. A linear algorithm to check for the existence of a rectangular dual of a planar triangulated graph. *Networks*, 17:307–317, 1987.
2. J. Bhasker and S. Sahni. A linear algorithm to find a rectangular dual of a planar triangulated graph. *Algorithmica*, 3:247–278, 1988.
3. Y.-T. Chiang, C.-C. Lin, and H.-I. Lu. Orderly spanning trees with applications to graph encoding and graph drawing. In *Proceedings of the 12th Annual ACM-SIAM Symposium on Discrete Algorithms*, pages 506–515, Washington, D. C., USA, 7–9 Jan. 2001. A revised and extended version can be found at `http://xxx.lanl.gov/abs/cs.DS/0102006`.
4. R. C.-N. Chuang, A. Garg, X. He, M.-Y. Kao, and H.-I. Lu. Compact encodings of planar graphs via canonical ordering and multiple parentheses. In K. G. Larsen, S. Skyum, and G. Winskel, editors, *Proceedings of the 25th International Colloquium on Automata, Languages, and Programming*, Lecture Notes in Computer Science 1443, pages 118–129, Aalborg, Denmark, 1998. Springer-Verlag.
5. H. de Fraysseix, P. Ossona de Mendez, and P. Rosenstiehl. On triangle contact graphs. *Combinatorics, Probability and Computing*, 3:233–246, 1994.
6. H. de Fraysseix, J. Pach, and R. Pollack. How to draw a planar graph on a grid. *Combinatorica*, 10:41–51, 1990.
7. U. Fößmeier, G. Kant, and M. Kaufmann. 2-visibility drawings of planar graphs. In S. North, editor, *Proceedings of the 4th International Symposium on Graph Drawing*, Lecture Notes in Computer Science 1190, pages 155–168, California, USA, 1996. Springer-Verlag.
8. X. He. On finding the rectangular duals of planar triangular graphs. *SIAM Journal on Computing*, 22:1218–1226, 1993.
9. X. He. On floor-plan of plane graphs. *SIAM Journal on Computing*, 28(6):2150–2167, 1999.
10. G. Jacobson. Space-efficient static trees and graphs. In *Proceedings of the 30th Annual Symposium on Foundations of Computer Science*, pages 549–554, Research Triangle Park, North Carolina, 30 Oct.–1 Nov. 1989. IEEE.
11. G. Kant. Drawing planar graphs using the canonical ordering. *Algorithmica*, 16(1):4–32, 1996.
12. G. Kant and X. He. Regular edge labeling of 4-connected plane graphs and its applications in graph drawing problems. *Theoretical Computer Science*, 172(1-2):175–193, 1997.
13. K. Koźmiński and E. Kinnen. Rectangular duals of planar graphs. *Networks*, 15(2):145–157, 1985.
14. K. A. Kózmiński and E. Kinnen. Rectangular dualization and rectangular dissections. *IEEE Transactions on Circuits and Systems*, 35(11):1401–1416, 1988.
15. Y. T. Lai and S. M. Leinwand. A theory of rectangular dual graphs. *Algorithmica*, 5(4):467–483, 1990.
16. K. Mailing, S. H. Mueller, and W. R. Heller. On finding most optimal rectangular package plans. In *Proceedings of the 19th Annual IEEE Design Automation Conference*, pages 263–270, 1982.

17. J. I. Munro and V. Raman. Succinct representation of balanced parentheses, static trees and planar graphs. In *Proceedings of the 38th Annual Symposium on Foundations of Computer Science*, pages 118–126, Miami Beach, Florida, 20–22 Oct. 1997. IEEE.
18. W. Schnyder. Planar graphs and poset dimension. *Order*, 5:323–343, 1989.
19. W. Schnyder. Embedding planar graphs on the grid. In *Proceedings of the First Annual ACM-SIAM Symposium on Discrete Algorithms*, pages 138–148, 1990.
20. S. Tsukiyama, K. Koike, and I. Shirakawa. An algorithm to eliminate all complex triangles in a maximal planar graph for use in VLSI floorplan. In *Proceedings of the IEEE International Symposium on Circuits and Systems*, pages 321–324, 1986.
21. K.-H. Yeap and M. Sarrafzadeh. Floor-planning by graph dualization: 2-concave rectilinear modules. *SIAM Journal on Computing*, 22(3):500–526, 1993.

Disconnected Graph Layout and the Polyomino Packing Approach

Karlis Freivalds[1], Ugur Dogrusoz[2], and Paulis Kikusts[1]*

[1] Institute of Mathematics and Computer Science, Univ. of Latvia, Riga, Latvia
{karlisf,paulis}@mii.lu.lv
[2] Computer Engineering Department, Bilkent Univ., Ankara, Turkey
ugur@cs.bilkent.edu.tr

Abstract. We review existing algorithms and present a new approach for layout of disconnected graphs. The new approach is based on polyomino representation of components as opposed to rectangles. The parameters of our algorithm and their influence on the drawings produced as well as a variation of the algorithm for multiple pages are discussed. We also analyze our algorithm both theoretically and experimentally and compare it with the existing ones. The new approach produces much more compact and uniform drawings than previous methods.

1 Introduction

Graphs model the complex information of a system of discrete objects and their relationship. *Graph layout* is the automatic positioning of the nodes and edges of a graph in order to produce an aesthetically pleasing drawing that is easy to comprehend [5,8].

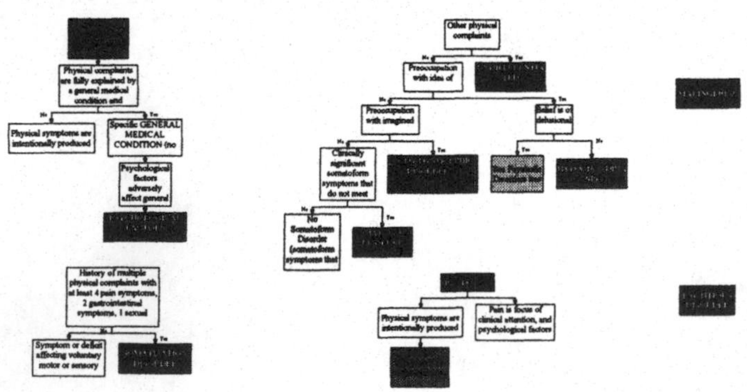

Fig. 1. An example of a disconnected graph.

* Research supported in part by NIST, Advanced Technology Program grant number 70NANB5H1162 and Tom Sawyer Software, Oakland, CA, USA.

P. Mutzel, M. Jünger, and S. Leipert (Eds.): GD 2001, LNCS 2265, pp. 378–391, 2002.
© Springer-Verlag Berlin Heidelberg 2002

Many graph layout and editing systems have been developed in the past [5, 7,11]. One essential aspect that has not been addressed sufficiently, is the layout of disconnected graphs; that is, the placement of the components (possibly consisting of a single isolated node) of a disconnected graph. Disconnected graphs occur rather frequently in real life applications either during the construction of a graph interactively or because of the nature of the application (Figure 1).

Most graph layout algorithms assume a graph to be connected and try to minimize the area needed for the resulting drawing. No matter how effective such an algorithm is, the space wasted overall could be arbitrarily large if the relative locations of disconnected objects of a graph are chosen by a naive, inefficient method.

Another key parameter here is the aspect ratio of the region (e.g., a window) within which the graph is to be displayed (Figure 2). When displaying a graph, the larger the wasted space is, the less visible objects will be, making the visualization process more difficult. Thus, a disconnected graph layout algorithm must strive for a packing of disconnected objects which respects the aspect ratio of the region in which it is to be displayed.

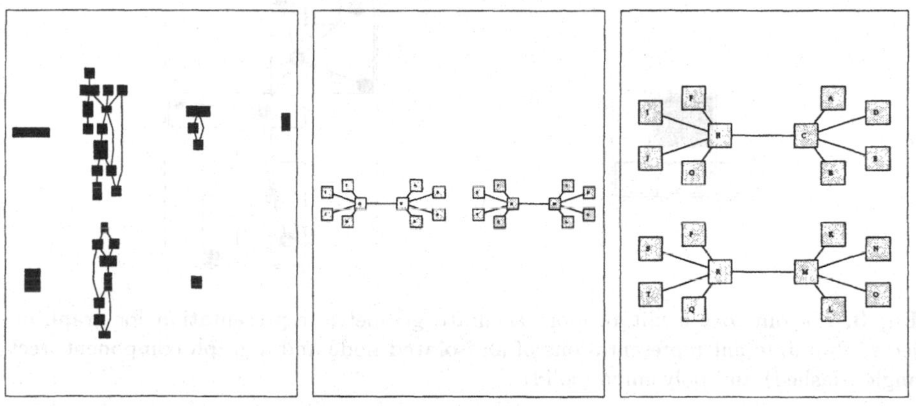

Fig. 2. How a naive disconnected graph layout algorithm can make inefficient use of the area **(left)**, and why the aspect ratio of the region in which the graph is to be drawn should be taken into account during disconnected graph layout **(middle and right)**.

In this paper, we review existing two-dimensional packing algorithms for the layout of disconnected graphs for a specified aspect ratio based on *strip-packing*, *tiling*, and *alternate-bisection* methodologies [6], and introduce a new algorithm that represents disconnected objects with polyominoes as opposed to rectangles. We also discuss the experimental results obtained and compare our algorithm with the previous ones. As expected, the new approach, which uses a more accurate representation for the objects, produces much more compact results. The drawings are also more aesthetically pleasing as the new approach places the objects more uniformly.

2 Definitions and Basics

Throughout the paper, the terms "graph object", or simply "object", are used interchangeably to denote a component or an isolated node of the graph to be laid out.

The tightest rectangle bounding the drawing of a graph object or the entire graph is said to be its *bounding rectangle.* The size of a graph's drawing is identified with its bounding rectangle's. The *aspect ratio* of a rectangle $R = (W, H)$ is equal to W/H.

Most layout algorithms represent graph objects with either points or rectangles in the plane. A *polyomino* is a geometric figure formed by joining unit squares at the edges. In our new approach we use polyominoes to represent graph objects (Figure 3).

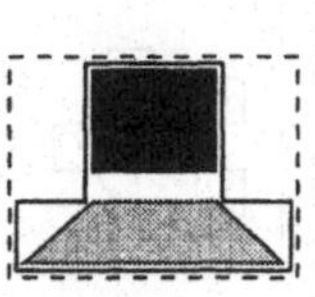

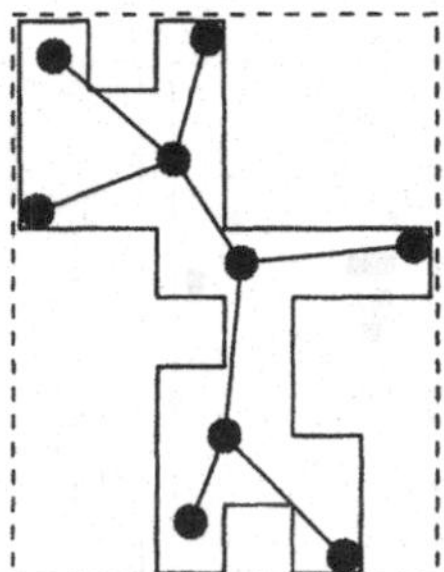

Fig. 3. Polyominoes facilitate more accurate geometric representation for graph objects. Two different representations of an isolated node and a graph component: rectangle (dashed) and polyomino (solid).

The task for a disconnected layout algorithm is to position a set of objects represented by rectangles or polyominoes with ordered dimensions (i.e., no rotations allowed) such that no pair of objects overlap and the area of the bounding rectangle of the drawing is minimized, respecting the aspect ratio of the region in which the graph is to be displayed. There has been extensive research done on two-dimensional packing of rectangles [3,1,4]. In the graph layout version of the problem, the user also specifies a *desired aspect ratio* for the resulting drawing so that the scaling that needs to be done before displaying the graph is minimal (Figure 2).

For an arbitrary list of n objects L_n, or simply L, let $\mathrm{A}^A(L)$ denote the area actually used by a particular algorithm A when applied to L. The *wasted space* is the unoccupied area of the packing: $\mathrm{WS}^A(L) = \mathrm{A}^A(L) - \sum_{i=1}^{n} \mathrm{A}_i$, where

A_i is the area of object L_i. Similarly, the *fullness* of a packing expresses, in percentage, how effectively the area is used by the packing algorithm: $\mathrm{F}^A(L) = 100 \cdot (\sum_{i=1}^{n} A_i)/\mathrm{A}^A(L)$. The *adjusted fullness* of a packing $\mathrm{AF}^A(L)$ ($\leq \mathrm{F}^A(L)$), expresses the fullness of a packing, in percentage, with respect to the desired aspect ratio. To be precise, it considers the additional area wasted when the final drawing is displayed in a region of desired aspect ratio DAR: $\mathrm{AF}^A(L) = \mathrm{F}^A(L) \cdot \frac{\mathrm{AR}^A(L)}{\mathrm{DAR}}$ where $\mathrm{AR}^A(L)$ is the aspect ratio of the packing produced by algorithm A when applied to objects L and we assume $\mathrm{AR}^A \leq \mathrm{DAR}$. Figure 4 illustrates this with an example, in which $\mathrm{F}^A(L) = \frac{A}{A+B}$, whereas $\mathrm{AF}^A(L) = \frac{A}{A+B+C}$.

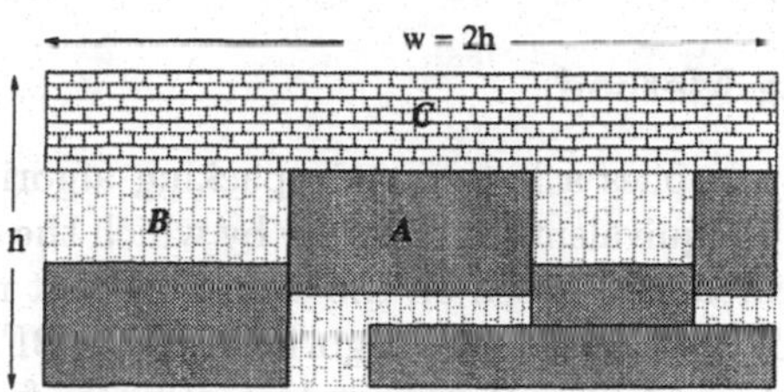

Fig. 4. Total area in which rectangles are packed is divided into three disjoint regions A (rectangles), B (wasted area), and C (additional area wasted when displayed in a region of aspect ratio 2).

Strip-Packing: One can find substantial literature on the design and analysis of algorithms for two-dimensional packing [1,3,4], the most popular version being *strip-packing.* In strip-packing, given a list of $n \geq 1$ rectangles $L_n = (R_1, \ldots, R_n)$, each having ordered dimensions (W_i, H_i), they are to be packed into a semi-infinite strip of unit width without any overlaps in order to minimize the height of the packing. This problem has applications in many areas including stock-cutting, two-dimensional storage problems, and resource-constrained scheduling in computer systems [2].

The most popular approach to strip-packing is the level algorithms. *First-Fit Decreasing Height* (FFDH) is a level algorithm, in which, at any point in the packing sequence, the next rectangle to be packed is placed left-justified on the first level on which it will fit. If none of the current levels will accommodate this rectangle, a new level is started. *Best-Fit Decreasing Height* (BFDH) is similar to FFDH except that the rectangles are packed, whenever possible, on current levels where they fit best.

For an arbitrary list of n rectangles L_n, all assumed to have width no greater than 1, $\mathrm{OPT}(L)$ denotes the minimum possible bin height within which rectangles in L can be packed.

Ordered One-Dimensional Packing: For a set of rectangles, *one-dimensional packing* or simply 1D packing along x-axis (y-axis) corresponds to the process of ordering these rectangles with respect to their x-coordinates (y-coordinates) without any overlaps to *minimize* the total width (height) of the bounding rectangle. If the current relative positions of rectangles are to be preserved in the packing, we call it *ordered 1D packing*. Ordered 1D packing of n objects can be performed in $O(n \log n)$ time [12].

3 Related Work

In this section, we review the existing algorithms for disconnected graph layout, which represent disconnected objects with rectangles. Detailed information on these algorithms may be found in [6].

3.1 Strip-Packing Method

This method directly applies a known strip-packing algorithm such as BFDH. The width of the strip (equivalently, the factor by which the rectangle dimensions are to be scaled) is calculated based on the desired aspect ratio, using the theoretical performance of the strip-packing algorithm. With BFDH, assuming object dimensions to be independent uniform random samples from the interval $[0, 1]$ and $\mathrm{OPT}(L) \approx n/4$, the expected value of adjusted fullness, $E[\mathrm{AF}^{BFDH}(L_n)]$, is shown to be 58.8 [6]. However, these calculations are based on the worst-case performance bounds and are rarely met in practice, making it not a particularly good "guess" for the bin width.

The algorithm is of $O(n \log n)$ time complexity.

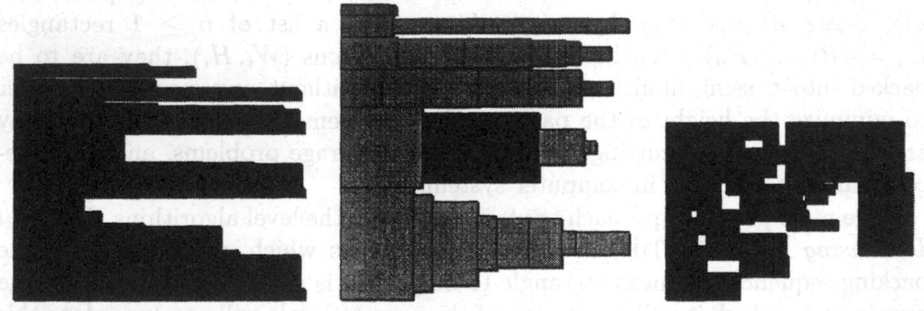

Fig. 5. The same graph laid out with strip-packing, tiling, and alternate-bisection methods, respectively for desired aspect ratio 1.0.

3.2 Tiling: Strip-Packing with Variable Width Strip

The tiling method eliminates the need to "guess" the right size strip by maintaining a bin whose width *dynamically* changes (i.e., increases). The algorithm starts by creating an initial level and placing the first rectangle in this level. It proceeds by determining whether the next rectangle in line should be added to one of the existing levels (the one which is the least utilized at the moment) or to a newly created level. The rectangle is tiled on one of the existing levels if there is enough room. Otherwise, a decision is made on whether the current strip width should be enlarged or a new level should be formed to keep the aspect ratio closer to the desired one.

In general, the tiling algorithm does not assume any particular ordering of the objects. However, experiments show that when graph objects are sorted in nonincreasing height, most compact drawings are obtained. Notice that when objects are processed in order of nonincreasing height, the algorithm turns into a variation of a strip-packing algorithm, BFDH to be more specific, where the strip width is dynamically increased as necessary to better fulfill the aspect ratio constraint.

The algorithm is of $O(n \log n)$ time complexity.

3.3 Alternate-Bisection Method

This divide-and-conquer method works by bisecting the disconnected objects of a graph alternately as follows. The objects are bipartitioned using a metric such as total area and objects in each partition are recursively laid out. The recursion continues until a partition consists of a small, constant number of objects (e.g., one) whose optimal layout becomes easy if not trivial. At the end of each recursive step, when placing the two embedded partitions relatively, the orientation is alternated. For instance, the last step would place the two already positioned partitions side by side (horizontally) if the four partitions in the previous step were placed one on top of the other (vertically) pairwise.

The theoretical analysis prove that the total area wasted by the algorithm for n objects, $W(n)$, is roughly $O(n^{1.41})$ [6], which is quite inefficient. However, when simple alternating ordered 1D packings are applied in each recursive step (e.g., objects in upper (lower) left partition are packed downwards (upwards), towards the horizontal separating axis in Figure 6), the experimental results show that much more compact results are obtained [6]. The overall time complexity of the algorithm is $O(n \log^2 n)$.

For independent, uniformly distributed random object dimensions, this algorithm will not favor one orientation over the other and yield "square-like" drawings. The desired aspect ratio can be respected by this algorithm by initially recursively partitioning the set of objects into two, one partition to be laid out with aspect ratio 1.0 and the other with $\text{DAR}(L) - 1.0$ $(= \frac{w-h}{h})$, assuming $\text{DAR}(L) = \frac{w}{h} > 1.0$ (Figure 6). Alternatively, the object dimensions can be scaled with respect to the desired aspect ratio as a preprocessing step, after which, the desired aspect ratio may be assumed to be 1.0.

3.4 Comparison of the Methods

In [6], experiments with graphs laid out with random aspect ratio and with random object dimensions are presented. Notice here that the graph objects

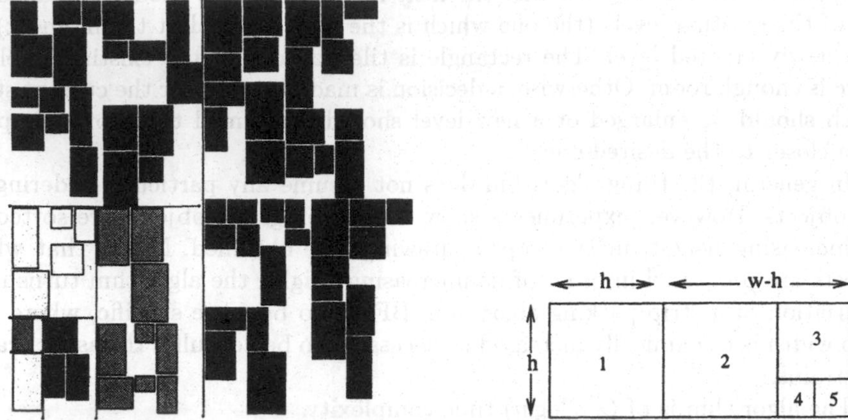

Fig. 6. An example of the alternate-bisection method; on one branch of the recursion, alternately partitioned objects are shown with separating lines and different colors **(left)**. An example of how the alternate-bisection method can be adapted to an arbitrary aspect ratio **(right)**.

are represented with rectangles and the area wasted by such representation is ignored. In the context of graph layout, it is argued that the object dimensions are not completely of uniform distribution since the two types of disconnected objects, isolated nodes and larger components, in most cases will be of highly varying dimensions. Experiments conducted with two groups of objects with dimensions uniformly distributed within each group but with different means are presented in Figure 7.

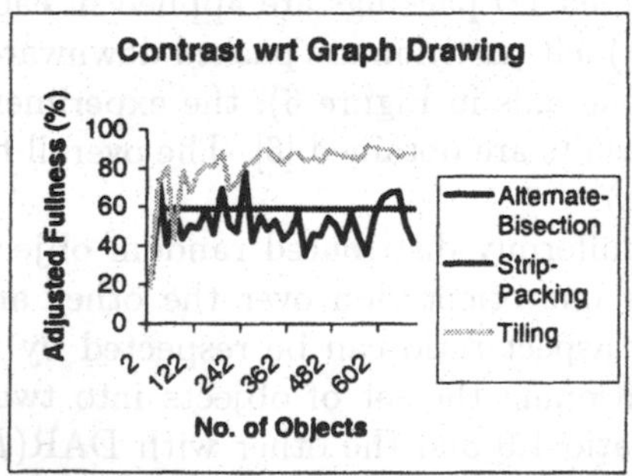

Fig. 7. Comparison of the performance of the three methods using a distribution model of dimensions that is more suitable in the context of graph layout.

In terms of execution time, both split-packing and tiling methods are superb since they are of $O(n \log n)$. The alternate-bisection method, on the other hand, gets a little slow as the number of objects are over a thousand. Considering that most interactive graph drawing applications will not consist of more than, say one hundred, disconnected objects, this method is also of practical value.

In terms of the quality of the packings produced, the experiments show that the tiling method clearly produces the most compact drawings. However, the results obtained from the alternate-bisection method tend to be more aesthetically pleasing since the objects are generally distributed more uniformly (Figure 5).

4 Polyomino Packing Approach

In this approach each graph object is represented by a polyomino. We define a polyomino as a finite set of $k \geq 1$ cells of the infinite planar square grid G that are fully or partially covered by the drawing of the object. If the case that an object is placed completely inside another one is not desirable, the definition can be modified and the uncovered grid cells that are completely bounded by the covered ones can be included as well.

Given a set of polyominoes $P_i, 1 \leq i \leq n$, packing them into a minimum area is clearly NP-hard, even when the polyominoes are restricted to rectangles [0]. Our heuristic algorithm for polyomino packing is a greedy one: it places the objects one by one, finding the optimal place for the new object, one at a time, with respect to the already placed ones. The optimal place for a polyomino is simply calculated as the grid cell G_{xy} located at (x, y) where the function $\max(|x|, |y|)$ is minimized over all grid cells. The cost function defines the order in which the cells are examined and this order is the same for all polyominoes.

A grid data structure is used to represent free grid cells, which are later marked as occupied as polyominoes are placed. In order to find the best place the algorithm PackPolyominoes looks sequentially through all cells in the increasing order of the cost function defined above. If an available spot (a set of unoccupied grid cells where the polyomino fits) is found, it is placed there and the corresponding grid cells are marked as occupied. When testing for intersections, we simply go through all polyomino cells and test whether each can be placed in a free grid cell.

Our experiments show that the quality of the packing depends very much on the order in which the polyominoes are processed. The best results are obtained when they are ordered and processed from the largest to the smallest, which conforms to the ordering in heuristic approaches for the bin packing [2] and strip-packing problems [3]. The size of each polyomino can be defined in several ways including its number of cells (i.e., area) and the perimeter of its bounding rectangle. Experiments show that both give similar results, so we choose the perimeter of the bounding rectangle for calculating the object sizes for the ease of implementation.

Here is a pseudo code of our algorithm:

algorithm PackPolyominoes($P_i, 1 \leq i \leq n$)
(1) sort $P_i, 1 \leq i \leq n$ in the order of nonincreasing size
(2) initialize the grid G using the sizes of $P_i, 1 \leq i \leq n$
(3) **foreach** polyomino P_i **do**

```
        calculate (x, y) such that the cost function is minimized
        while cannot place P_i in G centered at (x, y) do
            calculate next (x, y) using the cost function
        end while
        mark the cells in G covered by P_i as occupied
    end foreach
```

Figure 8 illustrates an example drawing produced using our algorithm for the component placement of a forest. Also see Figure 9 for the drawing produced by our algorithm for the graph in Figure 1.

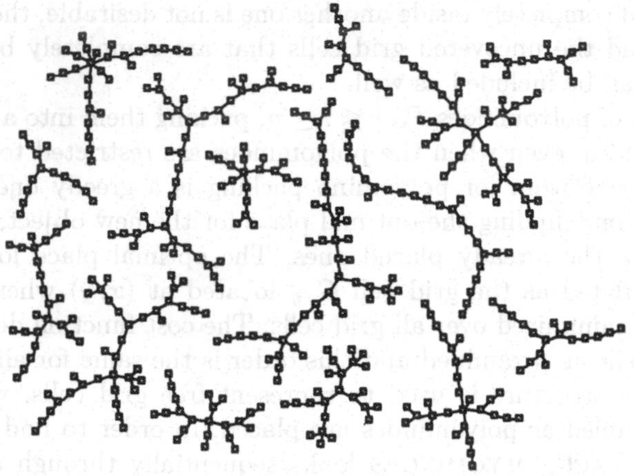

Fig. 8. An example packing produced by our algorithm.

4.1 Parameters

The grid step l is obviously the most significant parameter of this approach. We would like to guarantee that the average polyomino size s is not exceeding some constant c:

$$s = \frac{1}{n}\sum_{i=1}^{n}\lceil\frac{W_i}{l}\rceil\lceil\frac{H_i}{l}\rceil \le c$$

$$\Rightarrow \quad \frac{1}{n}\sum_{i=1}^{n}(\frac{W_i}{l}+1)(\frac{H_i}{l}+1) \le c$$

$$\Rightarrow \quad \sum_{i=1}^{n} W_i H_i + l\sum_{i=1}^{n}(W_i + H_i) + l^2 \le cnl^2$$

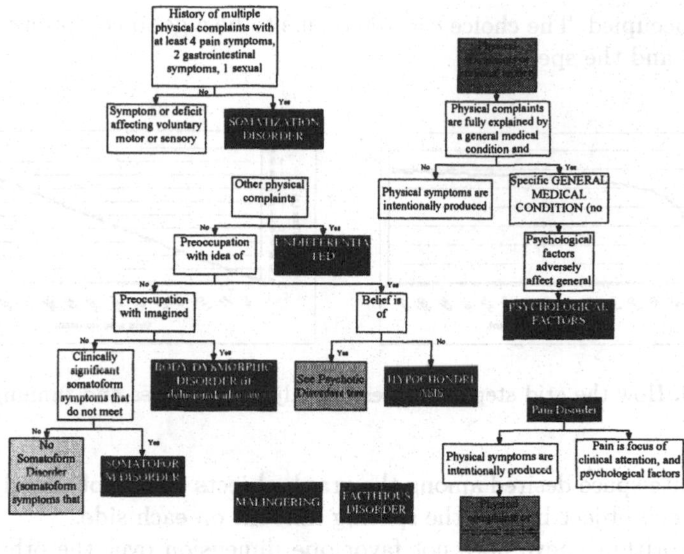

Fig. 9. The disconnected graph in Figure 1 laid out with the new algorithm (displayed with the same width to illustrate better usage of the area of aspect ratio 1.0).

Consequently, the grid step l can be calculated from the following quadratic equation:

$$(cn-1)\,l^2 - \sum_{i=1}^{n}(W_i + H_i)\,l - \sum_{i=1}^{n} W_i H_i = 0$$

With the average polyomino size $s \leq c$, the total area of all polyominoes does not exceed $n \cdot s$. Practical experiments show that the algorithm produces drawings of almost constant fullness (Figure 12), so the total packing area is also $O(n \cdot s)$.

To find a suitable place, each polyomino is tested for each cell. The test whether a polyomino fits in the specified place can be performed in $O(s)$ time in the worst case. Thus the complexity of the algorithm is $O(n^2 \cdot s^2)$. Since c and consequently s are constants this yields an $O(n^2)$ time overall, based on experimental results.

The value of the constant c must be selected carefully since its influence on the running time can be as much as $O(c^2)$. Figure 10 shows how the approximation quality paremeter c influences the adjusted fullness and the running time. For measurements, as a typical example a random forest of 300 trees of random order between 2 and 100 were generated (Figure 8 shows a smaller example of such a forest). The trees were laid out with a spring embedder algorithm similar to [10]. For small values of c, the adjusted fullness increases rapidly and converges towards 60%. The observed running time increases linearly with c. The difference from the theoretical bound of $O(c^2)$ can be explained by the observation that an occupied place is detected on average in constant time since almost all tested places are occupied. The choice $c = 100$ seems to be a good compromise between the quality and the speed.

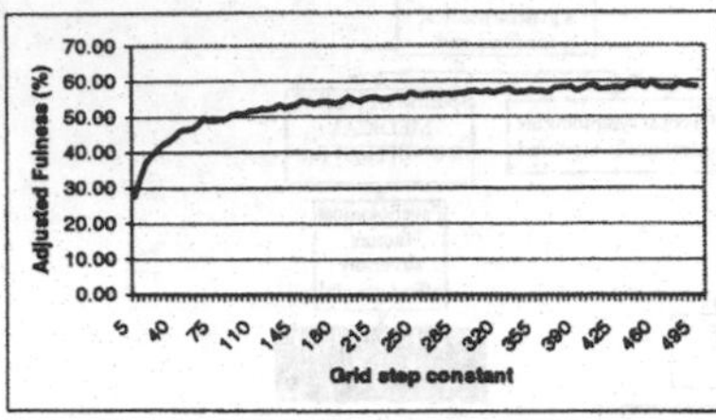

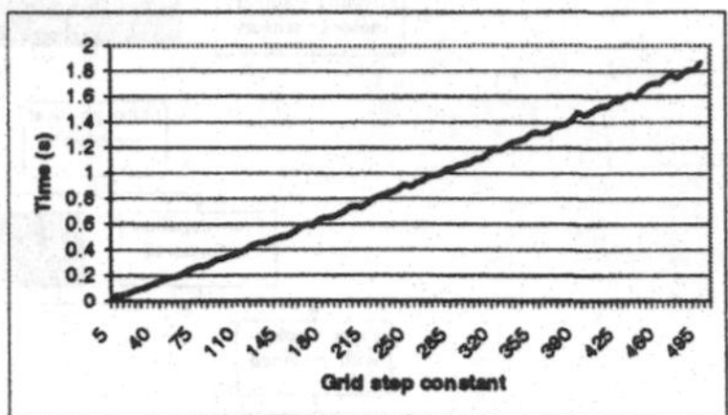

Fig. 10. How the grid step influences the adjusted fullness and running time.

The white space desired among the graph objects can be obtained by simply enlarging each object by half the spacing amount on each side.

The algorithm above does not favor one dimension over the other one and yields square-like drawings. In order to satisfy the desired aspect ratio DAR, one can simply take DAR as the unit grid step in x direction and 1 as the unit step in the y direction.

4.2 Packing in Multiple Pages

In certain applications, the graph is to be laid on multiple pages, and the task is to minimize the number of pages where the size of a page is defined in advance.

The approach is the same as above except the cost function is modified as $\max(x, y), x \geq 0, y \geq 0$, which defines the ordering starting from the corner of the page. Although such ordering lacks the nice central symmetry that we had in the original case, we have to modify the placement rule in order to better fill the sides of each page. We assume that each object separately fits in an empty page; otherwise an appropriate scaling should be performed. In algorithm PACKPOLYOMINOESINMULTIPAGES we start by fitting the first polyomino on the first page. If the current polyomino does not fit in the current page, the next page is tried until it is successfully placed. Similar to the original algorithm, the best results are obtained when the objects are sorted in their decreasing sizes. An optimization can be achieved by considering only those grid cells on the current page for which the bounding rectangle of the current polyomino is completely inside the page.

Here is a pseudo code of the multi page placement algorithm:

algorithm PACKPOLYOMINOESINMULTIPAGES($P_i, 1 \leq i \leq n$, *pageSize*)
sort $P_i, 1 \leq i \leq n$ in the order of nonincreasing size
initialize the grid G for the first page using *pageSize*
foreach polyomino P_i **do**
  set *pageNo* to 1
  while P_i not placed **do**

```
            calculate (x, y) such that the cost function is minimized
            while cannot place P_i on page pageNo centered at (x, y)
                    and (x, y) is within page boundaries do
                calculate next (x, y) using the cost function
            end while
            if P_i not placed then
                set pageNo to the next one
                if G not extended for page pageNo then
                    extend G for page pageNo
                end if
            end if
        end while
        mark those cells in G covered by P_i as occupied
    end foreach
```

5 Comparison with Previous Methods

We have compared our new method with the tiling and alternate-bisection methods discussed earlier. During the experiments, graphs that contained up to a thousand disconnected objects were used. Each object was assumed to be a star polygon with random number of corners in $[3 \ldots 8]$, each with random integer coordinates in $[1 \ldots 100]$, all independent and uniformly distributed. The value for the approximation quality constant c was taken to be 100. The desired aspect ratio was taken to be 1. For the previous methods the tightest rectangles bounding these polygons were used, whereas with our new approach, the smallest polyomino tightly bounding the polygons were used. Figure 11 shows a sample set of drawings produced by these methods for the same set of objects. The performance comparison of the methods is presented in Figure 12.

Clearly the new approach results in much more compact drawings. In terms of the execution time, it is slower but still easily within acceptable bounds given the fact that it is highly rare that a graph contains more than a few hundred disconnected objects.

6 Conclusion

In this paper, we reviewed existing algorithms and presented a new approach for layout of disconnected graphs. The new approach uses polyominoes as opposed to rectangles used in previous approaches for representation of isolated nodes and components, and produces much more compact and uniform drawings. The parameters of our algorithm and how they affect the drawings produced as well as a variation of the algorithm for multiple pages were discussed.

Acknowledgement. The authors wish to thank Cihad Baskoy for his help with the implementation and experimentation of certain algorithms.

Fig. 11. A sample from the random set of objects laid out with all three methods: tiling (left), alternate-bisection (right), and polyomino (middle) packing.

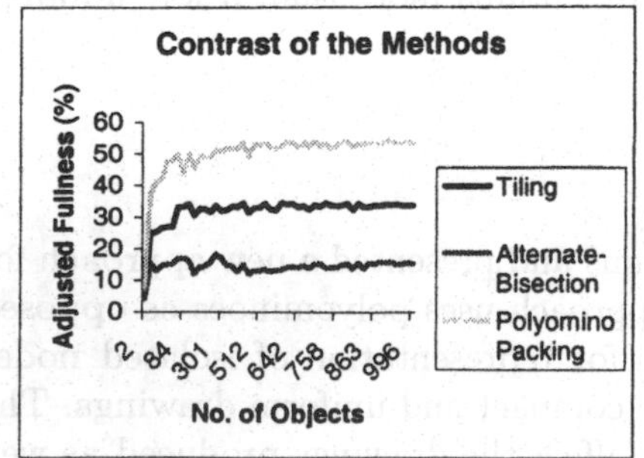

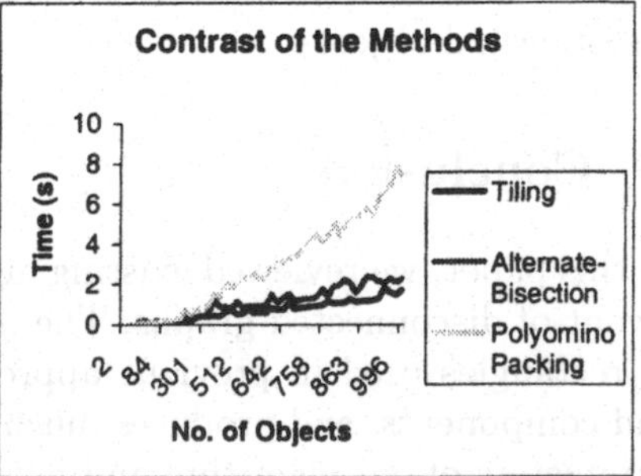

Fig. 12. Comparison of the new approach with the previous ones.

References

1. B. S. Baker, E. G. Coffman, and R. S. Rivest. Orthogonal packings in two dimensions. *SIAM Journal on Computing*, 9(4):846–855, November 1980.
2. E. G. Coffman, M. R. Garey, and D. S. Johnson. Approximation algorithms for bin packing: An updated survey. In G. Ausiello, M. Lucertini, and P. Serafini, editors, *Algorithm Design for Computer System Design*, pages 49–106. Springer-Verlag, New York, 1984.
3. E. G. Coffman, M. R. Garey, D. S. Johnson, and R. E. Tarjan. Performance bounds for level-oriented two-dimensional packing algorithms. *SIAM Journal on Computing*, 9(4):808–826, November 1990.
4. E. G. Coffman and P. W. Shor. Packings in two dimensions: Asymptotic average-case analysis of algorithms. *Algorithmica*, 9:253–277, 1993.
5. G. Di Battista, P. Eades, R. Tamassia, and I. G. Tollis. Algorithms for drawing graphs: an annotated bibliography. *Comput. Geom. Theory Appl.*, 4:235–282, 1994.
6. U. Dogrusoz. Algorithms for layout of disconnected graphs. *Information Sciences*, to appear.
7. U. Dogrusoz, M. Doorley, Q. Feng, A. Frick, B. Madden, and G. Sander. Toolkits for development of software diagramming applications. *IEEE Computer Graphics and Applications*, to appear.
8. U. Dogrusoz and G. Sander. Graph visualization. *ACM Computing Surveys*, to appear.
9. M. R. Garey and D. S. Johnson. *Computers and Intractability, A Guide to the Theory of NP-completeness.* Freeman, San Francisco, 1979.
10. T. Kamada and S. Kawai. An algorithm for drawing general undirected graphs. *Information Processing Letters*, 31:7–15, 1989.
11. P. Kikusts and P. Rucevskis. Layout algorithms of graph-like diagrams of GRADE windows graphic editors. In F.J. Brandenburg, editor, *Graph Drawing (Proc. GD '95)*, volume 1027 of *Lecture Notes in Computer Science*, pages 361–364. Springer-Verlag, 1995.
12. T. Lengauer. *Combinatorial algorithms for integrated circuit layout.* John Wiley & Sons, 1990.

Orthogonal Drawings of Plane Graphs without Bends
(Extended Abstract)

Md. Saidur Rahman[1], Mahmuda Naznin[1], and Takao Nishizeki[2]

[1] Department of computer Science and Engineering, Bangladesh University of Engineering and Technology (BUET), Dhaka-1000, Bangladesh.
saidur@cse.buet.edu, papri@cse.buet.edu

[2] Graduate School of Information Sciences, Tohoku University, Aoba-yama 05, Sendai 980-8579, Japan. nishi@ecei.tohoku.ac.jp

Abstract. In an orthogonal drawing of a plane graph G each vertex is drawn as a point and each edge is drawn as a sequence of vertical and horizontal line segments. A point at which the drawing of an edge changes its direction is called a bend. Every plane graph of the maximum degree at most four has an orthogonal drawing, but may need bends. A simple necessary and sufficient condition has not been known for a plane graph to have an orthogonal drawing without bends. In this paper we obtain a necessary and sufficient condition for a plane graph G of the maximum degree three to have an orthogonal drawing without bends. We also give a linear-time algorithm to find such a drawing of G if it exists.

Keywords: Graph, Algorithm, Graph Drawing, Orthogonal Drawing, Bend.

1 Introduction

Automatic graph drawings have numerous applications in VLSI circuit layout, networks, computer architecture, circuits schematics etc. For the last few years many researchers have concentrated their attention on graph drawings and introduced a number of drawing styles. Among these styles "orthogonal drawings" have attracted much attention due to their various applications, specially in circuit schematics, entity relationship diagrams, data flow diagrams etc. [DETT99]. An *orthogonal drawing* of a plane graph G is a drawing of G with the given embedding in which each vertex is mapped to a point, each edge is drawn as a sequence of alternate horizontal and vertical line segments, and any two edges do not cross except at their common end. A *bend* is a point where an edge changes its direction in a drawing. Every plane graph of the maximum degree four has an orthogonal drawing, but may need bends. For the cubic plane graph in Fig. 1(a) each vertex of which has degree 3, two orthogonal drawings are shown in Figs. 1(b) and (c) with 6 and 5 bends respectively. Minimization of

P. Mutzel, M. Jünger, and S. Leipert (Eds.): GD 2001, LNCS 2265, pp. 392–406, 2002.
© Springer-Verlag Berlin Heidelberg 2002

the number of bends in an orthogonal drawing is a challenging problem. Several works have been done on this issue [GT95, GT97, RNN99, T87]. In particular, Garg and Tamassia [GT97] presented an algorithm to find an orthogonal drawing of a given plane graph G with the minimum number of bends in time $O(n^{7/4}\sqrt{\log n})$, where n is the number of vertices in G. Rahman *et al.* gave an algorithm to find an orthogonal drawing of a given triconnected cubic plane graph with the minimum number of bends in linear time [RNN99].

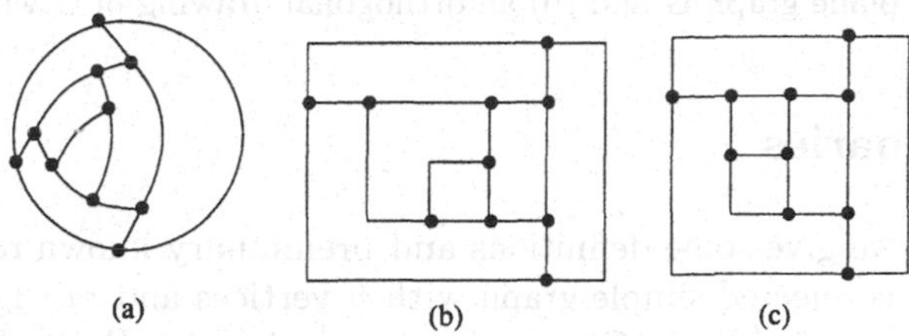

Fig. 1. (a) A plane graph G, (b) an orthogonal drawing of G with 6 bends, and (c) an orthogonal drawing of G with 5 bends.

In a VLSI floorplanning problem, an input is often a plane graph of the maximum degree 3 [L90, RNN00a, RNN00b]. Such a plane graph G may have an orthogonal drawing without bends. The graph in Fig. 2(a) has an orthogonal drawing without bends as shown in Fig. 2(b). However, not every plane graph of the maximum degree 3 has an orthogonal drawing without bends. For example, the cubic plane graph in Fig. 1(a) has no orthogonal drawing without bends, since any orthogonal drawing of the outer cycle of the graph needs at least four bends. Thus one may assume that there are four or more vertices of degree two on the outer cycle of G. It is interesting to know which classes of such plane graphs have orthogonal drawings without bends. However, no simple necessary and sufficient condition has been known for a plane graph to have an orthogonal drawing without bends, although one can know in time $O(n^{7/4}\sqrt{\log n})$ by the algorithm [GT97] whether a given plane graph has an orthogonal drawing without bends.

In this paper we obtain a simple necessary and sufficient condition for a plane graph G of the maximum degree 3 to have an orthogonal drawing without bends. The condition leads to a linear-time algorithm to find an orthogonal drawing of G without bends if it exists.

The rest of the paper is organized as follows. Section 2 describes some definitions and presents known results. Section 3 presents our results on orthogonal drawings of biconnected plane graphs without bends. Section 4 deals with orthogonal drawings of arbitrary (not always biconnected) plane graphs without bends. Finally Section 5 gives the conclusion.

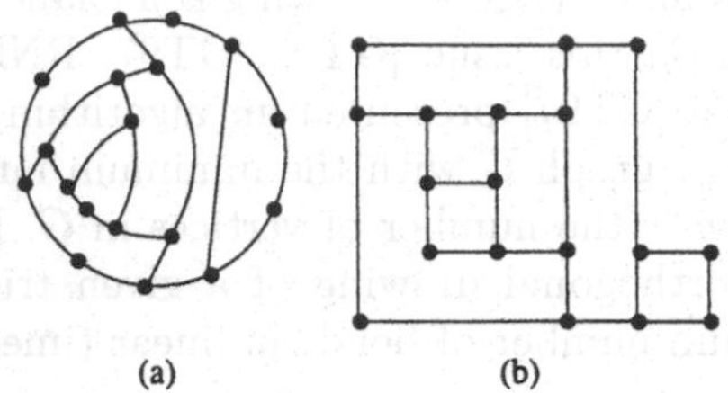

Fig. 2. (a) A plane graph G and (b) an orthogonal drawing of G without bends.

2 Preliminaries

In this section we give some definitions and preliminary known results.

Let G be a connected simple graph with n vertices and m edges. We denote the set of vertices of G by $V(G)$ and the set of edges by $E(G)$. The *degree* of a vertex v is the number of neighbors of v in G. We denote the maximum degree of graph G by $\Delta(G)$ or simply by Δ. The *connectivity* $\kappa(G)$ of a graph G is the minimum number of vertices whose removal results in a disconnected graph or a single vertex graph. We say that G is *k-connected* if $\kappa(G) \geq k$. We call a vertex of G a *cut vertex* if its removal results in a disconnected graph.

A graph is *planar* if it can be embedded in the plane so that no two edges intersect geometrically except at a vertex to which they are both incident. A *plane graph* G is a planar graph with a fixed planar embedding. A plane graph G divides the plane into connected regions called *faces.* We refer the *contour* of a face as a clockwise cycle formed by the edges on the boundary of the face. We denote the contour of the outer face of G by $C_o(G)$.

An edge of G which is incident to exactly one vertex of a cycle C and located outside C is called a *leg* of the cycle C. The vertex of C to which a leg is incident is called a *leg-vertex* of C. A cycle in G is called a k-legged cycle of G if C has exactly k legs in G.

An *orthogonal drawing* of a plane graph G is a drawing of G with the given embedding in which each vertex is mapped to a point, each edge is drawn as a sequence of alternate horizontal and vertical line segments, and any two edges do not cross except at their common end. A *bend* is a point where an edge changes its direction in a drawing. A *rectangular drawing* of a plane graph G is a drawing of G such that each edge is drawn as a horizontal or a vertical line segment, and each face is drawn as a rectangle. Thus a rectangular drawing is an orthogonal drawing in which there is no bends and each face is drawn as a rectangle. The following result is known on rectangular drawings.

Lemma 1. *Let G be a plane biconnected graph with $\Delta \leq 3$. Assume that four vertices of degree 2 on $C_o(G)$ are designated as the four corners of the outer rectangle. Then G has a rectangular drawing if and only if G satisfies the following two conditions [T84]:*

(r1) every 2-legged cycle contains at least two designated vertices, and
(r2) every 3-legged cycle contains at least one designated vertex.

Furthermore one can check in linear time whether G satisfies the condition above, and if G does then one can find a rectangular drawing in linear time [RNN98]. □

A cycle in G violating (r1) or (r2) is called a *bad cycle*: a 2-legged cycle is *bad* if it contains at most one designated vertex; a 3-legged cycle is *bad* if it contains no designated vertex.

A linear-time algorithm has been obtained in [RNN98] to find a rectangular drawing of a plane graph which has four designated corner vertices and satisfies the conditions in Lemma 1. We call it Algorithm **Rectangular-Draw** and use it in our orthogonal drawing algorithm in this paper.

For a cycle C in a plane graph G, we denote by $G(C)$ the plane subgraph of G inside C (including C). A bad cycle C in G is called a *maximal bad cycle* if $G(C)$ is not contained in $G(C')$ for any other bad cycle C' of G. We say that cycles C and C^* in a plane graph G are *independent* of each other if $G(C)$ and $G(C^*)$ have no common vertex. We now have the following lemma.

Lemma 2. *Let G be a biconnected plane graph of $\Delta \leq 3$, and let four vertices of degree 2 on $C_o(G)$ be designated as corners. Then the maximal bad cycles in G are independent of each other.* □

3 Orthogonal Drawings of Biconnected Plane Graphs

In this section we present our results on orthogonal drawings of biconnected plane graphs. From now on we assume that G is a biconnected plane graph with $\Delta \leq 3$ and there are four or more vertices of degree 2 on $C_o(G)$. The following theorem is the main result of this section.

Theorem 1. *Let G be a plane biconnected graph with $\Delta \leq 3$ and four or more vertices on $C_o(G)$. Then G has an orthogonal drawing without bends if and only if any 2-legged cycle in G contains at least two vertices of degree 2 and any 3-legged cycle in G contains at least one vertex of degree 2.* □

Note that Theorem 1 is a generalization of Lemma 1.

It is easy to prove the necessity of Theorem 1, as follows.

Necessity of Theorem 1. Assume that a plane biconnected graph G has an orthogonal drawing D without bends.

Let C be any 2-legged cycle. Then the drawing of C in D has at least four convex corners (of interior angle 90°). These convex corners must be vertices since D has no bends. The two leg-vertices of C may serve as two of the convex corners. However, each of the other convex corners must be a vertex of degree 2. Thus C must contain at least two vertices of degree 2.

Similarly we can show that any 3-legged cycle C in G contains at least one vertex of degree 2. □

In the rest of this section we give a constructive proof for the sufficiency of Theorem 1 and show that the proof leads to a linear-time algorithm to find an orthogonal drawing of a plane biconnected graph without bends if it exists.

Assume that G satisfies the condition in Theorem 1. We now need some definitions. Let C be a 2-legged cycle in G, and let x and y be the two leg vertices of C. We say that an orthogonal drawing $D(G(C))$ of the subgraph $G(C)$ is *feasible* if $D(G(C))$ has no bend and satisfies the following condition (f1) or (f2).

(f1) The drawing $D(G(C))$ intersects neither the first quadrant with the origin at x nor the third quadrant with the origin at y (after rotating the drawing and renaming the leg-vertices if necessary). (See Fig. 3.) Note that C is not always drawn by a rectangle.

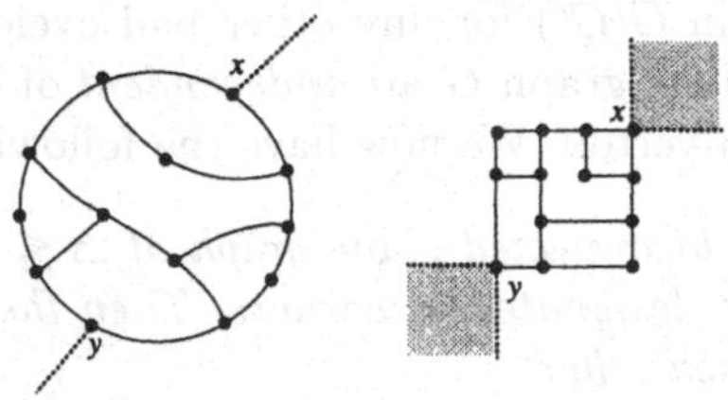

Fig. 3. Illustration of (f1) for a 2-legged cycle.

(f2) The drawing $D(G(C))$ intersects neither the first quadrant with the origin at x nor the fourth quadrant with the origin at y (after rotating the drawing and renaming the leg-vertices if necessary). (See Fig. 4.)

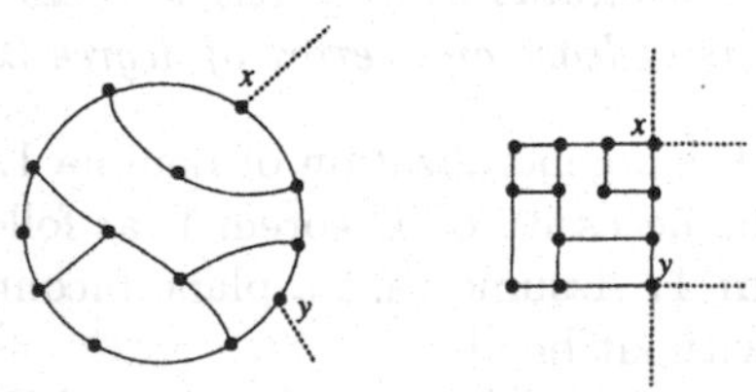

Fig. 4. Illustration of (f2) for a 2-legged cycle.

Let C be a 3-legged cycle in G, and let x, y and z be the three leg-vertices. One may assume that x, y and z appear clockwise on C. We say that an orthogonal drawing $D(G(C))$ of $G(C)$ is *feasible* if $D(G(C))$ has no bend and $D(G(C))$ satisfies the following condition (f3).

(f3) The drawing $D(G(C))$ intersects none of the following three quadrants: the first quadrant with origin at x, the fourth quadrant with origin at y, and the third quadrant with origin at z (after rotating the drawing and renaming the leg-vertices if necessary). (See Fig. 5.)

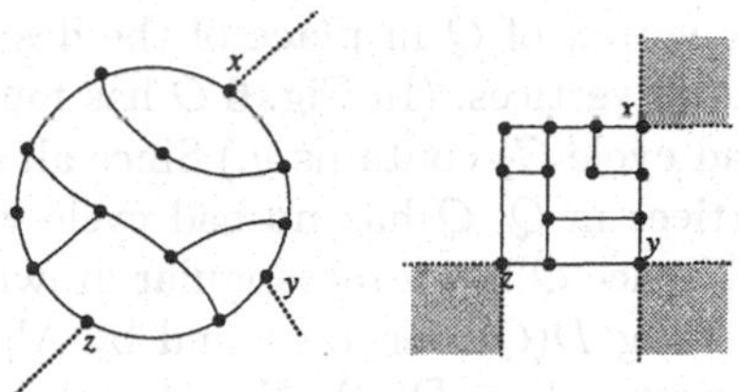

Fig. 5. Illustration of (f3) for a 3-legged cycle.

The conditions (f1), (f2) and (f3) imply that, in the drawing of $G(C)$, any vertex of $G(C)$ except leg-vertices is located in none of the shaded quadrants in Figs. 3, 4 and 5, and hence a leg incident to x, y or z can be drawn by a horizontal or a vertical line segments without edge-crossing as indicated by dotted lines in Figs. 3, 4 and 5.

We now have the following lemma.

Lemma 3. *Let G be a plane biconnected graph with $\Delta \leq 3$ and four or more vertices on $C_o(G)$, and assume that G satisfies the condition in Theorem 1, that is, any 2-legged cycle in G contains at least two vertices of degree 2 and any 3-legged cycle in G contains at least one vertex of degree 2. Then $G(C)$ has a feasible orthogonal drawing for any 2- or 3-legged cycle C in G.*

Proof. We give a recursive algorithm to find a feasible orthogonal drawing of $G(C)$. There are two cases to be considered.

Case 1: C is a 2-legged cycle.

Let x and y be the two leg-vertices of C, and let e_x and e_y be the legs incident to x and y, respectively. Since C satisfies the condition in Theorem 1, C has at least two vertices of degree 2. Let a and b be any two vertices of degree 2 on C. We now regard the four vertices x, y, a and b as the four designated corner vertices of C.

We first consider the case where $G(C)$ has no bad cycle with respect to the four designated vertices. In this case, by Lemma 1 $G(C)$ has a rectangular drawing D with the four designated corner vertices. Such a rectangular drawing D of $G(C)$ can be found by the algorithm **Rectangular-Draw** in [RNN98]. Since the outer cycle C of $G(C)$ is drawn as a rectangle in D, D satisfies Condition (f1) or (f2) and, in particular, x, y, a and b are the convex corners of the rectangular drawing of C. Since D is a rectangular drawing, D has no bend. Thus D is a feasible orthogonal drawing of $G(C)$.

We then consider the case where $G(C)$ has a bad cycle. Let $C_1, C_2, \cdots, C_l$ be the maximal bad cycles of $G(C)$. By Lemma 2 $C_1, C_2, \cdots, C_l$ are independent of each other. Construct a plane graph Q from $G(C)$ by contracting $G(C_i), 1 \leq i \leq l$, to a single vertex v_i, as illustrated in Figs. 6(a) and (b). Clearly Q is a plane biconnected graph with $\Delta \leq 3$. Every bad cycle C_i in $G(C)$ contains at most one designated vertex. If C_i contains a designated vertex, then we newly designate v_i as a corner vertex of Q in place of the designated vertex. Thus Q has exactly four designated vertices. (In Fig. 6 Q has four designated vertices a, b, x, and v_2 since the bad cycle C_2 contains y.) Since all maximal bad cycles are contracted to single vertices in Q, Q has no bad cycle with respect to the four designated vertices, and hence Q has a rectangular drawing $D(Q)$, as illustrated in Fig. 6(c). Such a drawing $D(Q)$ can be found by Algorithm **Rectangular-Draw**. Clearly there is no bend on $D(Q)$. The shrunken outer cycle of $G(C)$ is drawn as a rectangle in $D(Q)$, and hence $D(Q)$ satisfies conditions (f1) or (f2). If C_i is a 2-legged cycle, then v_i and the two legs e_{x_i} and e_{y_i} are embedded in $D(Q)$ as illustrated in Figs. 7(b) and 8(b) or as in their rotated ones, and C_i and the two legs e_{x_i} and e_{y_i} have the embeddings in Figs. 7(c) and 8(c) and their rotated ones. If C_i is a 3-legged cycle, then v_i and the three legs e_{x_i}, e_{y_i} and e_{z_i} are embedded in $D(Q)$ as illustrated in Fig. 9(b) or as in their rotated ones, and C_i and three legs e_{x_i}, e_{y_i} and e_{z_i} have the embeddings in Fig. 9(c) and their rotated ones. One can obtain a drawing $D(G(C))$ of $G(C)$ from the drawings of Q and $G(C_i)$ $1 \leq i \leq l$, as follows. Replace each v_i, $1 \leq i \leq l$, in $D(Q)$ with one of the feasible embeddings of $G(C_i)$ in Fig. 7(c), Fig. 8(c) and Fig. 9(c) and their rotated one that corresponds to the embedding of v_i with legs in $D(Q)$, and draw each leg of C_i in $D(G(C))$ by a straight line segment having the same direction as the leg in $D(Q)$, as illustrated in Fig. 6(d). We call this operation a *patching operation*.

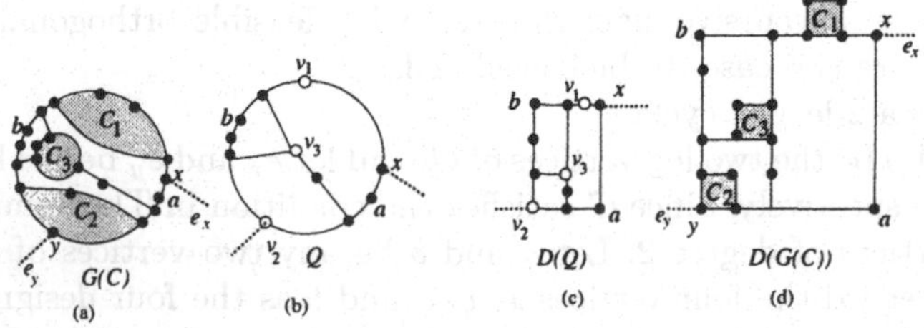

Fig. 6. Illustration for Case 1 where C has the maximal bad cycles C_1, C_2 and C_3.

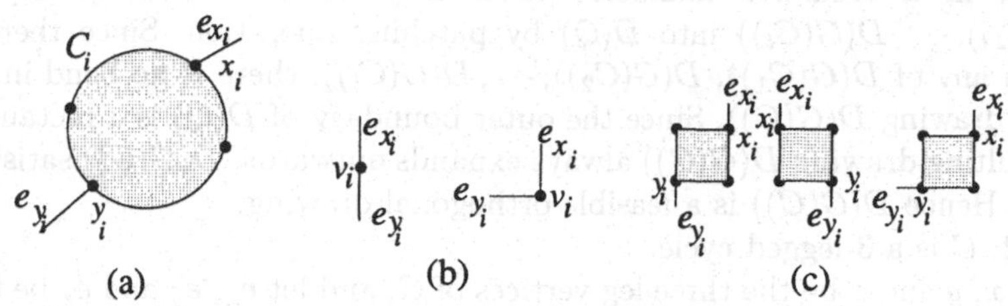

Fig. 7. (a) A 2-legged cycle C_i having a feasible orthogonal drawing satisfying (f1), (b) embeddings of a vertex v_i and two legs e_{x_i} and e_{y_i} incident to v_i, and (c) feasible orthogonal drawings of $G(C_i)$ with two legs.

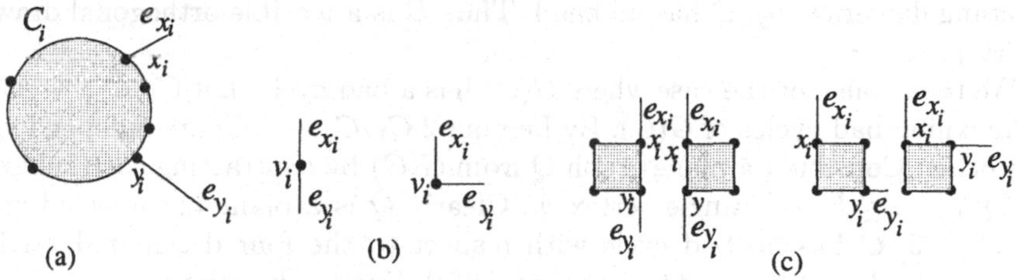

Fig. 8. (a) A 2-legged cycle C_i having a feasible orthogonal drawing satisfying (f2), (b) embeddings of a vertex v_i and two legs e_{x_i} and e_{y_i} incident to v_i, and (c) feasible orthogonal drawings of $G(C_i)$ with two legs.

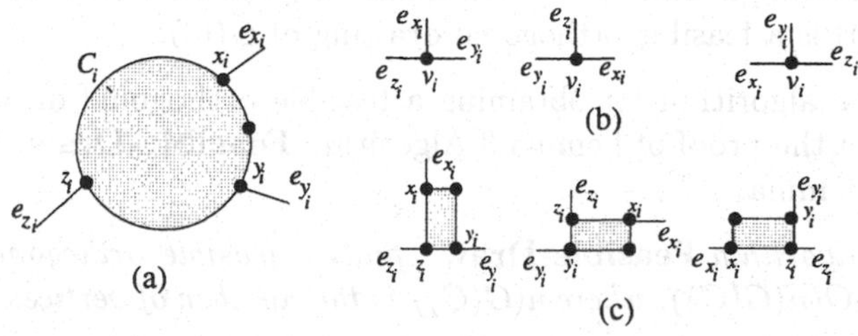

Fig. 9. (a) A 3-legged cycle C_i having feasible orthogonal drawings satisfying (f3), (b) embeddings of a vertex v_i and three legs e_{x_i}, e_{y_i} and e_{z_i} incident to v_i, and (c) feasible orthogonal drawings of $G(C_i)$ with three legs.

We find a feasible orthogonal drawing $D(G(C_i))$ of $G(C_i), 1 \leq i \leq l$, in a recursive manner. We then patch the drawings $D(G(C_1))$, $D(G(C_2)), \cdots, D(G(C_l))$ into $D(Q)$ by patching operation. Since there is no bend in any of $D(G(C_1))$, $D(G(C_2)), \cdots, D(G(C_l))$, there is no bend in the resulting drawing $D(G(C))$. Since the outer boundary of $D(Q)$ is a rectangle and the resulting drawing $D(G(C))$ always expands outwards, $D(G(C))$ satisfies (f1) or (f2). Hence $D(G(C))$ is a feasible orthogonal drawing.

Case 2: C is a 3-legged cycle.

Let x, y and z be the three leg-vertices of C, and let e_x, e_y and e_z be the legs incident to x, y and z, respectively. Since C satisfies the condition in Theorem 1, C has at least one vertex of degree 2. Let a be any vertex of degree 2 on C. We now regard the four vertices x, y, z and a as designated corner vertices.

We first consider the case where $G(C)$ has no bad cycle with respect to the four designated vertices. In this case by Lemma 1 $G(C)$ has a rectangular drawing D with the four designated vertices. Such a rectangular drawing D of $G(C)$ can be found by the algorithm **Rectangular-Draw.** Since the outer cycle C of $G(C)$ is drawn as a rectangle in D, D satisfies the condition (f3). Since D is a rectangular drawing, D has no bend. Thus D is a feasible orthogonal drawing of $G(C)$.

We then consider the case where $G(C)$ has a bad cycle. Let $C_1, C_2, \cdots, C_l$ be the maximal bad cycles of $G(C)$. By Lemma 2 $C_1, C_2, \cdots, C_l$ are independent of each other. Construct a plane graph Q from $G(C)$ by contracting each subgraph $G(C_i), 1 \leq i \leq l$, to a single vertex v_i. Clearly Q is a plane biconnected graph wih $\Delta \leq 3$, Q has no bad cycle with respect to the four designated vertices, and hence Q has a rectangular drawing $D(Q)$. Such a drawing can be found by Algorithm **Rectangular-Draw**. Clearly there is no bend on $D(Q)$. Since the outer cycle of Q is drawn as a rectangle in $D(Q)$, $D(Q)$ satisfies the condition (f3).

We then find a feasible orthogonal drawing $D(G(C_i))$ of $G(C_i), 1 \leq i \leq l$, in a recursive manner, and patch the drawings $D(G(C_1)), D(G(C_2)), \cdots, D(G(C_l))$ into $D(Q)$. Since there is no bend in any of $D(G(C_1)), D(G(C_2)), \cdots, D(G(C_l))$, there is no bend in the resulting drawing $D(G(C))$. Since the outer boundary of $D(Q)$ is a rectangle and $D(G(C))$ expands outwards, $D(G(C))$ satisfies (f3). Thus $D(G(C))$ is a feasible orthogonal drawing of $G(C)$. □

We call the algorithm for obtaining a feasible orthogonal drawing of $G(C)$ as described in the proof of Lemma 3 Algorithm **Feasible-Draw.** We now have the following lemma.

Lemma 4. *Algorithm* **Feasible-Draw** *finds a feasible orthogonal drawing of $G(C)$ in time $O(n(G(C))$, where $n(G(C))$ is the number of vertices in $G(C)$.* □

We are now ready to prove the sufficiency of Theorem 1; we actually prove the following lemma.

Lemma 5. *Let G be a plane biconnected graph with $\Delta \leq 3$ and four or more vertices of degree 2 on $C_o(G)$. If G satisfies the conditions in Theorem 1, then G has an orthogonal drawing without bends.*

Proof. Since there are four or more vertices of degree 2 on $C_o(G)$, we designate any four of them as (convex) corners.

Consider first the case where G does not have any bad cycle with respect to the four designated (convex) corners. Then by Lemma 1 there is a rectangular drawing of G. The rectangular drawing of G has no bends. Hence it is an orthogonal drawing $D(G)$ of G without bends.

Consider next the case where G has bad cycles. Let $C_1, C_2, \cdots, C_l$ be the maximal bad cycles in G. By Lemma 2 $C_1, C_2, \cdots, C_l$ are independent of each other. We contract each $G(C_i)$, $1 \le i \le l$, to a single vertex v_i. Let G^* be the resulting graph. Clearly, G^* has no bad cycle with respect to the four designated vertices, some of which may be vertices resulted from the contraction of bad cycles. By Lemma 1 G^* has a rectangular drawing $D(G^*)$, which can be found by the algorithm **Rectangular-Draw.** We recursively find a feasible orthogonal drawing of each $G(C_i)$, $1 \le i \le l$, by **Feasible-Draw.** Patch the feasible orthogonal drawings of $G(C_1), G(C_2), \cdots, G(C_l)$ into $D(G^*)$ by patching operations. The resulting drawing is an orthogonal drawing D of G. Note that $D(G^*)$ has no bend and $D(G(C_i))$, $1 \le i \le l$, has no bend. Furthermore, patching operation introduces no new bend. Thus D has no bend. □

We call the algorithm for obtaining an orthogonal drawing of a biconnected plane graph G described in the proof of Lemma 5 Algorithm **Bi-Orthogonal-Draw.** We now have the following theorem.

Theorem 2. *If G is a plane biconnected graph with $\Delta \le 3$, has four or more vertices of degree 2 on $C_o(G)$, and satisfies the condition in Theorem 1, then Algorithm* **Bi-Orthogonal-Draw** *finds an orthogonal drawing of G in linear time.* □

4 Orthogonal Drawings of Arbitrary Plane Graphs

In this section we extend our result on biconnected plane graphs in Theorem 1 to arbitrary (not always biconnected) plane graphs with $\Delta \le 3$ as in the following theorem.

Theorem 3. *Let G be a plane graph with $\Delta \le 3$. Then G has an orthogonal drawing without bends if and only if every k-legged cycle C in G contains at least $4 - k$ vertices having degree 2 in G for any k, $0 \le k \le 3$.*

The proof for the necessity of Theorem 3 is similar to the proof for the necessity of Theorem 1. In the rest of this section we give a constructive proof for the sufficiency of Theorem 3. We need some definitions.

We may assume that G is a plane connected graph of $\Delta \le 3$. We call a subgraph H of G a *biconnected component* of G if H is a maximal biconnected subgraph of G. We call a single edge (u, v) of G together with the vertices u and v a *weakly biconnected component* of G if either both u and v are cut vertices or one of u and v is a cut vertex and the other one is a vertex of degree one.

Let C be a cycle in G, and let v be a cut vertex of G on C. We call v an *out-cut vertex* for C if v is a leg-vertex of C in G, otherwise we call v an *in-cut vertex* for C. Any in-cut vertex for C is not a convex corner (having interior angle 90°) of the drawing of C in any orthogonal drawing of G; otherwise, the edge of G which is incident to v and is not on C could not be drawn as a horizontal or vertical line segment. Similarly, any out-cut vertex for C is not a concave corner (having interior angle 270°). Thus the orthogonal drawing of G must satisfy the following condition (f4).

(f4) Every in-cut vertex for any cycle is not a convex corner and every out-cut vertex is not a concave corner in the drawing of the cycle.

We now have the following lemmas.

Lemma 6. *Let G be a connected plane graph of $\Delta \leq 3$ satisfying the condition in Theorem 3. Then any biconnected component H of G has an orthogonal drawing which has no bends and satisfies (f4).* □

We call two subgraphs H_i and H_j of G are *disjoint* with each other if H_i and H_j have no common vertex. One can easily observe the following lemma.

Lemma 7. *Let G be a connected plane graph of $\Delta \leq 3$. Then the biconnected components in G are disjoint with each other.*

A *block* of a connected graph G is either a biconnected component or a weakly biconnected component of the graph. The blocks and cut-vertices in a connected graph G can be represented by a tree which is called the *BC-tree* of G. In the *BC*-tree of G every block is represented by a *B*-node and each cut vertex of G is represented by a *C*-node. The *BC*-tree of the plane graph $G(C_1)$ is depicted in Fig. 10(b), where each *B*-node is represented by a rectangle and each *C*-node is represented by a circle.

We call a cycle C in G a *maximal cycle* of G if $G(C)$ is not contained in $G(C')$ for any other cycle C' in G. Thus a maximal cycle is an outer cycle of a biconnected component of G. The graph G in Fig. 10(a) has two maximal cycles C_1 and C_2 drawn by thick lines. $G(C)$ is called a *maximal closed subgraph* of G if C is a maximal cycle of G. We now have the following lemma.

Lemma 8. *Let G be a connected plane graph of $\Delta \leq 3$ satisfying the condition in Theorem 3, and let C be a maximal cycle in G. Then $G(C)$ has an orthogonal drawing which has no bends and satisfies (f4).*

Proof. We give an algorithm for finding an orthogonal drawing of $G(C)$ which has no bends and satisfies (f4).

If $G(C)$ is a biconnected component of G, then by Lemma 6 $G(C)$ has an orthogonal drawing which has no bends and satisfies (f4). One may thus assume that $G(C)$ is not a biconnected component of G. Then $G(C)$ has some biconnected components and weakly biconnected components. By Lemma 7 the biconnected components of $G(C)$ are disjoint with each other. We can find an orthogonal drawing of a biconnected component which has no bend and satisfies (f4) by an algorithm similar to Algorithm **Bi-Orthogonal-Draw**. We can draw a weakly biconnected component by a horizontal or vertical line segment. It is thus remained to merge the drawings of biconnected components and weakly biconnected components without introducing new bends and edge crossings.

We construct a *BC*-tree of $G(C)$. Let B_0 be the node in the BC-tree corresponding to the biconnected component of $G(C)$ whose outer cycle is C. We consider the BC-tree of $G(C)$ as a rooted tree and regard B_0 as the root. Starting from the root we visit the tree by depth-first search and merge the orthogonal drawings of the blocks in the depth first-search order. Let $B_0, B_1, B_2, \cdots, B_b$ be the ordering of the blocks following a depth-first search order starting from B_0. The BC-tree of $G(C_1)$ of G in Fig. 10(a) is depicted in Fig. 10(b), where B_0 is the root of the tree and the other B-nodes are numbered according to a depth-first search order starting from B_0.

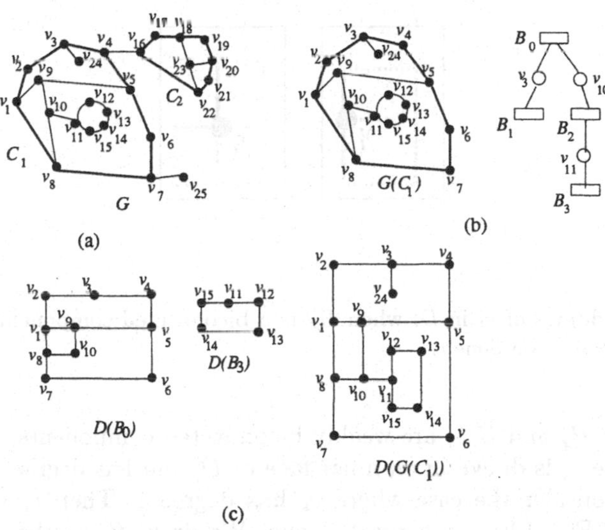

Fig. 10. (a) A plane graph G with two maximal cycles C_1 and C_2, (b) $G(C_1)$ and its BC-tree, (c) drawings of the two biconnected components B_0 and B_3 of $G(C_1)$ and the final drawing of $G(C_1)$.

We assume that we have obtained an orthogonal drawing D_i, which has no bends and satisfies (f4), by merging the orthogonal drawings of the blocks $B_0, \cdots, B_i$, and we are now going to obtain an orthogonal drawing D_{i+1}, which has no bends and satisfies (f4), by merging D_i with the orthogonal drawing of the block B_{i+1} with D_i. Let v_t be the cut-vertex corresponding to the C-node which is the parent of B_{i+1} in the BC-tree of $G(C)$. Let B_x be the parent of v_t

in the BC-tree. Then both B_x and B_{i+1} contain the vertex v_t, and D_i contains the drawing of B_x. We have the following three cases to consider.
Case 1: B_x is a biconnected component and B_{i+1} is a weakly biconnected component.

In this case B_{i+1} is an edge and will be drawn inside an inner face of the drawing D_i. Let C_f be the facial cycle of B_x. Then v_t is an in-cut vertex for C_f. Since we have obtained a feasible orthogonal drawing of B_x which has no bends and satisfies (f4), v_t is not drawn as convex corner in the drawing of C_f in $D(B_x)$, and hence the embedding of v_t in D_i is one of the two embeddings in Fig. 11 or a rotated one. We can draw B_{i+1} as a horizontal or a vertical line segment started from v_t as illustrated by dotted lines in Fig. 11. Thus we obtain the drawing D_{i+1}. Clearly no new bend is introduced in D_{i+1} and D_i may be expanded outwards to avoid edge crossings. In Fig. 10(c) the weakly biconnected component B_1 of edge (v_3, v_{24}) is merged to a biconnected component B_0 at vertex v_3.

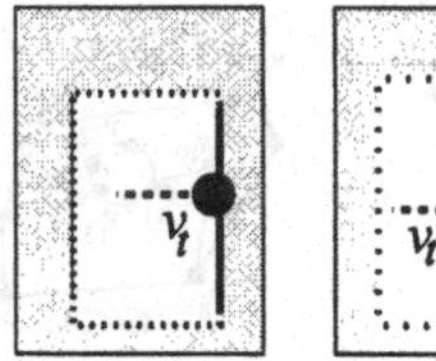

Fig. 11. Embeddings of v_t in D_i when B_x is a biconnected component and B_{i+1} is a weakly biconnected component.

Case 2: Both B_x and B_{i+1} are weakly biconnected components.

In this case v_t is drawn in an inner face of D_i and has degree 1 or 2 in D_i.

We first consider the case where v_t has degree 1. Then v_t in D_i has the embedding in Fig. 12(a) or a rotated one. We draw B_i as the dotted line in Fig. 12(a).

We next consider the case where v_t has degree 2 in D_i. Then v_t has degree 3 in $G(C)$, and let x, y, and z be the three neighbors of v_t in G. We may assume without loss of generality that edges (v_t, x) and (v_t, y) are already drawn in D_i and we now merge the drawing of the edge $(v_t, z) = B_{i+1}$ to D_i. It is evident from the drawing described above that (v_t, x) and (v_t, y) are drawn on a (horizontal or vertical) straight line segment. We draw the edge (v_t, z) as a dotted line as in Fig. 12(b).
Case 3: B_x is a weakly biconnected component and B_{i+1} is a biconnected component.

In this case v_t is drawn in D_i as the end of a horizontal or vertical line segment inside an inner face of D_i. Vertex v_t has degree 2 in B_{i+1} and is an

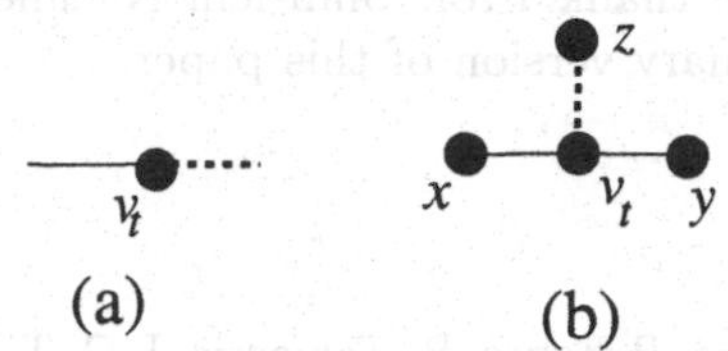

Fig. 12. Embedding of B_i when both B_x and B_{i+1} are weakly biconnected components

out-cut vertex for $C_o(B_{i+1})$. Hence by Lemma 6 v_t is not a concave corner of the drawing of $C_o(B_{i+1})$ in $D(B_{i+1})$. Therefore $D(B_{i+1})$ can be easily merged with D_i by rotating $D(G(B_{i+1}))$ 90° or 180° or 270° and expanding the drawing D_i if necessary. In Fig. 10(c) the orthogonal drawing of B_3 is merged to D_2 at vertex v_{11} where $D(B_3)$ has been rotated 90° and the drawing D_2 is expanded outwards. □

We call the algorithm described in the proof of Lemma 8 Algorithm **Maximal-Orthogonal-Draw**

We are now ready to give a proof for the sufficiency of Theorem 3.

Proof for Sufficiency of Theorem 3

We decompose G into maximal closed subgraphs and weakly biconnected components. We find an orthogonal drawing of each maximal closed subgraph by Algorithm **Maximal-Orthogonal-Draw.** Each weakly biconnected component can be drawn by a horizontal or a vertical line segment. Using a technique similar to one in the proof of Lemma 8 we merge the drawings of maximal closed subgraphs and weakly biconnected components in the outer faces of maximal closed subgraphs. The resulting drawing is an orthogonal drawing of G without bends. □

We call the algorithm described in the proof for the sufficiency of Theorem 3 Algorithm **No-bend-Orthogonal-Draw**. We now have the following theorem.

Theorem 4. *If G is a plane connected graph of $\Delta \leq 3$ and satisfies the condition in Theorem 3, then Algorithm* **No-bend-Orthogonal-Draw** *finds an orthogonal drawing of G without bends in linear time.* □

5 Conclusions

In this paper we established a necessary and sufficient condition for a plane graph G with the maximum degree at most 3 to have an orthogonal drawing without bends. We gave a linear-time algorithm to determine whether G has an orthogonal drawing without bends and find such a drawing of G if it exists. It is remained as a future work to establish a necessary and a sufficient condition for a plane graph of the maximum degree at most 4 to have an orthogonal drawing without bends.

Acknowledgment. We thank Prof. Shin-ichi Nakano for his comments and suggestions on a preliminary version of this paper.

References

[DETT99] G. Di. Battista, P. Eades, R. Tamassia, I. G. Tollis, *Graph Drawing: Algorithms for the Visualization of Graphs*, Prentice-Hall Inc., Upper Saddle River, New Jersey, 1999.

[GT95] A. Garg and R. Tamassia, *On the computational complexity of upward and rectilinear planarity testing*, Proc. of Graph Drawing'94, Lect. Notes in Computer Science, 894, pp. 99-110, 1995.

[GT97] A. Garg and R. Tamassia, *A new minimum cost flow algorithm with applications to graph drawing*, Proc. of Graph Drawing'96, Lect. Notes in Computer Science, 1190, pp. 201-206, 1997.

[L90] T. Lengauer, *Combinatorial Algorithms for Integrated Circuit Layout*, Wiley, Chichester, 1990.

[RNN00a] M. S. Rahman, S. Nakano and T. Nishizeki, *Box-rectangular drawings of plane graphs*, Journal of Algorithms, 37, pp. 363-398, 2000.

[RNN00b] M. S. Rahman, S. Nakano and T. Nishizeki, *Rectangular drawings of plane graphs without designated corners*, Proc. of COCOON'2000, Lect. Notes in Computer Science, 1858, pp. 85-94, 2000, also Comp. Geom. Theo. Appl., to appear.

[RNN98] M. S. Rahman, S. Nakano and T. Nishizeki, *Rectangular grid drawings of plane graphs*, Comp. Geom. Theo. Appl., 10(3), pp. 203-220, 1998.

[RNN99] M. S. Rahman, S. Nakano and T. Nishizeki, *A linear algorithm for bend-optimal orthogonal drawings of triconnected cubic plane graphs*, Journal of Graph Alg. and Appl., http://www.cs.brown.edu/publications/jgaa/, 3(4), pp. 31-62, 1999.

[T84] C. Thomassen, *Plane representations of graphs*, (Eds.) J.A. Bondy and U.S.R. Murty, Progress in Graph Theory, Academic Press Canada, pp. 43-69, 1984.

[T87] R. Tamassia, *On embedding a graph in the grid with the minimum number of bends*, SIAM J. Comput., 16, pp. 421-444, 1987.

Polar Coordinate Drawing of Planar Graphs with Good Angular Resolution

Christian A. Duncan[1] and Stephen G. Kobourov[2]

[1] Department of Computer Science
University of Miami
Coral Gables, FL 33124
duncan@cs.miami.edu

[2] Department of Computer Science
University of Arizona
Tucson, AZ 85721
kobourov@cs.arizona.edu

Abstract. We present a novel way to draw planar graphs with good angular resolution. We introduce the polar coordinate representation and describe a family of algorithms which use polar representation. The main advantage of using a polar representation is that it allows us to exert independent control over grid size and bend positions. Polar coordinates allow us to specify different vertex resolution, bend-point resolution and edge separation. We first describe a standard (Cartesian) representation algorithm (CRA) which we then modify to obtain a polar representation algorithm (PRA). In both algorithms we are concerned with the following drawing criteria: angular resolution, bends per edge, vertex resolution, bend-point resolution, edge separation, and drawing area.

The CRA algorithm achieves 1 bend per edge, unit vertex and bend resolution, $\sqrt{2}/2$ edge separation, $5n \times \frac{5n}{2}$ drawing area and $\frac{1}{2d(v)}$ angular resolution, where $d(v)$ is the degree of vertex v. The PRA algorithm has an improved angular resolution of $\frac{\pi}{4d(v)}$, 1 bend per edge, and unit vertex resolution. For the PRA algorithm, the bend-point resolution and edge separation are parameters that can be modified to achieve different types of drawings and drawing areas. In particular, for the same parameters as the CRA algorithm (unit bend-point resolution and $\sqrt{2}/2$ edge separation), the PRA algorithm creates a drawing of size $9n \times \frac{9n}{2}$.

1 Introduction

In the area of planar graph drawing there has been considerable interest in algorithms that produce readable drawings [4]. Among the many properties which contribute to the readability of planar graphs, edge smoothness, vertex resolution, bend-point resolution, angular resolution, and edge separation are of great importance. Edges are often drawn as straight-line segments connecting two vertices. An edge can also be drawn as a sequence of straight-line segments, in which case the smallest number of bends is desirable. An edge may also be drawn as a smooth curve. These three types of edges generally provide aesthetically pleasing drawings.

P. Mutzel, M. Jünger, and S. Leipert (Eds.): GD 2001, LNCS 2265, pp. 407–421, 2002.
© Springer-Verlag Berlin Heidelberg 2002

1.1 Definitions

A graph drawing has good *vertex resolution* if vertices cannot get arbitrarily close to one another, that is, if vertices are well distributed in the drawing. As a result, a great deal of research has been concentrated on graph drawing algorithms which place vertices on the integer grid such that the *drawing area* is proportional to the number of vertices n of the graph, typically $O(n) \times O(n)$. If there are bends in the edges, then the bend-points are also placed on the integer grid. The *bend-point resolution* of a graph refers to the minimum distance between two bends. The *edge separation* of a graph refers to the minimum distance between two edges that are sufficiently away from their endpoints (since incident edges can get arbitrarily close to each other near their common endpoint).

A graph drawing has good *angular resolution* if adjacent edges cannot form arbitrarily small angles. This is achieved by ensuring that the edges emanating from a given vertex "fan out" evenly around the vertex. Note, however, that good angular resolution cannot always be achieved while simultaneously guaranteeing straight-line edges and small sub-exponential drawing area [9]. By introducing bends in the edges, however, we can guarantee both good resolution and small drawing area.

1.2 Previous Work

Garg and Tamassia [5] consider the problem of drawing with good angular resolution, and Kant [8] shows how to create drawings with angular resolution of $\Theta(1/d(v))$ in an $O(n) \times O(n)$ area grid, using edges with at most three bends each. Gutwenger and Mutzel [7] describe an improved algorithm with better constant factors which produces very aesthetically pleasing drawings in a $(2n-5) \times (3n/2 - 7/2)$ grid with at least $2/d(v)$ angular resolution using at most three bends per edge. The algorithm of Goodrich and Wagner [6] requires one less bend per edge and guarantees angular resolution of $\Theta(1/d(v))$ for each vertex v, but at the expense of larger area, $(20n-48) \times (10n-24)$. Cheng, Duncan, Goodrich, and Kobourov [1] improve the above algorithm so that every edge has at most one bend while the angular resolution is $\Theta(1/d(v))$ for each vertex v and maximum area is $30n \times 15n$.

1.3 Our Results

We first present a new Cartesian representation algorithm (CRA) which improves the bounds of previous algorithms. In particular, CRA guarantees 1 bend per edge, unit vertex resolution, unit bend-point resolution, $\sqrt{2}/2$ edge separation, $5n \times \frac{5n}{2}$ drawing area, and $\frac{1}{2d(v)}$ angular resolution, where $d(v)$ is the degree of vertex v.

We then present a novel polar representation algorithm (PRA). The PRA algorithm also guarantees $\frac{\pi}{4d(v)}$ angular resolution, 1 bend per edge, and unit vertex resolution. The bend-point resolution and edge separation are parameters that can be modified to achieve different types of drawings and drawing areas.

In particular, for the same parameters as the CRA algorithm (unit bend-point resolution and $\sqrt{2}/2$ edge separation), the PRA algorithm creates a drawing of size $9n \times \frac{9n}{2}$. Note that in some situations the vertex resolution is more important than the bend-point resolution or the edge separation. In such situations, all of the previous algorithms perform poorly since they are designed to maintain constant resolution particularly between vertices and bend-points. Using the PRA algorithm, we can relax the bend-point resolution constraints and get significant improvements.

The PRA algorithm relies on a novel approach for representing bends and vertices. Traditionally, vertices and bend-points are restricted to lie on integer grid coordinates. One reason for this is that the points are defined by a pair of integers. In this way, all operations on the points (for example, shifting) are performed with integer arithmetic. At the drawing stage, the integer coordinates are mapped to pixels on the screen.

Another reason for placing vertices and bend-points on integer grid coordinates is that this approach guarantees good vertex resolution, good bend-point resolution, and good edge separation [1,6,7,8]. Rather than insisting that bend-points lie on integer grid coordinates, we propose an alternative approach which allows bend-points to be located on a grid represented by polar coordinates. We call this a *polar representation* approach because both the vertices and the bend-points are represented using polar coordinates.

At the exact moment of drawing the graph onto the screen, an algorithm using polar representation requires a rounding calculation to determine the exact pixel location for the bend-points. Note, however, that the traditional approach also uses a rounding calculation for scaling from the integer grid space to the pixel space.

The main advantage of using a polar representation is that it allows us to independently control grid size and bend positions. Polar coordinates allow us to specify different vertex resolution, bend-point resolution, and edge separation. We achieve this added flexibility at the expense of slightly increased storage for the graph representation. A Cartesian representation requires exactly two integers for each point while the polar representation requires up to five integers per point.

2 The CRA Algorithm

The Cartesian Representation Algorithm is a natural extension of the previous algorithms which guarantee good angular resolution [8,7,6,1]. For the remainder of this paper, when we say "graph" we mean a fully triangulated, undirected, planar graph. In our algorithm the vertices of the graph are inserted sequentially by their canonical ordering, generating subgraphs $G_1, G_2, \ldots, G_n$. The canonical ordering [3] for a planar graph G orders the vertices of G so that they can be inserted one at a time without creating any crossings. We define G_k at step i to be the graph induced by vertices $1, 2, \ldots, k$. Graph G_{k+1} is created from G_k by inserting the next vertex v_{k+1} in the canonical order. Before we show the details

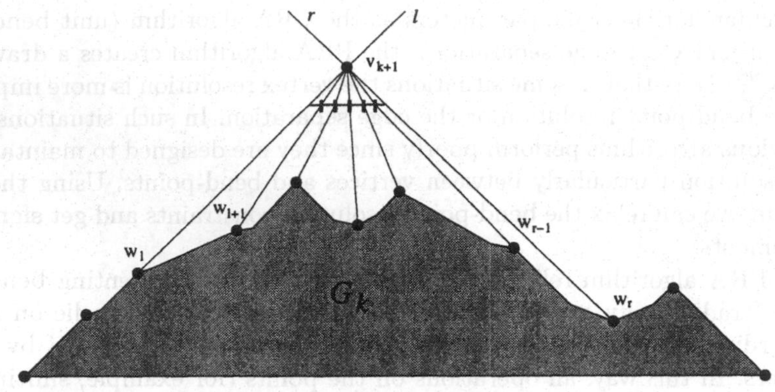

Fig. 1. Graph G_{k+1} after inserting v_{k+1}. The shaded part is G_k. Vertices w_l and w_r are the leftmost and rightmost neighbors of v_{k+1}. The horizontal line segment below v_{k+1} is the middle port region through which all the edges (v_{k+1}, w_i), $l < i < r$, are routed.

of our algorithm we need several definitions. Following the notation of [3], let $w_1 = v_1, w_2, \ldots, w_m = v_2$ be the vertices of the exterior face C_k of graph G_k in order. For a particular subgraph G_k and vertex v_{k+1}, we refer to w_l and w_r as the leftmost and rightmost neighbors of v_{k+1} on C_k, see Fig. 1.

2.1 Vertex Regions

In the immediate vicinity of every vertex there are two types of regions: *free regions* and *port regions*. The free and port regions alternate around the vertex, see Fig. 2(a). For each free region there is at most one edge passing through it to v. Each port region is bounded by a line segment with a number of ports and every edge inside the port region passes through a unique port. The number of ports in a port region is as small as possible. Define and name the six regions around v as follows:

There are three free regions M^f (between $-45°$ and $45°$), R^f (between $90°$ and $135°$), and L^f (between $-135°$ and $-90°$). There are also three port regions M^p (between L^f and R^f), L^p (between L^f and M^f), and R^p (between R^f and M^f).

The algorithm draws each edge in E by "routing" it through a port of one of the two vertices in a fashion similar to [1]. Each edge consists of two connected edge segments. One edge segment, the *port edge segment*, connects a vertex with one of its ports. The other segment, the *free edge segment*, connects a vertex to one of its neighbor's ports. For example, for an edge $e = (u, v)$, if we route e through the leftmost port in u's middle port region M^p, we would draw two line segments, see Fig. 2(b): the *port edge segment* would pass from u to the port, and the *free edge segment* would pass from the port to v. This method of

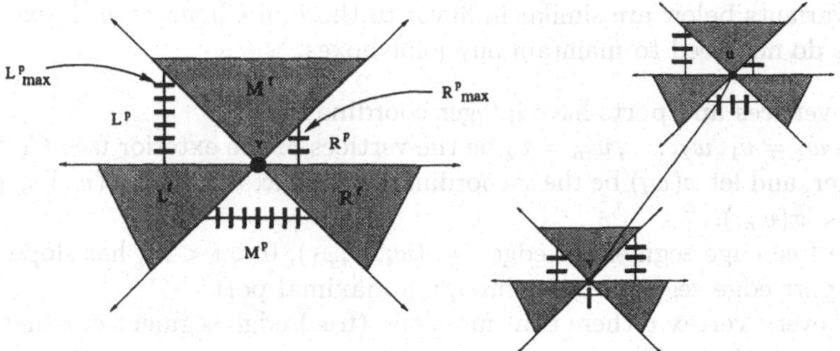

Fig. 2. Vertex regions and edge routing: the number of ports along each port region is determined by the number of edges that need to be routed through that port region. (a) The area around a vertex v is divided into 6 regions. The free regions are shaded and at most one free edge segment goes through each one of them. All the port segments use ports in the port regions of v. (b) Routing an edge $e = (u, v)$, where the port edge segment connects u to one of its ports and the free edge segment connects the port to v, going through one of v's free regions.

construction guarantees that the free edge segments always pass through free regions and that each port transmits at most one port edge segment.

We perform our construction in incremental stages, where each stage corresponds to the insertion of a new vertex. Observe that at each stage, for every vertex v except those on the external face, $w_1 = v_1, w_2, \ldots, w_m = v_2$, there are exactly three free edge segments. The remaining edges are connected to v via port segments. These remaining edges can be grouped into three classes based on which port region they are routed through, L^p, R^p, or M^p. Count the number of edges in each of these groups and let $d_l(v)$ be the number of port edge segments using port region L^p. Similarly, define $d_r(v)$ and $d_m(v)$ to be the number of port segments using port regions R^p and M^p. Observe that in the final stage, there are exactly three vertices on the exterior face, v_1, v_2, v_n, and then $\sum_{v \in V}(d_l(v) + d_r(v) + d_m(v)) = |E|$. That is, for every edge, there is a corresponding port and free edge segment.

For a vertex v we define the *maximal right port* R^p_{max} as follows. Let v have coordinates (v_x, v_y). Then the R^p_{max} of v has coordinates $(v_x + d_r(v) + 1, v_y + d_r(v))$ if $d_r(v) > 0$ and (v_x, v_y) otherwise. We define the *maximal left port* L^p_{max} of v in a similar fashion, see Fig. 2(a).

2.2 Invariants of the CRA Algorithm

By design, our algorithm is incremental with n stages, where each stage corresponds to the insertion of the next vertex in the canonical order. Thus it is natural to define several key invariants to be maintained at every stage. The four invariants below are similar in flavor to those of Cheng *et al* [1] except that here we do not need to maintain any joint boxes.

1. All vertices and ports have integer coordinates.
2. Let $w_1 = v_1, w_2, \ldots, w_m = v_2$ be the vertices of the exterior face C_k of G_k in order, and let $x(w_i)$ be the x-coordinate of vertex w_i. Then $x(w_1) < x(w_2) < \ldots < x(w_m)$.
3. The free edge segment of edge $e = (w_i, w_{i+1})$, $0 < i < m$, has slope ± 1 and e's port edge segment goes through a maximal port.
4. For every vertex v there is at most one (free) edge segment crossing each of its free regions. All other edge segments are port edge segments.

2.3 Vertex Shifting

In the algorithms that maintain good angular resolution with the aid of vertex joint boxes [1,6], every time a new vertex is inserted, already placed vertices need to be shifted a great deal so that the joint box can fit amongst them. The amount of shifting required is typically of the order of the degree of the vertex. Invariably this leads to large constants behind the $O(n) \times O(n)$ area, e.g. $(20n - 48) \times (10n - 24)$ in [6] and $30n \times 15n$ in [1]. In our algorithm we never need to shift any vertex by more than five grid units allowing us to draw G in a $5n \times \frac{5n}{2}$ grid. When a new vertex v is inserted, we must create enough space so that the leftmost w_l and rightmost w_r neighbors of v can "see" v through their respective maximal port regions. Note that the previous R^p_{max} port of w_l and L^p_{max} of w_r were used at an earlier stage. Thus, we must create an additional port along the R^p region of w_l. Similarly, additional space is necessary along the L^p region of w_r.

In order to create more space we need to move w_l and w_r. We also have to ensure that the four invariants and the planarity of the graph are maintained. This is achieved by shifting the "shifting set" of the vertex as well as the vertex itself. Using the definition of de Fraysseix *et al* [3], define the *shifting set* $M_k(w_i)$ for a vertex w_i on the external face of G_k to be a subset of the vertices of G such that:

1. $w_j \in M_k(w_i)$ iff $j \geq i$
2. $M_k(w_1) \supset M_k(w_2) \supset \ldots \supset M_k(w_m)$
3. Let $\delta_1, \delta_2, \ldots, \delta_m > 0$; if we sequentially translate all vertices in $M_k(w_i)$ by distance δ_i to the right ($i = 1, 2, \ldots, m$), then the embedding of G_k remains planar.

These shifting sets can be defined recursively. Let w_l and w_r be the leftmost and rightmost neighbors of v on C_k. Then construct $M_{k+1}(w_i)$ recursively as follows:

$$M_{k+1}(w_i) = M_k(w_i) \cup v_{k+1}, \text{ for } i \leq l,$$

$$M_{k+1}(v_{k+1}) = M_k(w_{l+1}) \cup v_{k+1},$$
$$M_{k+1}(w_j) = M_k(w_j), \text{ for } j \geq r.$$

For convenience, define a *right-shift of m units* for a vertex w_i as shifting $M_k(w_i)$ by m units to the right so that all ports for every vertex in $M_k(w_i)$ also shift except the ports in the L^p region of w_i. Define a *left-shift of m units* for vertex w_i as shifting $M_k(w_{i+1})$ by m units to the right so that all ports for every vertex in $M_k(w_{i+1})$ also shift. Note that in a left-shift, we also shift the ports in the R^p region of w_i by m units to the right.

2.4 CRA Overview

The CRA algorithm constructs the graph one vertex at a time, by creating the graphs $G_1, G_2, \ldots, G_n$. Constructing G_i, $1 \leq i \leq 3$ is straightforward, so assume that G_k has been constructed with exterior face $C_k = (v_1 = w_1, w_2, \ldots, w_m = v_2)$. Suppose we have embedded G_k with exterior face C_k. To construct G_{k+1}, let v_{k+1} be the next vertex in the canonical ordering and recall that w_l and w_r are, respectively, the leftmost and rightmost neighbors of v_{k+1} on the exterior face C_k.

Recall that $d_r(w_l)$ is the current number of port edge segments using R^p of w_l, and that $d_l(w_r)$ is the current number of port edge segments using L^p of w_r. There are two cases to consider:

- **case (a)** $d_r(w_l) = 0$, see Fig. 3(a).
- **case (b)** $d_r(w_l) > 0$, see Fig. 3(b).

In case (a) perform a left-shift of 2 units on w_l in order to free space for a port in the R^p region of w_l. In case (b) perform a left-shift of 1 unit on w_l. Similarly, if $d_l(w_r) = 0$ then perform a right-shift of 2 units on w_r. Otherwise perform a right-shift of 1 unit on w_r.

Insert v_{k+1} at the intersection of lines l and r, where l is the line with slope $+1$ through w_l's maximal right port and r is the line with slope -1 through w_r's maximal left port, see Fig. 1. In the case where lines l and r do not intersect in a grid point it suffices to shift all the elements in $M_k(w_r)$ one additional unit to the right.

The edges from v_{k+1} to w_l and w_r are routed through w_l's maximal right port and w_r's maximal left port, respectively. The remaining edges go from v_{k+1} to vertices w_i, $l < i < r$. Let $(v_{k+1}(x), v_{k+1}(y))$ be the coordinates of vertex v_{k+1}. Let w_a be the rightmost vertex such that $w_a(x) < v_{k+1}(x)$. Before placing the M^p region of v_{k+1} it is necessary to ensure that there are enough ports on it that can be used to connect v_{k+1} to $w_l, w_{l+1}, \ldots, w_r$. The M^p region is a horizontal line segment with $1, 3, \ldots, 2m+1$ ports when the line segment is $1, 2, \ldots, m$ grid units below v_{k+1}. It is necessary to find how many w_i's are to the left and right of v_{k+1} in order to find exactly how far below vertex v_{k+1}, the region M^p must be placed. These numbers are $a - l$ and $r - a - 1$ respectively (from the definition of w_a above). Then place the M^p region of v_{k+1} so that its y coordinate is equal to $v_{k+1}(y) - \max(a - l, r - a - 1)$. As shown in the next section the M^p region can be placed correctly, that is, placed so that it does not intersect the old graph G_k. After determining the location of M^p the edges are

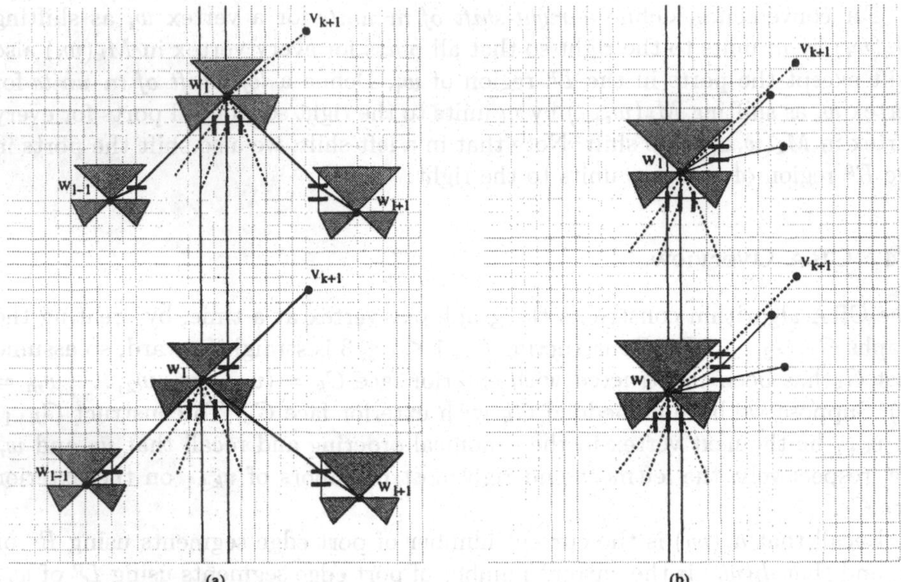

Fig. 3. Adding the current vertex v_{k+1}. Here w_l is the leftmost neighbor of v_{k+1} on the exterior face of G_k. (a) If $d_r(w_l) = 0$, then we need to shift w_l two grid units to the left. (b) If $d_r(w_l) > 0$, then it suffices to shift w_l only one unit to the left. Note that the shifting set $M_k(w_l)$ also shifts with w_l.

routed from v_{k+1} to $w_{l+1}, w_{l+2}, \ldots, w_a$ starting from the leftmost port of M^p of v_{k+1}. Similarly, route the edges from v_{k+1} to $w_{r-1}, w_{r-2}, \ldots, w_{a+1}$ starting from the rightmost port of M^p of v_{k+1}.

3 Correctness of the Algorithm

The algorithm works correctly if all four invariants are maintained. We show that free edge segments always remain in free edge regions and that there is at most one free edge segment per free region. We then need to bound the drawing area required by the algorithm and show that good angular resolution is maintained. Finally, we have to bound the number of bends created and analyze the running time. We leave the detailed proofs for the full version of the paper and present brief proof sketches instead.

Lemma 1. *Free edge segments in free regions remain in the free regions.*

Proof Sketch: Recall that there are three types of free regions: M^f, L^f, R^f. Let us first look at free edge segments in the M^f regions. Edge segments in the M^f regions are created by a vertex v dominating another vertex w. But whenever v dominates w, w is added to the shifting set for v and is only shifted when v is shifted. Therefore, the slope of the free edge segment of the edge connecting v

and w remains constant and the free edge segment remains within M^f. As the two remaining cases are symmetric, without loss of generality, let us examine the case when the free edge segment lies in the L^f region. This implies that the slope of the free edge segment is between 0 and +1. Since shifting only moves vertices farther apart in the x-direction, the slope can only get closer to 0, thus remaining in L^f. □

Lemma 2. *Every free edge segment passes through a free region which contains no other edges.*

Proof Sketch: When a new vertex $v = v_{k+1}$ is inserted there are two types of new edges added: the *outside edges* between v and the outside neighbors, w_l and w_r, and the *inside edges* between v and w_i where $l < i < r$. In both cases the new edge is routed through a port creating one free edge segment and one port edge segment. A free edge segment of an outside edge has slope either +1 or −1 by construction; therefore it lies inside the free regions L^f and R^f of vertex v. Since v is a new vertex, there are no other segments inside these two free regions.

Dealing with the inside edges is more complex. We first need to show that there is sufficient space between the vertices on the exterior face of G_k and the new vertex v. Second, we need to show that v has enough ports in its middle port region M^p for each of the vertices on G_k that it is connected to. Third, we need to show that the free edge segments of the inside edges remain inside their free regions. We begin by showing that for every vertex w_i, $l < i < r$,

- vertex w_i lies below M^p, the middle port region for v, and
- we can assign a unique port along the M^p port region of v, such that the edge segment connecting w_i to that port fits inside w_i's middle free region M^f

First consider the vertices $w_l, w_{l+1}, \ldots, w_a$. In the worst case, they have monotonically increasing y-coordinates. Intuitively, this is the worst case because the area of $G_{k+1} - G_k$ is the smallest and hence it is more difficult to ensure that there is enough space for the M^p region of v. Using invariant (3) we can show that $w_a(y) \leq v(y) - (a - l)$, since all the edges connecting consecutive vertices on the outer face have both port and free edge segments, thus "freeing" at least one grid point distance between v and w_a for every w_i. Similarly, $w_{a+1}(y) \leq v(y) - (r - (a + 1))$.

The same argument implies that there are at least as many ports along the middle port region M^p of v as there are vertices w_i, $l < i < r$. We route the inside edges for v as follows: starting with the leftmost port on the middle port region M^p of v, we assign consecutive ports to $w_{l+1}, w_{l+2}, \ldots, w_a$ and route the corresponding free edge segments from w_i to the ports. The port edge segments for the inside edges connect the same ports to v. Similarly, starting with the rightmost port on the middle port region M^p of v we assign consecutive ports to $w_{r-1}, w_{r-2}, \ldots, w_{a+1}$ and route the corresponding free edge segments from w_i to the ports. The port edge segments for the the inside edges connect the same ports to v.

Combining the above observations, it can also be shown that for every vertex w_i, $l < i < r$, the free edge segment of its edge to v fits inside the middle free region M^f of w_i. □

Lemma 3. *If G_k maintains invariants one through four, then G_{k+1} maintains invariants one through four.*

Proof Sketch: By definition of the shifting set, invariants one and two hold, see [6]. By construction of the algorithm, invariant three holds as well. Also by construction every edge inserted has a port edge segment and a free edge segment. By lemmas 1 and 2 invariant four also holds. □

Lemma 4. *The angular resolution for vertex $v \in G$ as produced by the algorithm is $1/2d(v)$, where $d(v)$ is the degree of vertex v.*

Proof Sketch: The worst angle is achieved between a free edge segment for some edge f and a port edge segment for some edge e, where f is located at the boundary of its free region and e is the neighboring port edge segment. There are six possible cases but the argument is the same for all of them, so without loss of generality consider the case in Fig. 4. Let v be the vertex and $d(v) = d$ its degree. Also let s and t be the lengths as shown in Fig. 4. Let θ be the angle between f and e, and x the number of ports as shown in the figure. Note that all vertices have three edges connected to them via free edge segments. Then the number of ports in any port region is at most $d - 3$. From the figure, $\tan(\theta) = t/(s - t)$ and hence $\arctan(t/(s - t)) = \theta$. But

$$\frac{t}{s-t} = \frac{\sqrt{2}/2}{\sqrt{2}(x+1) - \sqrt{2}/2} = \frac{1}{2x+1}$$

Using the Maclaurin expansion for $\arctan(y)$, where $y < 1$ we have

$$\arctan(y) = y - y^3/3 + y^5/5 - ...$$

Since $1/(2x + 1) < 1$, and $x \leq d - 3$ this yields $\theta \geq 1/2d$ which completes the proof. □

Theorem 1. *For a given planar graph G, the algorithm produces in $O(n)$ time a planar embedding with grid size $5n \times 5n/2$, using at most one bend. The angular resolution for every vertex v of G is $1/2d(v)$.*

Proof Sketch: Since every edge has only two segments, there can be at most one bend per edge. Chrobak and Payne [2] show how to implement the algorithm of De Fraysseix, *et al.* [3] in linear time. Their approach can be easily extended to our algorithm. By invariants three and four and by lemma 4 the angular resolution is at most $1/2d(v)$.

It remains to show that the drawings produced by the algorithm fit on the $5n \times 5n/2$ grid. Every time we insert a vertex v_k, we increase the grid size by at most 5 units, which implies that the width of the drawing is at most $5n$. The final drawing fits inside an isosceles triangle with sides of slope $0, +1, -1$. The width of the base is $5n$ and so the height is less than $5n/2$. □

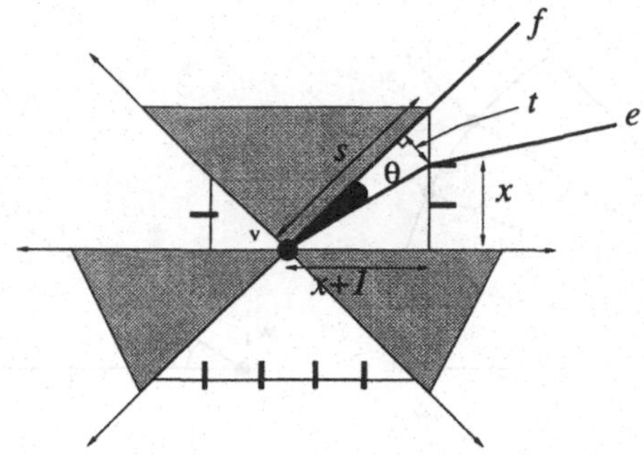

Fig. 4. The minimum angle between two edges adjacent to vertex v is proportional to the degree d of the vertex. Using our algorithm the angle cannot be smaller than $2/d$.

4 The PRA Algorithm

In this section, we introduce a novel approach to representing bends and vertices. Rather than insisting that bends lie on integer grid coordinates, we propose an alternative approach which allows bends to be located on a grid represented by polar coordinates. Using a polar representation allows us to independently control the grid size and edge bend positions. We begin by considering the polar representation in general and then present the PRA algorithm that uses the new approach.

A point p in the polar grid system is represented by a set of integers. For the vertices we only need two integers (p_x, p_y). For the bend-points we may need up to five integers. We shall see in the PRA algorithm that these five integers need not be explicitly stored for every bend-point. In general, a bend-point is given by:

- (p_x, p_y), the origin of the polar system
- p_r, the radius of the circle around the origin (p_x, p_y)
- p_d and p_n, the angle (p_θ) of the circle where the point is located, i.e., $p_\theta = 2\pi p_n/p_d$. For convenience, we consider $p_\theta = 0$ to be the vertical direction.

The PRA algorithm places vertices at integer grid coordinates, thus guaranteeing unit vertex resolution. As it is based on the CRA algorithm it also uses only 1 bend per edge. The main difference in the two algorithms is in the placement of the bend-points. In the PRA algorithm, bend-points will be placed on a circle around the vertex (rather than on a straight-line segment). Therefore, the origin, (p_x, p_y) for each bend-point need not be explicitly stored – it suffices to store the origin of the vertex that the bend-point is associated with. Similarly, groups of bend-points around a given vertex will have the same radius and hence each of the bend-points need not explicitly store p_r. Since the points will be evenly spaced in a port region, the values for p_θ need also not be explicitly stored for each bend-point.

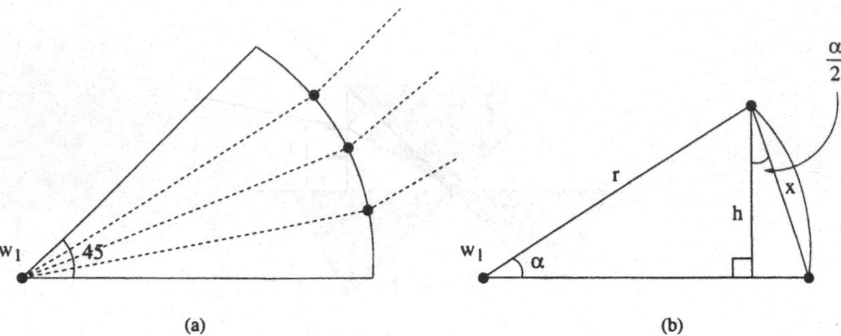

Fig. 5. Vertex w_l is the left-most neighbor of the next vertex v_{k+1} along the exterior face of G_k. The $d_r(w_l)$ ports of w_l are evenly spaced on the arc of a circle of radius $2d_r(w_l)$ bounded by the middle free region M^f and the right free region R^f. (a) An example of the layout for the R^p region with $d_r(w_l) = 3$. (b) The distance x between two adjacent ports or a port and an adjacent free region can be computed given the radius of the circle and the angle between the edges connecting the ports to w_l: $x = 2r \sin \frac{\alpha}{2}$.

Consider the leftmost neighbor, w_l, of the next vertex in the canonical order, v_{k+1}. The ports are evenly spaced in the R^p region for w_l, Fig. 5(a). The length of the straight-line segment separating two bend-points or a bend-point and an adjacent free region can be computed as follows. Consider the example in Fig. 5(b). We would like to compute the length x in terms of the radius of the circle and the angle between the two line segments connecting consecutive ports to w_l. From basic trigonometry, the angle between h and x is $\alpha/2$. We can express h in terms of r and α: $h/r = \sin \alpha$ and we can express x in terms of h and α: $h/x = \cos \alpha/2$. Combining the two expressions we obtain

$$x = \frac{h}{\cos \frac{\alpha}{2}} = \frac{r \sin \alpha}{\cos \frac{\alpha}{2}} = \frac{2r \sin \frac{\alpha}{2} \cos \frac{\alpha}{2}}{\cos \frac{\alpha}{2}} = 2r \sin \frac{\alpha}{2}.$$

Assume we have inserted $v_1, v_2, \ldots, v_k$ and have a drawing of G_k with exterior face C_k. Consider inserting the next vertex v_{k+1} in the canonical order. Let w_l and w_r be the leftmost and rightmost neighbors of v_{k+1} on the exterior face C_k. Define f_b and f_e to be the *bend-point resolution* and *edge separation* respectively. Observe that in the standard Cartesian representation algorithms $f_b = 1$ and $f_e = \sqrt{2}/2$. Let $d_r(w_l)$, respectively $d_l(w_r)$, be the number of port edge segments using R^p of w_l, respectively L^p of w_r. When inserting v_{k+1}, the degrees for w_l and w_r affect the amount of shifting necessary to ensure proper resolution. As the cases for $d_r(w_l)$ and $d_l(w_r)$ are symmetrical, we shall concentrate on $d_r(w_l)$. There are two cases to consider:

- **case (a)** $d_r(w_l) = 0$ prior to insertion
- **case (b)** $d_r(w_l) \geq 1$ prior to insertion

In **case (a)** we insert the first edge in the port region R^p between the two free regions R^f and M^f of w_l. We place the port in the middle of the arc of a circle connecting R^f and M^f. Since there are no other bends yet in R^p we are only concerned with maintaining the edge separation. We need to place the port sufficiently away from the vertex w_l. Consider the relationship between the radius of the circle and the edge separation, see Fig. 5.

The edge separation $f_e = x = 2r \sin \frac{\alpha}{2}$. But since there is only one port and it is in the middle of the arc, $\alpha = \pi/8$. We are interested in the radius necessary to achieve the edge separation f_e which is given by

$$r = \frac{f_e}{2 \sin \frac{\alpha}{2}} = \frac{f_e}{2 \sin \frac{\pi}{16}} < \frac{4f_e}{\sqrt{2}} = 2\sqrt{2}f_e.$$

Since we maintain that the vertices are at integer coordinates and the radii are also integers, then the minimum radius required in case (a) is

$$r < \lceil 2\sqrt{2}f_e \rceil.$$

In **case (b)** we insert an additional port in the port region R^p which already has at least one port. In this case, we must ensure that both the edge separation f_e and bend-point resolution f_b are preserved. In this case the radius required is given by:

$$\max\{\lceil \frac{f_e}{2 \sin \frac{\pi}{8(d_r(w_l)+1)}} \rceil, \lceil \frac{f_b}{2 \sin \frac{\pi}{8(d_r(w_l)+1)}} \rceil\}.$$

Typically, $f_b \geq f_e$, so we can assume that the bend-point resolution determines the radius in case (b). Using this together with the fact that $\sin \alpha > 0.97\alpha$ for $\alpha < \pi/8$, the minimum radius required is

$$r < \lceil \frac{f_b}{2 \sin \frac{\pi}{8(d_r(w_l)+1)}} \rceil < \lceil \sqrt{2} f_b (d_r(w_l) + 1) \rceil$$

Summing over all vertices in the graph, the sum of the radii used for the right port regions, R, yields:

$$R = \sum_{v_i \in V : d_r(v_i)=1} \lceil 2\sqrt{2}f_e \rceil + \sum_{v_i \in V : d_r(v_i)>1} \lceil \sqrt{2} f_b (d_r(v_i) + 1) \rceil. \quad (1)$$

With R we bounded the number of shifts required because of "right" neighbors. Similarly, we can define L, the shifts necessary due to "left" neighbors:

$$L = \sum_{v_i \in V : d_l(v_i)=1} \lceil 2\sqrt{2}f_e \rceil + \sum_{v_i \in V : d_l(v_i)>1} \lceil \sqrt{2} f_b (d_l(v_i) + 1) \rceil. \quad (2)$$

L and R bound the number of shifts required due to left and right neighbor visibility. Note, however, that if we shift by the minimum amount required by the f_e and f_b parameters, the location of the next vertex v_{k+1} may not be at integer coordinates. We can guarantee that v_{k+1} is placed on the integer grid

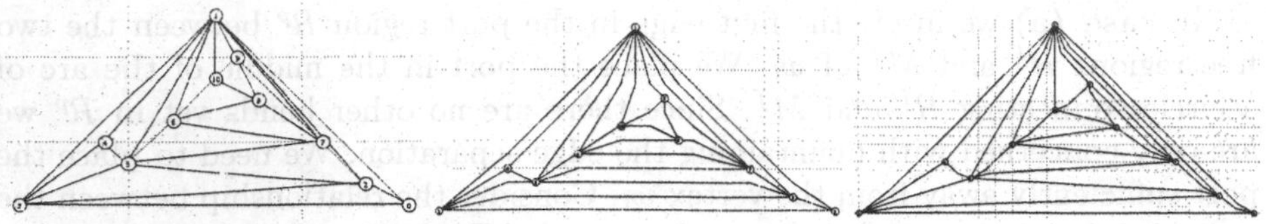

Fig. 6. A graph with 11 vertices drawn using (a) the canonical ordering on the 10×19 grid; (b) the CRA algorithm on the 14×29 grid; (c) the PRA algorithm on the 23×45 grid.

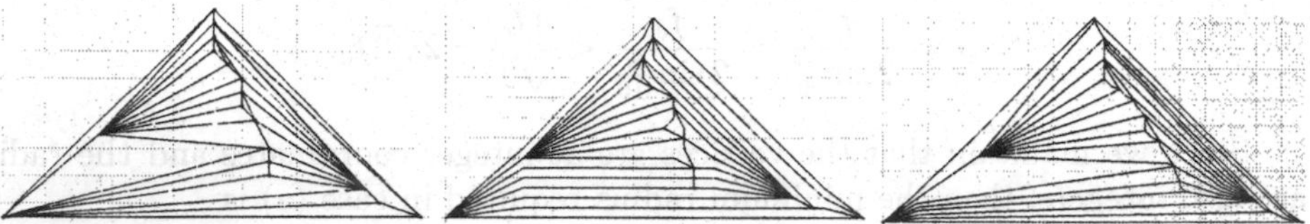

Fig. 7. A graph with 17 vertices drawn using (a) the canonical ordering on the 16×31 grid; (b) the CRA algorithm on the 21×41 grid; (c) the PRA algorithm on the 43×85 grid.

by performing some additional shifts. By shifting at most 3 more units, we are guaranteed to find an integer location for v_{k+1}. Then the total shifting required is at most $L + R + 3n$. Since the final drawing fits inside an isoceles right-angle triangle, the total area required for the drawing is $(L + R + 3n) \times (\frac{L+R+3n}{2})$.

In order to compare the PRA algorithm to the CRA algorithm, we evaluate equations 2 and 1 using two sets of parameters, Table 1. In all three cases the algorithms guarantee at most one bend per edge. The PRA algorithms place all the vertices on grid points and each bend-point is determined by at most five integer polar coordinates.

Table 1. Fixing specific values for the vertex resolution f_v, bend-point resolution f_b, and edge separation f_e allows us to compare the PRA and CRA algorithms.

Algorithm	f_v	f_b	f_e	drawing area	resolution
CRA	1	1	$\sqrt{2}/2$	$5n \times 5n/2$	$1/2d(v)$
PRA1	1	1	$\sqrt{2}/2$	$9n \times 9n/2$	$\pi/4d(v)$
PRA2	1	1/2	1/2	$7n \times 7n/2$	$\pi/4d(v)$

5 Conclusion and Open Problems

In this paper we present two algorithms for drawing planar graphs with good angular resolution while maintaining small drawing area. Other drawing criteria optimized by the algorithms include number of bends, vertex resolution, bend-point resolution, and edge separation. The first algorithm, CRA, is a traditional algorithm in which vertices and bend-points are represented using Cartesian coordinates. It improves on the best known simultaneous bounds for the six drawing criteria. In the PRA algorithm vertices and bend-points are represented using polar coordinates. It is based on the CRA algorithm but allows for independent control over the grid size and bend positions.

Using a polar coordinate representation yields slightly worse area bounds compared to the CRA algorithm, see Fig 6 and Fig. 7. We believe, however, that the PRA approach is more promising. The angular resolution of the PRA algorithm is better and it provides greater control over the drawing process.

The PRA bounds presented in this paper can be further improved. Using two integers to represent the radius (similar to the way the angles are currently represented) will most likely result in smaller drawing area. Our current estimates indicate that certain (small) values of edge separation and bend-point resolution yield grids of size $4n \times 2n$. It is likely that when using only one bend per edge, the best angular resolution will be achieved for vertex regions in which each of the port and free regions have angles $\pi/3$ rather than a combination of $\pi/4$ and $\pi/2$. The biggest challenge, however, to the success of the PRA algorithm deals with the three potential shifts needed to align a new vertex onto an integer grid. If we can reduce this bottleneck, we feel that the PRA algorithm can significantly surpass the bounds of the CRA algorithm.

References

1. C. C. Cheng, C. A. Duncan, M. T. Goodrich, and S. G. Kobourov. Drawing planar graphs with circular arcs. *Discrete and Computational Geometry*, 25:405–418, 2001.
2. M. Chrobak and T. Payne. A linear-time algorithm for drawing planar graphs. *Inform. Process. Lett.*, 54:241–246, 1995.
3. H. de Fraysseix, J. Pach, and R. Pollack. How to draw a planar graph on a grid. *Combinatorica*, 10(1):41–51, 1990.
4. G. Di Battista, P. Eades, R. Tamassia, and I. G. Tollis. *Graph Drawing: Algorithms for the Visualization of Graphs.* Prentice Hall, Englewood Cliffs, NJ, 1999.
5. A. Garg and R. Tamassia. Planar drawings and angular resolution: Algorithms and bounds. In *Proc. 2nd European Symposium on Algorithms*, pages 12–23, 1994.
6. M. T. Goodrich and C. G. Wagner. A framework for drawing planar graphs with curves and polylines. In *Proc. 6th Symposium on Graph Drawing*, pages 153–166, 1998.
7. C. Gutwenger and P. Mutzel. Planar polyline drawings with good angular resolution. In *Proc. 6th Symposium on Graph Drawing*, pages 167–182, 1998.
8. G. Kant. Drawing planar graphs using the canonical ordering. *Algorithmica*, 16:4–32, 1996. (special issue on Graph Drawing, edited by G. Di Battista and R. Tamassia).
9. S. Malitz and A. Papakostas. On the angular resolution of planar graphs. *SIAM J. Discrete Math.*, 7:172–183, 1994.

On Polar Visibility Representations of Graphs

Joan P. Hutchinson*

Department of Mathematics and Computer Science
Macalester College
St. Paul, MN 55105, USA
hutchinson@macalester.edu

Abstract. We introduce polar visibility graphs, graphs whose vertices can be represented by arcs of concentric circles with adjacency determined by radial visibility including visibility through the origin. These graphs are more general than the well-studied bar-visibility graphs and are characterized here, when arcs are proper subsets of circles, as the graphs that embed on the plane with all but at most one cut-vertex on a common face or on the projective plane with all cut-vertices on a common face. We also characterize the graphs representable using full circles and arcs.

1 Introduction

Visibility graphs are now a well-established area of graph drawing [10]. Much has been written about their importance and application; however, they continue to pique the imagination of mathematicians with their intrinsic appeal and intriguing questions [2]. There has been a natural progression from bar-visibility graphs (BVGs) [11,17] to rectangles [1,3,4,6], from bars with visibilities in the plane to those on the sphere and cylinder [12,13], on a (flat) torus [8], or on the Möbius band [5]. These rectilinear representations are natural ones for most applications; however, we turn instead to the realm of polar representations with arcs of circles and radial visibility. In many ways circular representations and related polar coordinates are equally natural and in some contexts more applicable than rectilinear ones. With this change of perspective we can and do represent a class of graphs, larger than with bars in the plane, though ultimately constrained by the real projective plane. Thus with a planar representation of arcs of circles nonplanar graphs are drawn in a natural way, resulting in diagrams often reminiscent of time-exposed shots of the North Star and surrounding stars.

We introduce the layout of graphs as polar visibility graphs (PVGs) using arcs of concentric circles (arcs that are proper subsets of a circle) with radial visibility, including visibility through the origin, the center of all the concentric circles. These graphs, though arising naturally from visibility in the plane, correspond to graphs embedded on the (real) projective plane, the nonorientable surface of Euler characteristic 1. PVGs are characterized as the planar graphs that can be

* Research supported in part by NSA Grant #MDA904-99-1-0069.

P. Mutzel, M. Jünger, and S. Leipert (Eds.): GD 2001, LNCS 2265, pp. 422–434, 2002.
© Springer-Verlag Berlin Heidelberg 2002

drawn in the plane with all but at most one cut-vertex on a common face plus the graphs that can be embedded on the projective plane with all cut-vertices on a common face. We also consider the variation in which full circles are allowed along with arcs, and characterize the graphs so representable (CVGs) in terms of their block-cutpoint tree.

2 Background

Just as visibility wider than along a line is required for BVGs, we ask that radial visibility in PVGs be available through a nondegenerate cone. Define a (nondegenerate) *cone* in the plane to be a 4-sided region of positive area with two opposite sides being arcs of circles, centered at the origin, and the other two sides, possibly intersecting, being radial line segments on lines through the origin. Thus, both $\{(r,\theta) : 1 \le r \le 2, 0 \le \theta \le \pi/6\}$ and also

$$\{(r,\theta) : 0 \le r \le 1, 0 \le \theta \le \frac{\pi}{6} \text{or} \pi \le \theta \le \frac{7\pi}{6}\} = \{(r,\theta) : -1 \le r \le 1, 0 \le \theta \le \frac{\pi}{6}\}$$

are considered to be cones, respectively, not containing and containing the origin. Given a set of arcs, all centered at the origin, two of these arcs a_1 and a_2 are said to be *radially visible* if there is a cone that intersects only these two arcs and whose two circular ends are subsets of the two arcs; the same definition holds for visibility between an arc and a circle and between two circles. A graph is called a *polar visibility graph* if its vertices can be represented by arcs, including endpoints, of circles centered at the origin, having pairwise disjoint relative interiors, so that two vertices are adjacent if and only if the corresponding arcs are radially visible; see Figure 1a below for a PVG representation of K_6. If this model is used, but without visibility through the origin, the graphs arising are one of the cylindrical types characterized in [13]. Note that for a 2-connected graph (a graph without cut-vertices) there is no loss in taking arcs as proper subsets of circles since a full circle can be cut down to a smaller arc, leaving the same visibilities. Arcs in a PVG layout spanning more than half its circle will provide interesting variations, full circles even more. We use graph theoretic terminology as in [15], topological notions as in [9], and algorithmic ideas following the BVG presentation in [10].

Similarly a graph is called a *circular visibility graph* if its vertices can be represented by arcs and circles with radial visibility between arcs and circles determining edges as for PVGs. When possible we prefer, but do not require, arcs over circles; that is, in a layout we will decrease a circle to become a proper arc if no additional visibilities are introduced. We shall see that some planar and projective planar graphs with cut-vertices on an arbitrary number of faces are CVGs, but not PVGs, but that these faces must be nested appropriately. Figure 1b shows such a planar CVG. In that layout the inner circle contains one arc; if instead, it contained four mutually visible arcs, encircling the origin and forming a K_5, the example becomes a nonplanar CVG.

Note that in a PVG or CVG layout of a graph G, we may draw each arc and circle on a distinct circle, and we may take these circles to have radii $1, 2, \ldots, n$

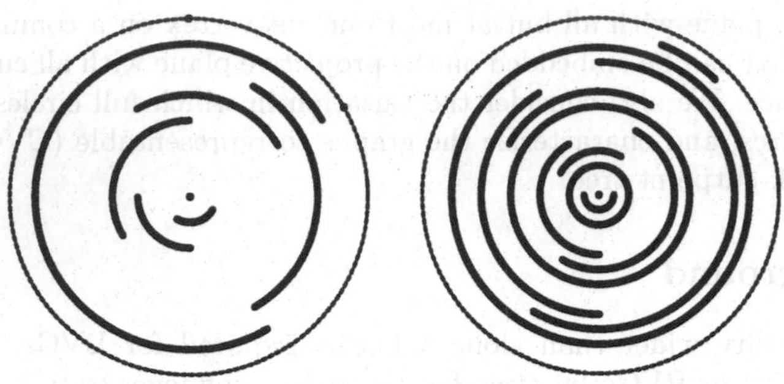

Fig. 1. **a.** K_6 **b.** A circular visibility layout

where $n = |V(G)|$. This naturally leads to another layout of the graph in a disk of radius n+1 and centered at the origin by inverting each circle and arc through the circle of radius $(n+1)/2$. That is, each point with polar coordinates (r, θ), $0 < r < n+1$, is mapped by the inversion to the point $(n+1-r, \theta)$. This inversion preserves circles, arcs, and the angles defining these arcs. If the original layout was L, we denote this inverted layout by $\mathbb{I}(L)$.

Recall that the (real) projective plane can be obtained by taking a circular disk and identifying opposite (or antipodal) points. Thus if we identify opposite points of the circle of radius $n+1$, we create a projective plane. Two arcs in $\mathbb{I}(L)$ (or an arc and a circle or two circles) that were previously radially visible in a cone, not containing the origin, are still radially visible, and a pair visible in a cone through the origin are now visible in a "generalized cone" that crosses the boundary of the projective plane, reemerging on the other side. The coordinates of such a generalized cone are given by $\{(r,\theta), r^* \leq r \leq n+1 \text{or} -(n+1) \leq r \leq -s^*, \theta_1 \leq \theta \leq \theta_2\}$ where r^*, s*, $\theta_1 < \theta_2$ are constants, $0 \leq r^*, s^* < n+1$. In addition, the interior of no two of these new cones intersect. Fig. 2a shows the inverted layout of K_6 on the projective plane with dashed lines indicating a conical area of visibility, and in 2b we see an embedding of K_6 created by shrinking each arc to a vertex. The first proposition is then clear since each inverted arc and circle on the projective plane can be replaced by a single vertex. Then the visibility cones can each be shrunk and transformed to a set of nonintersecting edges on the projective plane.

Proposition 1. *A PVG or a CVG embeds on the projective plane. Recall that a graph G is said to embed on a surface S if it can be drawn there without any edge crossings, and that each maximal connected component of* $S \backslash \{V(G), E(G)\}$ *is called a face of the embedding (we do not require that the faces be simply connected).*

Theorem 1. *A graph Gis a PVG if and only if either a) G has an embedding in the plane with all but at most one cut-vertex on a common face, or else b) G has an embedding on the projective plane with all cut-vertices on a common face.*

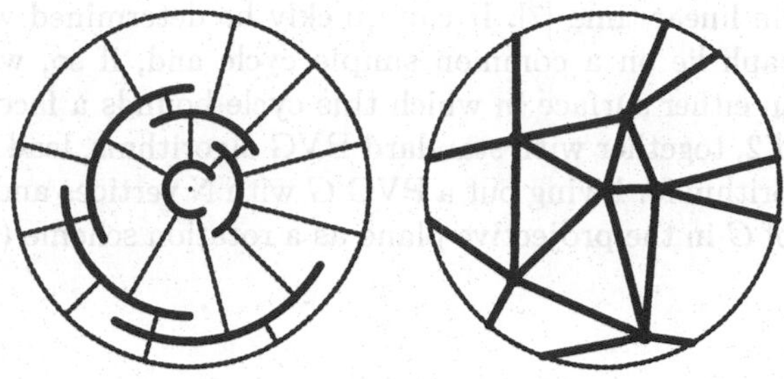

Fig. 2. a. $\mathbb{I}(K_6)$ and $\mathbb{I}(K_6)^*$ on the projective plane **b.** For $G = K_6$, $\mathbb{I}(L_G) = (\mathbb{I}(L))_G$

Note that condition (a) allows for the representation of planar graphs that are not BVGs; for example $K_{2,3}$ with three additional vertices of degree 1 appended, one each to a vertex of degree two, is a PVG. Similarly K_4+4e (K_4 plus a pendant vertex and edge at each vertex) is a PVG (see Fig. 3); these are the smallest graphs that are not BVGs. Condition (b) also allows for more planar graphs; for example, two vertices joined by three internally disjoint paths of length three (i.e., three edges each) plus six vertices of degree 1, each adjacent to a different vertex of degree two, satisfies (b), but not (a).

Every graph G can be decomposed into its blocks and their connecting cut-vertices (a *block* is either an edge or a 2-connected subgraph; see [15]), and these connections determine a tree, called the *block-cutpoint tree* of the graph, BC(G). This tree has a vertex for each block and for each cut-vertex of G, and two vertices of BC(G) are adjacent if and only if they correspond to an incident cut-vertex and block. We call a block *planar* if it represents a planar graph.

Theorem 2. *A graph G is a CVG if and only if* BC(G) *consists of a path $P = (e_1, e_2, \ldots, e_{2k+1})$, $k \geq 0$, with e_{2i} representing a planar block, $i = 1, \ldots, k$, so that*

1. a) e_1 is also incident with one additional (nonempty) block representing a (2-connected) projective planar graph, or
b) e_1 is also incident with one or more (nonempty) planar blocks, and

2. e_{2k+1} is also incident with an arbitrary tree structure T so that $T \cup \{e_{2k+1}\}$ represents a planar graph that can be drawn in the plane with all cut-vertices, except possibly for that representing e_{2k+1}, on a common face.

When $k = 0$, these conditions reduce to those of Theorem 1. On the other hand, it may be that each cut-vertex of G, represented by $e_1, e_3, \ldots, e_{2k+1}$, lies on a different face, as in Figure 1b. This example is the first of an infinite family of CVGs with an increasing number of cut-vertices, all on different faces; the family is obtained by nesting repeatedly the same pattern of arcs and circles. Most of the details of the proofs of Theorems 1 and 2 are included below.

As described in [10], planar layouts and the block-cutpoint tree of a graph can be determined in linear time. Projective planar graphs can also be recognized and embedded in linear time [7]. It can quickly be determined whether all cut-vertices of a graph lie on a common simple cycle and, if so, whether there is an embedding in either surface in which this cycle bounds a face. The proofs of Theorems 1 and 2, together with standard BVG algorithms, lead to a $\mathcal{O}(\mathrm{NE}^2) = \mathcal{O}(N^3)$time algorithm for laying out a PVG G with N vertices and E edges, given an embedding of G in the projective plane as a rotation scheme (defined below), as in [7,9].

3 Main Results on PVGs

We develop theory that will also allow extension to CVGs. We focus on simple graphs and their characterizations as in Theorems 1 and 2. Thus we say that two arcs are radially visible if there is at least one maximal cone providing mutual visibility; however, we can also obtain more precise results by keeping track of multiple and even self-visibility between arcs and circles.

First we need more precise topological and geometric definitions. Consider a PVG or CVG layout L of a graph G and its inverse layout on the projective plane, $\mathbb{I}(L)$. We let L^* (respectively, $\mathbb{I}(L)^*$) denote the visibility depiction obtained by shrinking each maximal visibility cone of L (resp., $\mathbb{I}(L)$) to a distinct line segment by reducing its angles$b_1 \leq \theta \leq b_2$ to some constant $\theta = b$, $b_1 < b < b_2$; strict inequality ensures distinct visibility segments. For $G = K_6$, $\mathbb{I}(L)^*$ is shown in Fig. 2a.

Also let L_G (resp., $(\mathbb{I}(L))_G$) denote the graph obtained from L^* (resp., $\mathbb{I}(L)^*$) by shrinking each arc to a vertex, consisting of one point, and transforming each visibility line segment to an edge that intersects no other edge except possibly at the origin (resp., an edge that intersects no other edge on the projective plane). If L^* or $\mathbb{I}(L)^*$ contains a circle, it is replaced by a point as vertex. Thus $(\mathbb{I}(L))_G$ is a graph embedded on the projective plane, see Fig. 2b. Note that L^*, $\mathbb{I}(L)^*$, $\mathbb{I}(L_G)$, and $(\mathbb{I}(L))_G$ have visibility segments and edges for each distinct, maximal visibility cone so that multiple edges and loops may be present in these depictions; however, a pair of multiple edges will not form an embedded digon with empty interior.

Note that the complement of the arcs, circles, and lines of L^* divide up the plane into faces; similarly $\mathbb{I}(L)^*$ divides up the projective plane. One face of L^* is the exterior face, possibly containing the origin; this exterior face is the one in which most cut-vertices of a PVG and their blocks can be placed. We say that an arc or circle of a layout L lies on the exterior face if it lies on the exterior face of L^*. We use the following combinatorial description of an embedded graph and of a PVG or CVG layout. If a graph is embedded on any surface, then for each vertex there is naturally defined a cyclic rotation of its neighbors, given by the order, say clockwise, of its edges in the embedding; such a collection of rotations, one for each vertex, is called a *rotation scheme.* (See, for example, [16] where it is shown that an embedding is equivalent to a rotation scheme.) Such a description is generally used for algorithms on embedded graphs [9]. Similarly, given a PVG layout L in the plane and its inverse layout $\mathbb{I}(L)$ in the projective plane, one can define the *arc-rotation scheme* to be the set of cyclic rotations of neighbors about each arc of its visibilities to other arcs; note that the rotations at the arcs of L and of $\mathbb{I}(L)$ are inverses of each other. We say that an embedding of a PVG graph G in the plane or on the projective plane and its polar visibility layout L are *equivalent* if the arc-rotation scheme of $\mathbb{I}(L)$, when translated into a set of vertex-neighbor cycles, yields the rotation scheme of the embedded graph; see Figs. 1, 2. Given a circle in a CVG layout L or $\mathbb{I}(L)$, the neighbors divide into two cyclic rotations of the inner and outer visibilities, called the *circle-rotation scheme.* Then a drawing of a CVG and its layout L are equivalent if the arc/circle-rotation schemes of $\mathbb{I}(L)$ agree with those of the embedded graph.

It is not hard to see the following, by bending or straightening corresponding BVG and PVG layouts.

Proposition 2. *A connected graph has a PVG layout with no visibilities through the origin if and only if the graph is a BVG.*

A PVG layout with no visibilities through the origin contains arcs in sectors, alternating about the origin, so that some can be reflected through the origin, leaving the layout in two quadrants, and this can be straightened to form a BVG.

Of course there are planar graphs with layouts as PVGs including visibilities through the origin and with cut-vertices represented on the exterior face. Note that whenever there are visibilities through the origin in a layout L, then the equivalent graph $(\mathbb{I}(L))_G$ is embedded on the projective plane. It turns out that in some PVG layouts there is a (sneaky) hiding place for a cut-vertex and its connecting blocks, but the resulting graphs turn out to be planar. In a PVG layout we call an arc a^* a *long arc* if its angular span is greater than π. Suppose $a^* = \{(r^*, \theta), 0 \leq \theta \leq \pi + x\}$, for some $0 < x < \pi$. Then the cone defined by $C(a^*) = \{(r, \theta), -r^* \leq r \leq r^*, 0 \leq \theta \leq x\}$ is an area in which interior arcs can see the arc a^* and possibly no others; see Fig. 3.

In preparation for CVG layouts, we require special PVG layouts. Let a^* be a long-arc at radius 1, spanning $\theta_1 \leq \theta \leq \theta_1 + \pi + x$for some $x > 0$. Arcs a^* and b^* are called a *long-arc pair at the origin* if they are mutually visible, together they span 2π, and if b^* lies at radius $r^* > 1$, no arcs intersect the *long-arc cone* $\{(r, \theta) : 0 \leq r < r^*, \theta_1 + \pi + x < \theta < \theta_1 + 2\pi\}$. (For example, when $r^* = 2$, no arcs can meet the designated cone.) Similarly if a^* is a long-arc at the outermost radius $n = |V(G)|$, spanning $\theta_2 \leq \theta \leq \theta_2 + \pi + y$ for some $y > 0$, then a^* and b^* are a *long-arc pair at infinity* if they are mutually visible, together span 2π, and if b^* lies at radius $r^* < n$, no arcs intersect the long-arc cone $\{(r, \theta) : r^* < r < n, \theta_2 + \pi + y < \theta < \theta_2 + 2\pi\}$; see Figs. 1a, 3. Notice that in long-arc pairs the long arc at radius 1 or at radius n could be extended to form a full circle without changing visibilities.

Here are the building-block results needed for the PVG characterization.

Proposition 3. *Let G be laid out as a PVG L including a long arc a^* that represents a cut-vertex x^*, not lying on the exterior face, and let B be a block of G incident with x^* and whose representation lies within $C(a^*)$ in L. Then G is*

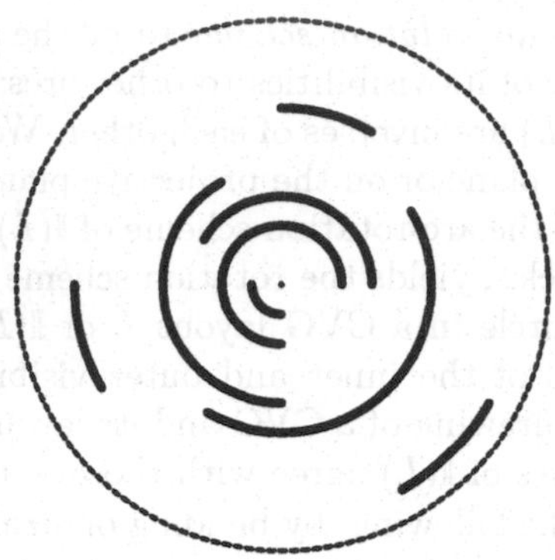

Fig. 3. K_4+4e.

a planar graph and can be drawn in the plane with one face including all vertices whose arcs lie on the exterior face of L.

Proof. The proof consists of observing that in $\mathbb{I}(L)$ and in $(\mathbb{I}(L))_G$ the representation of a block B incident with x^* lies within a (noncontractible) sector of the projective plane that divides the space into two contractible (planar) regions.

The next result is the necessary topological argument needed to characterize PVGs; it is a contraction proof similar to that of [14] and [8]. This result is carried out for multigraphs, those embedded with no digon face with empty interior except for two special faces. For these graphs we achieve layouts with a one-to-one correspondence between distinct, maximal visibility cones and edges of G. If x is a vertex of a PVG, we let a_x denote its arc in the layout, and conversely x_a is the vertex corresponding to an arc a.

Proposition 4. *(i) Let G be a loopless 2-connected plane multigraph, let F be a face in the embedding, and let c be a vertex of G. Suppose G has at most two digon faces, possibly F and, when c does not lie on F, possibly one incident with c. Then $G' = G$ plus a loop at c has a PVG layout L' in which all vertices of F are represented on the exterior face of L' and (a_c, a_d) is a long-arc pair at the origin for some neighbor d of c. In addition, G' has an embedding on the projective plane that is equivalent to L'.*

(ii) Let G be a loopless 2-connected plane multigraph with v_1 and v_2 designated, distinct vertices and with no digon face. Then $G' = G$ plus a loop at v_2 has a PVG layout L' with v_i represented by arc a_i, $i = 1, 2$, with (a_1, b_1) a long-arc pair at infinity, and with (a_2, b_2) a long-arc pair at the origin, where for $i = 1$ and 2, arc b_i corresponds to some neighbor of v_i. Also G' has an embedding on the projective plane that is equivalent to L'.

(iii) Let G be a loopless 2-connected multigraph with a 2-cell embedding on the projective plane, with F a face in the embedding, and with no digon face except possibly for F. Then G has a PVG layout L that is equivalent to the embedding of G with exterior face corresponding to F.

(iv) *Let G be a loopless 2-connected multigraph with a 2-cell embedding on the projective plane, with no digon face, and with v_1 a designated vertex. Then G has a PVG layout L, equivalent to the embedding of G, in which v_1 is represented by arc a_1 with (a_1, b_1) a long-arc pair at infinity and with arc b_1 corresponding to some neighbor of v_1.*

Proof. Sketch of proof of (i). For most cases (when c does not lie on F), the proof is by induction on n.

We can always find a nonloop, nonmultiple edge $e = (x, y)$ of G so that G with e contracted, G/e, is 2-connected, loopless, embedded on the plane, and F is still bounded by at least two edges. G/e satisfies the inductive hypothesis and so has PVG layout L_e, equivalent to an embedding of G/e plus a loop on the projective plane. If the contraction combines vertices x and y into new vertex x^*, let a^* be its representation in L_e, at say radius r. Because the embeddings of G/e and L_e are equivalent, the lines of visibility to arcs representing vertices adjacent to x in G are consecutive in the rotation of visibility lines about a^* in L_e. Then, when a^* is not one of the special long arcs at the origin, it can be replaced by two arcs a_x and a_y at radii $r - 0.5$ and $r + 0.5$ (or vice versa), representing vertices x and y of G, so that their visibilities give all edges incident with x and y and preserve the arc-rotations at x and y in G; see Fig 4. This alteration gives the desired PVG layout L for G. The argument is similar, though a bit more intricate, when a^* is part of the long-arc pair at the origin. The proofs for (ii–iv) are analogous.

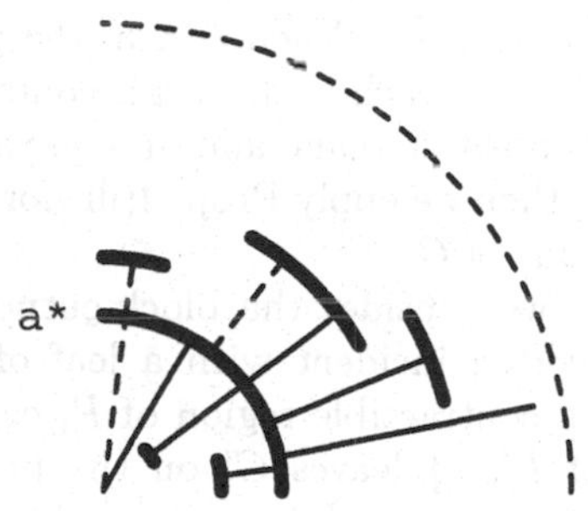

Fig. 4. An arc and its neighbors.

We then obtain the following.

Proposition 5. *If G has a PVG layout L, then the embedding $(\mathbb{I}(L))_G$ of G on the projective plane has cut-vertices on at most two faces. If the embedding has cut-vertices on two faces, then on one face there is only one cut-vertex, represented in L by a long arc.*

Corollary 1. *If G has a PVG layout L with a long arc a^*, representing a cut-vertex x^* and not lying on the exterior face, then G has a planar embedding with all cut-vertices except for x^* lying on a common face.*

Theorem 3. *A simple planar graph G has a PVG representation if it has a planar embedding with all but at most one cut-vertex on a common face.*

Proof. (Sketch) Assume that G is not a BVG and so can be drawn in the plane with cut-vertices lying on the exterior face F_1 and an additional cut-vertex c lying on $F_2 \neq F_1$. Consider the block-cutpoint tree BC(G) of G; c may lie on several blocks, but at least one, call it B_0, contains a cut-vertex $c' \neq c$ lying on F_1. Both of the faces F_i are bounded by a facial walk W_i, and each W_i contains a unique simple subcycle C_i, lying in B_0 and containing c' and c, respectively. If G has cut vertices $c_1, \ldots, c_i$ lying on F_1, we label the blocks other than B_0 incident with $c_1, \ldots, c_i$ $B_1, B_2, \ldots, B_j$, and the blocks $D_1, \ldots, D_k$ incident with c. Then we prove by induction on j that there is a PVG layout L of G with F_1 represented by the exterior face of L, with (a_c, a_d) a long-arc pair at the origin for some neighbor d of c, and with the blocks incident with c represented within $C(a_c)$.

Theorem 4. *If a simple graph has an embedding on the projective plane with all cut-vertices on a common face, then it is a PVG.*

Proof. Let G have an embedding on the projective plane P with all cut-vertices on a common face F. We prove by induction on $n = |V(G)|$ that G has a PVG layout with arcs representing cut-vertices on the exterior face and with its embedding equivalent to that of G. When $n < 5$, the graph has a BVG layout and so a PVG one by Prop. 2; each such graph containing a cycle also has a 2-cell embedding on the projective plane and an equivalent PVG layout.

If G has no cut-vertex, then we apply Prop. 4(iii) for graphs on the projective plane to get the PVG layout of G.

If G has a cut-vertex, we consider the block-cutpoint tree T of G, and, if possible, let c be a cut-vertex incident with a leaf of T with that leaf-block planar and embedded in a contractible region of P; call this block B. Deleting the vertices and edges of $B \backslash \{c\}$ leaves G' on the projective plane with face F now a face F', containing all remaining cut-vertices. By induction G' has a PVG layout L' that is equivalent to G' and with exterior face representing F'. Then there is a BVG layout of B with the bar representing c bottommost and extending the width of the layout, and by Prop. 2 B has a corresponding PVG layout L_B. Then a_c in L_B can be inserted as a subarc of a_c on the exterior face of L' so that L_B together with L' gives the desired layout of G.

Otherwise every leaf-block B is embedded in a noncontractible region of P and contains a noncontractible cycle in its embedding. If blocks B and B' are two such leaves, they must intersect at a cut-vertex c since every pair of noncontractible cycles on P intersects. If there are additional blocks, there are additional leaves which must also all meet at c so that T is a star $K_{1,i}$ with the non-leaf vertex of T representing c, the only cut-vertex of G, and each block is embedded in a wedge of P, all wedges meeting at, say, the origin. Such a graph is planar with one cut-vertex c and so by Theorem 3 is a PVG. □

Proof. of Theorem 1 for simple graphs. By Theorems 3 and 4 the graphs described are PVGs. Conversely if L is a layout of a PVG G, then G has an embedding on the projective plane by Prop. 1 with embedding $(\mathbb{I}(L))_G$. If L has no visibility through the origin, then by Prop. 2 G is a BVG and so embeds in the plane with all cut-vertices on a common face. Otherwise, if L contains a long arc, satisfying the conditions of Cor. 1, then G embeds in the plane with all but one cut-vertex on a common face. Otherwise G embeds in the projective plane with all cut-vertices on a common face by Prop 5. □

4 Results on CVGs

As the example in Fig. 2b and its extensions demonstrate, cut-vertices on many faces can be achieved using circles in layouts. We characterize CVGs in this section, as given in Theorem 2.

Suppose G has a layout L with circles $c_1, c_2, \ldots, c_k$ at radii $r_1 < r_2 < \cdots < r_k$ and with no circle replaceable by an arc so that the same visibilities are achieved. The circles c_i divide up the plane into annular regions and one projective planar region; note that neither the interior of c_1, denoted $\text{int}(c_1)$, nor the exterior of c_k, $\text{ext}(c_k)$, is empty in L since neither circle can be replaced by an arc. Then the corresponding vertices $v_1, v_2, \ldots, v_k$ of G are cut-vertices, and G is the union of graphs whose layouts lie in the annular regions plus the innermost region: $G = G_1 \cup G_2 \cup \ldots \cup G_k \cup G_{k+1}$ where G_1 is the subgraph whose layout in L lies on $c_1 \cup \text{int}(c_1)$, G_{k+1} lies on $c_{k+1} \cup \text{ext}(c_{k+1})$, and for $i = 2, \ldots, k$, G_i lies on the annulus given by $c_{i-1} \cup c_i \cup \{\text{int}(c_i) \cap \text{ext}(c_{i-1})\}$. Thus $G_2, \ldots, G_{k+1}$ are each planar. In addition for $i = 2, \ldots, k$ G_i is 2-connected since each block of G_i contains some vertices adjacent to v_{i-1} and some to v_i. Thus the block-cutpoint tree for G, BC(G), contains a path of $2k-1$ vertices, representing consecutively $v_1, G_2, v_2, \ldots, G_k, v_k$. What sorts of graphs are possible for G_1 and for G_{k+1}, and what additional tree structure in BC(G) is possible at the two ends of this path?

Consider G_1, laid out on $c_1 \cup \text{int}(c_1)$, with c_1 opened up to become an arc $a_1$so that this is a PVG layout of G_1. If G_1 is planar, by Prop. 5 and its proof, G_1 can have at most one additional cut-vertex, not on the exterior face but represented by a long arc a^* at radius 1. If there is no long arc a^* besides a_1, then v_1 may be attached to an arbitrary positive number of planar blocks. If there is a long arc $a^* \neq a_1$, then each block represented between a^* and a_1 sees these two arcs and so there is only one block lying in this annular region. Inside and attached to a^* may be any number $i_a \geq 0$ of 2-connected, planar graphs, but in any case, BC(G) has attached to the path-end v_1 either $i_1 > 0$ leaves or else one additional block vertex b, representing part or all of G_1, then a vertex for a^* that is also adjacent to $i_a > 0$ vertices of degree one. (Thus the latter case corresponds to having v_3 represented by c_1 and v_1 by a^*.) If G_1 is not planar, by Prop. 5 and Cor. 1 it is 2-connected so that the path of BC(G) is extended at v_1 by one additional vertex representing G_1.

The layout for the planar graph G_{k+1} lies in the infinite region, $c_k \cup \text{ext}(c_k)$. In this layout of G_{k+1} the circle c_k can be opened up to a long arc with empty interior to form a PVG layout; by Prop. 5 G_{k+1} has all its cut-vertices on a common face, the exterior face, and so can have arbitrarily many cut-vertices with arbitrarily many connected blocks, provided all cut-vertices lie on the infinite face. Thus attached to v_k in BC(G) is any tree representing a planar graph with all cut-vertices, except possibly for v_k, on a common face. These remarks prove the necessity of Theorem 2.

Lemma 1. *Let L be a layout of a PVG G with n vertices and with a long-arc pair at infinity or at the origin (or both). Then L can be laid out as a CVG with a circle on the exterior face at radius n or a circle about the origin at radius 1 (or both).*

As noted in Section 3, a long arc at radius 1 or at n can be extended to a full circle, changing no visibilities.

Proof. Proof of the sufficiency of Thm. 2. Suppose G has BC(G) satisfying (1a) and (2) so that BC(G) is $(b_0, e_1, e_2, \ldots, e_{2k+1}, T)$ where for $i = 1, \ldots, k$, each e_{2i-1} represents a cut-vertex v_i of G, each e_{2i} represents a 2-connected planar graph, b_0 is a 2-connected projective planar graph, and T represents a plane graph with all cut-vertices on a face F. Such a graph embeds on the projective plane; in the layout each cut-vertex v_i will be represented by a circle c_i.

By Prop. 4(iv) the projective planar subgraph of G corresponding to b_0 has a PVG layout L'_0 with the arc a_1 representing v_1 in a long-arc pair at infinity with some neighbor of v_1. By Lemma 1 L'_0 can be changed to the CVG L_0 so that a_1 becomes a circle surrounding L_0. By Prop. 4(ii) the planar subgraph of G corresponding to e_2 can be represented as a PVG L'_1 with a_1, representing v_1, part of a long-arc pair at the origin and with a_2, representing v_2, part of a long-arc pair at infinity. By Lemma 1 L'_1 can be changed to the CVG L_1 so that a_1 and a_2 each become circles inside and surrounding L_1 respectively. Then L_1 is joined with L_0 by identifying the two copies of the circle a_1, placing L_1 wholly outside of L_0. This process of expansion can be repeated for $e_4, \ldots, e_{2k}$. Finally by Prop. 4(i) T can be laid out as a PVG with v_k represented by a_k, part of a long-arc pair at the origin. Again by Lemma 1 a_k can be extended to a full circle inside of T's layout and can be identified with the circle representing a_k on the exterior of the layout previously constructed. In this way G is laid out.

If BC(G) satisfies (1b) and (2), it can be laid out similarly, only differing within c_1.

Since v_1 is incident with one or more planar blocks, we can lay these out in radial segments within c_1. Each planar block can be represented as a BVG with v_1 represented top-most and a neighbor bottom-most, then as a PVG via Prop. 2, and then inserted with v_1's arc as a subarc of c_1 within a distinct wedge of, say, $0 \leq \theta \leq \pi$, giving the desired visibilities. Thus in all cases the graph can be laid out as a CVG. □

5 Concluding Thoughts

It is clear that more complex graphs can be achieved in the polar visibility model by allowing visibility through the origin and diagonally across the boundary of a disc with antipodal points identified; call such a layout a doubly polar visibility layout and the resulting graphs doubly polar visibility graphs (DPVGs). These naturally lead to graphs that embed on the Klein bottle, the nonorientable surface of Euler characteristic 0. Analogous proofs to those given on the projective plane give the following results.

Proposition 6. *a) A DPVG embeds on the Klein bottle.*
b) If G has a layout L as a DPVG with no long arcs, then G contains no cut-vertex.
c) If a 2-connected graph G has an embedding on the Klein bottle, then G is a DPVG and has an equivalent doubly polar visibility layout. It seems that a DPVG that is neither a BVG nor a PVG can have at most two cut-vertices, represented by a long arc about the origin and at infinity.

Acknowledgements. The author wishes to thank Alexandru Burst, Alice Dean, Michael McGeachie, Bojan Mohar, William Owens, and Stan Wagon for useful conversations and insightful examples concerning this work.

References

1. Bose, P., Dean, A., Hutchinson, J., Shermer, T. On rectangle visibility graphs In: North, S. (ed.): Proc. Graph Drawing '96. Lecture Notes in Computer Science, Vol. 1190. Springer, Berlin (1997) 25–44
2. Chang, Y-W., Jacobson, M. S., Lehel, J., West, D. B. The visibility number of a graph (preprint)
3. Dean, A., Hutchinson, J. Rectangle-visibility representations of bipartite graphs Discrete Applied Math. 75 (1997) 9–25
4. —— Rectangle-visibility layouts of unions and products of trees J. Graph Algorithms and Applications 2 (1998) 1–21
5. Dean, A. Bar-visibility graphs on the Möbius Band In: Marks, J. (ed.): Proc. Graph Drawing 2000. (to appear)
6. Hutchinson, J., Shermer, T., Vince, A. On Representations of some Thickness-two graphs Computational Geometry, Theory and Applications 13 (1999) 161–171
7. Mohar, B. Projective planarity in linear time J. Algorithms 15 (1993) 482–502
8. Mohar, B., Rosenstiehl, P. Tessellation and visibility representations of maps on the torus Discrete Comput. Geom. 19 (1998) 249–263
9. Mohar, B., Thomassen, C. Graphs on Surfaces Johns Hopkins Press (to appear)
10. O'Rourke, J. Art Gallery Theorems and Algorithms Oxford University Press, Oxford (1987)
11. Tamassia, R., Tollis, I. G. A unified approach to visibility representations of planar graphs Discrete and Computational Geometry 1 (1986) 321–341
12. —— Tesselation representations of planar graphs In: Medanic, J. V., Kumar, P. R. (eds.): Proc. 27th Annual Allerton Conf. on Communication, Control, and Computing. (1989) 48–57

13. ____ Representations of graphs on a cylinder SIAM J. Disc. Math. 4 (1991) 139-149
14. Thomassen, C. Planar representations of graphs In: Bondy, J. A., Murty, U. S. R. (eds.): Progress in Graph Theory. (1984) 43–69
15. West, D. Introduction to Graph Theory Prentice Hall, Upper Saddle River, NJ (1996)
16. White, A. Graphs, Groups and Surfaces revised ed. North-Holland, Amsterdam (1984)
17. Wismath, S. Characterizing bar line-of-sight graphs In: Proc. 1st Symp. Comp. Geom. ACM (1985) 147–152

Tulip

Auber David

LaBRI-Université Bordeaux 1, 351 Cours de la Libération, 33405 Talence, France
auber@labri.u-bordeaux.fr, fax:+33 5 56 84 66 69

1 Short Description

This paper briefly presents some of the most important capabilities of a graph visualization software called Tulip[1]. This software has been developed in order to experiment tools such as clustering, graph drawing and metrics coloring for the purpose of information visualization. The main Tulip's characteristics are: a graph model which allows clustering with data sharing, and a general property evaluation mechanism that makes the most part of the software reusable and easily extendable. This software has been written in C++ and uses the Tulip graph library, the OpenGL library and the QT library[5]. The current version is fully usable and enables to visualize graphs with about 500.000 elements on a personal computer.

2 Areas of Application

Tulip has been designed to enable manipulation and visualization of huge graphs. Thus, its areas of application are those where visual-analyse of such kind of data structures must be addressed. We are currently using it in some works about the visualization of metabolic pathways and for the visualization of the interactions between proteins. Some others direct applications are the visualization of: hyperlinks(WWW), file system structures or computer networks.

3 Layout Algorithms and Layout Features

The Tulip's layout engine makes it possible to connect dedicated algorithms to each properties of the displayed layout. For nodes and bends positions it exists about ten graph drawing algorithms. The most important of them are:

- A variant of the so-called Sugiyama algorithm. It manages general graphs with fixed elements size[2],
- A 3D general graph drawing algorithm,
- The tree walker algorithm[4],
- The 3D cone tree layout[1],
- An implementation of the famous force directed approach[3].

[1] Tulip, software is under GPL license and is available on the world wide web at the URL: `http://dept-info.labri.fr/~auber/projects/tulip`.

P. Mutzel, M. Jünger, and S. Leipert (Eds.): GD 2001, LNCS 2265, pp. 435–437, 2002.
© Springer-Verlag Berlin Heidelberg 2002

For the others properties(sizes, colors, shapes), Tulip also includes dedicated algorithms that enables to compute them automatically. Most of these algorithms are based on the combinatorical properties of graphs. They are used to highlight important elements of the graph in order to help the user in his research.

4 Architecture

The Tulip software has been written in C++ and is currently available for the Linux operating system. It uses the Qt library for its graphics user interface and the OpenGl library for the 3D displaying. The extensions are supported by the so-called plug-in mechanism which allows to receive dynamically new algorithms (selection, graph drawing, import, export, metric, etc ...).

5 Interfaces

The Tulip software interfarce enables to manipulate the following parts of the visualized graph:

- Cluster and graph structure: The user can add or remove elements (nodes or edges) of a graph, or of a cluster, by a simple click on the visualization window. Furthermore, it can graphically select parts of a graph, or of a cluster, to create a new cluster.
- Cluster and graph properties: The layout property of a graph, or of a cluster, can be directly modified with the visualization window. Another window called "property editing" enables to modify the values of other properties and to add or to remove a property from a graph or a cluster.
- Clusters hierarchy : With the cluster-tree window, the user can add or remove clusters in the hierarchy and change their positions in it.

The interface also enables to navigate in the graph. The available operations are: rotations, zoom, and translations.

6 Screenshots

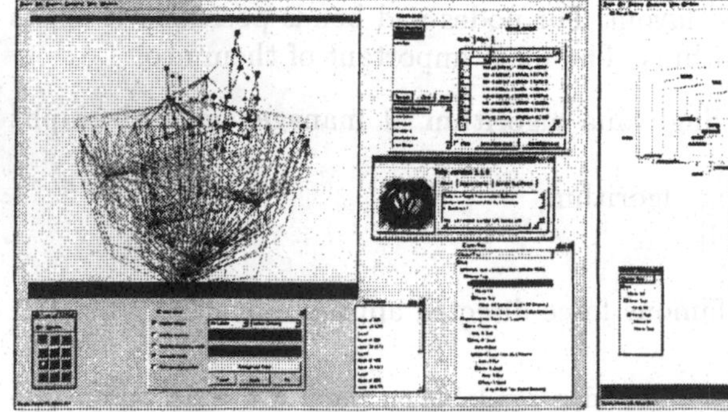

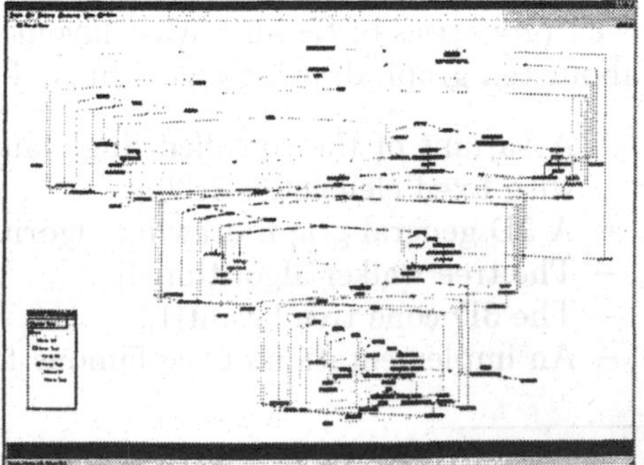

References

1. J.D.Mackinlay G.G.Robertson and SK.Card. Cone trees:animated 3d visualizations of hierarchical ingormation. In *CHI'91*, pages 189–194. ACM Press, 1991.
2. S.Tagawa K.Sugiyama and M.Toba. Methods for visual understanding of hierarchical system structures. *IEEE Transactions on system, Man and Cybernetics.*, 11:109–125, 1981.
3. M.Kaufmann and D.Wagner. *Drawing Graphs.* LNCS 2025. Springer-Verlag, 2001.
4. Edward M. Reingold and John S. Tilford. Tidier drawings of trees. *IEEE Transactions on Software Engineering*, 7(2):223–228, March 1981.
5. Trolltech. Qt the crossplatform c++ gui framework. `http://www.trolltech.com/`.

The ILOG JViews Graph Layout Module

Georg Sander and Adrian Vasiliu

ILOG SA, 9, rue de Verdun - BP 85, 94253 Gentilly Cedex, FRANCE
{sander|vasiliu}@ilog.fr, http://jviews.ilog.com

1 Short Description

The ILOG JViews Component Suite is a set of pure Java components for building sophisticated interactive commercial Web-based user interfaces. Besides the graphics, interaction, and animation framework, the graph layout module is one of the key components.

2 Areas of Application

Applications of the ILOG JViews Component Suite range from Telecom (network display) over Geographic Information Systems (cartographic maps) to Resource Scheduling (Gantt charts and workflow diagrams). It contains a broad set of layout algorithms for the automatic arrangement of different kinds of diagrams such as PERT charts, process and workflow diagrams, WAN and LAN networks, and supply chain diagrams.

3 Layout Algorithms and Layout Features

The ILOG JViews graph layout module includes the following features:

- **Topological Mesh Layout:** Designed for biconnected graphs as they occur in database and knowledge engineering. It arranges outer nodes on a circle and tries to expose symmetries of inner nodes [4].
- **Spring Embedder Layout:** A force-directed algorithm [1] that distributes nodes evenly in a specified area.
- **Uniform Length Edges Layout:** Another force-directed algorithm that arranges the graph such that all edges have approximately the same length.
- **Tree Layout:** An incremental algorithm based on [5] to arrange trees. Besides the classical tree style, it includes a tip-over mode (children are arranged sequentially instead of in parallel to optimize the area) and a radial mode (nodes are placed in circles/ellipses around the root).
- **Hierarchical Layout:** Flow layout based on partitioning into layers, with support for various edge routing styles [3], connection ports and constraints.
- **Bus Layout:** Designed for bus topologies in networking and telecommunications.
- **Circular Layout:** It displays ring and star network topologies interconnected in a tree structure for telecommunication applications.

P. Mutzel, M. Jünger, and S. Leipert (Eds.): GD 2001, LNCS 2265, pp. 438–439, 2002.
© Springer-Verlag Berlin Heidelberg 2002

- **Grid Layout:** Arranges disconnected nodes in rows, columns, or on a grid. It is also utilized to arrange the connected components of graphs.
- **Link Layout:** An edge routing algorithm that doesn't move the nodes. It supports orthogonal and direct edge shapes and offers two modes: a fast combinatoric algorithm for short edges and a classical obstacle/maze routing for long edges similar to the PCB and VLSI routing algorithms [2].
- **Automatic Label Placement:** A fast simulated annealing technique to place labels at nodes and edges while avoiding overlaps.
- **Nested Layout:** The layout algorithms can be combined in multiple ways to arrange nested graphs (that is, graphs with nodes that contain graphs).

4 Architecture

The ILOG JViews Graph Layout framework contains extensive functionality (a generic graph model, 13 layout algorithms, notification mechanisms, graph filters, etc.). The framework is a set of extensible Java classes that can easily be integrated into customer applications. ILOG also provides ready-to-use components built upon the graph layout module, such as the ILOG JViews for Workflow Applications or the ILOG Telecom Graphic Objects for Java.

5 Screenshots

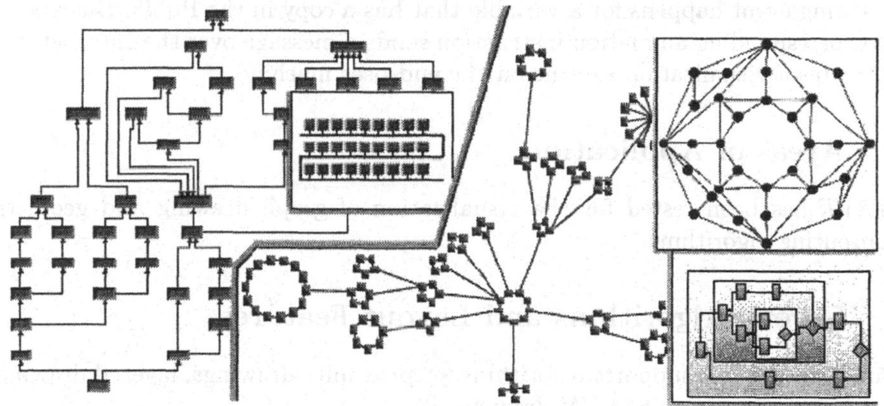

References

1. P. Eades. A heuristic for graph drawing. *Congressus Numeratium*, 42:149–160, 1984.
2. D.T. Lee, C.D. Yang, and C.K. Wong. Rectilinear paths among rectilinear obstacles. *Discrete Appl. Math.*, 70:185–215, 1996.
3. G. Sander. A fast heuristic for hierarchical manhattan layout. In *Proc. Symposium on Graph Drawing, LNCS 1027*, pages 447–458, 1996.
4. A. Vasiliu. A topological graph unfolding algorithm. In *Proc. COMES'93, Douai, France*, pages 73–76, 1993.
5. J.Q. Walker II. A node positioning algorithm for generalized trees. *Software Pract. and Exper.*, 20(7):685–705, 1990.

WAVE*

Emilio di Giacomo and Giuseppe Liotta

Dipartimento di Ingegneria Elettronica e dell'Informazione,
Università degli Studi di Perugia,
{digiacomo,liotta}@diei.ing.unipg.it

1 Short Description

WAVE (http://lis.ing.unipg.it/wave) is a system for algorithm visualization over the Internet designed with a novel paradigm, called *Publication-driven* approach [1,2]. The Publication-driven approach separates the task of executing the algorithm from that of running its visualization and thus it makes it possible to easily distribute such two tasks over the Internet. The idea behind the approach is as follows: The algorithm code runs on the developer machine, while the variables which are the subject of the animation are copied on the end-user machine in a suitable structure, called *Public Blackboard*. The algorithm code on the developer side is automatically enriched with a set of *animation instructions*, each corresponding to an event that is relevant for the animation. When an interesting event happens for a variable that has a copy in the Public Blackboard, the corresponding animation instruction sends a message over the Internet, that activates a visualization routine on the end-user machine.

2 Areas of Application

WAVE has been tested for the visualization of graph drawing and geometric computing algorithms.

3 Layout Algorithms and Layout Features

Currently, WAVE supports algorithms for proximity drawings, layered drawings, dominance drawings, and HV-drawings [3].

4 Architecture

WAVE is a system distributed over the Internet. The algorithm to be visualized is executed by the *Algorithm server* on the algorithm developer's machine. The animation is displayed on an *Animation applet* running inside a Web browser on the end-user's machine. The *WAVE main server* acts as bridge between them.

* Research supported in part by the CNR Project "Geometria Computazionale Robusta con Applicazioni alla Grafica ed al CAD", the project "Algorithms for Large Data Sets: Science and Engineering" of the Italian Ministry of University and Scientific and Technological Research (MURST 40%).

P. Mutzel, M. Jünger, and S. Leipert (Eds.): GD 2001, LNCS 2265, pp. 440–441, 2002.
© Springer-Verlag Berlin Heidelberg 2002

4.1 Programming Language

WAVE is implemented in the Java Language.

4.2 Operating System

WAVE has been tested on Windows, Linux and Unix system.

5 Interfaces

WAVE has a graphical interface that allows user to graphically edit input for the algorithm, and to see its animation.

6 Screenshot

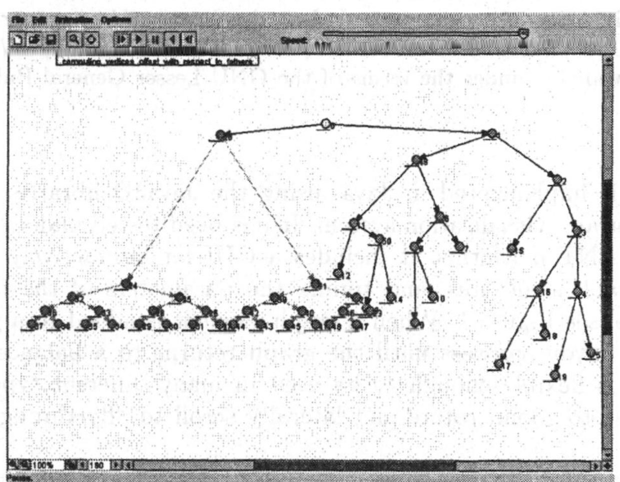

Fig. 1. Animating an algorithm for layered drawings of binary trees with WAVE.

References

1. C. Demetrescu, I. Finocchi, and G. Liotta. Visualizing Algorithms over the Web with the Publication-driven Approach. In D. Wagner, editor, *Workshop on Algorithm Engineering(Proc. WAE '00)*, Lecture Notes Comput. Sci. Springer-Verlag, 2000.
2. C. Demetrescu, E. Di Giacomo, I. Finocchi, and G. Liotta. Visualizing Geometric Algorithms with WAVE: System Demonstration. In *Fall Workshop on Computational Geometry, Stony Brook NY 2000*, 2000.
3. G. Di Battista, P. Eades, R. Tamassia, and I. G. Tollis. *Graph Drawing.* Prentice Hall, Upper Saddle River, NJ, 1999.

WilmaScope – An Interactive 3D Graph Visualisation System

Tim Dwyer and Peter Eckersley

Department of Computer Science, University of Melbourne,
Parkville, Victoria 3052, Australia,
dwyer@cs.usyd.edu.au, pde@cs.mu.oz.au,
http://www.wilmascope.org

Abstract. This is a brief description of the WilmaScope interactive 3D graph visualisation system. Wilma features clustering of related groups of nodes, a GUI for editing graphs and adjusting the force layout parameters and a CORBA interface for creating and interacting with graphs remotely from other programs. It has also been used to construct 3D UML Class and Object models as part of a usability study. Wilma is freely available under the terms of the GNU Lesser General Public License[1].

Wilmascope is implemented in Java, using the Java3D libraries to provide a portable 3D graph visualisation system. It has been tested under Linux, Solaris and Windows NT platforms. It includes a GUI for interactively creating and modifying graph structures, and tuning the parameters of the force directed layout engine, see Figure 1. When a graph element is added to a graph Wilma animates the layout process until the graph returns to a balanced state. The basic engine is flexible enough to be used in many different types of graphs. Users are able to zoom, rotate and laterally scroll to "fly through" 3D graph structures.

A usability study was conducted on a 3D UML Class modelling tool based on the Wilma engine with very encouraging results[1].

Wilma also includes an API which allows other programs to generate complex graphs which can be displayed in real-time. The API is accessible via a CORBA interface, and we have implemented a number of clients in Python[2] to create example graphs. The CORBA interface features callbacks allowing remote programs to not only generate graphs but also to interact with them via a point and click interface. For example, a Wilma client could provide a bridge to a web browser such as Netscape, graphing the history of visited pages and when a node in the history graph is clicked, reloading that page into the browser.

Wilma supports a powerful hierarchical graph model, supporting nested clusters such that each cluster has its own force parameters. Force parameters for a cluster or the root graph may be adjusted in real time and the affects on the

[1] http://www.gnu.org/copyleft/lesser.html

[2] A high-level, interpreted, object-oriented language; http://www.python.org

P. Mutzel, M. Jünger, and S. Leipert (Eds.): GD 2001, LNCS 2265, pp. 442–443, 2002.
© Springer-Verlag Berlin Heidelberg 2002

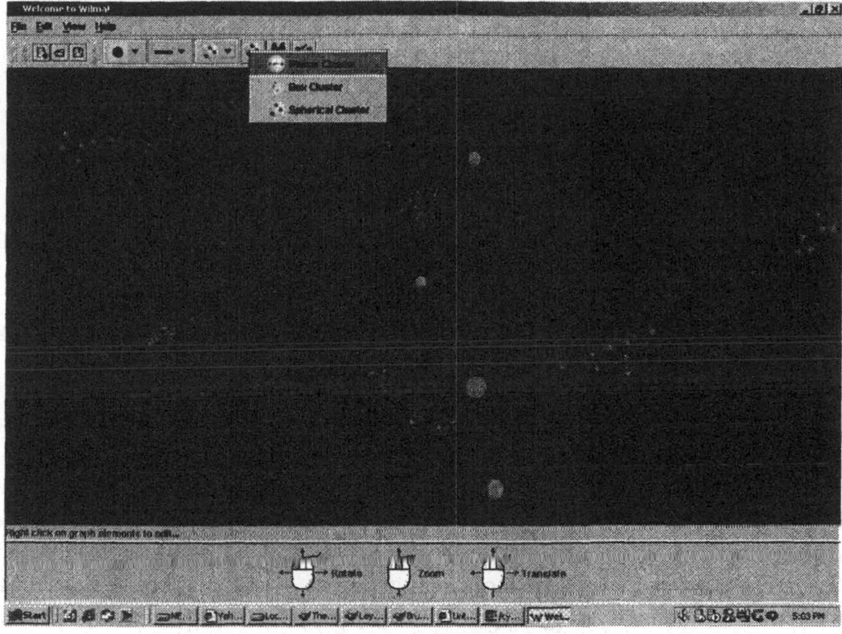

Fig. 1. A screenshot of Wilma in action.

layout are visible instantly. Well clustered graphs will also be arranged faster since internode forces need only be calculated for each cluster. Clusters may also be collapsed or expanded by the user to elide or show detail providing an efficient way to navigate large nested graphs.

Within clusters it is also possible to define constraint systems such as confining nodes in the cluster to a plane. An attempt is made to satisfy the constraint through the use of a force attracting nodes to the ideal location. An equal force is exerted back on the constraint system causing the whole cluster to rotate such that the constraints may be satisfied within the cluster while forces external to the cluster affect the cluster's orientation. This can be used to obtain sophisticated 3D layouts such as Cone Trees within the force directed system.

References

1. T. Dwyer. Three dimensional uml using force directed layout. Technical report, Department of Computer Science, The University of Melbourne, Parkville, Australia, 3052. 2000.

Exploration and Visualization of Computer Networks: Polyphemus and Hermes*

Gabriele Barbagallo[1], Andrea Carmignani[1], Giuseppe Di Battista[1], Walter Didimo[2], and Maurizio Pizzonia[1]

[1] Dipartimento di Informatica e Automazione, Università di Roma Tre, Via della Vasca Navale 79, 00146 Roma, Italy.
{barbagal,carmigna,gdb,pizzonia}@dia.uniroma3.it

[2] Dipartimento di Ingegneria Elettronica e dell'Informazione, Università di Perugia, Via G. Duranti 93, 06125 Roma, Italy. didimo@diei.unipg.it

1 Short Description

POLYPHEMUS and HERMES [1] are systems for exploring and visualizing computer networks at different abstraction levels. Today's Internet is organized as *Autonomous Systems* (*AS*), each of them groups a set of networks under a single administrative authority (e.g. a private company). HERMES visualizes Internet as a set of ASes and interconnections between them. POLYPHEMUS is a complementary system that is able to look inside an AS and to visualize its networks at the level of the routers and their physical links.

2 Areas of Application

The two systems presented in this paper are targeted to Internet service providers and to content delivery companies which need to have a clear view of the Internet structure in planning and designing their infrastructure.

3 Layout Algorithms and Layout Features

The drawing-server of POLYPHEMUS and the middle-tier server of HERMES both encapsulate a drawing engine that computes maps of the networks inside a single AS or draws parts of the ASes interconnection graph. The drawing engine is based on the GDToolkit library and has the following main features. Its basic drawing convention is the podevsnef model for orthogonal drawings having vertices of degree greater than four. However, since the graphs handled by HERMES often have many vertices (ASes) of degree one connected with the same vertex, the podevsnef model is enriched with new features for representing such vertices. The adopted algorithm implements a modified version of the standard topology-shape-metrics approach [1]. The user performs her exploration interactively and each step enriches the current map with new ASes and connections.

* Research supported in part by the Murst Project: "Algorithms for Large Data Sets". HERMES is available at http://www.dia.uniroma3.it/~hermes

P. Mutzel, M. Jünger, and S. Leipert (Eds.): GD 2001, LNCS 2265, pp. 444–445, 2002.
© Springer-Verlag Berlin Heidelberg 2002

4 Architecture and Interfaces

For the two systems we have used different architectures. Since HERMES uses publicly available data, it is targeted to offer a service to any Internet client. On the other hand, POLYPHEMUS requires to access private routing information. Hence, it is conceived as a local application that has to be installed and used by single users.

POLYPHEMUS is based on a two tiers architecture with a client and a drawing-server. The client (a Java application) allows the user to discover the network exploiting the knowledge of the routers that uses the Ospf routing protocol. Discovery is performed by accessing the Management Information Base (MIB) of the routers, with the Snmp protocol. Visualization of the the topology shows AS routers, networks grouped in Ospf areas, and inter-area relationships. For each item a rich set of information can be displayed.

HERMES has a three tiers architecture. The user interacts with a top-tier client which forwards the user requests to a middle-tier server. The server translates the requests into queries to a repository (the bottom-tier). Using the top-tier the user can interactively explore and visualize the ASes interconnections, and several information about ASes and their routing policies. HERMES handles a large repository that is updated off-line from a plurality of sources [2]: APNIC, ARIN, RIPE, RADB, etc. The data in the repository are used by HERMES to construct the ASes interconnection graph.

5 Screenshots

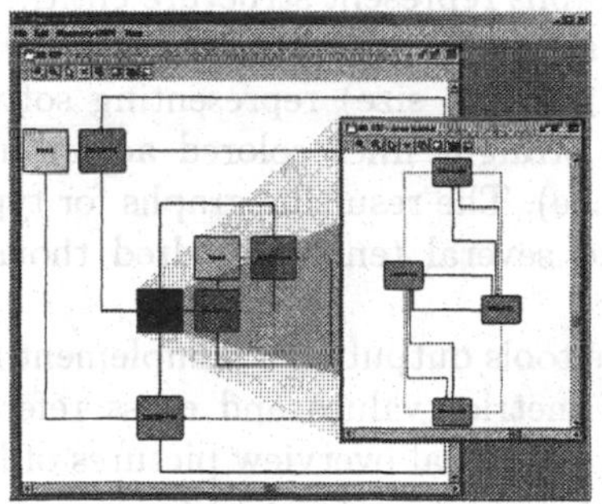

(a) Polyphemus

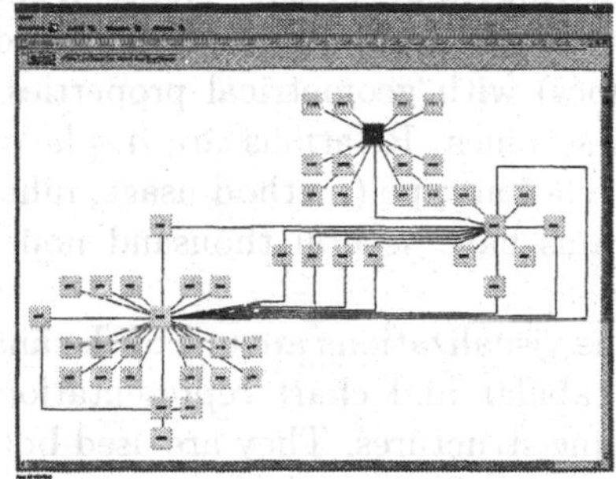

(b) Hermes

References

1. A. Carmignani, G. Di Battista, W. Didimo, F. Matera, and M. Pizzonia. Visualization of the autonomous systems interconnections with hermes. In J. Marks, editor, *Graph Drawing (Proc. GD '00)*, volume 1984 of *Lecture Notes Comput. Sci.*, pages 150–163. Springer-Verlag, 2000.
2. Merit Network, Inc. Radb database services. On line. http://www.radb.net.

CrocoCosmos

Claus Lewerentz, Frank Simon, and Frank Steinbrückner

Software & Systems Engineering Research Group
BTU Cottbus, D-03013 Cottbus, Germany
www.software-systemtechnik.de

1 Short Description

The ***CrocoCosmos*** tool was developed as part of the ***Crocodile*** tool set for the analysis and visualization of large object-oriented software systems. The context of our research is to support maintenance and re-engineering processes in an appropriately automated way for large programs. One aspect is program comprehension through the visualization of program structures on an architectural level. Thus, CrocoCosmos is not a general graph drawing tool but serves a dedicated purpose in a specific graph drawing application.

2 Areas of Application

Our approach to visualize large object-oriented programs (10^6 LOC, 10^3 – 10^4 classes, 10^2 subsystems) on the basis of extracted structure and metrics data uses attributed 3D graphs. Nodes in such graphs represent structure entities like classes or packages. They are visualized by simple geometric objects (as spheres or cubes) with geometrical properties (as color or size) representing software metrics values. Relations are displayed as straight lines colored according to their relation type (method usage, inheritance). The resulting graphs for typical programs have several thousand nodes and several ten to hundred thousand edges.

The visualizations are one of the analysis tools output that complement multiple tabular and chart representations of metrics values and cross reference browsing structures. They are used both to get initial overview pictures of large systems as well as exploring particularly interesting parts of systems in detail. The first results from several case studies that were done together with industrial software developers are very encouraging [2]. The 3D visualization proved to be a very effective means to quickly detect and explain typical design weaknesses and to give restructuring recommendations on the basis of simple visual patterns.

3 Layout Algorithms and Layout Features

A central idea for drawing these graphs is the use of a generic similarity and distance concept that allows to calculate metric distances for each pair of nodes [1]. The distances may be calculated from arbitrary common property sets of the

P. Mutzel, M. Jünger, and S. Leipert (Eds.): GD 2001, LNCS 2265, pp. 446–447, 2002.
© Springer-Verlag Berlin Heidelberg 2002

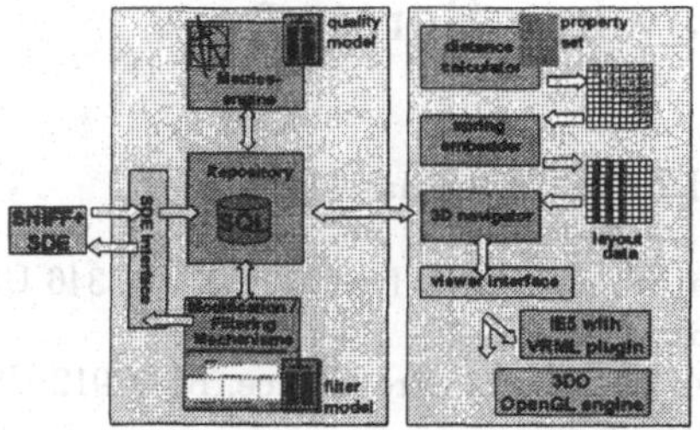

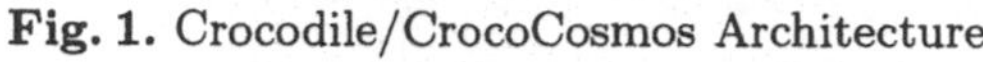
Fig. 1. Crocodile/CrocoCosmos Architecture

Fig. 2. Visualization of a program with 1.200 classes and 220.000 lines of Java code

nodes. To calculate a layout of the graph, the metric distances between nodes are used as weights of the edges in the complete graph. The 3D graph layout algorithm is based on a spring-embedding method that is combined with simulated annealing techniques to control the iteration processes. It takes the edge weights and produces a layout which approximately preserves the node distances on an ordinal scale in the 3D space. In the resulting graph layouts, that contain all nodes and the edges for the structural relations as call or inheritance relations the spatial relationships are meaningful and can be interpreted with respect to the problem domain.

4 Architecture

The Crocodile tool consists of a parsing frontend, that fills a relational database containing all structural and metrics data. CrocoCosmos is the visualization backend, consisting of a distance calculator, the spring-embedder, and a high-performance 3D display engine (cf. Figure 1). Except the display engine which is implemented directly on top of OpenGL, the system is written in Java.

5 Screenshot

Figure 2 shows an example depicting more than 1.200 classes of a Java program with more than 25.000 use relationships between them. The layout reflects distances according to similarities in usage relations between classes. Interaction and navigation mechanisms allow for a user-driven exploration of the graphs using 3D display devices.

References

1. F. Simon, S. Löffler, and C. Lewerentz. Distance-based cohesion measuring. In *Proc. 2nd Conference on Software Measurement (FESMA'99)*, pages 69–84, 1999.
2. F. Simon, F. Steinbrückner, and C. Lewerentz. Anpassbare explorierbare virtuelle Informationsräume. In *Proc. 3rd GI Workshop on Software Reengineering*, 2001.

The Graph Drawing Server*

Stina Bridgeman[1] and Roberto Tamassia[2]

[1] Department of Computer Science, Colgate University, Hamilton, NY 13346 USA
stina@cs.colgate.edu
[2] Center for Geometric Computing, Brown University, Providence, RI 02912 USA
rt@cs.brown.edu

1 Short Description

There are many obstacles in the way of someone wishing to make use of existing graph drawing technology — software installation and data conversion can be time-consuming and may be prohibitively difficult for the casual or novice user, and software may be limited to a particular platform or provided interface. The Graph Drawing Server (GDS) [2] seeks to remove many of these obstacles by providing a graph drawing and translation service with an easy-to-use web-based interface[1]. A user needs only a commonly-available web browser to access a variety of algorithms, without having to install any additional software or do any format translations (once the data is in one of many supported formats).

GDS has received over 62,000 requests from 43 countries since June 1996.

2 Areas of Application

Potential uses of GDS include:
- drawing user-supplied graphs,
- translating between graph descriptions in different formats,
- performing experimental comparisons of graph drawing algorithms,
- testing and developing new graph drawing algorithms,
- demonstrating graph drawing algorithms in educational settings,
- creating a database of real-life user graphs for experimental studies, and
- rapid prototyping of software containing a graph drawing component.

3 Layout Algorithms and Layout Features

The algorithms available through GDS come from a variety of sources — the Graph Drawing Server does not replace existing graph drawing packages, but rather brings them to a wider audience.

* Research supported in part by the NSF under grants CCR-9732327 and CDA-9703080, and by the U.S. Army Research Office under grant DAAH04-96-1-0013. Work done while the first author was at Brown University.

[1] http://loki.cs.brown.edu/geomNet/gds/

P. Mutzel, M. Jünger, and S. Leipert (Eds.): GD 2001, LNCS 2265, pp. 448–450, 2002.
© Springer-Verlag Berlin Heidelberg 2002

GDS currently supports several general-purpose algorithms for orthogonal, hierarchical, and force-directed drawings, plus specialized algorithms for binary trees and series-parallel digraphs. New algorithms can be easily incorporated into the service. Output can be generated in a variety of formats including GIF and Postscript, so that drawings can easily be included in papers and presentations.

4 Architecture

The Graph Drawing Server is a component of GeomNet [1], a distributed geometric computing system providing access to computational geometry and graph drawing algorithms over the Internet. It utilizes the web services framework, where users connect to services running on remote machines rather than installing software packages locally.

The GDS system consists of a network of individual servers, each supporting some subset of the available algorithms. Users can send requests to any server, and that server will coordinate the others as needed to complete the request.

Individual GDS servers have a layered object-oriented architecture which facilitates adding new applications and new functionality. Current work involves adding support for enhanced security and for "cooperative computing", where client and server can negotiate the best distribution of computation in order to satisfy the request as quickly as possible.

4.1 Programming Language

The server component of the Graph Drawing Server is implemented in Java. Individual algorithms, translation filters, and interfaces may be implemented in any language (e.g. C, C++, Java, perl, and, for interfaces, HTML).

4.2 Operating System

GDS servers will run on any operating system for which a Java virtual machine exists; individual algorithms, translation filters, and interfaces may be limited to certain platforms because of language or environment requirements.

5 Interfaces

Users interact with GDS by sending a request consisting of the graph to be drawn, the algorithm to be run and values for its parameters, the format of the input, and the desired format for the output. Requests can be sent using a graph editor applet or a set of forms-based web pages, allowing casual or novice users to make use of the service immediately. For more advanced applications, user programs can make socket or HTTP connections directly to the service. This allows drawing algorithms to be used by any program, regardless of the platform and language of program and algorithm.

The authors would like to thank Gill Barequet, Christian Duncan, Ashim Garg, and Michael Goodrich for their contributions to GDS and GeomNet, and Rob Mason for the creation of the graph editor applet.

References

1. G. Barequet, S. Bridgeman, C. Duncan, M. Goodrich, and R. Tamassia. GeomNet: Geometric computing over the Internet. *IEEE Internet Computing*, 3(2):21–29, 1999.
2. S. Bridgeman, A. Garg, and R. Tamassia. A graph drawing and translation service on the WWW. *Internat. J. Comput. Geom. Appl.*, 9(4/5):419–446, 1999.

Drawing Database Schemas with DBdraw

Giuseppe Di Battista[1], Walter Didimo[2], Maurizio Patrignani[1], and Maurizio Pizzonia[1]

[1] Dipartimento di Informatica e Automazione, Università di Roma Tre, Via della Vasca Navale 79, 00146 Roma, Italy. {gdb,patrigna,pizzonia}@dia.uniroma3.it
[2] Dipartimento di Ingegneria Elettronica e dell'Informazione, Università di Perugia, Via G. Duranti 93, 06125 Roma, Italy. didimo@diei.unipg.it

1 Short Description

DBdraw is an application that allows the user to automatically produce drawings of database schemas according to a drawing standard that is well accepted by the database community. The drawing engine of DBdraw is based on the GDToolkit library.

2 Areas of Application

DBdraw is clearly targeted to developers and maintainers of databases. It may give a valuable help in building first release documentation as well as in facing the lack of documentation of old systems.

3 Layout: Algorithms and Features

The drawing standard adopted by DBdraw represents the tables of the schema with boxes, and table attributes with distinct stripes inside each table. Links connecting attributes of two different tables represent referential constraints or join relationships, and may attach arbitrarily to the left or to the right side of the stripes representing the attributes. The drawing algorithm is inspired by the topology-shape-metric approach. Briefly, it consists of four steps:

Constrained Planarization. A planarization is performed on the graph underlying the database schema in order to obtain a planar embedding consistent with the specific sequence of attributes of the table. Dummy vertices of degree four are introduced to replace crossings, see Figure 1(a).

U-Turns Assignment. In our drawing convention each link may monotonically follow in the left-to-right direction or may perform one or more u-turns. In this step a (possibly empty) sequence of u-turns is associated with each link trying to minimize their total number, see Figure 1(b).

Orthogonalization. This step associates an orthogonal shape to the schema. Each u-turn is replaced with two bends of 90 degrees. The shape is such that links approach tables horizontally, see Figure 1(c).

Constrained Compaction. The output of this step is a complete drawing of the database schema. The length of the link and the size of the vertices are

P. Mutzel, M. Jünger, and S. Leipert (Eds.): GD 2001, LNCS 2265, pp. 451–452, 2002.
© Springer-Verlag Berlin Heidelberg 2002

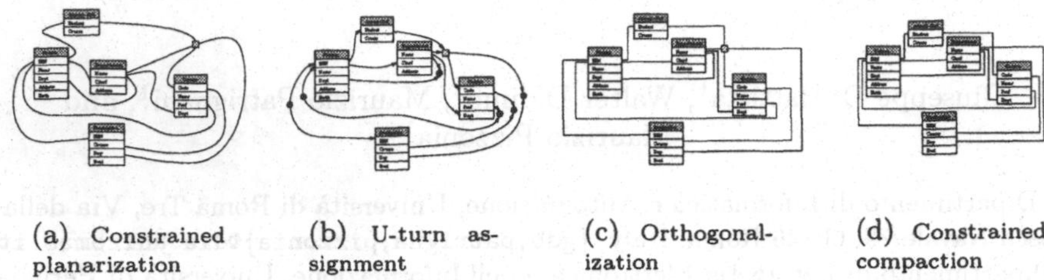

(a) Constrained planarization (b) U-turn assignment (c) Orthogonalization (d) Constrained compaction

Fig. 1. Illustration of the main steps of the DBdraw algorithm.

computed, keeping as small as possible the area and the total link length. The adopted technique allows us to exactly specify the incidence point of each link on the tables involved in the link. Dummy vertices introduced in the Constrained Planarization step are removed, see Figure 1(d).

4 Architecture and Interfaces

The architecture of DBdraw takes advantage of the capability provided by the Windows® operating system for importing database schemas and producing an output that may be rapidly inserted into database documentation. The user specifies a database, DBdraw extracts the schema from the database, computes the drawing and inserts it in a Microsoft® Word document whose name is again specified by the user. The drawing is stored in a vectorial format and may be further modified. Currently only Microsoft® Access databases (.mdb files) are supported. The drawing engine is written in C++ and it is based on the GDToolkit graph drawing library. The other parts of the system are written in Visual Basic.

5 Screenshots

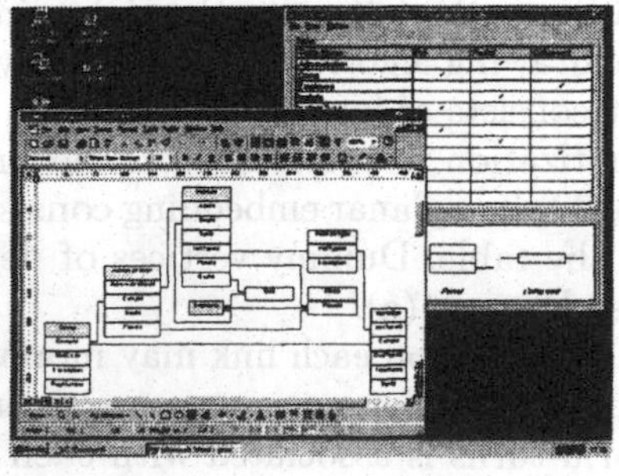

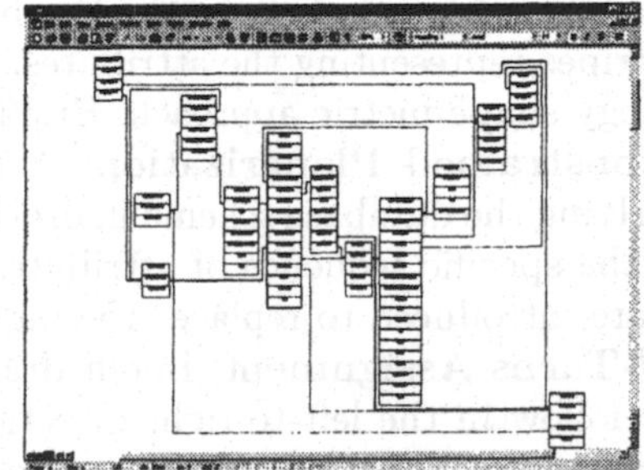

References

1. G. Di Battista, W. Didimo, M. Patrignani, and M. Pizzonia. Drawing relational schemas. In W. C. de Leeuw and R. van Liere, editors, *Data Visualization 2000*, Eurographics. Springer-Verlag Wien, 2000.

yFiles: Visualization and Automatic Layout of Graphs

Roland Wiese, Markus Eiglsperger, and Michael Kaufmann

Wilhelm-Schickard-Institut für Informatik, Universität Tübingen,
Sand 13, 72026 Tübingen, Germany
{wiese, eiglsper, mk}@informatik.uni-tuebingen.de

1 Short Description

yFiles is a Java-based library for the visualization and automatic layout of graph structures. Included features are data structures, graph algorithms, diverse layout and labeling algorithms and a graph viewer component.

The graph viewer architecture adheres to the model-view-control design pattern. The *view* component itself is a Java-Swing based component that can easily be added to any application GUI. It supports features like zooming, scrolling, layout morphing and different levels of detail rendering.

yFiles was designed to be easily integratable into any Java-based application that either needs a viewer component or layout algorithms for graph structures or both.

By now, *yFiles* is a commercial product distributed by yWorks GmbH. An evaluation version of *yFiles* can be obtained from the authors or the yWorks web page *www.yWorks.de*.

2 Areas of Application

An extensive and liable visualization system is crucial in application areas such as software engineering, database management and database modelling, WWW-visualization, bioinformatics, business process engineering and networking. *yFiles* has been successfully employed to all of these domains and, thus, has already proved its high flexibility and usefulness.

3 Layout Algorithms and Layout Features

Currently *yFiles* includes graph layout algorithms for the following styles: layered, tree, force-directed, circular-radial and orthogonal. These algorithms are tuned variants of published algorithms.

Besides layout algorithms which assign coordinates to edge paths and nodes, *yFiles* also supports the automatic assignment of edge and node label coordinates.

Both layout and labeling algorithms can be customized to a high degree. Customization can be done by either setting layout parameters or by exchanging certain stages of an algorithm by custom code.

P. Mutzel, M. Jünger, and S. Leipert (Eds.): GD 2001, LNCS 2265, pp. 453–454, 2002.
© Springer-Verlag Berlin Heidelberg 2002

4 Architecture

4.1 Programming Language

yFiles is written in Java.

4.2 Operating System

The library runs on all Java2 platforms, currently including Linux, Solaris, HPUX, MacOS X, and Microsoft Windows (95/98/2000/NT).

5 Interfaces

The main *yFiles* interface is provided at the Java application programming level. The API allows seamless integration of graph viewer and/or -layout components in any Java-based application.

Additional support for the widely used GML file format allows data exchange with non-Java applications and other graph drawing tools.

6 Screenshots

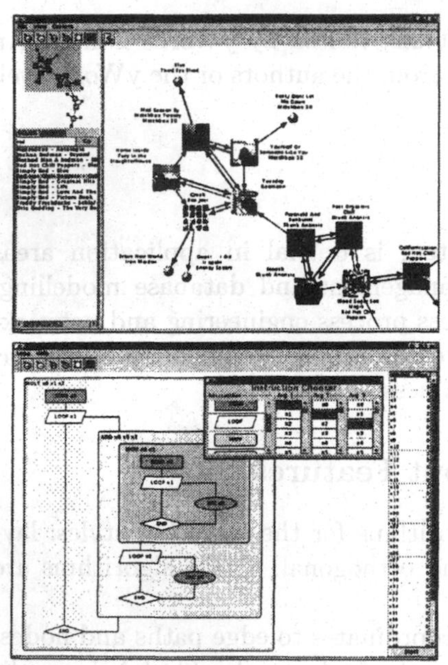

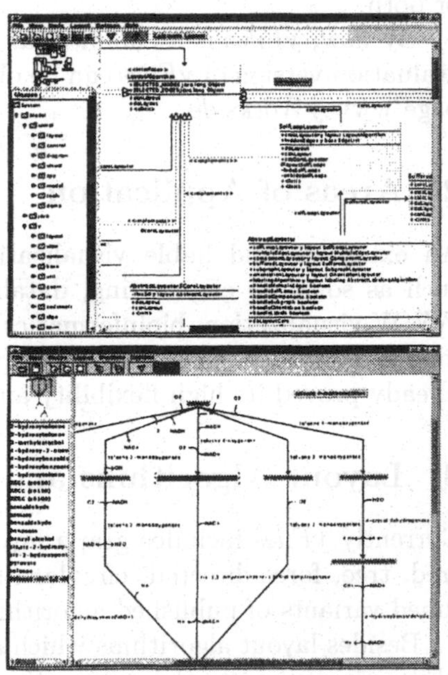

Fig. 1. Applications built upon yFiles: a product browser, a UML tool, a visual programming editor and a biochemical pathways browser.

BioPath*

Franz J. Brandenburg, Michael Forster, Andreas Pick,
Marcus Raitner, and Falk Schreiber

Universität Passau, Innstraße 33, 94032 Passau
{brandenb,forster,pick,raitner,schreiber}@fmi.uni-passau.de

1 Short Description

Biochemical processes in organisms are considered as very large networks consisting of reactants, products and enzymes with interconnections representing reactions and regulation. Examples are given by the well known *Boehringer poster* [4] and the *Biochemical Pathways* atlas [5]. These networks are very complex and grow fast by the steady progress of knowledge in life sciences.

BioPath is the result of the *Electronic Biochemical Pathways Project* [1,3], a joint work of research groups at the universities of Erlangen, Mannheim, Passau and Spektrum Verlag, Heidelberg. *BioPath* provides a convenient electronic access to biochemical reactions at a high level of detail and explores all advantages of an electronic version over a printed one. It uses an innovative algorithm for drawing directed graphs [7] based on *Graphlet* [2].

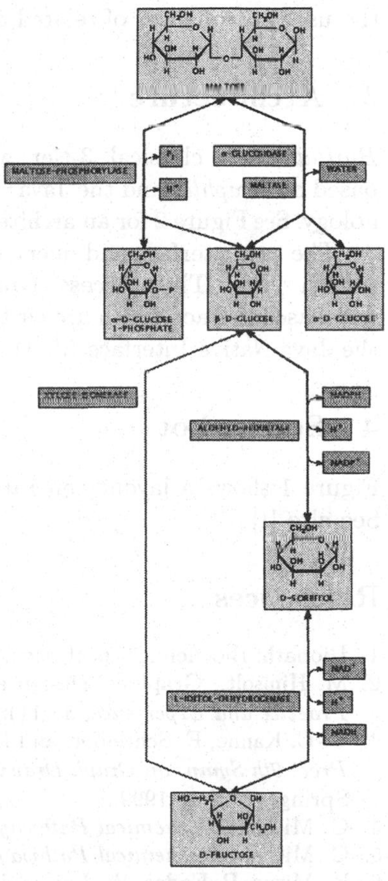

Fig. 1. Example

2 Layout Algorithm and Layout Features

The state of the art in the visualization of biochemical reaction networks are manually produced drawings, as they appear in textbooks, on the poster [4] or in electronic information systems. *BioPath* is the first tool with a dynamic visualization of pathways, which meet the following requirements:

* Supported in part by the German Ministry of Education and Research (BMBF)

P. Mutzel, M. Jünger, and S. Leipert (Eds.): GD 2001, LNCS 2265, pp. 455–456, 2002.
© Springer-Verlag Berlin Heidelberg 2002

Local. Components of the reactions should be placed in the established drawing style of biochemistry as in textbooks.

Global. All reactions should be placed according to their temporal order.

Context sensitive navigation and views. The drawing should maintain the mental map when the granularity of the provided information changes by a new view [6].

As biochemical pathways can be represented by graphs, the visualization of the objects and their connections is a typical graph drawing problem. However, common algorithms are insufficient to represent pathways according to the established conventions of biology and chemistry. Our new customized algorithm produces hierarchical layouts of directed graphs taking node sizes and constraints into account. The sizes of the nodes enforce a new layering strategy. This leads to compact drawings. Using constraints we can draw distinguished paths differently, e. g. the citrate cycle as a real cycle, and we preserve the mental map of the user in sequences of related drawings. The algorithm is described in [7].

3 Architecture

BioPath is a classical 3-tier web application based on *Graphlet* and the Java™ Servlet Technology. See Figure 2 for an architecture overview.

The web interface and query engine are written in Java. They access *Graphlet* and the database interface which are written in C++ via the Java Native Interface (JNI).

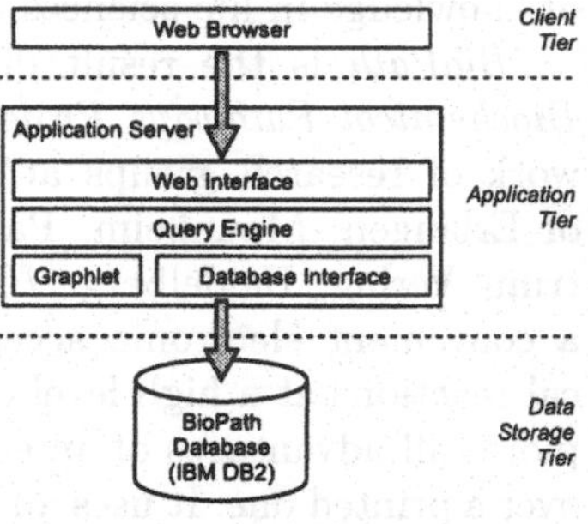

Fig. 2. *BioPath* Architecture

4 Screenshot

Figure 1 shows a layout generated by *BioPath*. See also [1].

References

1. Biopath: Biochemical pathways. `http://biopath.fmi.uni-passau.de`, 2000.
2. M. Himsolt. Graphlet: Design and implementation of a graph editor. *Software - Practice and Experience*, 30(11):1303–1324, 2000.
3. C.-C. Kanne, F. Schreiber, and D. Trümbach. Electronic Biochemical Pathways. In *Proc. 7th Symp. on Graph Drawing (GD'99)*, volume 1731 of *LNCS*, pages 418–419. Springer Verlag, 1999.
4. G. Michal. *Biochemical Pathways (Poster)*. Boehringer Mannh., Penzberg, 1993.
5. G. Michal. *Biochemical Pathways*. Spektrum Akad. Verlag, Heidelberg, 1999.
6. K. Misue, P. Eades, W. Lai, and K. Sugiyama. Layout adjustment and the mental map. *J. of Visual Languages and Computing*, 6:183–210, 1995.
7. F. Schreiber. *Visualisierung biochemischer Reaktionsnetze*. PhD thesis, Universität Passau. 2001.

Graph Visualization API Library for Application Builders

François Bertault, Wendy Feng, Uli Fößmeier,
Gabe Grigorescu, and Brendan Madden

Tom Sawyer Software, 1625 Clay Street, Sixth Floor, Oakland, CA 94612
Tel: (510) 208-4370
Fax: (510) 208-4371
{bertault,wfeng,uli,gabe,bmadden}@tomsawyer.com
www.tomsawyer.com

1 Short Description

Founded in 1991, Tom Sawyer Software produces quality graph-based architectures for application developers. These technologies include graph management, graph layout, graph diagramming, and graph visualization technologies. This software is growing in scope both architecturally and functionally and is packaged as flexible and well-documented library technology that enables universities, governments, and companies to produce graph drawing applications very quickly and with high quality.

2 Areas of Application

Tom Sawyer component technology is used in a large number of different application domains. Our customers build applications for bioinformatics, telecommunications, enterprise software and electronic commerce, IT infrastructure, software engineering, knowledge management, engineering design, and financial analysis.

3 Layout Algorithms and Layout Features

Tom Sawyer Software provides five different layout styles: Hierarchical layout, Orthogonal layout, Circular layout and Symmetric layout (1). Each of this style emphasizes a particular property in the graph to be displayed. Each style is highly customizable with a large set of options such as spacing between nodes, edge routing styles, specific support of undirected graphs and clustering, placement constraints. Tom Sawyer provides also some advanced complexity management features such as hiding, folding and nesting of graphs. Different layout styles can be combined in the same drawing.

P. Mutzel, M. Jünger, and S. Leipert (Eds.): GD 2001, LNCS 2265, pp. 457–458, 2002.
© Springer-Verlag Berlin Heidelberg 2002

4 Architecture and Interfaces

Tom Sawyer Software produces the Graph Editor Toolkit in specific versions for Pure Java and for Microsoft Windows MFC and ActiveX. The Graph Editor Toolkit acts as a graph window presentation system for an application with customizable graph sources, customizable user interface, and customizable graphics. The Graph Editor Toolkit is designed to become an integrated part of a customer application.

Tom Sawyer Software produces the Graph Layout Toolkit in specific versions for C, C++, JNI Java, and ActiveX with a C++-based kernel, and the company is introducing a new Graph Layout Toolkit for Pure Java. The Graph Layout Toolkit, with its graph model management, serves as a very solid infrastructure for the Graph Editor Toolkit

Our portable architectures include API versions that are entirely portable without graphics and versions that include comprehensive API layers with graphics.

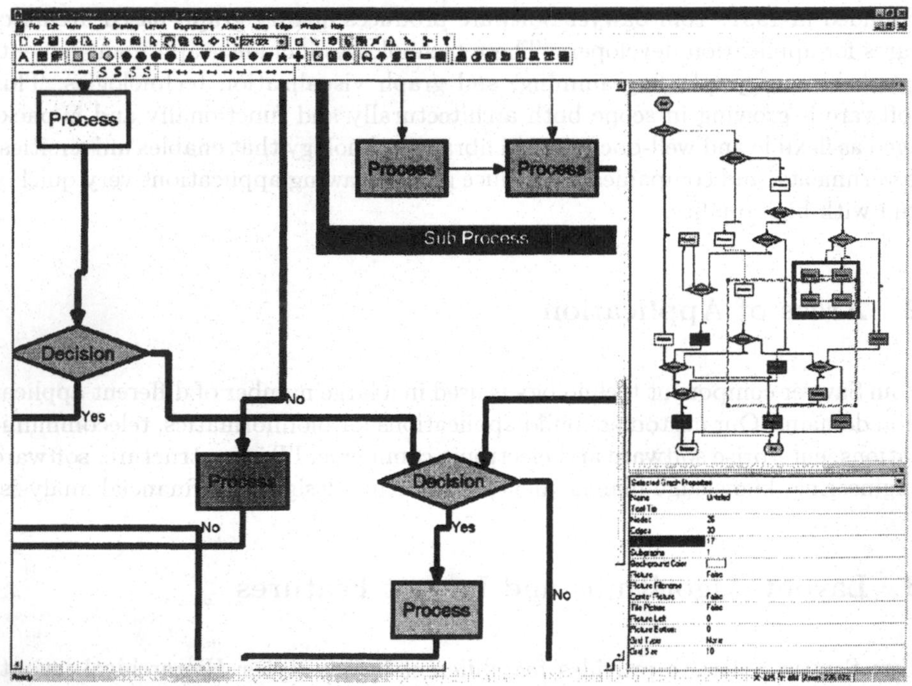

Fig. 1. Drawings with the symmetric and circular layout styles obtained using the Windows Graph Editor.

JGraph – A Java Based System for Drawing Graphs and Running Graph Algorithms

Jay Bagga[1] and Adrian Heinz[2]

[1] Ball State University, Muncie IN 47306, USA
[2] Ontario Systems, Muncie, IN 47304, USA

1 Introduction

JGraph is a Java based system for drawing graphs and for running graph algorithms. A number of well-known algorithms are provided, including those for planarity testing and drawing planar graphs on a grid. The algorithms can be run with an animation feature where the user can see the intermediate steps as the algorithm executes. The system is extensible in that new algorithms can be easily added.

The JGraph project was conceived and led by Jay Bagga. Several of his graduate students have made contributions. Aaron Nall designed the original version and the graphical user interface. The most recent version has been managed by Adrian Heinz who has added many new features and has also implemented algorithms for planarity testing and drawing of planar graphs on a grid.

This system was initially created for students in computer science, mathematics, and other related disciplines who are learning graph theory and using graph algorithms. It can be used by anyone with an interest in graph drawing and graph algorithms.

To draw a planar graph on a grid, we use the Shift Method of de Fraysseix et. al. [1,2]

2 Architecture

2.1 Programming Language

The source code is written in Java, and we have used Java Swing components.

2.2 Operating System

The code can be compiled to run on multiple platforms. We have been running it on PCs with MS-WINDOWS 95/98/2000 and JDK 1.3 installed.

P. Mutzel, M. Jünger, and S. Leipert (Eds.): GD 2001, LNCS 2265, pp. 459–460, 2002.
© Springer-Verlag Berlin Heidelberg 2002

3 Features

The graphical user interface is designed so that a user can draw graphs on the screen almost as one would on paper. The vertices and edges are drawn by clicking and dragging the mouse. Properties such as colors of vertices and edges, labels, and weights can be modified. Vertices and edges can be easily moved to any position. A number of example graphs are provided. Some graph operations such as rotation are also provided. Many well-known graph algorithms are available. These include depth-first, breadth-first search, minimum spanning tree, shortest path, blocks-finding, and planarity-testing. Also included is an implementation of an algorithm for drawings of planar graphs on a grid.

New algorithms can be easily added to the system. Algorithms can be selected from a menu. Running the algorithm in a new window preserves the original graph. If the animated running algorithm option is selected, the intermediate steps are shown in the window as the algorithm runs.

The figure below shows a maximal planar graph drawn on a grid by the JGraph software.

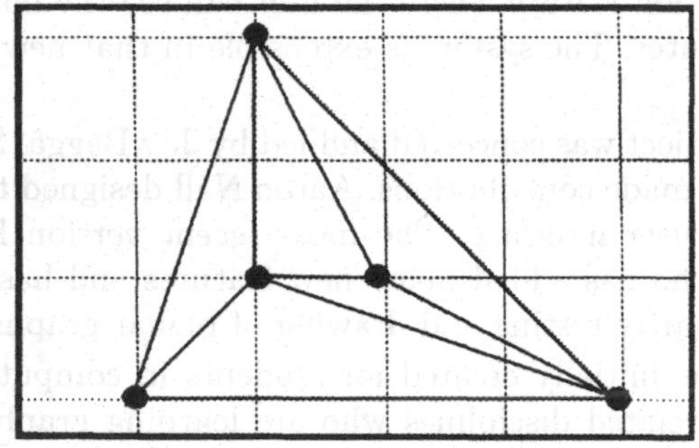

4 Conclusion

The JGraph system has been used as a learning tool. A version of this system is available on the web (`http://www.cs.bsu.edu/homepages/gnet`.) This is an ongoing project. New features and algorithms will continue to be added. Suggestions and comments are greatly welcome. The first author can be contacted at `jbagga@bsu.edu`.

References

1. H. de Fraysseix, J. Pach and R. Pollack, "How to draw a planar graph on a grid," Combinatorica **10**(41-51), 1990.
2. Shin-ichi Nakano, "Planar Drawings of Plane Graphs," IEICE Trans. Fundamentals **E00-A**(1-9), 2000.

Caesar Automatic Layout of UML Class Diagrams

Carsten Gutwenger[1], Michael Jünger[2], Karsten Klein[1], Joachim Kupke[1], Sebastian Leipert[1], and Petra Mutzel[3]

[1] Research Center caesar, Friedensplatz 16, D-53111 Bonn, Germany
{gutwenger|kklein|kupke|leipert}@caesar.de
[2] Universität zu Köln, Pohligstr. 1, D-50969 Köln, Germany
mjuenger@informatik.uni-koeln.de
[3] Technische Universität Wien, Favoritenstr. 9–11 E186, A-1040 Wien, Austria
mutzel@ads.tuwien.ac.at

1 Short Description

UML diagrams have become increasingly important in the engineering and reengineering processes for software systems. Of particular interest are UML class diagrams whose purpose is to display class hierarchies (generalizations), associations, aggregations, and compositions in one picture. The combination of hierarchical and non-hierarchical relations poses a special challenge to a graph layout tool. Commercial software typically uses Sugiyama-style methods, see, e.g., [1] that cannot properly distinguish between hierarchical and non-hierarchical relations.

The caesar graph drawing group develops and implements a library of algorithms and data structures for graph drawing. Special emphasis is on the layout of UML class diagrams.

2 Areas of Application

The software addresses users of UML-class diagrams such as software engineers and data base engineers.

3 Layout Algorithms and Layout Features

Our UML-class diagram layout tool applies state-of-the-art techniques for planarization, planar drawing, and compaction. In contrast to existing tools, the hierarchical and non-hierarchical relations are neither treated alike nor as separate tasks in a two-phase process as in, e.g., Seemann [2].

For a clear visualization of the specific combination of hierarchical and non-hierarchical components in UML Class Diagrams we put special emphasis on meeting a balanced mixture of the following aesthetic criteria: Crossing minimization, bend minimization, uniform direction within each hierarchical component, no nesting of one hierarchy within another, orthogonal layout, merging of multiple inheritance edges, good edge labeling.

P. Mutzel, M. Jünger, and S. Leipert (Eds.): GD 2001, LNCS 2265, pp. 461–462, 2002.
© Springer-Verlag Berlin Heidelberg 2002

4 Architecture

The caesar C++ library for graph drawing is an object oriented system using its own implementations of efficient data structures and algorithms.

The library runs on Microsoft and Unix systems. The Microsoft Visio Add-On operates on Windows NT/2000/98/Me/XP using Visio 2000 or Visio 2002.

5 Interfaces

The caesar library provides a generic layout interface that is independent of the drawing environment. The library currently provides an interface to Visio 2000/2002, a diagramming software by Microsoft.

6 Screenshot

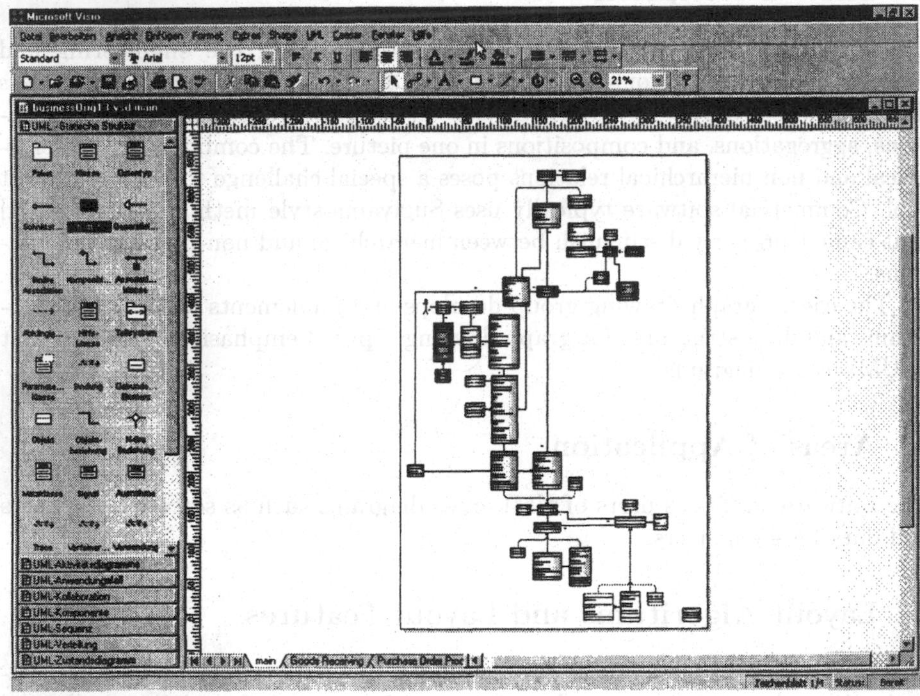

References

1. Rational Rose, 2001. Rational Software Corporation.
2. J. Seemann. Extending the Sugiyama algorithm for drawing UML class diagrams. In G. Di Battista, editor, *Proc. Graph Drawing 1997 (GD '97)*, volume 1353 of *Lecture Notes in Computer Science*, pages 415–424. Springer, 1998.

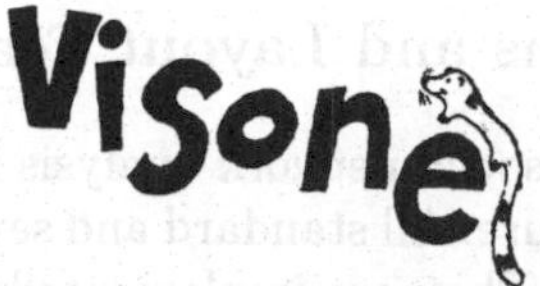

Software for Visual Social Network Analysis*

Michael Baur, Marc Benkert, Ulrik Brandes, Sabine Cornelsen, Marco Gaertler, Boris Köpf, Jürgen Lerner, and Dorothea Wagner

Department of Computer & Information Science,
University of Konstanz, 78457 Konstanz, Germany.
www.visone.de

1 Short Description

We are developing a social network tool that is powerful, comprehensive, and yet easy to use. The unique feature of our tool is the integration of network analysis and visualization. In a long-term interdisciplinary research collaboration, members of our group had implemented several prototypes to explore and demonstrate the feasibility of novel methods. These prototypes have been revised and combined into a stand-alone tool which will be extended regularly.

2 Areas of Application

Social network analysis is a subdiscipline of the social sciences using graph-theoretic concepts to understand and explain social structure. Its methods are also applied, e.g., to financial networks, citation networks, and Web graphs.

Over the last few years, interest in methods for visual analysis of social networks has risen substantially, and several novel approaches have been devised. In response to numerous requests, we are developing this software specifically to allow non-specialist users in the social sciences to apply innovative and advanced methods with ease and accuracy.

The tool is intended for research and teaching in social networks, with special emphasis on visual means of exploring and communicating network data and analyses. In contrast to other tools common in the social sciences, ours is entirely visual. Technicalities are either transparent, or expressed in simpler terms. Initial feedback indicates that users who often regard data exploration and analysis as complicated and unnerving enjoy the playful nature of visual interaction.[1]

* Supported by Deutsche Forschungsgemeinschaft (DFG) under grant BR 2158/1-1.

[1] We are grateful to Steven Corman, Jürgen Grote, Patrick Kenis, Jörg Raab, Volker Schneider, and many participants of the 21st Social Networks Conference (Sunbelt XXI) and of the Summer School on the Analysis of Political and Managerial Networks (POLNET) who provided valuable feedback on preliminary versions.

P. Mutzel, M. Jünger, and S. Leipert (Eds.): GD 2001, LNCS 2265, pp. 463–464, 2002.
© Springer-Verlag Berlin Heidelberg 2002

3 Layout Algorithms and Layout Features

A central line of research in social network analysis is the investigation of prominent actors in a social structure. All standard and several more specialized prominence indices are provided. They are implemented using novel algorithms and, due to our newly introduced normalization scheme, can easily be compared with each other. Future versions are likely to support also various forms of cluster analysis.

Two visualization methods specifically designed to convey the result of such prominence analyses (see screenshots below) are implemented using our own, recently improved algorithms. In addition, we provide a simple spring embedder, a modified spectral layout algorithm, and some simple layout adjustment routines. Future versions shall include, e.g., multidimensional scaling.

4 Architecture

The user interface of visone is a graphical editor tailored to social networks with specialized components for analysis and visualization. The editor uses terminology consistent with the social network literature, and provides different data views in a way that is intuitive for social network analysts.

For interoperability, we support a number of data formats common in social network analysis. Publication quality export is available in PostScript and SVG (which can be converted into, e.g., PDF or JPEG). A batch mode is anticipated.

The program is written in C++, making extensive use of LEDA, the Library of Efficient Data Types and Algorithms from Algorithmic Solutions GmbH. It is available for systems running Linux, Solaris, or Windows, free of charge for academic purposes.

5 Screenshots

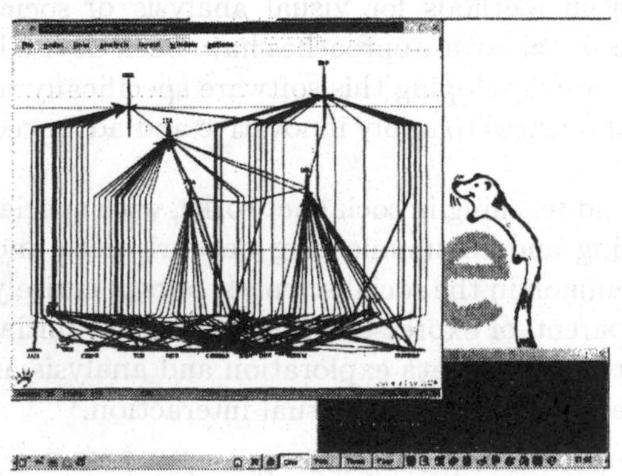

Status visualization
(Screenshot)

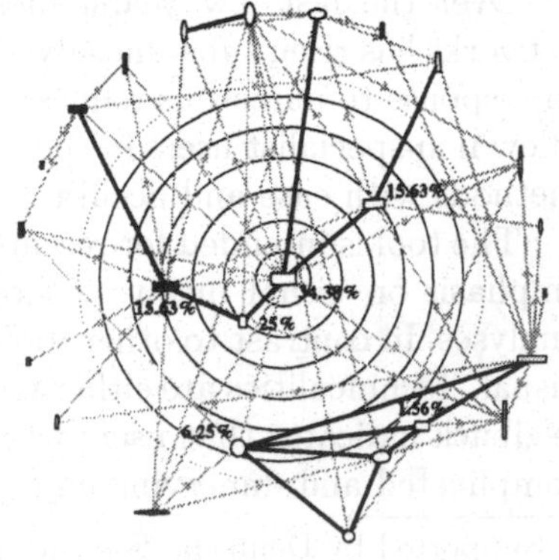

Centrality visualization
(Publication export)

Generating Schematic Cable Plans Using Springembedder Methods

Ulrich Lauther and Andreas Stübinger

Siemens AG, CT SE 6, Munich
{Ulrich.Lauther,Andreas.Stuebinger}@mchp.siemens.de

Motivation and Area of Application. For documentation of communication networks like telephon networks normaly there are two different representations: a *ground plan* showing the exact coordinates of all net elements (man holes, trenches, cables cabinets, exchanges, ...), usually stored and managed in a GIS system, and a *schematic plan* containing the same information, but drawn in a schematic way, not properly scaled, but in an easier to understand way. Another kind of schematic maps are often used to display connections in a urban public transportation system: the arrangement of train stations shows some similiarity to geographic coordinates, but the main importance is to show an easy to understand drawing of the connection possibilities by using a more orthogonal drawing.

In telecomunication GIS systems it is possible to handle both views of the net: the ground plan, and a schematic plan. However, this implies a high burdon on keeping both plans consistent up to date. Every time a connection is inserted, updated or deleted in the ground plan it has to be re-done in the schematic map, too. Therfore, updateing both plans is a very tedious task and very prone to errors.

This burdon is resolved by an automatic schematic map generation algorithm described in this paper. Instead of updating two different views in lockstep, only one map, the ground plan, is updated manually. The schematic map then is generated automaticly by SCHEMAP.

Input data. The input for SCHEMAP consists of the ground plan, containing two different but highly interwoven graphs: a *track graph*, describing man holes and duct elements, and a *cable graph* describing the electrical components of the telecomunication network. The cable graph is embedded inside the track graph as all cables have to lie inside some trenches. In the case of an air cable a dummy duct is created in the track graph.

Layout Algorithms and Constraints to honor. The track layout is done by a modified springembedder based on [1], the cable layout by a specialized sorting algorithm and the track node layout (i.e. cable nodes inside a track node) by a combination of pattern router and Lee router.

To get a schematic view of the input graph the algorithm has to take the following constraints into account:

P. Mutzel, M. Jünger, and S. Leipert (Eds.): GD 2001, LNCS 2265, pp. 465–466, 2002.
© Springer-Verlag Berlin Heidelberg 2002

Similarity of location. Crossings in the input have to be preserved; however, no new crossings may be added. Also, north/south and east/west relations have to be preserved.

Orthogonal. An "as orthogonal as possible" drawing should be achieved.

Sizes for nodes and edges. To draw cable edges inside the track edges a track edge has to have some size. This implies that track nodes get a rectangle box having some size to connect a track edge.

Results. In figure 1 on the left a typical input for SCHEMAP is shown. Only the track graph is visible, all the cables are hidden, as there are no coordinates given for the cable graph elements. The boxes near the track edges represent textlabels. On the right of figure 1 the output of SCHEMAP is shown. Nearly all track edges are orthogonal, all cable edges are sorted.

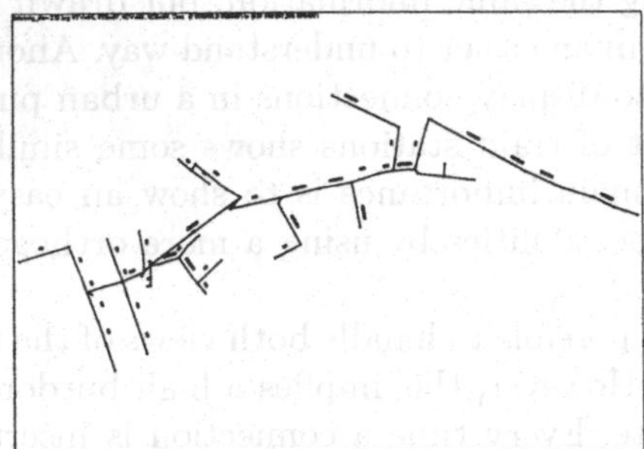
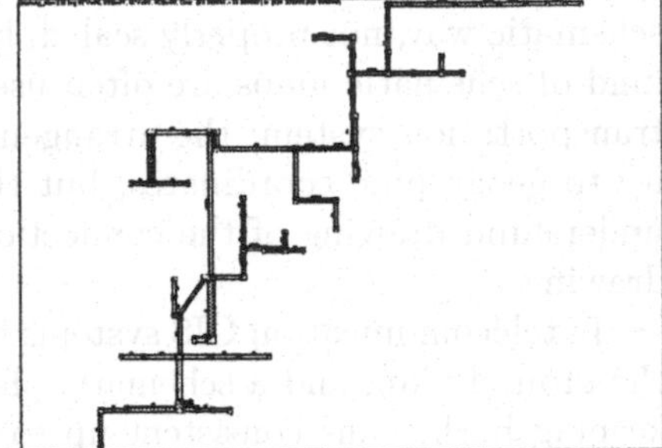

Fig. 1. Left: Typical input ground plan, Right: Resulting schematic view. To show the cables in more detail different colors are chosen.

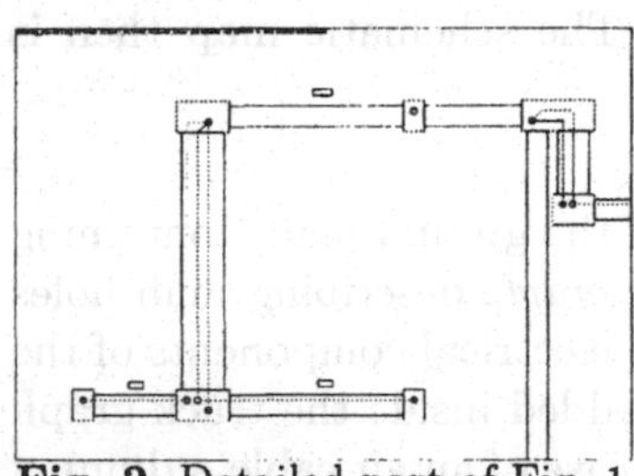

Fig. 2. Detailed part of Fig. 1

The drawings are normaly not perfect in every sense, i.e. normaly there are some minor modifications to be done by hand. However, SCHEMAP achieves its goal pretty good: the schematic drawings show the input graph in a more schematic way respecting all the given constraints and it is fast enough to be used in daily work. SCHEMAP is implemented in C++ using the TURBO library (see [3]) and is integrated as module SCHEMAP into NetMinister-OSP (see [2]).

References

1. T.M.J. Fruchterman, E.M. Reingold: *Graph Drawing by Force-Directed Placement*, Software — Practice and Experience, Vol. 21, pp. 1129 – 1164, 1991.
2. *NetMinister-OSP*, Order No. A50001-N12-P14-1-7600, Siemens AG, 2000.
3. U. Lauther: *TURBO – In Projekten schneller zum Ziel*, Software@Siemens, 1999.

SugiBib

Holger Eichelberger

Würzburg University, eichelberger@informatik.uni-wuerzburg.de

1 Short Description

UML (Unified Modeling Language) class diagrams, which are widely used for specifying aspects of object-oriented software systems, can be laid out by our tool SugiBib. In 1998 J. Seemann described how to apply the Sugiyama algorithm to class diagrams and the first version of SugiBib was implemented as a masters thesis.

2 Areas of Application

SugiBib can be used as a class browser, an online rendering engine and as a plugin layout component for external programs in order to calculate the layout of UML class diagrams specified as UMLscript [1] files.

3 Layout Algorithms and Layout Features

The algorithm identifies the edges of a pseudo-hierarchy (usually the inheritance edges), calculates a semantic ordering, inserts compound nodes due to incremental layout, association classes and annotations, inserts additional edges to reflect containment of nodes and removes reflective associations. Then the algorithm by Seemann is applied and after the calculation of coordinates all composite nodes inserted in the preparation steps are expanded. Nested nodes are currently laid out by a frame layout approach which is applied to the entire input graph.

4 Architecture

SugiBib is a framework which was designed to implement a general, highly configurable, component-based version of the Sugiyama algorithm [3]. The components can be combined in different sequences to implement other layout algorithms. Nodes and edges of the framework are parametrized by their individual graphical information.

4.1 Programming Language

SugiBib is a pure Java framework which can be compiled and executed on all Java 2 platforms. The graphical frontend is implemented in AWT and in Swing. The current implementation is tested on the JDK version 1.3.1

P. Mutzel, M. Jünger, and S. Leipert (Eds.): GD 2001, LNCS 2265, pp. 467–468, 2002.
© Springer-Verlag Berlin Heidelberg 2002

4.2 Operating System

Since SugiBib is written in Java, it runs on every platform that supportes by Java.

5 Interfaces

As a framework SugiBib represents an open interface. New components are be implemented by subclassing existing classes. Components are used in sequence, i.e. a component is called by its constructor and its input is the preceeding component itself. Further implementations will provide XMI (XML Metadata Interchange) and GXL (Graph Exchange Language) beside UMLscript[1] as input languages. Graph descriptions in UMLscript can be generated from Java source code by an appropriate compiler.

6 Screenshots

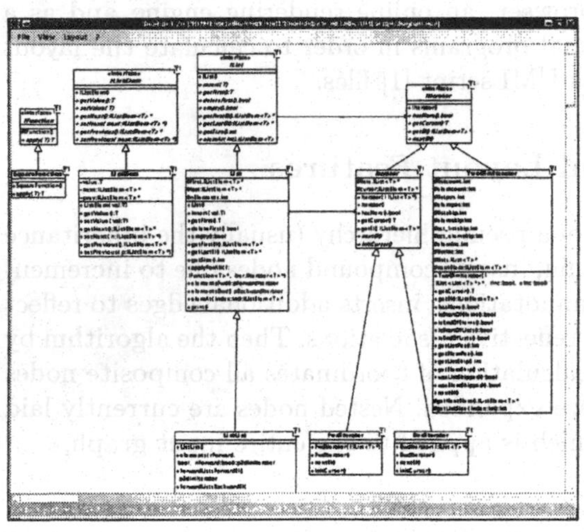

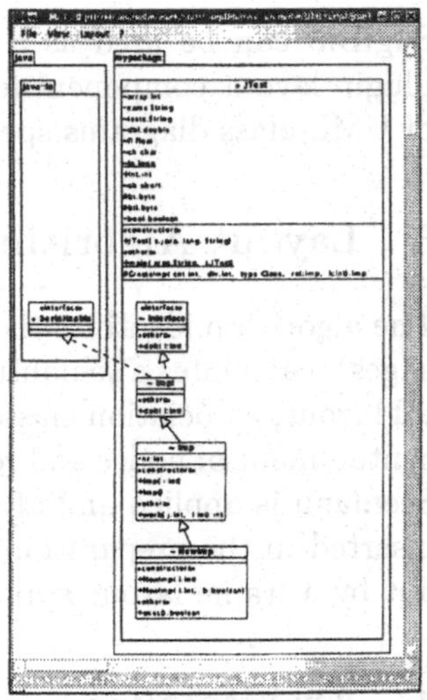

References

1. H. Eichelberger, J. Wolff von Gudenberg: UMLscript Sprachspezifikation, Technical Report No. 272, University of Wuerzburg, February 2001
2. J. Seemann: Extending the Sugiyama Algorithm for Drawing UML Class Diagrams: Towards Automatic Layout of Object-Oriented Software Diagrams, *Lecture Notes in Computer Science*, LNCS 1353 G. DiBattista (Editor), 414-423, 1998
3. K. Sugiyama, S. Tagawa, M. Toda: Methods for Visual Understanding of Hierarchical System Structures, *IEEE Transactions on Systems, Man, and Cybernetics*, SMC-11(2):109-125, February 1981

Knowledge Index Manager

Jean Delahousse and Pascal Auillans

Mondeca (http://www.mondeca.com)

1 Short Description

Mondeca is the editor of KIM[1], a software solution for content and knowledge organization, which allows companies to: access quickly and precisely data and documents available in the enterprise and/or on the Web; federate scattered document resources; index and organize document content managed in different applications; build knowledge bases; customize access to information according to each user profile.

2 Areas of Application

Mondeca software is used for legal, medical, technical documentation, expert database, human resources management, marketing survey, mapping of financial structures, cultural publications, sports database... In all those applications KIM acts as an organizer and navigation tool to access external documents and data. Hierarchical and non-hierarchical semantical links between Topics of the index enable to build hierarchical organization of information but also of knowledge database. Organization vocabulary is customized depending on the needs of the users.

Mondeca largest clients include such companies as EDF, TNO-FEL (Dutch center for applied research), Supercomputer Center of San Diego University.

3 Layout Algorithms and Layout Features

Hierarchical drawings lead to the conception of a new orientation driven planarization algorithm which produces an upward drawing with at most one bend per edge. This polynomial algorithm may probably be enhanced into a linear time one by a suitable modification of our planarity algorithm, relaxing the DFS condition on the traversal tree.

4 Architecture

The software relies on three main components : Information Network Repository, Information Network Navigation Engine for edition, filtering and publication, Graphic Server[2] built on Pigale library from EHESS.

[1] free demo at http://semantopic.mondeca-publishing.com/semantopic/pregen/html/anonymous/

[2] freely available at ftp://pr.cams.ehess.fr/pub/pigale.tar.gz

P. Mutzel, M. Jünger, and S. Leipert (Eds.): GD 2001, LNCS 2265, pp. 469–470, 2002.
© Springer-Verlag Berlin Heidelberg 2002

4.1 Programming Language

C++, Java (EJB, J2EE), Pliant[3] and XML (SVG, Topic Maps, XSLT).

4.2 Operating System

Windows NT, Linux/Unix with J2EE application server and Oracle.

5 Interfaces

XML for publication, XTM and RDF (available 2002) for database interchange.

6 Screenshots

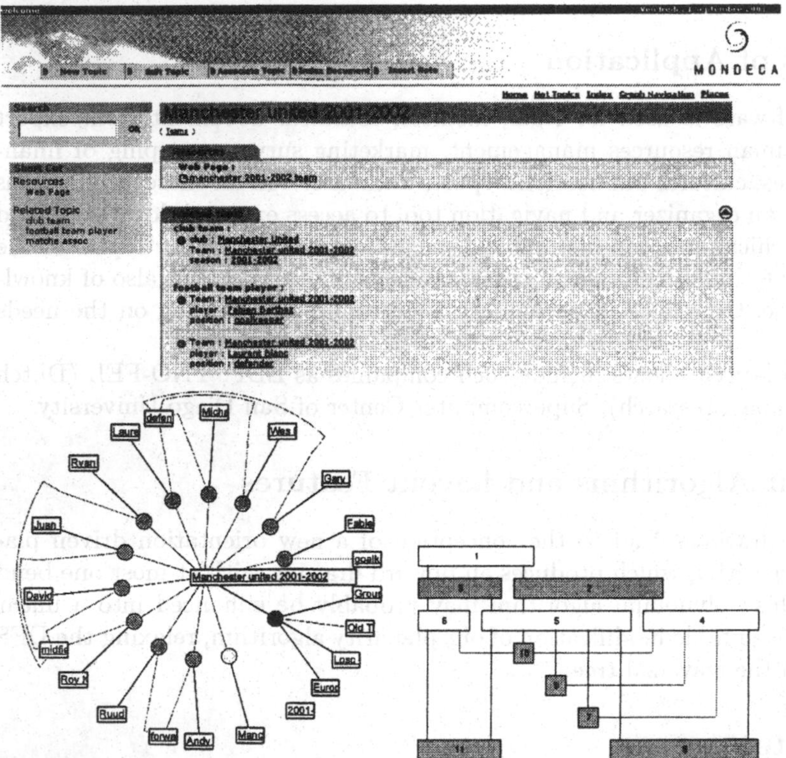

Fig. 1. Screenshot of the centered topic screen, of a hierarchical and a polar drawing.

[3] http://pliant.cx

Planarity Testing of Graphs on Base of a Spring Model

Günter Hotz and Steffen Lohse

Universität des Saarlandes – Fachbereich Informatik, Saarbrücken, Germany
http://www-hotz.cs.uni-sb.de

1 Short Description

It is well known that planar embeddings of 3-connected graphs are uniquely determined up to isomorphy of the induced complex of nodes, edges and faces of the plane or the 2-sphere [1]. Moreover, each of the isomorphy classes of these embeddings contains a representative that has a convex polygon as outer border and has all edges embedded as straight lines. We fixate the outer polygon of such embeddings and regard each remaining edge e as a spring, its resilience being $|e|^k$ ($|e|$ euclidean length of e, $k \in \mathbb{R}, 1 < k < \infty$). For 3-connected graphs, exactly one power-balanced embedding for each k exists, and this embedding is planar if and only if the graph with the fixated border polygon has a planar embedding inside that very polygon. For $k = 1$ or $k = \infty$, some faces may be collapsed; we call such embeddings *quasi-planar* [2]. It is possible to decide the planarity of any graph embedding in linear time [3]. The motivation for this result was to develop a planarity test that simultaneously with the decision process constructs a concrete planar embedding. This algorithm should work in three steps:

1. Choose any circle of the graph and embed it as a convex polygon in the plane or as an equator of the 2-sphere.
2. Fixate this circle and hang in the rest of the graph. Substitute each edge by a spring of a fixed norm k and compute its power-balanced position.
3. Decide if the computed embedding can be decomposed into two planar flanks. The graph is planar if and only if this is possible. On the sphere, these flanks should be represented by the two hemispheres belonging to the fixated circle.

If the input graph originates from a hierarchical definition of a chip layout system (e.g. [4]), that definition gives the base for an efficient algorithm to compute this embedding by use of a multi grid method [5].

The algorithm presented here assumes the graph to be 3-connected – but it does not make special assumptions on its definition. We only use the isomorphy theorem mentioned. The stepwise construction of the embedding reflects the *single step method* of iterative solutions of equations. Some steps produce quasi-planar embeddings which in later steps will be unfolded to planar embeddings by hanging in more springs. This requires some complex data structures, the handling of which is responsible for the high worst case runtime bound $O(n^2 \log n)$ of the algorithm. The proof of this bound assumes a worst case situation which

P. Mutzel, M. Jünger, and S. Leipert (Eds.): GD 2001, LNCS 2265, pp. 471–472, 2002.
© Springer-Verlag Berlin Heidelberg 2002

seems never to appear. We assume that a more thorough analysis will come out with a $O(n \log n)$ bound.

We ran tests on randomly generated planar graphs with up to 6000 nodes and compared the running time with the algorithms Fáry [6] ($O(n^2)$) and Schnyder [7] ($O(n)$) implemented in [8]. In all examples, our algorithm was in the worst case and the average case faster than Fáry, and Schnyder always was faster than our algorithm. Our algorithm shows complexity jumps for graphs of size n at 2^n, suggesting a possible additional speedup. The pictures produced by our algorithm, representing the planar embeddings on the sphere, have a certain aesthetical attraction.

The detailed description of the algorithm can be found in the Master's Thesis of the second of the authors [9]. It includes a C++ implementation for MS Windows and UNIX.

2 Screenshots

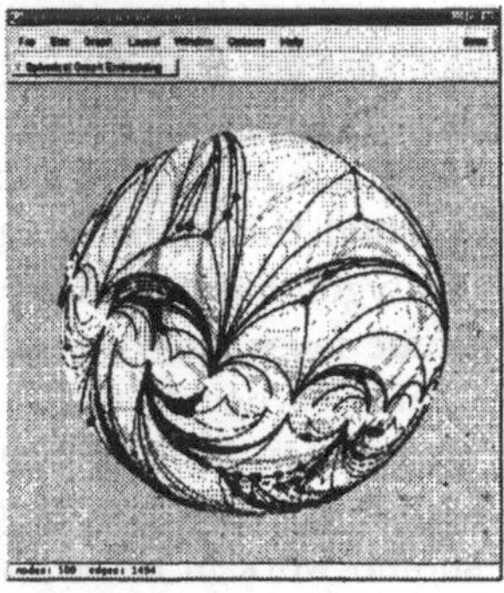

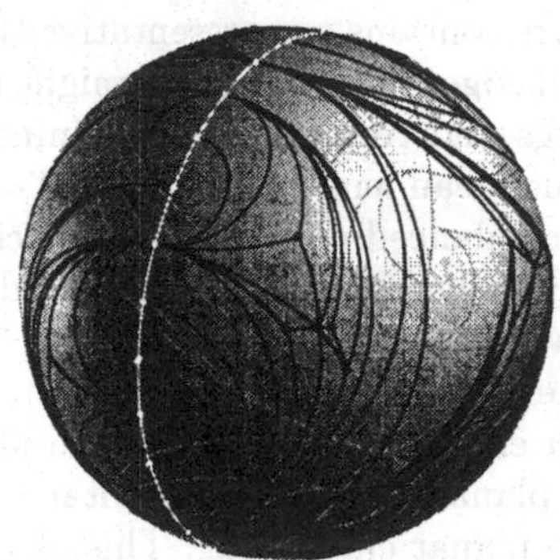

References

1. Hotz, G.: Einbettung von Streckenkomplexen in die Ebene. Math. Annalen 167, 214–223 (1966)
2. Becker, B., Hotz, G.: On the Optimal Layout of Planar Graphs with Fixed Boundary. SIAM Journal on Computing 16(5), 946–972 (1987)
3. Becker-Groh, U., Hotz, G.: Ein Planaritätstest für planar-konvexe Grapheneinbettungen mit lin. Komplexität. Beiträge zur Algebra u. Geom. 18, 191–200 (1984)
4. Hotz, G., Reichert, A.: Hierarchischer Entwurf komplexer Systeme. Chapter 8 in: Wegener, I. (Ed.): Highlights aus der Informatik. Springer-Verlag (1996)
5. Osthof, H.-G.: Optimale Grapheinbettungen und ihre Anwendungen. Dissertation. Universität des Saarlandes (1990)
6. Fáry, I.: On Straight Line Representing of Planar Graphs. Acta Sci. Math. 11, 229–233 (1948)
7. Schnyder, W.: Embedding Planar Graphs on the Grid. Proc. 1st ACM-SIAM Symp. on Discrete Alg., 138–148 (1990)
8. Mehlhorn, K., Näher, S., Seel, M., Uhrig, C.: The LEDA User Manual (Version 3.7). Max-Planck-Institut für Informatik, Saarbrücken (1999)
9. Lohse, S.: Ein Planaritätstest und Einbettungsverfahren für Graphen. Diplomarbeit. Fachbereich Informatik, Universität des Saarlandes (2001)

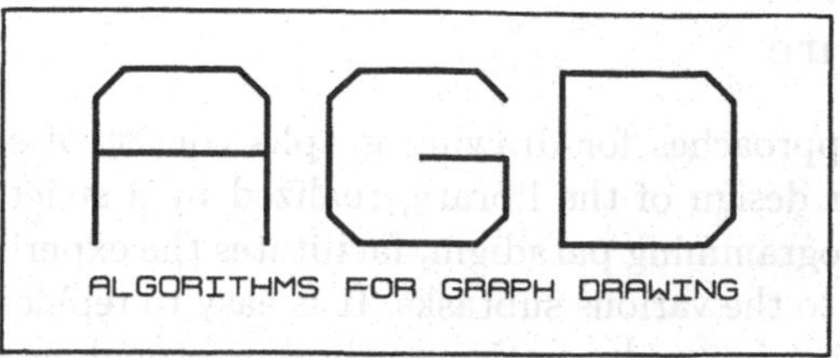

A Library of Algorithms for Graph Drawing*

Carsten Gutwenger[1], Michael Jünger[2], Gunnar W. Klau[3], Sebastian Leipert[1], Petra Mutzel[3], and René Weiskircher[3]

[1] Stiftung caesar, Bonn, Germany
[2] Universität zu Köln, Germany
[3] Technische Universität Wien, Austria

1 Short Description

The AGD library provides algorithms, data structures, and tools to create geometric representations of graphs and aims at bridging the gap between theory and practice in the area of graph drawing. It consists of C++ classes and is built on top of the library of efficient data types and algorithms LEDA; an optional add-on to AGD requires ABACUS, a framework for the implementation of branch-and-cut algorithms, and contains implementations of exact algorithms for many *NP*-hard optimization problems in algorithmic graph drawing.

The fully documented library is freely available for non-commercial use at `http://www.ads.tuwien.ac.at/AGD`. The site also contains an online manual, links to AGD related papers, and contact information.

2 Layout Algorithms and Layout Features

The library contains a large number of state-of-the-art drawing algorithms for many of which implementations can only be found in AGD. Figure 1 shows UML-diagrams of three major components of the library: (a) planar drawing algorithms, (b) hierarchical drawing algorithms, and (c) the planarization method. Among the highlights in the latest version are implementations of the Kandinsky- and the Giotto-algorithm, new heuristics and exact algorithms for two-dimensional compaction, and new strategies for crossing minimization based on a linear-time implementation of SPQR-trees.

* Partially supported by DFG-grants Ju204/7-3, Mu1129/3-1, and Na 303/1-3. In addition to the authors, the following persons have contributed to AGD: D. Alberts, D. Ambras, R. Brockenauer, C. Buchheim, M. Elf, S. Fialko, K. Klein, G. Koch, T. Lange, D. Lüttke-Hüttmann, S. Näher, T. Ziegler

P. Mutzel, M. Jünger, and S. Leipert (Eds.): GD 2001, LNCS 2265, pp. 473–474, 2002.
© Springer-Verlag Berlin Heidelberg 2002

3 Architecture

Most successful approaches for drawing graphs consist of several phases. The open and modular design of the library, realized by a strict compliance to the object-oriented programming paradigm, facilitates the experimentation with different approaches to the various subtasks. It is easy to replace or extend existing modules by plugging in an alternative method or providing a derived solution.

3.1 Programming Language and Operating Systems

The library is written in C++ and can be downloaded for Linux, Solaris, and Windows platforms and different compilers.

4 Interfaces

AGD implements a generic layout interface that is independent of a fixed drawing component. This partition makes it easy to integrate the library's functionality within application programs. However, several interface implementations are already included in the library; most notably, two demo programs that use the graph editor in LEDA provide instant access to the functionality of the base part and the optional add-on of AGD. These interfaces can be easily extended and adapted. Furthermore, due to the file- and socket-based AGD server interface, the usage of AGD is not restricted to C++ applications.

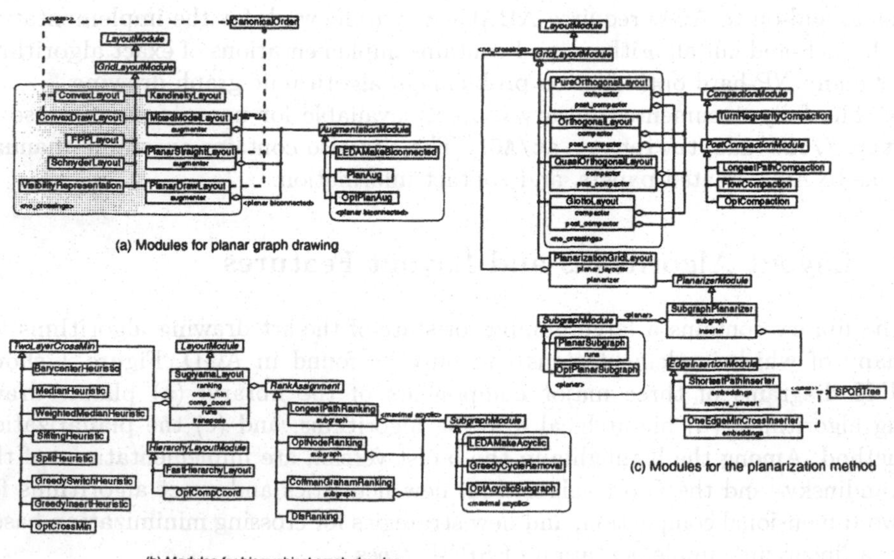

Fig. 1. UML-diagrams of selected AGD modules

Industrial Plant Drawer*

Walter Didimo[1], Maurizio Patrignani[2], and Maurizio Pizzonia[2]

[1] Università di Perugia, Perugia, Italy
[2] Università di Roma Tre, Rome, Italy

1 Short Description

Industrial Plant Drawer is a prototypical system especially designed to cope with the complexity of the requirements of industrial plant schemas. It produces orthogonal drawings in which the size of the nodes is assigned according to the symbol associated with them. Also, it allows the user to choose a set of nodes to be placed on the border of the drawing. Labels, colors, and other graphic features are dealt with. Created as a GDToolkit demo, Industrial Plan Drawer grew into an independent system.

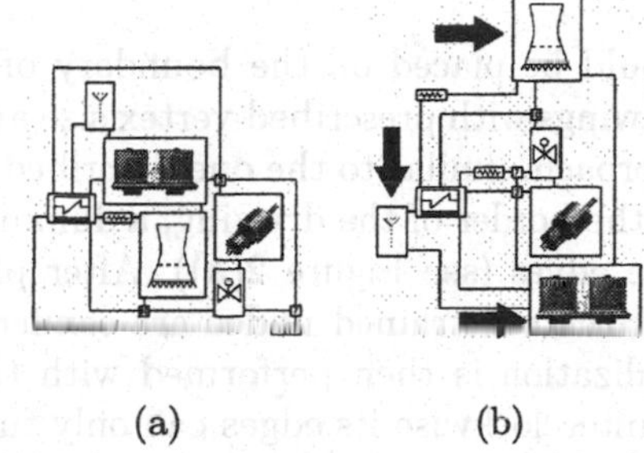

Fig. 1. Two drawings generated with Industrial Plant Drawer. In (a) some of the vertices are replaced with symbols of appropriate dimensions. In (b) the same graph where the highlighted nodes are constrained to be on the border.

2 Areas of Application

Engineers who design, maintain, and manage industrial plants extensively use diagrammatic representations which are necessary to understand the complexity of the industrial systems they deal with. In such schemas each node represents a component of the plant, and each edge represents flow of material or control information. They are usually very large, difficult to modify or update, and produced by hand in the design phase of the life-cycle of the plant.

3 Layout Algorithms and Layout Features

Even though industrial plants may be represented as graphs, they are particularly challenging for the currently available graph drawing tools. One of the features that most affect the readability of industrial schemas is concerned with the size of each symbol that should reflect the role it plays in the system or its real dimensions. Also, devices that interface the system with other systems

* Work partially supported by: "Progetto Algoritmi per Grandi Insiemi di Dati: Scienza e Ingegneria", MURST Programmi di Ricerca di Rilevante Interesse Nazionale.

P. Mutzel, M. Jünger, and S. Leipert (Eds.): GD 2001, LNCS 2265, pp. 475–476, 2002.
© Springer-Verlag Berlin Heidelberg 2002

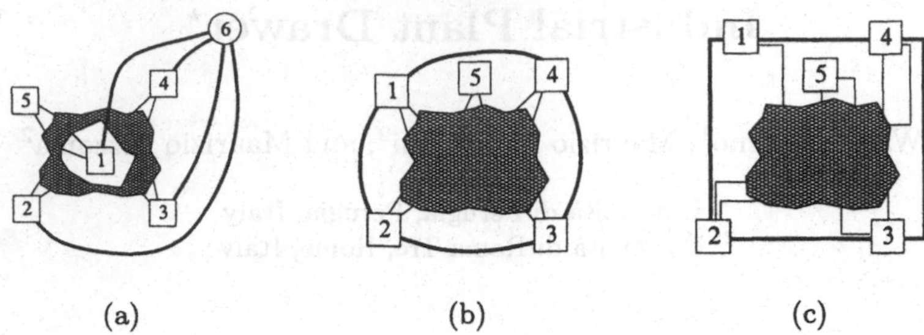

Fig. 2. This figure illustrates the procedure that our system uses to constrain some nodes (1,2,3, and 4 in the picture) on the boundary of the drawing.

should be placed on the boundary of the drawing (see Figure 1). Orthogonal drawings with prescribed vertex size are produced with a topology-shape-metrics approach, similar to the one described in [1]. Since some nodes need to be placed on the border of the drawing, a dummy node is connected to them with uncrossable edges (see Figure 2(a)). After planarization the dummy node is removed and the constrained nodes are connected in a cycle as in Figure 2(b). Orthogonalization is then performed with the constraint that walking on such cycle counterclockwise its edges can only turn to the left. This forces the shape of the external face to be a rectangle (see Figure 2(c)). Then compaction is performed as described in [1]. The edges of the rectangle are removed before visualization.

Industrial Plant Drawer is written in C++ language, using the GDToolkit graph drawing library [2] and LEDA [3], and runs on all the operating systems supported by these libraries.

4 Interfaces

Industrial Plant Drawer is extremely friendly. A graph is inserted through the graph editor or by loading it from a file. The user can customize the drawing through dialog boxes that pop up by clicking on the nodes and edges. Each node can be associated with a plant component chosen from a collection. Furthermore, the user can specify whether the node should be constrained to be on the boundary of the drawing. Each edge may be directed, colored, labeled, etc.

References

1. G. Di Battista, W. Didimo, M. Patrignani, and M. Pizzonia. Orthogonal and quasi-upward drawings with vertices of prescribed size. In J. Kratochvil, editor, *Graph Drawing (Proc. GD '99)*, volume 1731 of *Lecture Notes Comput. Sci.* Springer-Verlag, 1999.
2. Gdtoolkit 3.0: An object-oriented library for handling and drawing graphs, 1999. Third University of Rome, http://www.dia.uniroma3.it/~gdt.
3. K. Mehlhorn and S. Näher. *LEDA: A Platform for Combinatorial and Geometric Computing.* Cambridge University Press, New York, 1998.

Pajek – Analysis and Visualization of Large Networks

Vladimir Batagelj[1] and Andrej Mrvar[2]

[1] Department of mathematics, FMF, University of Ljubljana
[2] Faculty of Social Sciences, University of Ljubljana

1 Short Description

`Pajek` (spider, in Slovene) is a program package, for Windows (32 bit), for analysis and visualization of *large networks* (having thousands of vertices). It is freely available, for noncommercial use, at its home page:
`http://vlado.fmf.uni-lj.si/pub/networks/pajek/`

We started the development of Pajek in November 1996. The main goals in the design of Pajek are [1]:

- to support abstraction by (recursive) factorization of a large network into several smaller networks that can be treated further using more sophisticated methods;
- to provide the user with some powerful visualization tools;
- to implement a selection of efficient (subquadratic) algorithms for analysis of large networks.

Besides ordinary networks Pajek supports also 2-mode networks and temporal networks.

Pajek is essentially a collection of procedures based on 6 data types: network, partition, cluster, vector, permutation, and hierarchy. These procedures are available through the main window menus. Frequently used sequences of operations can be defined as macros. This allows also the adaptations of Pajek to special groups of users.

We developed efficient algorithms for determining main parts in acyclic networks, cores, counting triads, and for pattern (subnetwork) searching [2,3,4]. Pajek contains also some data analysis procedures such as clustering and blockmodeling.

Pajek is still under development. The latest version is available for download at its home page.

2 Areas of Application

Pajek was applied by researchers in different areas: social network analysis [2], chemistry (organic molecule), genealogies [5], Internet networks, citation networks, diffusion networks (AIDS, news), data-mining (2-mode networks), ...

P. Mutzel, M. Jünger, and S. Leipert (Eds.): GD 2001, LNCS 2265, pp. 477–478, 2002.
© Springer-Verlag Berlin Heidelberg 2002

It is also used at universities (Ljubljana, Rotterdam, Irvine, The Ohio State University, Penn State, Madrid...) as a support in courses on network analysis.

Together with Wouter de Nooy from University of Rotterdam we wrote a course book *Exploratory Social Network Analysis With Pajek.*

3 Layout Algorithms and Layout Features

Special emphasis is given to automatic generation of network layouts. Several standard algorithms for automatic graph drawing are implemented: spring embedders (Kamada-Kawai and Fruchterman-Reingold), layouts determined by eigenvectors (Lanczos algorithm), drawing in layers (genealogies and other acyclic structures), fish-eye views and block (matrix) representation.

These algorithms were modified and extended to enable additional options: drawing with constraints (optimization of the selected part of the network, fixing some vertices to predefined positions, using values of lines as similarities or dissimilarities), drawing in 3D space. Pajek also provides tools for manual editing of graph layout.

Properties of vertices/lines (given as data or computed) can be represented using colors, sizes and/or shapes of vertices/lines.

4 Architecture

Pajek is implemented in Delphi and runs on Windows 32 operating systems.

5 Interfaces

Pajek supports also some non-native input formats: UCINET DL files; chemical MDLMOL and BS; and genealogical GEDCOM.

The layouts can be exported in the following output graphic formats that can be examined by special 2D and 3D viewers: Encapsulated PostScript, Scalable Vector Graphics, VRML, MDLMOL, and Kinemages.

References

1. Batagelj, V., Mrvar, A. (1998): Pajek – A Program for Large Network Analysis. *Connections*, **21 (2)**, 47-57.
2. Batagelj, V., Mrvar, A. (2000): Some Analyses of Erdős Collaboration Graph. *Social Networks*, **22**, 173-186.
3. Batagelj, V., Mrvar, A. (2001): A Subquadratic Triad Census Algorithm for Large Sparse Networks with Small Maximum Degree. *Social Networks*, **23**, 237-243.
4. Batagelj, V., Mrvar, A., Zaveršnik M. (1999): Partitioning Approach to Visualization of Large Graphs. Kratochvil, J. (Ed.) Graph Drawing. 7th International Symposium, GD'99, Štiřin Castle, Czech Republic, Proceedings. *Lecture Notes in Computer Science*, 1731. Springer-Verlag, Berlin/Heidelberg, 90-97.
5. White, D.R., Batagelj, V., Mrvar, A. (1999): Analyzing Large Kinship and Marriage Networks with Pgraph and Pajek. *Social Science Computer Review*, **17 (3)**, 245-274.

GLIDE

Kathy Ryall

MERL, 201 Broadway, Cambridge, MA 02139, USA, ryall@merl.com

1 Short Description

The GLIDE (Graph Layout Interactive Diagram Editor) system (first presented at GD'96 [4]) improves on general constraint-based approaches to drawing and layout by supporting only a small set of "macro" constraints, known as VOFs (visual organization features). They are specifically suited to graph drawing, contributing both aesthetic and semantic information. To date, GLIDE remains the only interactive graph drawing tool to support the use of VOFs or similar constructs. More recent work [1][3] has begun to investigate the importance of visual organization in graph drawing.

2 Areas of Application

GLIDE is intended for interactive drawing of small graphs, most often used in publications or presentations. By exploiting the advanced techniques that have been developed by both the graph-drawing or constraint-based-layout communities, GLIDE supports the exquisite symmetries, spacings, and alignments that graphic designers utilize in professional-grade work.

3 Layout Algorithms and Layout Features

GLIDE is based on constraints, but ones that are designed specifically for drawing graphs, not general graphics. These "macro" constraints, or *Visual Organization Features (VOFs)* [2], and their application in GLIDE are described more fully in [5]. In GLIDE, VOFs can be applied and removed interactively. Furthermore, the tool enforces syntactic constraints, such as preventing nodes overlapping other nodes/edges. The VOF and syntactic constraints are enforced by a generalized spring algorithm.

GLIDE converts each VOF instance into a set of constraints. The physical simulation of the resulting mass-spring model is continuously animated, indicating to the user the influence of the chosen VOFs. The user may move nodes and groups of nodes while the simulation proceeds in order to aid the system in finding better global solutions to the implicit constraint-satisfaction problem. The use of a constraint-satisfaction scheme (mass-spring simulation) that is intuitive and predictable, rather than one better at finding global solutions, is deliberate. GLIDE is intended to support a user and the computer in jointly solving the layout problem. For this purpose predictability, simplicity, and the compelling nature of the animation are more important than global optimality.

P. Mutzel, M. Jünger, and S. Leipert (Eds.): GD 2001, LNCS 2265, pp. 479–480, 2002.
© Springer-Verlag Berlin Heidelberg 2002

4 Architecture

Glide is written in C and Tcl/Tk, and currently runs under most flavors of Unix.

5 Interfaces

Figure 1 depicts GLIDE's interface, displaying a graph layed out using a variety of VOFs — *Symmetry, Alignment, HubShape.* GLIDE supports other standard functionality (e.g. ability to change fonts, colors, shapes, etc). The VOFs we support provide a natural and powerful vocabulary for users to express easily the desired characteristics of a graph layout; our intuitive constraint-satisfaction method allows for a collaborative interaction between user and computer. An informal comparison with commercial drawing packages shows GLIDE to be markedly superior for drawing small, aesthetic graphs.

6 Screenshot

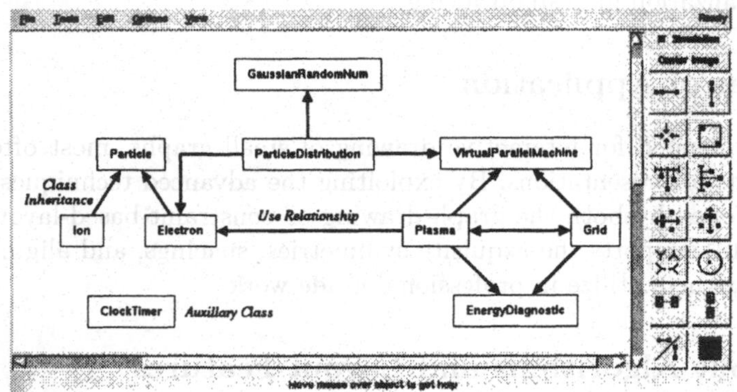

Fig. 1. The GLIDE interface

References

1. Stina Bridgeman and Roberto Tamassia. A user study in similarity measures for graph drawing. In *Proc. of Graph Drawing 2000*, pages 19–30, 2001.
2. Corey Kosak, Joe Marks, and Stuart Shieber. Automating the layout of network diagrams with specified visual organization. *IEEE Transactions on Systems, Man, and Cybernetics*, 24(3):440–454, March 1994.
3. Helen Purchase, Jo-Anne Allder, and David Carrington. User preference of graph layout aesthetics: A uml study. In *Proc. of Graph Drawing 2000*, pages 5–18, 2001.
4. Kathy Ryall, Joe Marks, and Stuart Shieber. An interactive system for drawing graphs. In *Proc. of Graph Drawing 96*, pages 387–393, September 1996.
5. Kathy Ryall, Joe Marks, and Stuart Shieber. An interactive constraint-based sytem for drawing graphs. In *Proc. of UIST 97*, pages 97–104, Banff, Alberta, October 1997.

ViSta*

Rodolfo Castelló, Rym Mili, and Ioannis G. Tollis

[1] Department of Computer Science, Institute of Technology and Higher Education of Monterey, Mexico
[2] Department of Computer Science, The University of Texas at Dallas, Richardson, TX 75083-0688, USA

1 Short Description

ViSta is a tool suite designed to support the requirements specification phase of reactive systems. It includes a template wizard, a graphical editor and a statechart visualization tool. The template wizard guides the user through the steps necessary for the extraction of relevant information from a textual description of requirements. This information is stored in a database that is used by the statechart visualization tool to automatically generate statechart layouts. The statechart visualization tool offers a framework that combines hierarchical drawing, labeling, and floorplanning techniques.

2 Areas of Application

Statecharts are widely used for the requirements specification of reactive systems. This notation captures the requirements attributes that are concerned with the behavioral features of a system, and models these features in terms of a hierarchy of diagrams and states. The usefulness of statecharts depends primarily on their readability, that is the capability of the drawing to convey the meaning quickly and clearly. Several visualization tools for the specification of reactive systems are available in the market [2,3]. Even though these tools are helpful in organizing designers' thoughts, they are mostly sophisticated small scale graphical editors, and therefore are severely inadequate for the modeling of complex reactive systems. Specifically, hand made diagrams quickly become unreadable when the specification complexity and size increase. Therefore computer assistance is of paramount importance for the graphical representation of complex reactive systems.

3 Layout Algorithms and Layout Features

In our approach, a statechart is treated as a graph. Nodes in the graph correspond to states, and arcs correspond to transitions between states. Our algorithmic framework [1] improves the readability of the drawings by focusing

* Research supported in part by Sandia National Labs and by the Texas Advanced Research Program under grant number 009741-040.

P. Mutzel, M. Jünger, and S. Leipert (Eds.): GD 2001, LNCS 2265, pp. 481–482, 2002.
© Springer-Verlag Berlin Heidelberg 2002

on some of the most important aesthetic criteria for graph drawing, namely, *edge-crossing reduction*, *edge-bend reduction*, *labeling*, and *optimization of the drawing area*. It is based on several techniques that include hierarchical drawing, labeling, and floorplanning. The use of layout algorithms implies that even when a minor modification is performed on a drawing (e.g., addition of a node or a transition), the layout algorithms re-orders the nodes in such a way that the aesthetic criteria are met. Hence, the structure of the resulting drawing could be very different from the original one. ViSta offers a feature that can be used to preserve the *mental map* (i.e., the structure) of statecharts. This is important since the designer/specifier may find mistakes/omissions and hence may need to delete and/or insert states and/or transitions without loosing the mental map.

4 Architecture

ViSta (version 2.3) was developed in JAVA 1.2 and uses JAVA's Swing API. It consists of four components: the *template wizard*, the *statechart graphical editor*, the *statechart visualization tool*, and the *database*.
The data stored in the database is summarized in a *decomposition tree*. This structure reflects the decomposition of superstates into substates. A node in the decomposition tree includes the following information: its name; its width and height; the coordinates of its origin point; a pointer to its parent; the list of its children; its decomposition type (e.g., *AND*, *OR* or *leaf*); the list of incoming arcs; the list of outgoing arcs; a list of attributes; and finally its aliases.
The template wizard is used when a textual description of requirements exists, and the user wants ViSta to generate the drawings. The user inputs a textual description by either opening an existing document or creating a new file. Then s/he selects information from the textual document and dynamically introduces it into a set of templates.
The user may decide to directly draw statecharts using the graphical editor. In the graphical editor, statechart elements are described using *item-forms* (e.g., state-form, transition-form). The information is immediately stored in the database, and is used by ViSta to generate (or update) the graphical, template and structured descriptions.

References

1. R. Castelló, R. Mili, and I. G. Tollis. An algorithmic framework for visualizing statecharts. In Joe Marks, editor, *Graph Drawing (Proceedings GD'00)*, pages 139–149. Springer-Verlag, 2001. Lecture Notes in Computer Science 1984.
2. D. Harel et al. Statemate: A working environment for the development of complex reactive systems. *IEEE Transactions on Software Engineering*, 16(4):403–414, May 1990.
3. Artisan Software Tools. Real-time studio: The rational alternative. Available from Artisan Software Tools over the Internet.
http://www.artisansw.com/rtdialogue/pdfs/rational.pdf. Accessed November 2000.

Graphviz – Open Source Graph Drawing Tools

John Ellson[1], Emden Gansner[2], Lefteris Koutsofios[2], Stephen C. North[2], and Gordon Woodhull[2]

[1] Lucent Technologies
[2] AT&T Labs Research

1 Short Description

Graphviz is a heterogeneous collection of graph drawing tools containing batch layout programs (**dot**, **neato**, **fdp**, **twopi**); a platform for incremental layout (**Dynagraph**); customizable graph editors (**dotty**, **Grappa**); a server for including graphs in Web pages (**WebDot**); support for graphs as COM objects (**Montage**); utility programs useful in graph visualization; and libraries for attributed graphs. The software is available under an Open Source license. The article[1] provides a detailed description of the package.

The Graphviz software began with a precursor of **dot** in 1988, followed by **neato** in the early 90's. The features expanded greatly over the years, driven by user request. Graphviz became Open Source in 2000, and was recently distributed on about 500,000 CDROMs as an add-on package for the SUSE Linux release, and is redistributed by Debian, Mandrake, SourceForge, and soon OpenBSD.

2 Areas of Application

Thanks to the variety of components available and its open, "toolkit" design, Graphviz supports a wide variety of applications. The foremost application is probably presentation layouts, such as including graphs in papers. As stream processors, the Graphviz tools can be used as co-processes with interactive components to provide dynamic layouts for debuggers, process monitors, program analysis software, etc. Graphviz tools have been adopted as a visualization service by the W3C Resource Description Framework XML project at MIT, and the Doxygen software engineering system.

3 Layout Algorithms and Layout Features

At present, Graphviz offers 3 batch layout algorithms: hierarchical, symmetric and circular, each allowing extensive parameterization. There is also an incremental hierarchical layout, with plans to provide incremental versions of all layouts. A distinguishing feature of the layouts is their support of a rich graphics model for nodes, and many output formats, such a PostScript, SVG, HPGL, JPEG, etc.

P. Mutzel, M. Jünger, and S. Leipert (Eds.): GD 2001, LNCS 2265, pp. 483–484, 2002.
© Springer-Verlag Berlin Heidelberg 2002

4 Architecture

The Graphviz architecture follows the Unix "toolkit" model, having multiple open layers, including C libraries, scripting language interfaces, stream processors and editors with GUIs. This provides the most flexibility and opportunities for reuse.

4.1 Programming Language and Operating System

The base software uses C, C++ and Java. APIs are also available for perl and tcl. The **dotty** editor is customized using the **lefty** scripting language. Graphviz runs on most versions of Unix and Windows.

5 Interfaces

By its nature, Graphviz has many levels of interfaces, from programming language APIs to customizable editors to command-line tools to servers.

6 Screenshot

The figure below shows **Histograph**, a 100-line C++ program using **Montage** and **Dynagraph**. It provides a clickable nonlinear history display as a browser feature.

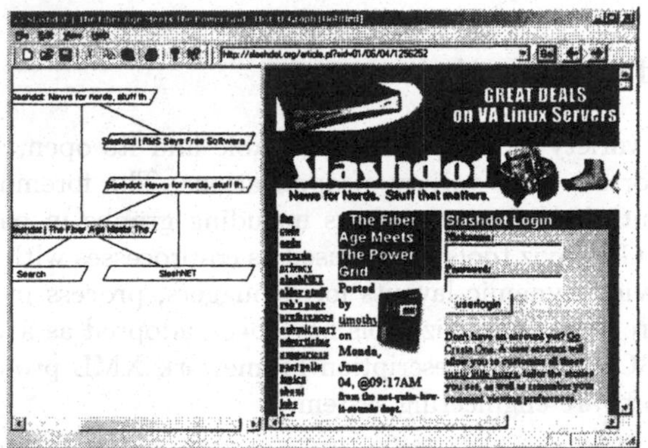

References

1. E.R. Gansner and S.C. North. An open graph visualization system and its applications to software engineering. *Software – Practice and Experience*, 30:1203–1233, 2000.

Exchanging Graphs with GXL

Andreas Winter

Universität Koblenz-Landau, Institut für Softwaretechnik
D-56016 Koblenz, Postfach 201602
winter@uni-koblenz.de
http://www.gupro.de/winter

Abstract. GXL (Graph eXchange Language) is designed to be a standard data exchange format for graph-based tools. GXL is defined as an XML sublanguage, which offers support for exchanging instance graphs together with their appropriate schema information in a uniform format. Formally, GXL is based on typed, attributed, directed, ordered graphs which are extended by concepts to represent hypergraphs and hierarchical graphs. Using this general graph model, GXL offers a versatile support for exchanging nearly all kinds of graphs.

1 Motivation and Background

A great variety of software tools relies on graphs as internal data representation. A standardized language for exchanging those graphs offers a first step in improving *interoperability* between these tools. In software reengineering, for instance, various graph-based tools are used. These include *extractors* (e. g. scanner, parser), *abstractors* (e. g. query tools, structure recognition tools, slicing tools etc.), and *visualizer* (e. g. graph and diagram visualizer, code browser). Currently, these tool components are used more or less independently. [29] gives an overview on existing combinations of tools used in various reengineering projects. Using a common graph interchange format, these tools can be composed to build a genral and powerful reengineering workbench.

The development of *GXL (Graph eXchange Language)* originally started to support data interoperability between reengineering tools. But since GXL was developed as a general format for describing graph structures, it is applicable in further areas of tool interoperability. Especially, GXL is used to support interoperability between graph transformation systems [45] or graph visualization systems. Now, the work on GXL aims at offering a *general exchange format for graph-based tools.*

Exchanging graphs with GXL deals with both, *instance graphs* and their corresponding *graph schemas.* Firstly, GXL offers a versatile support for exchanging all kinds of data based on *typed, attributed, directed, ordered graphs* including *hypergraphs* and *hierarchical graphs.* Secondly, GXL offers means for exchanging graph schemas representing the graph structure i. e. the definition of node and edge classes, their attribute schemas and their incidence structure. Both, instance graphs and graph schemas, are exchanged by XML documents (Extended Markup Language) [47].

P. Mutzel, M. Jünger, and S. Leipert (Eds.): GD 2001, LNCS 2265, pp. 485–500, 2002.
© Springer-Verlag Berlin Heidelberg 2002

After a short survey of the genealogy of GXL in section 2, the major concepts of GXL to exchange instance graphs are introduced in section 3. The language definition of GXL is given by its XML document type definition (DTD) in section 3.4. Section 4 describes the exchange of graph schemas. The current and intended usage of GXL is summarized in section 5.

More information on GXL can be found in [29] and [27]. Up-to-date information including tutorials and further GXL documents are collected at `http://www.gupro.de/GXL`.

2 Genealogy of GXL

GXL originated in a merger of *GRAph eXchange format (GraX)* [10], *Tuple Attribute Language (TA)* [26], and the graph format of the *PROGRES* graph rewriting system [41]. The graph model resulting from this merger was supplemented by additional concepts to handle hierarchical graphs and hypergraphs. Furthermore, GXL includes ideas from common exchange formats used in reengineering, including *ATerms* [46], *Relation Partition Algebra (RPA)* [36], and *Rigi Standard Format (RSF)* [52]. Further features from XML-based exchange of graph transformation systems, developed by groups in Barcelona, Berlin, Budapest, and Kent [22] were included into GXL. The development of GXL was also influenced by various formats used in graph drawing e. g. *daVinci* [13], *GML/Graphlet* [18], *GRL* [33] *XGMML* [53], and *GraphXML* [25]. Thus, GXL covers most of the important graph formats. It can be viewed as a generalization of these formats. The genealogy of GXL is depicted in figure 2.

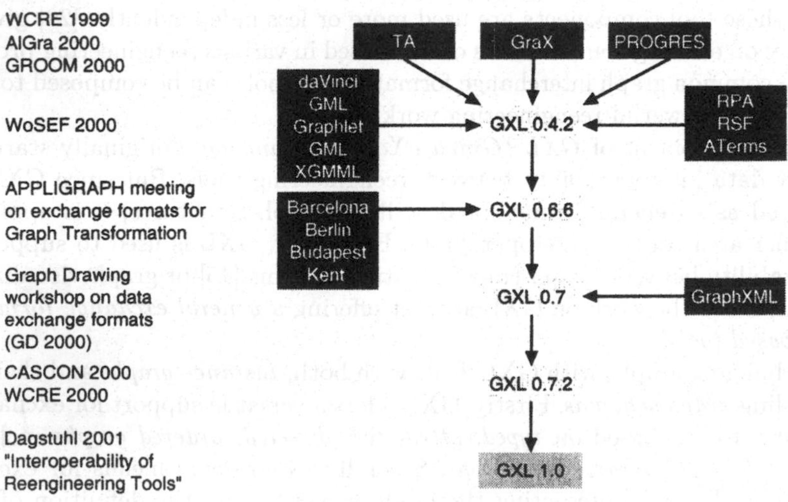

Fig. 1. Genealogy of GXL

The development of GXL was advanced during various conferences and workshops since 1998. First efforts on defining a general exchange format for reengineering data were made at WCRE 1998 [49] and at CASCON 1998 [5]. Approaches for graph-based exchange formats were discussed during meetings at WCRE 1999 [50], AlGra 2000 [1], and GROOM 2000 [20]. These discussions resulted in the first version of GXL, which was presented at the ICSE 2000 workshop on standard exchange formats (WoSEF 2000) [43]. Subsequent versions were discussed and compared to similar approaches from related areas at the APPLIGRAPH meeting for exchange formats for graph transformation systems [22] and the Graph Drawing 2000 workshop on exchange formats [16]. Improvements of these versions were presented in CASCON 2000 tutorials [28] and workshops [44] and during the WCRE 2000 exchange formats workshop [31].

GXL (version 1.0) was ratified as *standard exchange format* in software reengineering at the Dagstuhl Seminar "Interoperability of Reengineering Tools" in January 2001 [7]. Current work deals with gathering experiences with GXL version 1.0 and providing tool support for working with GXL.

3 Exchanging Graphs

Due to their mathematical foundation and algorithmic power, graphs are a common data structure in software engineering. Different graph models e. g. directed graphs, undirected graphs, node attributed graphs, edge attributed graphs, node typed graphs, edge typed graphs, ordered graphs, relational graphs, acyclic graphs, trees, etc. or combinations of these graph models are utilized in many software systems. To support interoperability of graph-based tools, the underlying graph model has to be as rich as possible to cover most of these graph models.

Such a common graph model is given by *typed, attributed, directed, ordered graphs (TGraphs)* [9], [10]. TGraphs are *directed* graphs, whose nodes and edges may be *attributed* and *typed*. Each type can be assigned an individual attribute schema specifying the possible attributes of nodes and edges. Furthermore, TGraphs are *ordered*, i. e. the node set, the edge set, and the sets of edges incident to a node have a total ordering. This ordering gives modeling power to describe sequences of objects (e. g. parameter lists) and facilitates the implementation of deterministic graph algorithms. In applying TGraphs to the other graph models, not all properties of TGraphs have to be used to their full extent. These graph models can be viewed as specializations of TGraphs. Exchanging typed, attributed, directed, ordered graphs or their specializations with GXL is introduced in section 3.1

To offer support for *hypergraphs* and *hierarchical graphs*, TGraphs were extended by n-ary edges and by nodes and edges containing lower level graphs. GXL language constructs for exchanging hypergraphs and hierarchical graphs are sketched in section 3.2 and 3.3. The complete GXL language definition is given in section 3.4 in terms of a XML document type definition.

3.1 Exchanging Typed, Attributed, Directed, Ordered Graphs

The object diagram (cf. [40]) in figure 2 shows a node and edge typed, node and edge attributed, directed, ordered graph representing a program fragment on ASG (abstract syntax graph) level. Function *main* calls function $a = max(a, b)$ in line *8* and function $b = min(b, a)$ in line *19*. The functions *main*, *max* and *min* are represented by nodes of type *Function*. These nodes are attributed with the function name. *FunctionCall* nodes represent the calls of functions *max* and *min*. They are associated to the caller by *isCaller* edges and to the callee by *isCallee* edges. *isCaller* edges are attributed with a line attribute showing the line number which contains the call. Input parameters (represented by *Variable* nodes that are attributed with the variable name) are associated by *isInput* edges. The ordering of parameter lists is given by ordering the incidences of *isInput* edges pointing to *FunctionCall* nodes. The first edge of type *isInput* incident to function call *v2* (modeling the call of *max(a,b)*) comes from node *v6* representing variable *a*. The second edge of type *isInput* connects to the second parameter *b* (node *v7*). The incidences of *isInput* edges associated with node *v3* model the reversed parameter order. Output parameters are associated to their function calls by *isOutput* edges.

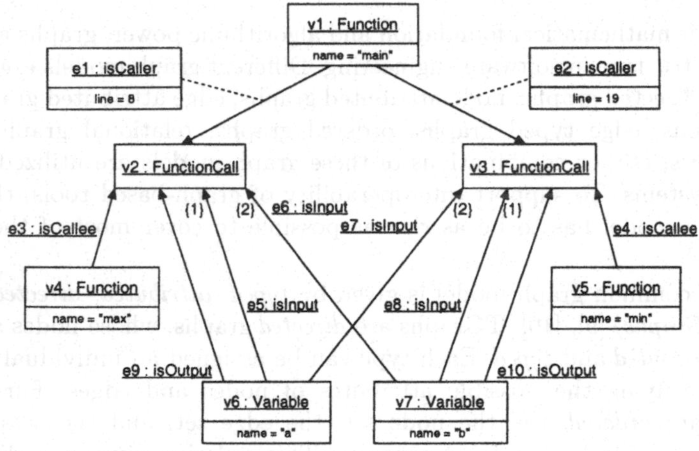

Fig. 2. typed, attributed, directed, ordered graph

Exchanging graphs like the one in figure 2 requires language constructs for representing nodes, edges and their incidence relation. Furthermore, support for describing type information, attribute values, and ordering information is needed.

Figure 3 depicts the graph from figure 2 as GXL document. The complete grammar for these documents is given in section 3.4. XML documents start with specifying the XML version and the underlying document type definition, here "gxl.dtd". The body of a GXL document is enclosed in <**gxl**> tags. The GXL

```
<?xml version = "1.0" ?>
<!DOCTYPE gxl SYSTEM "gxl.dtd" >
<gxl>
<graph id = "simpleGraph"
              edgeids = "true">
   <type xlink:href = "schema.gxl"/>
   <node id = "v1" >
      <type xlink:href =
      "schema.gxl#Function"/>
      <attr name = "name" >
         <string>main</string>
      </attr>
   <node id = "v2" >
      <type xlink:href =
      "schema.gxl#FunctionCall"/>
   </node>
   <node id = "v3" >
      <type xlink:href =
      "schema.gxl#FunctionCall"/>
   </node>
   <node id = "v4" >
      <type xlink:href =
      "schema.gxl#Function"/>
      <attr name = "name" >
         <string>max</string>
      </attr>
   </node>
   <node id = "v5" >
      <type xlink:href =
      "schema.gxl#Function"/>
      <attr name = "name" >
         <string>min</string>
      </attr>
   </node>
   <node id = "v6" >
      <type xlink:href =
      "schema.gxl#Variable"/>
      <attr name = "name" >
         <string>a</string>
      </attr>
   </node>
   <node id = "v7" >
      <type xlink:href =
      "schema.gxl#Variable"/>
      <attr name = "name" >
         <string>b</string>
      </attr>
   </node>
   <edge id = "e1"
      from = "v2" to = v1"/>
      <type xlink:href =
      "schema.gxl#isCaller"/>
      <attr name = "line" >
         <int>8</int>
      </attr>
   </edge>
   <edge id = "e2"
      from = "v3" to = v1"/>
      <type xlink:href =
      "schema.gxl#isCaller"/>
      <attr name = "line" >
         <int>19</int>
      </attr>
   </edge>
   <edge id = "e3"
      from = "v4" to = v2"/>
      <type xlink:href =
      "schema.gxl#isCallee"/>
   </edge>
   <edge id = "e9"
      from = "v6" to = v2"
      <type xlink:href =
      "schema.gxl#isOutput"/>
   </edge>
   <edge id = "e5"
      from = "v6" to = v2"
      toorder = "1"/>
      <type xlink:href =
      "schema.gxl#isInput"/>
   </edge>
   <edge id = "e6"
      from = "v7" to = v2"
      toorder = "2"/>
      <type xlink:href =
      "schema.gxl#isInput"/>
   </edge>
   <edge id = "e7"
      from = "v6" to = v3"
      toorder = "2"/>
      <type xlink:href =
      "schema .gxl#isInput"/>
   </edge>
   <edge id = "e8"
      from = "v7" to = v3"
      toorder = "1"/>
      <type xlink:href =
      "schema.gxl#isInput"/>
   </edge>
   <edge id = "e9"
      from = "v6" to = v2"
      <type xlink:href =
      "schema.gxl#isOutput"/>
   </edge>
   <edge id = "e10"
      from = "v7" to = v3"
      <type xlink:href =
      "schema.gxl#isOutput"/>
   </edge>
</graph>
</gxl>
```

Fig. 3. GXL representation of graph from figure 2

document in figure 3 contains one graph, enclosed in <**graph**> tags, with an unique identifier "simpleGraph". The graph refers to its associated graph schema (cf. section 4) stored in file schema.gxl. GXL supports both, graphs with edges having a unique object identifier, and graphs with unnamed edges. The attribute edgeids = "true" indicates uniquely named edges.

Nodes and edges of a given graph are exchanged by <**node**> and <**edge**> elements which can be addressed by their identifier attribute. Incidence information of edges including edge orientation is stored in from and to attributes within <**edge**> tags. Ordering of incidences is also modeled here. Attributes fromorder and toorder represent the position of an edge in the incidence list of its start and target node. Node and edge types are represented by links pointing to the appropriate schema information. This link is enclosed in <**type**> tags.

<**node**> and <**edge**> elements may additionally contain further attribute information. <**attr**> elements describe attribute name and value. Like OCL [48], GXL provides <**bool**>, <**int**>, <**float**>, and <**string**> attributes. Furthermore, enumeration values (<**enum**>) and URI-references (<**locator**>) pointing to ex-

ternally stored objects are supported. Attribute values might be sub structured. Here, GXL offers composite attributes like sequences (<**seq**>), sets (<**set**>), multi sets (<**bag**>), and tuples (<**tup**>).

3.2 Exchanging Hypergraphs

In addition to graphs, GXL provides the exchange of *hypergraphs*. *Hypergraphs* [3] are graphs with n-ary edges (hyperedges) with arbitrary n. Hyperedges represent n-ary relations. GXL provides the exchange of typed, attributed, directed and undirected, ordered hypergraphs.

Figure 4 shows a hypergraph in UML notation, modeling the function call $a = max(a, b)$ by a 5-ary hyperedge of type *FunctionCall2*. The diamond, representing the hyperedge, is connected by undirected lines (tentacles) to its related *Function*- and *Variable*-nodes. These tentacles are marked with roles, identifying *caller*, *callee*, *input*, and *output*. Numbers describing the order of incidences of tentacles according the hyperedge, indicate the ordering of parameters. Like edge *e1* in figure 2, the hyperedge is attributed with a *line* attribute.

The GXL representation of this hyperedge is given in figure 5. Hyperedges are represented by <**rel**> elements (relation). Like <**node**> and <**edge**> elements, <**rel**> elements can contain type (<**type**>) and attribute (<**attr**>) information. Tentacles, which point to the related graph objects (**target**), are represented by <**relend**> subelements (relation end). Roles of tentacles are stored in **role** attributes. Incidences according to the hyperedge are exchanged by **startorder** attributes. The ordering of tentacles according their target objects can be modeled by **endorder** attributes. Directed or undirected hyperedges and tentacles are distinguished by attributes **isdirected** and **direction** (cf. the GXL DTD in section 3.4).

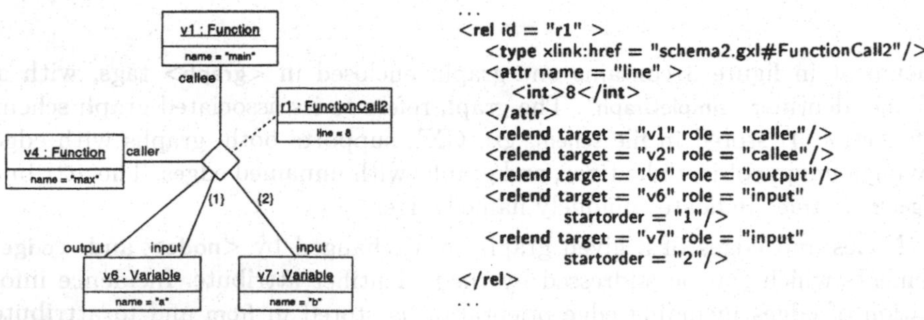

Fig. 4. Hypergraph

Fig. 5. GXL representation

Edges can be viewed as 2-ary hyperedges. Thus, in GXL, edge information can be represented by *binary hyperedges*. Since graphs with (binary) edges are widespread in software engineering and most applications deal with graphs instead of hypergraphs, GXL offers both, the element <**rel**> for hyperedges, and, as a shortcut for binary hyperedges, <**edge**> elements.

3.3 Exchanging Hierarchical Graphs

Graphs gather their popularity from their mathematical foundation and their visual capabilities to express complex contexts. However, due to their size, large graphs become bulky and difficult to understand. This complexity can be reduced by structuring graphs. Parts of graphs representing related objects can be grouped together to form encapsulated, higher level structures. *Hierarchical graphs* [4] support structuring graphs by grouping and encapsulation.

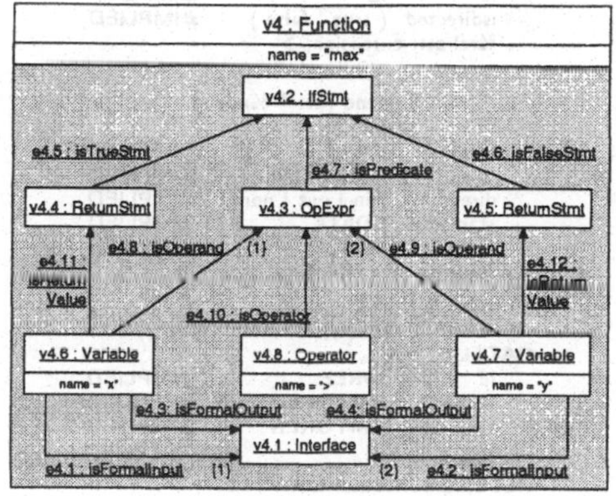

Fig. 6. Hierarchical Graph

```
...
<node id = "v4" >
  <type xlink:href = "schema.gxl#Function" />
  <attr name = "name" >
    <string>max</string>
  </attr>
  <graph id = "max" >
    <type xlink:href = "asg.gxl" >
    <node id = "v4.1" >
      <type xlink:href = "asg.gxl#Interface" >
    </node>
    ...
    <edge id = "e4.12"
      from = "v4.7" to = v4.5"/>
      <type xlink:href =
        "asg.gxl#isReturnValue"/>
    </edge>
  </graph>
</node>
...
```

Fig. 7. GXL representation

Figure 6 depicts a hierarchical graph. Node *v4* (cf. figure 2) contains a subgraph of type *asg* (abstract syntax graph) representing the implementation of function *max*. The GXL representation in figure 7 shows this subgraph as **<graph>** subelement of node *v4*. Subgraphs associated to edges or hyperedges are exchanged analogously (cf. the GXL DTD in section 3.4).

This representation for hierarchical graphs works for those hierarchical graphs with strong ownership for each graph object. This representation also permits edges and hyperedges crossing the boundaries of graph hierarchies. Those edges are contained in the least-common-ancestor-graph.

Since no general model for hierarchical graphs exists so far (cf. [4]), GXL provides further support for exchanging graph hierarchies. Alternatively, references to subgraphs and their elements can be exchanged using **<locator>** attributes pointing to their appropriate GXL representation. Further support might be offered in a next GXL version by graph-valued attributes or by special edges, representing hierarchy.

3.4 GXL Document Type Definition

The language features of GXL for exchanging typed, attributed, directed, ordered graphs, hypergraphs, and hierarchical graphs are summarized in a concep-

```
<!– extensions –>
<!ENTITY % gxl-extension         "" >
<!ENTITY % graph-extension       "" >
<!ENTITY % node-extension        "" >
<!ENTITY % edge-extension        "" >
<!ENTITY % rel-extension         "" >
<!ENTITY % value-extension       "" >
<!ENTITY % relend-extension      "" >
<!ENTITY % gxl-attr-extension    "" >
<!ENTITY % graph-attr-extension  "" >
<!ENTITY % node-attr-extension   "" >
<!ENTITY % edge-attr-extension   "" >
<!ENTITY % rel-attr-extension    "" >
<!ENTITY % relend-attr-extension "" >

<!– attribute values –>
<!ENTITY % val " locator | bool | int | float | string |
                 enum | seq | set | bag | tup
                 % value-extension;" >

<!– gxl –>
<!ELEMENT gxl (graph* %gxl-extension;) >
<!ATTLIST gxl
    xmlns:xlink CDATA          #FIXED
                               "www.w3.org/1999/xlink"
    %gxl-attr-extension; >

<!– type –>
<!ELEMENT type EMPTY>
<!ATTLIST type
    xlink:type  (simple)       #FIXED "simple"
    xlink:href  CDATA          #REQUIRED >

<!– graph –>
<!ELEMENT graph (type? , attr* ,
                 ( node | edge | rel )*
                 %graph-extension;) >
<!ATTLIST graph
    id          ID             #REQUIRED
    role        NMTOKEN        #IMPLIED
    edgeids     ( true | false ) "false"
    hypergraph  ( true | false ) "false"
    edgemode    ( directed | undirected |
                defaultdirected | defaultundirected)
                "directed"
    %graph-attr-extension; >

<!– node –>
<!ELEMENT node (type? , attr*, graph*
                %node-extension;) >
<!ATTLIST node
    id          ID             #REQUIRED
    %node-attr-extension; >

<!– edge –>
<!ELEMENT edge (type?, attr*, graph*
                %edge-extension;) >
<!ATTLIST edge
    id          ID             #IMPLIED
    from        IDREF          #REQUIRED
    to          IDREF          #REQUIRED
    fromorder   CDATA          #IMPLIED
    toorder     CDATA          #IMPLIED
    isdirected  ( true | false ) #IMPLIED
    %edge-attr-extension; >

<!– rel –>
<!ELEMENT rel (type? , attr*, graph*, relend*
               %rel-extension;) >
<!ATTLIST rel
    id          ID             #IMPLIED
    isdirected  ( true | false ) #IMPLIED
    %rel-attr-extension; >

<!– relend –>
<!ELEMENT relend (attr* %relend-extension;) >
<!ATTLIST relend
    target      IDREF          #REQUIRED
    role        NMTOKEN        #IMPLIED
    direction   ( in | out | none) #IMPLIED
    startorder  CDATA          #IMPLIED
    endorder    CDATA          #IMPLIED
    %relend-attr-extension; >

<!– attr –>
<!ELEMENT attr (type?, attr*, (%val;)) >
<!ATTLIST attr
    id          IDREF          #IMPLIED
    name        NMTOKEN        #REQUIRED
    kind        NMTOKEN        #IMPLIED >

<!– locator –>
<!ELEMENT locator EMPTY >
<!ATTLIST locator
    xlink:type  (simple)       #FIXED "simple"
    xlink:href  CDATA          #IMPLIED >

<!– attribute values –>
<!ELEMENT bool (#PCDATA) >
<!ELEMENT int (#PCDATA) >
<!ELEMENT float (#PCDATA) >
<!ELEMENT string (#PCDATA) >
<!ELEMENT enum (#PCDATA) >
<!ELEMENT seq (%val;)* >
<!ELEMENT set (%val;)* >
<!ELEMENT bag (%val;)* >
<!ELEMENT tup (%val;)* >
```

Fig. 8. GXL Document Type Definition

tual model defining the graph model supported by GXL. The GXL graph model is completely described at `http://www.gupro.de/GXL/` (graph model) with its graph structure part and its attribute part.

Since GXL is a XML sublanguage, the GXL graph model had to be transcribed into a XML document type definition (DTD) or an appropriate XML schema definition. To keep GXL simple and less verbose, this translation was done manually. The resulting DTD (cf. figure 8, a commented version is given at `http://www.gupro.de/GXL` (DTD)) requires only 18 XML elements. In contrast, an appropriate DTD generated with IBMs XMI Toolkit [30] according the XML Metadata Interchange (XMI) principles for developing DTDs [35, section 3] requires 66 elements for the GXL core and and additional 63 elements for XMI and Corba related aspects.

4 Exchanging Graph Schemas

Graphs only offer a plain structured means for describing objects (nodes) and their interrelationship (edges, hyperedges). Graphs have no meaning of their own. The meaning of graphs corresponds to the context in which they are used and exchanged. The application and interchange context determines

- which node, edge, and hyperedge classes are used,
- which relations exist between nodes, edges, and hyperedges of given classes,
- which attribute structures are associated to nodes, edges, and hyperedges,
- which graph hierarchies are supported, and
- which additional constraints (like ordering of incidences, degree-restrictions etc.) have to be complied.

This schematic data can be described by *conceptual modeling techniques.* Class diagrams offer a suited declarative language to define graph classes with respect to a given application or interchange context [10].

4.1 Describing Graph Classes by Class Diagrams

In GXL graph classes are defined by UML class diagrams [40]. Figure 9 shows a graph schema defining classes of graphs like the one given in figure 2. Node classes (*FunctionCall*, *Function*, and *Variable*) are defined by classes. Edge classes (*isCallee*, *isInput*, and *isOutput*) are defined by associations. Attributed edge classes (*isCaller*) are described by associated classes. Like classes, they contain the associated attribute structures. The orientation of edges is depicted by a filled triangle (cf. [40, p. 155]. Since most of the available UML tools do not offer this UML construct, directed arrows (depicting visibility in original UML) can be used alternatively (cf. figure 13). Multiplicities denote degree restrictions. Ordering of incidences is indicated by the keyword {ordered}.

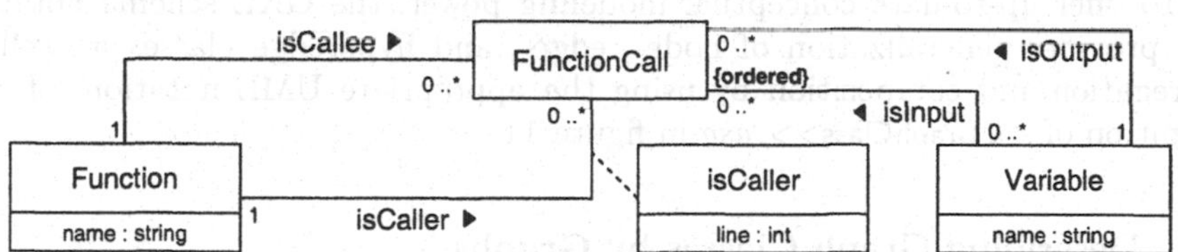

Fig. 9. Graph - Schema

In a similar way, UML class diagrams offer language constructs to specify classes of hypergraphs and hierarchical graphs. Figure 10 shows a class diagram defining hypergraphs like the one in figure 4. Classes of hyperedges are defined by n-ary associations depicted by a diamond. This diamond is connected by links to the related node classes. These links can be annotated by multiplicity information to demand cardinalities, and by names indicating the role of

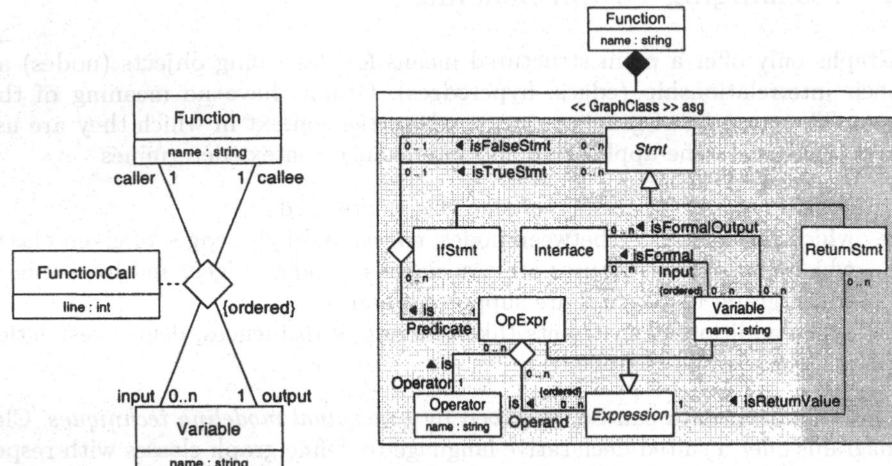

Fig. 10. Hypergraph-Schema **Fig. 11.** Hierarchical Graph-Schema

participating classes. The keyword {ordered} demands ordering of tentacles in appropriate instance graphs.

The definition of hierarchical graphs in UML requires an additional language construct representing graph classes themselves. The stereotype <<GraphClass>> distinguishes classes representing graph classes from classes defining node classes. <<GraphClasses>> compose classes and associations to define the graph schema on the next level of hierarchy. UML provides a nested notation to represent these subschemas within the <<GraphClass>> classes. Strong ownership of subgraphs to graph elements is expressed by composition (filled diamond). Figure 11 defines a graph class for hierarchical graphs like the one, depicted in figure 6. Nodes of Class Function contain graphs of graph class *asg*.

To offer up-to-date conceptual modeling power, the GXL schema notation also provides generalization of node-, edge-, and hyperedge classes as well as aggregation and composition by using the appropriate UML notation (cf. the definition of <<GraphClass>> *asg* in figure 11.

4.2 Describing Graph Classes by Graphs

Since UML class diagrams are structured information themselves, they may be represented as graphs as well. For exchanging graph schemas in GXL, UML class diagrams are transfered into equivalent graph representations. Thus, instance graphs and schemas are exchanged with the *same type of document*, i.e. XML documents matching the GXL DTD (cf. section 3.4).

Figure 12 depicts the transformation of the class diagram in figure 9 into a node and edge typed, node and edge attributed, directed graph. Node-, edge- and hyperedge-classes, attributes and their domains are modeled by nodes of suitable node types. Their attributes describe further properties. Interrelationships between surrogates of these classes are represented by edges of proper types. Attribute information is associated with surrogates of node, edge, and hyperedge classes and associations by *hasAttribute* and *hasDomain* edges. *from* and *to* edges model incidences of associations including their orientation. Multiplicities of associations are stored in limits-attributes. The boolean attribute isOrdered indicates ordered incidences.

GXL documents, representing instance graphs to a given graph schema, refer to those nodes of the equivalent schema graph representing node classes (*NodeClass*), edge classes (*EdgeClass*) and hyperedge classes (*RelClass*). The nodes representing these class definitions in figure 12 which are referred by the graph in figure 2 are shaded.

Class diagrams defining hypergraphs or hierarchical graphs can be transformed into graphs analogously. Each class diagram defining a GXL graph schema can be transformed into a graph (schema graph) matching a suited schema, representing GXL schema graphs. Schema graphs are instances of the *GXL metaschema* (cf. section 4.3). They are exchanged like all instance graphs (cf. section 3) referring to a GXL schema graph. Since the schema graph, representing the GXL metaschema, is an instance of itself, it is exchanged by a self referring GXL document.

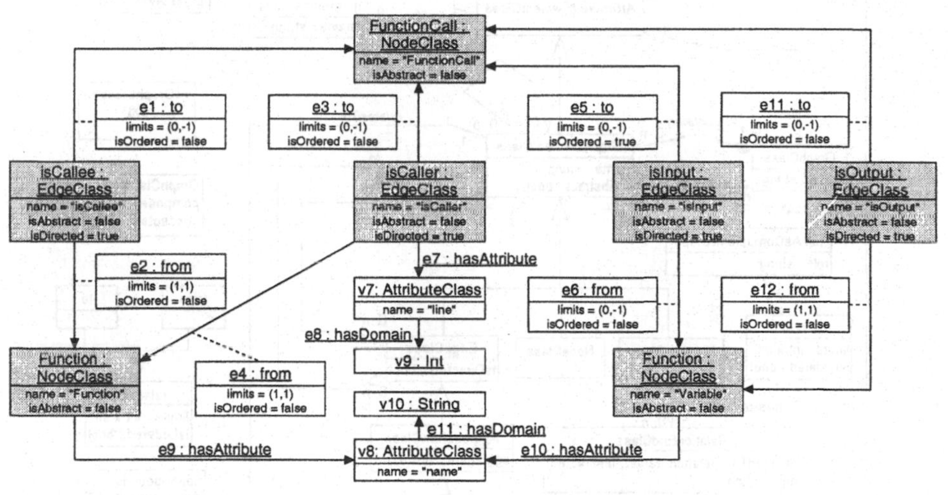

Fig. 12. Graph - Schema (schema graph)

In contrast to the strategy proposed by XML Meta Data Interchange (XMI) [35], GXL schemas are *not* exchanged by XML documents according to the Meta Object Facility (MOF) [34]. XMI/MOF offers a general, but very verbose format for exchanging UML class diagrams as XML streams. Next to its exaggerated verbosity, which contradicts the requirement for exchange formats of as compact as possible documents, the XMI/MOF approach requires *different types of documents* for representing schema and instance graphs. Especially in applications dealing with schema information on instance level (e. g. in tools for editing and analyzing schemas), this leads to the disadvantage of different documents representing the same information, one on instance level (as XML document) and one on schema level (as XML DTD). The GXL approach treats schema and instance information in exactly the same way. Schema and instance graphs are exchanged according the same DTD given in figure 8.

4.3 GXL Metaschema

The GXL metaschema defines the set of graphs representing correct GXL schema graphs. The class diagram in figure 13 shows the graph part of the GXL metaschema. This graph class provides constructs to define classes of graph elements i. e. nodes (NodeClass), edges(EdgeClass), and hyperedges (RelClass) including their interrelationships. Their attributes distinguish abstract classes, classes of directed or undirected edges or hyperedges or contain multiplicity constraints etc.

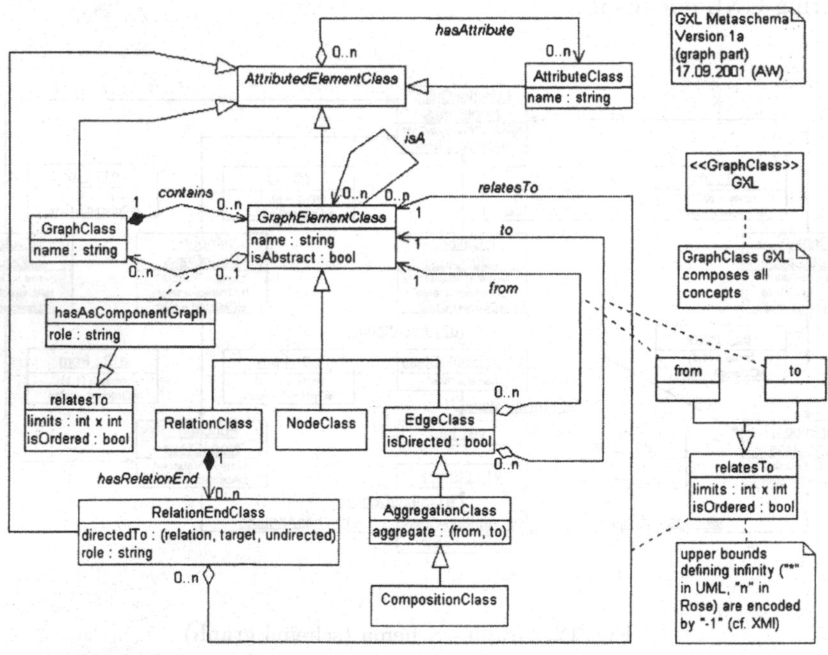

Fig. 13. GXL Metaschema (graph part)

To express associated attribute structures GraphElementClasses can be connected to attribute structures (AttributedElementClass). The definition of attribute structures supports the structured attributes used in GXL including the definition of default values (cf. the complete GXL metaschema, including its attribute and value parts, at `http://www.gupro.de/GXL/` (meta schema)). Generalization is provided for all GraphElementClasses by isA edges. GraphElementClasses containing lower level graphs are associated to the representation of the lower level GraphClass representation by contains edges. The GraphClass contains those node-, edge-, and hyperedge classes representing its structure. Aggregation (AggregationClass) and composition (CompositonClass) are modeled by specializations of EdgeClasses.

5 Using GXL

At the Dagstuhl seminar on "Interoperability of Reengineering Tools" GXL version 1.0 was ratified as the *standard exchange format* in software reengineering [7]. Currently, various groups in software (re)engineering are implementing GXL import and export facilities to their tools (e.g. Bauhaus [2], Columbus [11], CPPX [6], Fujaba [14], GUPRO [23], PBS [37], RPA (Philips Research), PROGRES [38], Rigi [39], Shrimp [42]) Others are going to implement tools to support working with GXL. For instance, a framework for GXL Converters [15] and a XMI2GXL translator [54] are being developed at Univ. BW München. Further activities deal with providing graph query machines (GReQL, Univ. Koblenz) to GXL graphs or GXL-based graph databases (Univ. Aachen).

An important feature of GXL is its support for exchanging schema information. Based on this capability, reference schemas for certain standard applications in reengineering are currently under development. These activities address reference schemas for data reverse engineering (DRE, Univ. Namur, Paderborn, Victoria), the Dagstuhl Middle Model [32] or abstract syntax graph models for C++ [6], [12].

Furthermore, groups developing graph transformation tools (e.g. GenSet [17], PROGRES [38]) or graph visualization tools (e.g. GVF [24], Shrimp [42], yFiles [55]) already use GXL or pronounced to use GXL. At University of Toronto, GXL is applied within an undergraduate software engineering course to create a graph editor/layouter [8].

GXL also serves as foundation to define further graph oriented exchange formats. Thus, GXL defines the graph part in the exchange format GTXL (Graph Transformation eXchange Language) [21], [45]. Activities in the graph drawing community also deal with the development of an exchange format for graphs including layout information [16]. There is evidence of combining the structure part of GXL with the graph layout part and the modularization part of GraphML [19] to form a general and comprehensive graph exchange format.

6 Conclusion

The previous sections gave a short introduction in the GXL Graph eXchange Language version 1.0 and its current application.

Summarizing, GXL offers an already widely used XML sublanguage for interchanging typed, attributed, directed, ordered graphs including hypergraphs and hierarchical graphs together with their appropriate schemas. By focusing on graph structure, GXL contributes the core for defining a family of special suited graph exchange formats.

Acknowledgment. I would like to thank the GXL co-authors Richard C. Holt, Andy Schürr, and Susan Elliott Sim for various fruitful discussions on the development of GXL, Jürgen Ebert, Bernt Kullbach, and Volker Riediger for many interesting discussions on TGraphs and GXL, and Kevin Hirschmann for implementing the GUPRO related GXL tools. Thanks to all users of GXL, who currently applying and testing GXL 1.0 in their tools. Their experience will be an important aid to improve GXL.

References

1. Workshop on Algebraic and Graph-Theoretic Approaches in Software Reengineering, Koblenz, February 28, 2000. `http://www.uni-koblenz.de/~winter/AlGra/algra.html` (14.9.2001).
2. Bauhaus: Software Architecture, Software Reengineering, Program Understanding. `http://www.informatik.uni-stuttgart.de/ifi/ps/bauhaus/` (1.9.2001).
3. C. Berge. *Graphs and Hypergraphs*. North-Holland, Amsterdam, 2 edition, 1976.
4. G. Busatto. An Abstract Model of Hierarchical Graphs and Hierarchical Graph Transformation (current draft). `http://www.informatik.uni-bremen.de/~giorgio/papers/phd-thesis.ps.gz` (16.9.2001).
5. Data Exchange Group, Conclusions from Meeting at CASCON 1998,30. Nov 1998. `http://plg.uwaterloo.ca/~holt/sw.eng/exch.format` (14.9.2001).
6. CPPX: Open Source C++ Fact Extractor. `http://swag.uwaterloo.ca/~cppx/` (1.9.2001).
7. J. Ebert, K. Kontogiannis, J. Mylopoulos: Interoperability of Reengineering Tools. `http://www.dagstuhl.de/DATA/Reports/01041/` (18.4.2001), 2001.
8. S. Easterbrook. CSC444F: Software Engineering I (Fall term 2001), University of Toronto. `http://www.cs.toronto.edu/~sme/CSC444F/` (15.9.2001), 2001.
9. J. Ebert and A. Franzke. A Declarative Approach to Graph Based Modeling. In E. Mayr, G. Schmidt, and G. Tinhofer, editors. *Graphtheoretic Concepts in Computer Science, LNCS 903*. Springer, Berlin, pages 38–50. 1995.
10. J. Ebert, B. Kullbach, and A. Winter. GraX – An Interchange Format for Reengineering Tools. In *[50]*, pages 89–98. 1999.
11. R. Ferenc, F. Magyar, Á. Beszédes, Á. Kiss, and M. Tarkiainen. Columbus - Tool for Reverse Engineering Large Object Oriented Software Systems. In *Proceedings SPLST 2001, Szeged, Hungary (*http://www.inf.u-szeged.hu/~ferenc/research/ferencr_columbus.pdf, *(1.9.2001))*, pages 16–27. June 2001.
12. R. Ferenc, S. Elliott Sim, R. C. Holt, R. Koschke, and T. Gyimòthy. Towards a Standard Schema for C/C++. To appear in *8th Working Conference on Reverse Engineering*. IEEE Computer Soc., 2001.
13. M. Fröhlich and M. Werner. daVinci V2.0.x Online Documentation. `http://www.tzi.de/~davinci/docs/` (18.4.2001), June 1996.
14. Fujaba: From UML to Java and back again. `http://www.uni-paderborn.de/cs/fujaba/` (1.9.2001).

15. GCF - a GXL Converter Framework.
http://www2.informatik.unibw-muenchen.de/GXL/triebsees/ (1.9.2001).
16. Workshop on Data Exchange Formats, Graph Drawing 2000.
http://www.cs.virginia.edu/~gd2000/gd-satellite.html (14.9.2001), 2001.
17. GenSet: Design Information Fusion.
http://www.cs.uoregon.edu/research/perpetual/dasada/Software/GenSet/ (1.9.2001).
18. The GML File Format.
http://www.infosun.fmi.uni-passau.de/Graphlet/GML/ (18.4.2001).
19. The GraphML File Format. **http://www.graphdrawing.org/graphml/** (31.8.2001), 2001.
20. 7-ter Workshop des GI-Arbeitskreises GROOM, UML - Erweiterungen und Konzepte der Metamodellierung, 4.-5. April 2000, Universität Koblenz-Landau. **http://www2.informatik.unibw-muenchen.de/GROOM/META/** (14.9.2001).
21. Graph Transformation System Exchange Language.
http://tfs.cs.tu-berlin.de/projekte/gxl-gtxl.html (18.08.2001).
22. First APPLIGRAPH meeting on GXL (graph exchange language) and GTXL (graph transformation exchange language) Paderborn (September 5-6, 2000). **http://tfs.cs.tu-berlin.de/projekte/gxl-gtxl/paderborn.html** (11.9.2001).
23. GUPRO: Generic Understanding of Programs. **http://www.gupro.de/** (1.9.2001).
24. GVF - Graph Visualization Framework. **http://www.cwi.nl/InfoVisu** (1.9.2001).
25. I. Herman and M. S. Marshall. Graph XML – An XML based graph interchange format. Report INS-0009, CWI, Amsterdam, April 2000.
26. R. C. Holt. An Introduction to TA: The Tuple-Attribute Language.
http://plg.uwaterloo.ca/~holt/papers/ta.html (18.4.2001), 1997.
27. R. C. Holt and A. Winter. A Short Introduction to the GXL Software Exchange Format. In *[51]*, pages 299–301. 2000.
28. R. C. Holt and A. Winter. Software Data Interchange with GXL: Introduction and Tutorial, CASCON 2000.
http://www.cas.ibm.com/archives/2000/workshops/descriptions.shtml\#16 (15.9.2001), November 13-16, 2000.
29. R. C. Holt, A. Winter, and A. Schürr. GXL: Toward a Standard Exchange Format. In *[51]*, pages 162–171. 2000.
30. XMI Toolkit 1.15. **http://alphaworks.ibm.com/tech/xmitoolkit** (1.9.2001).
31. K. Kontogiannis. Exchange Formats Workshop. In *[51]*, pages 277–301. 2000.
32. T. Lethbridge, E. Plödereder, S. Tichelar, C. Riva, and P. Linos. The Dagstuhl Middle Level Model (DMM). internal note, 2001.
33. F. Newbery Paulish. *The Design of an Extendible Graph Editor, LNCS 704*. Springer, Berlin, 1991.
34. Meta Object Facility (MOF) Specification.
http://www.omg.org/technology/documents/formal/mof.htm (2.9.2001), March 2000.
35. XML Meta Data Interchange (XMI) Specification.
http://www.omg.org/technology/documents/formal/xmi.htm (1.9.2001), November 2000.
36. R. Ommering, L. van Feijs, and R. Krikhaar. A relational approach to support software architecture analysis. *Software Practice and Experience*, 28(4), pages 371–400, April 1998.
37. PBS: The Portable Bookshelf. **http://swag.uwaterloo.ca/pbs/** (1.9.2001).

38. A Graph Grammar Programming Environment - PROGRES. `http://www-i3.informatik.rwth-aachen.de/research/projects/progres/` (1.9.2001).
39. RIGI: a visual tool for understanding legacy systems. `http://www.rigi.csc.uvic.ca/` (1.9.2001).
40. J. Rumbaugh, I. Jacobson, and G. Booch. *The Unified Modeling Language Reference Manual.* Addison Wesley, Reading, 1999.
41. A. Schürr, A. J. Winter, and A. Zündorf. PROGRES: Language and Environment. In H. Ehrig, G. Engels, H.-J. Kreowski, and G. Rozenberg, editors. *Handbook on Graph Grammars*, volume 2. World Scientific, Singapore, pages 487–550. 1999.
42. ShriMP Views: simple Hierarchical Multi-Perspective. `http://www.shrimpviews.com/` (1.9.2001).
43. S. Elliot Sim, R. C. Holt, and R. Koschke. Proceedings ICSE 2000 Workshop on Standard Exchange Format (WoSEF). Technical report, Limerick, 2000.
44. S. Elliott Sim. Software Data Interchange with GXL: Implementation Issues, CASCON 2000. `http://www.cas.ibm.com/archives/2000/workshops/descriptions.shtml\#17` (14.9.2001), November 13-16, 2000.
45. G. Taenzer. Towards Common Exchange Formats for Graphs and Graph Transformation Systems. In *Proceedings UNIGRA satellite workshop of ETAPS'01.* 2001.
46. M. van den Brand, H. A. de Jong, P. Klint, and P. A. Olivier. Efficient annotated Terms. *Software: Practice and Experience*, 30(3), pages 259–291, March 2000.
47. Extensible Markup Language (XML) 1.0. W3c recommendation, W3C XML Working Group, `http://www.w3.org/XML/` (17.4.2001), February 1998.
48. J. B. Warmer and A. G. Kleppe. *The Object Constraint Language : Precise Modeling With UML.* Addison-Wesley, 1998.
49. *5th Working Conference on Reverse Engineering.* IEEE Computer Soc., 1998.
50. *6th Working Conference on Reverse Engineering.* IEEE Computer Soc., 1999.
51. *7th Working Conference on Reverse Engineering.* IEEE Computer Soc., 2000.
52. K. Wong. RIGI User's Manual, Version 5.4.4. `http://www.rigi.csc.uvic.ca/rigi/rigiframe1.shtml?Download` (18.4.2001), 30. June 1998.
53. Extensible Graph Markup and Modeling Language). `http://www.cs.rpi.edu/~puninj/XGMML/` (19.8.2001), 2001.
54. XIG - An XSLT-based XMI2GXL-Translator. `http://ist.unibw-muenchen.de/GXL/volk/` (1.9.2001).
55. yFiles - Interactive Visualization of Graph Strucutres. `http://www-pr.informatik.uni-tuebingen.de/yfiles/` (1.9.2001).

GraphML Progress Report*
Structural Layer Proposal

Ulrik Brandes[1], Markus Eiglsperger[2], Ivan Herman[3], Michael Himsolt[4], and M. Scott Marshall[3]

[1] Department of Computer & Information Science, University of Konstanz.
Ulrik.Brandes@uni-konstanz.de

[2] Wilhelm Schickard Institute for Computer Science, University of Tübingen.
eiglsper@informatik.uni-tuebingen.de

[3] Centrum voor Wiskunde en Informatica. {ivan|scott}@cwi.nl

[4] DaimlerChrysler Research. Michael.Himsolt@daimlerchrysler.com

Abstract. Following a workshop on graph data formats held with the 8th Symposium on Graph Drawing (GD 2000), a task group was formed to propose a format for graphs and graph drawings that meets current and projected requirements.
On behalf of this task group, we here present GraphML (Graph Markup Language), an XML format for graph structures, as an initial step towards this goal. Its main characteristic is a unique mechanism that allows to define extension modules for additional data, such as graph drawing information or data specific to a particular application. These modules can freely be combined or stripped without affecting the graph structure, so that information can be added (or omitted) in a well-defined way.

1 Introduction

Graph drawing tools, like all other tools dealing with relational data, need to store and exchange graphs and associated data. Despite several earlier attempts to define a standard, no agreed-upon format is widely accepted and, indeed, many tools support only a limited number of custom formats which are typically restricted in their expressibility and specific to an area of application.

Motivated by the goals of tool interoperability, access to benchmark data sets, and data exchange over the Web, the Steering Committee of the Graph Drawing Symposium started a new initiative with an informal workshop held in conjunction with the 8th Symposium on Graph Drawing (GD 2000) [1]. As a consequence, an informal task group was formed to propose a modern graph exchange format suitable in particular for data transfer between graph drawing tools and other applications.

On behalf of this group we propose GraphML (Graph Markup Language), an XML format that takes a unique approach to represent graphs and graph drawings by specifying

* The latest information on GraphML is maintained on the GraphML homepage [2].

P. Mutzel, M. Jünger, and S. Leipert (Eds.): GD 2001, LNCS 2265, pp. 501–512, 2002.
© Springer-Verlag Berlin Heidelberg 2002

1. core elements to describe graph structures together with
2. an extension mechanism that allows to independently build application-specific graph data formats on top of them.

In particular, such extensions can be freely combined or ignored without affecting the graph data itself. Thus, drawing information can be added to an application-specific format, and graphs can be extracted from foreign application data. These features seem to be essential requirements for today's and future graph data formats, since graph models are ubiquitous and there will certainly be no agreement on a single general format across all disciplines.

This report is organized as follows. In Sect. 2, we outline the guidelines used in the design of GraphML. The core of the language is described in Sect. 3 and in Sect. 4 we outline how to add non-structural data and thus bind GraphML to specific applications. We conclude with future plans in Sect. 5.

2 Usage Scenarios and Design Goals

A modern graph exchange format cannot be defined in a monolithic way, since graph drawing services are used as components in larger systems and Web-based services are emerging. Graph data may need to be exchanged between such services, or stages of a service, and between graph drawing services and systems specific to areas of applications.

The typical usage scenarios that we envision for the format are centered around systems designed for arbitrary applications dealing with graphs and other data associated with them. Such systems will contain or call graph drawing services that add or modify layout and graphics information. Moreover, such services may compute only partial information or intermediate representations, for instance because they instantiate only part of a staged layout approach such as the topology-shape-metrics or Sugiyama frameworks. We hence aimed to satisfy the following key goal.

> The graph exchange format should be able to represent arbitrary graphs with arbitrary additional data, including layout and graphics information. The additional data should be stored in a format appropriate for the specific application, but should not complicate or interfere with the representation of data from other applications.

GraphML is designed with this and the following more pragmatic goals in mind:

- *Simplicity*: The format should be easy to parse and interpret for both humans and machines. As a general principle, there should be no ambiguities and thus a single well-defined interpretation for each valid GraphML document.
- *Generality*: There should be no limitation with respect to the graph model, i.e. hypergraphs, hierarchical graphs, etc. should be expressible within the same basic format.

- *Extensibility*: It should be possible to extend the format in a well-defined way to represent additional data required by arbitrary applications or more sophisticated use (e.g., sending a layout algorithm together with the graph).
- *Robustness*: Systems not capable of handling the full range of graph models or added information should be able to easily recognize and extract the subset they can handle.

There was no arguing that the format be based on XML (eXtensible Markup Language) [7] to stay compatible with other emerging standards such as, e.g., SOAP (Simple Object Access Protocol) [8], but also to enable use of the many widely supported tools for parsing and handling XML-formatted data. Another principal decision was to conceptually separate different layers of information, such as graph structure, application data, topology, geometry, or graphics. Figure 1 sketches the conceptual units of our design.

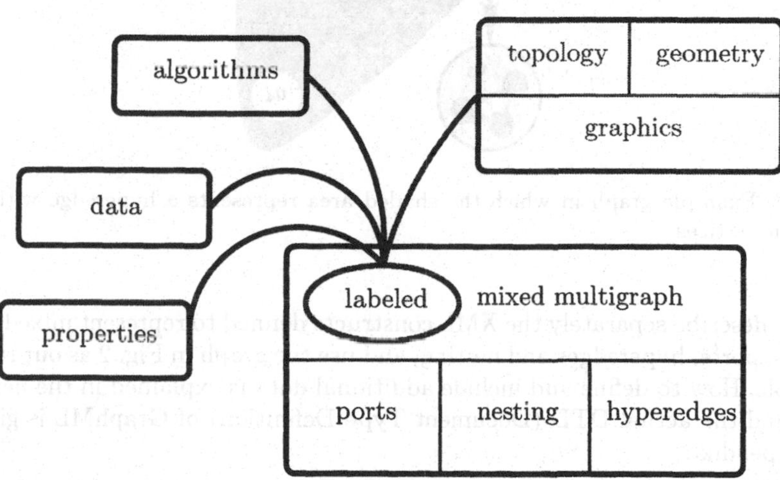

Fig. 1. The basic graph model of GraphML are labeled mixed multigraphs with optional node ports, hyperedges, and nesting. Graph drawing information is planned to be separated into topological and geometric, with a graphics layer on top. Like any other associated data, it will be encapsulated in a special tag

To date, the GraphML group has specified the structural layer, i.e. the core elements of the format describing the incidence structure of arbitrary graphs, and an extension mechanism to add non-structural data. Using this extension mechanism, it is possible to define application-specific modules that can be added to the structural layer common to all variants thus created. Information relevant to graph drawing services will later be defined within one or more such modules.

3 Structural Layer

In this section, we describe how graphs are represented in GraphML. This part of GraphML is called the structural layer and constitutes the essence of the format. The fundamental graph model underlying our design is a labeled mixed multigraph, which may, but need not, include node ports, hyperedges, and nested graphs. When these concepts are not supported by an application, they can be easily identified and may simply be ignored without losing the remaining structural information.

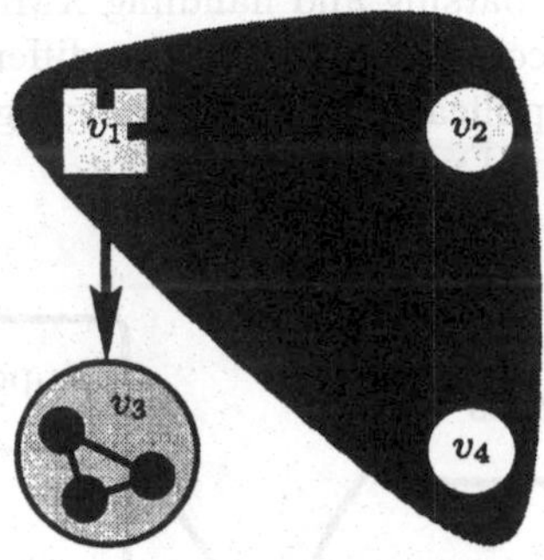

Fig. 2. Example graph in which the shaded area represents a hyperedge with three incident vertices

We describe separately the XML constructs defined to represent mixed multigraphs, ports, hyperedges and nesting, and use the graph in Fig. 2 as our running example. How to define and include additional data is explained in the next section, and the actual DTD (Document Type Definition) of GraphML is given in the appendix.

3.1 Mixed Multigraphs

A *mixed multigraph* is a graph which may contain both directed and undirected edges and may have loops and multi-edges. The representation chosen for GraphML is a simple list if nodes and edges. GraphML defines XML tags `<graph>`, `<node>`, and `<edge>` for this purpose.

- A GraphML document may contain any number of `<graph>`s.
- The mandatory XML attribute `edgedefault` of `<graph>` specifies whether `<edge>`s are directed or undirected by default. The optional XML attribute `directed` of `<edge>` can be used to overwrite the default.
- An `<edge>` refers to a source and a target node, regardless of whether it is directed or not.
- A `<graph>` tag may contain `<node>`s and `<edge>`s in any order.

Tools supporting only mixed multigraphs may describe and view the example graph as shown in Fig. 3. Note, however, that they may also choose to represent the features they do not support in a different way (e.g., by adding a dummy node for each hyperedge). Although there is a single well-defined interpretation for each GraphML document, no prescriptions are made on the representation of more complex graph models when only basic elements are supported.

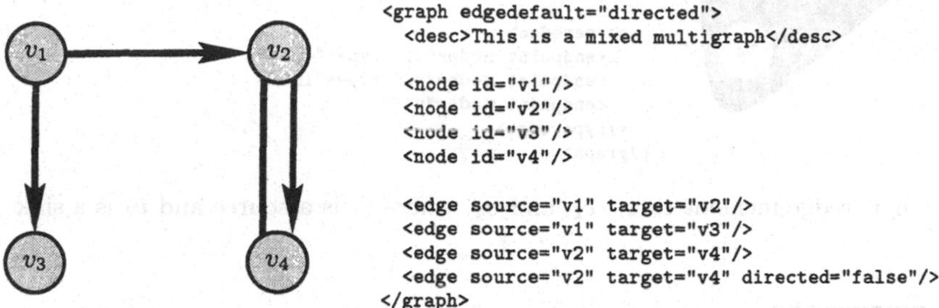

Fig. 3. The graph of Fig. 2 as represented in the most basic layer of GraphML

3.2 Ports

A *port* is a subset of the incidence relations of a node and can be viewed as a part of a node to which edges may attach. In electrical circuits for instance ports can be legs of a chip, and in graph drawing they may be used to specify points at which edges connect to a node. In GraphML <port>s appear as nested subelements of <node>s, and <edge>s may specify <port>s they attach to for both of their endpoints. See Fig. 4 for an example.

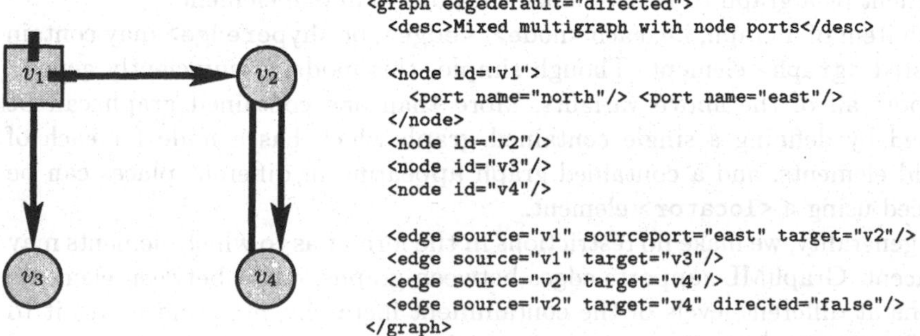

Fig. 4. Ports have local names at vertices, and can be referenced separately by edges

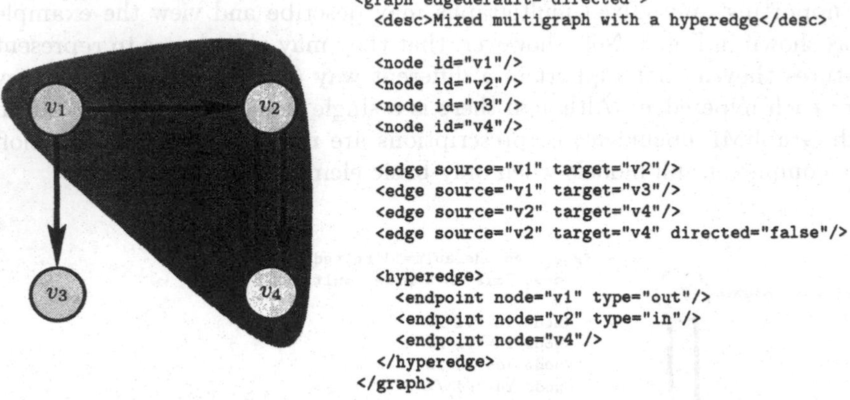


```
<graph edgedefault="directed">
  <desc>Mixed multigraph with a hyperedge</desc>

  <node id="v1"/>
  <node id="v2"/>
  <node id="v3"/>
  <node id="v4"/>

  <edge source="v1" target="v2"/>
  <edge source="v1" target="v3"/>
  <edge source="v2" target="v4"/>
  <edge source="v2" target="v4" directed="false"/>

  <hyperedge>
    <endpoint node="v1" type="out"/>
    <endpoint node="v2" type="in"/>
    <endpoint node="v4"/>
  </hyperedge>
</graph>
```

Fig. 5. A hyperedge incident to v_1, v_2, and v_4, where v_1 is a source and v_2 is a sink

3.3 Hypergraphs

A *hyperedge* is a subset of nodes, together with a classification of these nodes into inputs, outputs, or neither of the two. A GraphML tag **<hyperedge>** may therefore contain any number of **<endpoint>**s which, in turn, refer to **<node>**s, but also classify these nodes using the XML attribute **type**. See Fig. 5 for the addition of hyperedges to a mixed multigraph.

Hyperedges are generalizations of edges, and edges could hence be represented as hyperedges. We have chosen to separate the two concepts for the benefit of applications that do not support hyperedges. By using two different XML tags, parsers can easily distinguish the two cases and invoke special treatment of **<hyperedge>**s if needed. Such applications may choose, e.g., to represent hyperedges using dummy nodes or to ignore them altogether.

3.4 Nested Graphs

A *nested graph* is a graph occurring in an element of another graph. There are many models of hierarchical graphs, e.g., allowing more than one nested graph per element or a graph to be contained in more than one element.

Each item of a graph, i.e. each **<node>**, **<edge>**, or **<hyperedge>** may contain one nested **<graph>** element. Though simple, this model is sufficiently general to support all of the above variants. More than one contained graph can be expressed by defining a single contained graph which has a node for each of the child elements, and a contained graph appearing in different places can be referenced using a **<locator>** element.

For generality, we make no restrictions in the format as to which elements may be adjacent. GraphML supports edges between graphs, edges between elements of graphs at different levels of the containment hierarchy, etc., and leaves it to the application to detect inconsistencies with respect to its own model.

See Figure 6 for an example of a nested mixed multigraph.

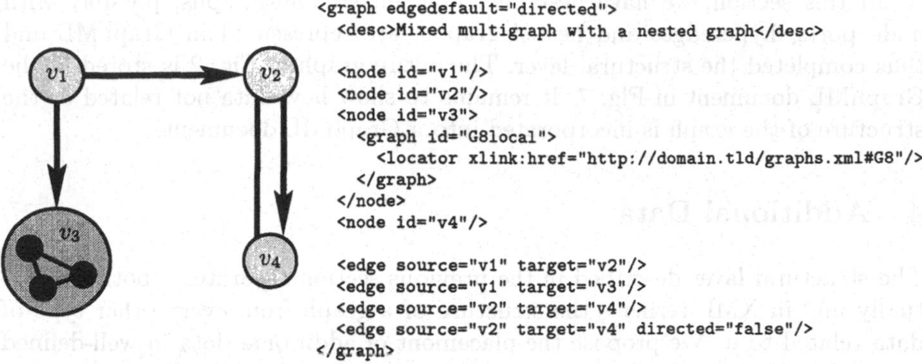


```
<graph edgedefault="directed">
  <desc>Mixed multigraph with a nested graph</desc>

  <node id="v1"/>
  <node id="v2"/>
  <node id="v3">
    <graph id="G8local">
      <locator xlink:href="http://domain.tld/graphs.xml#G8"/>
    </graph>
  </node>
  <node id="v4"/>

  <edge source="v1" target="v2"/>
  <edge source="v1" target="v3"/>
  <edge source="v2" target="v4"/>
  <edge source="v2" target="v4" directed="false"/>
</graph>
```

Fig. 6. A nested graph (located in another document)

```
<?xml version="1.0" standalone="no"?>
<!DOCTYPE graphml SYSTEM "graphml.dtd">
<graphml>
  <graph edgedefault="directed">
    <desc>The entire example graph</desc>

    <node id="v1">
      <port name="north"/>
      <port name="east"/>
    </node>
    <node id="v2"/>
    <node id="v3">
      <graph id="G8">
        <locator xlink:href="http://domain.tld/graph.xml#G8"/>
      </graph>
    </node>
    <node id="v4"/>

    <edge source="v1" sourceport="east" target="v2"/>
    <edge source="v1" target="v3"/>
    <edge source="v2" target="v4"/>
    <edge source="v2" target="v4" directed="false"/>

    <hyperedge>
      <endpoint node="v1" port="north" type="out"/>
      <endpoint node="v2" type="in"/>
      <endpoint node="v4"/>
    </hyperedge>
  </graph>
</graphml>
```

Fig. 7. A complete GraphML document representing the graph of Fig. 2

In this section, we have described how mixed multigraphs, possibly with node ports, hyperedges and nested graphs, are represented in GraphML and thus completed the structural layer. The entire graph of Fig. 2 is stored in the GraphML document in Fig. 7. It remains to show how data not related to the structure of the graph is incorporated into a GraphML document.

4 Additional Data

The structural layer described in the previous section separates – both conceptually and in XML terms – the structure of a graph from every other type of data related to it. We propose the placement of additional data in well-defined locations without prescribing the representation of of the data. These locations are defined with the help of `<key>` and `<data>` tags. Furthermore, we propose means to structure and type the content of `<data>` tags.

4.1 Unstructured Data

Data labelings are considered to be (partial) functions that assign values in an arbitrary range to elements of the graph (which usually have the same type). Edge weights, for instance, can be viewed as a function from the set of edges into, say, the real numbers. For each such function, GraphML requires a `<key>`, providing a name and a domain (via the XML attribute `for`) for the class of labels. The optional content of a `<key>` tag is used as the function's default value.

Each element in the GraphML structural layer – except for `<locator>` – may contain any number of `<data>` tags, representing data values assigned to the corresponding graph item. A `<data>` tag refers to its `<key>`, i.e. the function, for which it provides a value, which in turn is defined by its content. If no `<data>` tag is present for an element in the domain of a given `<key>`, the default value is assumed.

Unless explicitly defined otherwise, the range of data labels is not restricted (i.e. `#PCDATA`).

4.2 Structured Data

More structured content can be defined by replacing the XML content model shared by `<key>` and `<data>` tags with a self-defined one. The mechanism proposed for such variations mimicks the W3C Recommendation for XHTML Modularization [5], which is currently implemented with DTDs only. The W3C is working on an implementation using XML Schema (see [9] for a primer), which we will adapt as soon as it becomes a Recommendation.

The DTD implementation is as follows. A parameter entity `\%GRAPHML.data.content`, initially defined to be `\#PCDATA`, is used to specify the content model of `<key>` and `<data>` tags. A GraphML extension module, say EXT, would define new XML tags to represent data specific to the corresponding application and a parameter entity `\%EXT.data.content` specifying a content model for this module.

Recall that there may be more than one module. A driver file therefore defines the GraphML variant resulting from a specific combination of modules by importing them into GraphML and overwriting the content models of `<key>` and `<data>` with a combination of the content models defined in the modules. For details and examples see the GraphML homepage [2] and the XHTML Modularization Tutorial [6].

In effect, arbitrary variants of GraphML can be created by specifying valid XML for the content of `<key>` and `<data>` tags in extension modules, but the structural definition of the graph remains unaltered, regardless of the structure and type of data associated with it.

4.3 Typed Data

A second issue important for parsers and human interpretation alike is the typing of data labels. While the mechanism outlined in the previous subsection ensures that the XML content of `<data>` tags is any valid data item defined in an extension module, we suggest a way to infer the intended data type prior to having read all `<data>` tags that refer to the same `<key>`.

Each GraphML tag has an associated parameter entity in its list of XML attributes. This parameter is empty by default, but can be overwritten by extensions, thus adding new XML attributes to GraphML tags. In particular, one can define new XML attributes for `<key>`s. A GraphML extension module can therefore define an XML attribute with enumeration type values that correspond to meaningful types to content. It is likely that, when modularization via XML Schema becomes available, part of the type-checking can be delegated down to the XML level.

In summary, extensions can define any XML format to describe additional data and they can also provide typing information for parsers to deal with data in an efficient and type-safe manner. Multiple extensions can be combined freely, and extensions need not be supported by a parser to correctly extract contained graphs, since extensions are confined to subtrees of the Document Object Model (DOM) rooted at `<data>` tags and therefore cannot interfere with the structural layer.

5 Summary and Future Work

We have proposed GraphML, an XML format for the exchange of graph data. GraphML has been designed to be simple, general, robust, and, in particular, extensible. Application developers may define GraphML variants to include application-specific data, thereby making full use of XML's capabilities, but the design of GraphML ensures that such extensions are transparent to systems unaware of the extension. A tutorial on extending GraphML is in preparation.

Up-to-date information on GraphML is available on the Web [2], together with experimental parsers and graph editors using them. After a public review period, the specification will be finalized and published in full. To avoid even further diversification of formats, we aim towards the integration of GraphML with GXL (Graph Exchange Language) [3].

GraphML lays the ground towards an exchange format for graph visualizations. Future work will concentrate on defining GraphML extension modules for various applications, in particular graph drawing and information visualization. An important option considered in our preliminary designs for such modules is to make use of SVG (Scalable Vector Graphics) [4] in some form or another.

Acknowledgments. Many people have contributed to the current proposal. We would especially like to thank Stephen North and Roberto Tamassia for their continuing support and expertise, Jürgen Lerner and Sascha Meinert for implementing experimental parsers, John Punin for translating the GraphML DTD into XML Schema, Giuseppe Liotta for running the graph format panel discussion at GD 2001, and Petra Mutzel for including it into the program as well as giving us the opportunity to publish this progress report in the proceedings. We would also like to express our thanks to all members active or passive of the GraphML mailing list for their support of the project. In addition to all those mentioned before these are Vladimir Batagelj, Anne-Lise Gros, Carsten Gutwenger, David Jensen, Serban Jora, Michael Kaufmann, Guy Melançon, Maurizio Patrignani, Tim Pattison, Matthew Phillips, John Punin, Susan Sim, Adrian Vasiliu, Vance Waddle, and Andreas Winter.

References

1. Ulrik Brandes, M. Scott Marshall, and Stephen C. North. Graph data format workshop report. In Joe Marks, editor, *Proceedings of the 8th International Symposium on Graph Drawing (GD 2000)*, volume 1984 of *Lecture Notes in Computer Science*, pages 407–409. Springer, 2001.
2. GraphML homepage. `http://www.graphdrawing.org/graphml/`.
3. Ric Holt, Andy Schürr, Susan Sim, Andreas Winter. GXL – Graph eXchange Language. `http://www.gupro.de/GXL/`. Also refer to the article in this volume.
4. W3C. Scalable Vector Graphics (SVG). `http://www.w3.org/Graphics/SVG`.
5. W3C. XHTML™ 1.1 – Module-based XHTML. `http://www.w3.org/TR/2001/REC-xhtml11-20010531/`.
6. W3C. XHTML Modules and Markup Languages. `http://www.w3.org/MarkUp/Guide/xhtml-m12n-tutorial/`.
7. W3C. Extensible Markup Language (XML). `http://www.w3.org/XML/`.
8. W3C. XML Protocol Activity. `http://www.w3.org/2000/xp/`.
9. W3C. XML Schema Part 0: Primer. `http://www.w3.org/TR/xmlschema-0/`.

A GraphML Document Type Definition

The following is a simplified version of the proposed Document Type Definition (DTD) for GraphML. The omitted details are relevant only for extension module developers and given on the GraphML homepage [2].

```
<!-- documents -----------------------------------------------------------
  GraphML documents start with an optional content description,
  followed by the declaration of any number of keys and a sequence of graphs.
  ------------------------------------------------------------------------->
<!ELEMENT graphml ((desc)?,(key)*,(graph)*)>


<!-- comments ------------------------------------------------------------
  A description element contains human-readable text
  describing the content of the element it appears in.
  ------------------------------------------------------------------------->
<!ELEMENT desc (#PCDATA)>


<!-- remote definitions --------------------------------------------------
  A locator may be used instead of other content of a graph or data element
  to refer to the location of the actual definition of the enclosing item's
  content.
  ------------------------------------------------------------------------->
<!ELEMENT locator EMPTY>
<!ATTLIST locator
  xmlns:xlink CDATA    #FIXED    "http://www.w3.org/TR/2000/PR-xlink-20001220/"
  xlink:href  CDATA    #REQUIRED
  xlink:type  (simple) #FIXED    "simple"
>


<!-- graphs --------------------------------------------------------------
  A graph contains an optional description, keys local to this graph,
  and either a locator indicating that the graph is defined elsewhere,
  or lists of nodes, (hyper)edges, and data associated with the graph
  (in any order).
  A graph may be identified using the "id" attribute.  The mandatory attribute
  "edgedefault" indicates whether edges are directed or undirected by default;
  this can be overwritten locally by every edge.
  ------------------------------------------------------------------------->
<!ELEMENT graph ((desc)?,(key)*,((((data)|(node)|(edge)|(hyperedge))*)|(locator)))>
<!ATTLIST graph
  id          ID                      #IMPLIED
  edgedefault (directed|undirected) #REQUIRED
>


<!-- nodes ---------------------------------------------------------------
  Each node in a graph has to have a (unique) id.
  It may contain a description, followed by a sequence of ports and node
  data in any order, and may contain another graph.  Alternatively,
  it can be defined in another location, including a different file.
  Ports are identified by a name which does not have to be unique throughout
  the document, but within a node.  They can be nested hierarchically.
  ------------------------------------------------------------------------->
<!ELEMENT node ((desc)?,((((data)|(port))*,(graph)?)|(locator)))>
<!ATTLIST node id ID #REQUIRED>

<!ELEMENT port ((desc)?,((data)|(port))*)>
<!ATTLIST port name NMTOKEN #REQUIRED>



<!-- edges ---------------------------------------------------------------
  Similiar to nodes, edges may contain a separate description, followed by
  any number of edge data and, potentially, a nested graph.
  An edge must refer to a source and a target node, and may specify ports
  it attaches to.  However, such ports are not implicitly created and must
  therefore be defined at the corresponding node.
  Using the attribute "directed", the default value defined for the enclosing
  graph can be overwritten.
```

```
  ---------------------------------------------------------------------------->
<!ELEMENT edge ((desc)?,(data)*,(graph)?)>
<!ATTLIST edge
  id         ID            #IMPLIED
  source     IDREF         #REQUIRED
  sourceport NMTOKEN       #IMPLIED
  target     IDREF         #REQUIRED
  targetport NMTOKEN       #IMPLIED
  directed   (true|false)  #IMPLIED
>


<!-- hyperedges ---------------------------------------------------------------
  Since the number of nodes incident to a hyperedge is arbitrary, they are
  not referred to via attributes of hyperedge.  Rather, a child element
  endpoint is created for each incident node, which refers to the node and,
  optionally, the port it is incident to. For each incidence, it can be
  specified separately whether the node is a source ("out"), a target ("in"),
  or neither ("undir").
  ---------------------------------------------------------------------------->
<!ELEMENT hyperedge ((desc)?,((data)|(endpoint))*,(graph)?)>
<!ATTLIST hyperedge id ID #IMPLIED>

<!ELEMENT endpoint ((desc)?)>
<!ATTLIST endpoint
  id   ID                  #IMPLIED
  node IDREF               #REQUIRED
  port NMTOKEN             #IMPLIED
  type (in|out|undir) "undir"
>


<!-- additional data ----------------------------------------------------------
  Additional data can be attached to any GraphML item by inserting data tags.
  To distinguish different sorts of data, those of the same sort refer
  to a common key tag.  A key may specify the domain it is valid for,
  and may contain a default value for that domain.  A key can thus be seen
  as the declaration of an array, non-default values of which are defined
  by the respective element.
  Extension modules may overwrite the common content model of key and data
  and add new attributes to keys to provide data type information.
  ---------------------------------------------------------------------------->
<!ENTITY % GRAPHML.key.attrib "">
<!ENTITY % GRAPHML.data.content "(#PCDATA)">

<!ELEMENT key %GRAPHML.data.content;>
<!ATTLIST key
  id  ID                                                  #REQUIRED
  for (graph|node|edge|hyperedge|port|endpoint|all) "all"
  %GRAPHML.key.attrib;
>

<!ELEMENT data %GRAPHML.data.content;>
<!ATTLIST data
  key IDREF #REQUIRED
  id  ID    #IMPLIED
>
```

Graph-Drawing Contest Report

Therese Biedl[1] and Franz J. Brandenburg[2]

[1] Institute of Computer Research
University of Waterloo
Waterloo, Ontario, Canada N2L 3G1
biedl@magnolia.math.uwaterloo.ca
[2] Universität Passau
94030 Passau, Germany
brandenb@informatik.uni-passau.de

Abstract. This report describes the Eight Annual Graph Drawing Contest, held in conjunction with the Ninth Graph Drawing Symposium in Vienna, Austria. The purpose of the contest is to monitor and challenge the current state of the graph-drawing technology and to display artistic work related to graph drawing.

1 Introduction

Text descriptions of the four categories for the 2001 contest were available via the World Wide Web (WWW) and announced with the Graph Drawing Symposium. The data of the challenge graphs was provided in GML format. Only fourteen separate submissions from eight teams were received, seven teams had at least one German team member. The winners were selected by the jury members Therese Biedl, Franz J. Brandenburg, Peter Eades and Joe Marks. The winning entries are described below, and are available under `http://www.infosun.fmi.uni-passau.de/GD/GD2001`.

2 Winning Submissions

2.1 Category A

The graph for Category A has an intimate relation to the Graph Drawing Symposia. It is the GD2000 self-citation graph. There is a node for every paper in the proceedings of GD94 to GD2000, and an arc if a paper refers to another GD paper. The citations are restricted only to the proceedings, thus the data is not suited for an analysis and ranking of GD contributions. The data was created by Susanne Lenz, Passau, using Graphlet.

The graph has 311 nodes and 647 edges. It has 52 isolated nodes and 5 isolated edges. There was one erroneous arc (GD94/143, GD98/423) which should have been reversed. There are four small cycles by mutual references in the GD94 and GD95 proceedings.

P. Mutzel, M. Jünger, and S. Leipert (Eds.): GD 2001, LNCS 2265, pp. 513–521, 2002.
© Springer-Verlag Berlin Heidelberg 2002

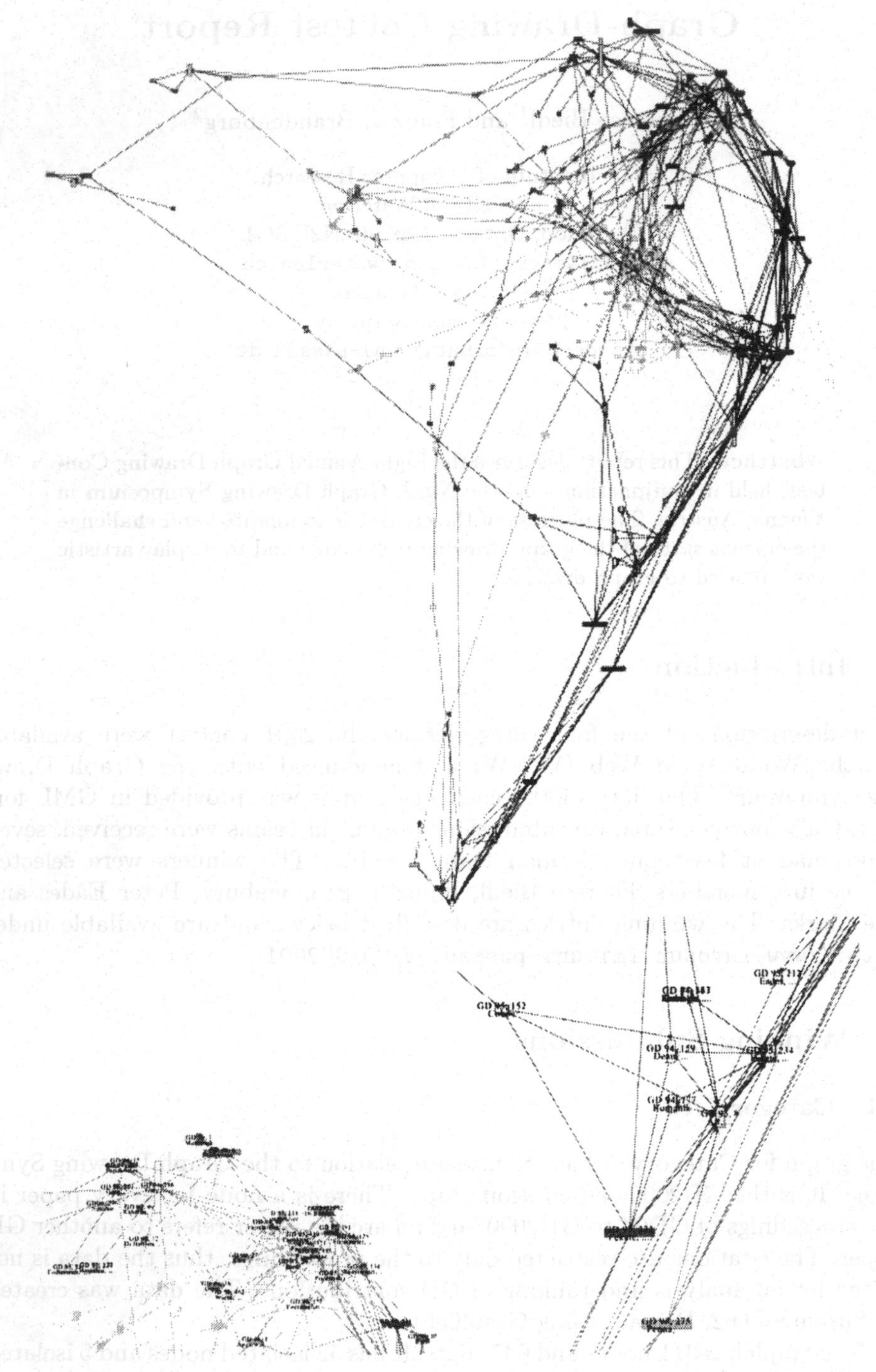

Fig. 1. Three-dimensional drawing of GD proceedings papers. The layout clusters dense subgraphs which in turn correspond to topics considered in graph drawing.

The winning entry is a poster submitted by Ulrik Brandes and Marco Gärtler from the University of Konstanz. The poster shows eight two-dimensional projections of a three-dimensional layout with depth cues. The projections correspond to views through the faces of an enclosing octahedron. Each paper is represented by a rectangle, where height and width indicate the number of citations received and made. Positions are determined by three eigenvectors of a generalized Laplacian matrix of the underlying undirected graph without manual postprocessing. The layout clearly shows that the citation network clusters around topics typically considered in graph drawing (Fig. 1), and how these are connected to each other. The poster also offers another perspective common in citation network analysis, in which just one eigenvector is used for the x-coordinate, and an index ranking papers by authority [4] is used for the y-coordinate (Fig. 2).

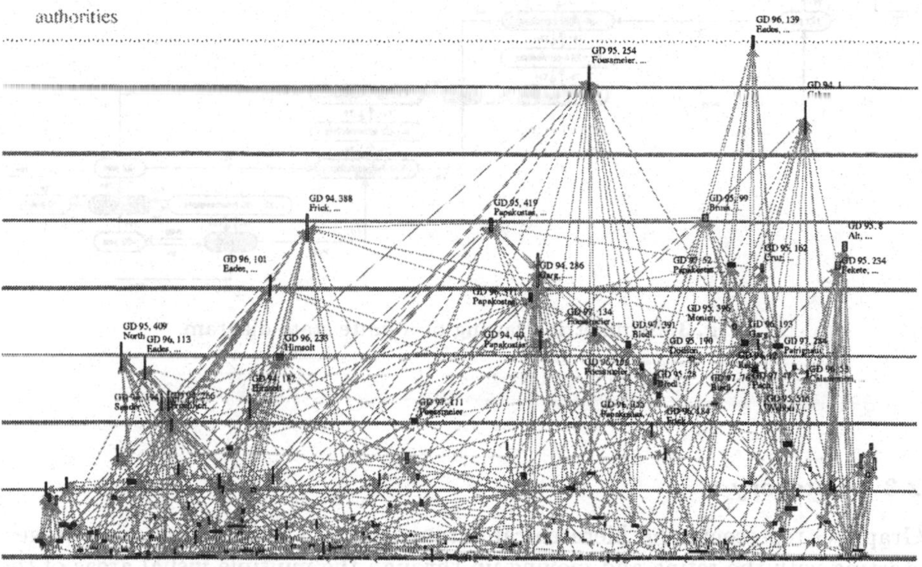

Fig. 2. GD proceedings papers ranked by authority with topical clustering in horizontal direction. The most authoritative paper relative to this data set is [2].

2.2 Category B

Graph B was this year's "easier challenge". It represents a finite state diagram from an industrial application with 18 nodes and 37 arcs. Two nodes have self-loops and some nodes have parallel arcs in each direction. The nodes and arcs are labeled. The six entries on Graph B were quite similar. Each detected and displayed the symmetries of the two biconnected components. The winning entry by Carsten Gutwenger, Karsten Klein, Joachim Kupke, and Sebastian Leipert, Stiftung caesar, Bonn, perfectly reflects the structure of the diagram. They observed that the graph is 4-planar when the multiple edges are deleted. The drawing was computed using Tamassia's bend minimization algorithm, first adding a "+60 pos" dummy node and choosing an embedding with maximal outer face containing the node "0 pos". The outcome of the algorithm was not symmetric. Symmetry is obtained by spitting some edges and inserting dummy nodes at expected bends. The final drawing Fig. 3 was obtained by flipping the right biconnected components using Microsoft Visio.

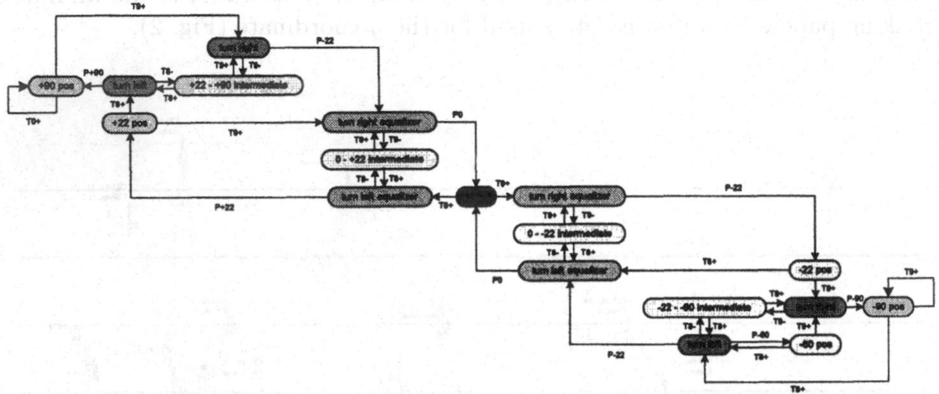

Fig. 3. Symmetric drawing of a finite state diagram.

2.3 Category C

Graph C represents the hierarchical arrangement of visual processing stages, starting with the retina and moving up through the multiple visual areas of the brains. The graph was given together with a prescribed layering of the nodes in 14 layers. It was contributed by Therese Biedl adapting the data from [3].

The winning entry by Roland Wiese, University of Tübingen, was produced automatically with the HierarchicLayouter of yFiles-1.4 using the non-default settings AsIsLayerer (applied to manually layered input sketch), minimalLayerDistance=100, minimalEdgeDistance=10 and weightHeuristic = MEDIAN_HEURISTIC. In the final drawing, nodes and their outgoing edges were colored by a random rainbow color scheme.

2.4 Category D

Category D is the free or artistic category which combines arts and graphs.

The jury selected two out of the four entries. The symposium attendees were asked for a ranking "by applause", which ended in a tie-break.

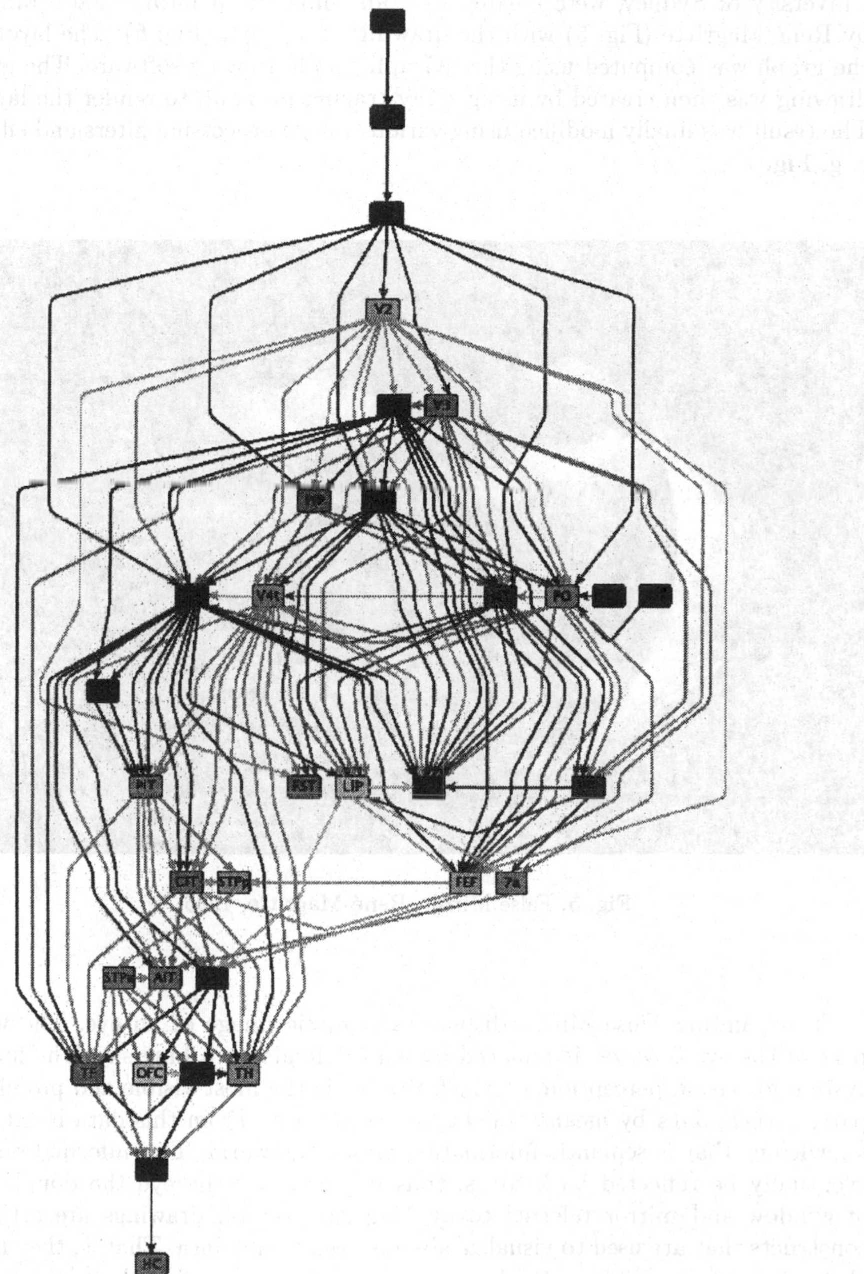

Fig. 4. A hierarchical drawing of the graph with a predefined hierarchy.

D1. The images of the information visualization by Carsten Friedrich from the University of Sydney, were created by combining the painting "False Mirror" by René Magritte (Fig. 5) with the drawing of a graph (Fig. 6). The layout of the graph was computed using the jjGraph graph drawing software. The graph drawing was then created by using a ray tracing program to render the layout. The result was finally modified using various image processing filters and effects, e. g. Fig. 7.

Fig. 5. False Mirror, René Magritte, 1935.

The painting False Mirror displays a realistic image of an eye. The white part of the eye however is replaced by white clouds on a blue sky. The human system for visual perception, of which the eye is the most visible and prominent part, gathers data by means of light hitting the eye. From this data it extracts knowledge, that is semantic information about the world. This information will eventually be reflected back to us, thus imposing on the eye the double role of window and mirror referred to by Magritte. Graph drawings are artificial constructs that are used to visualize abstract relational data. That is, they make abstract data accessible to the human eye and by that accessible to perception and understanding. It lies in the nature of these visualizations that even if they try to communicate this data as true as possible they can never achieve this goal completely. The visualization has to convert the data into a metaphor and set it into a context which alters the way we perceive it. In the background picture we

Fig. 6. 3D drawing of a graph.

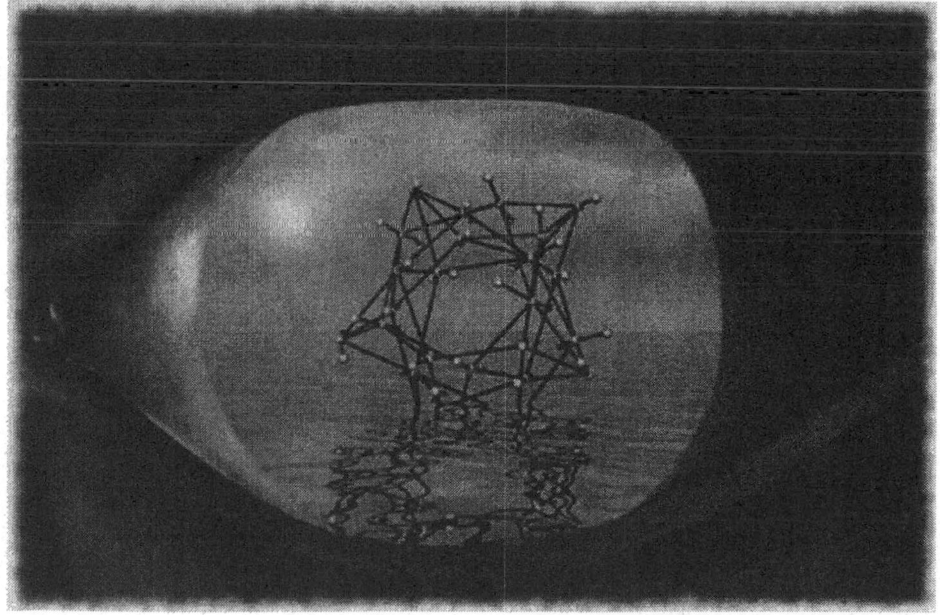

Fig. 7. Composition of Fig. 5 and Fig. 6.

see a graph interpreted as a three-dimensional object and set into a scenery which reflects the graph, highlights certain parts, obscures or hides others, but overall stays mainly neutral. Apart from the objective physical settings and biological mechanisms, perception is always highly subjective. Whatever a person perceives will always be altered by the special idiosyncratic circumstances. This effect is of course exponentiated when we try to perceive how somebody else perceives. The sequence of drawings shows interpretations of these effects.

D2. The *Vienna ferris wheel graph*, the GD'2001 logo, was designed by Merijam Percan from the University of Cologne. The logo was created as a three dimensional representation of a graph, representing the vertices by spheres and the edges by cylinders. The drawing of the graph of Vienna's Ferris Wheel was generated using PovRay3.1*g* and animated using gifmerge. The entry consists of a picture and a gif-animation. The scene contains an island with a lighthouse, a house, birds and trees. Some of these objects were taken from `www.povworld.de` and modified and composed to the picture. Merijam Percan expresses her thanks to Martin Gruber for providing the necessary hardware and his advice.

Fig. 8. Scene with Vienna ferris wheel graph.

3 Observation and Conclusion

The high quality and originality of this year's contributions, particularly for the challenge graphs, demonstrates the capabilities of graph drawing. There were 573 visitors on the web site of the GD competition, but too few entries. For next year's GD competition we will try to attract members inside and outside the graph drawing community to contribute to the graph drawing competition and shall open new categories.

Acknowledgments. Sponsorship for this context was provided by COMSORT, Hunt Valley, MD, USA, DaimlerChrysler AG, Stuttgart, Germany and MERL-Mitsubishi Electric Research Laboratories, Cambridge, Mass.

References

1. F.J. Brandenburg, U. Brandes, M. Himsolt and M. Raitner, Graph-Drawing Contest Report, *Proc. 8th International Symposium on Graph Drawing (GD '2000)* vol. 1984 *LNCS* (2000) 410–418.
2. P. Eades, A. Symvonis, and S.H. Whitesides, Two algorithms for three-dimensional orthogonal graph drawing. *Proc. 4th International Symposium on Graph Drawing (GD '96)*, vol. 1190 *LNCS* (1996) 139–154.
3. R. Sekular and R. Blake, Perception, McGraw Hill, 1994.
4. J. M. Kleinberg, Authoritative sources in a hyperlinked environment. *J. Assoc. Comput. Mach.*, 46 (1999) 604–632.

Author Index

Lecture Notes in Computer Science

For information about Vols. 1–2194
please contact your bookseller or Springer-Verlag

Vol. 2195: H.-Y. Shum, M. Liao, S.-F. Chang (Eds.), Advances in Multimedia Information Processing – PCM 2001. Proceedings, 2001. XX, 1149 pages. 2001.

Vol. 2196: W. Taha (Ed.), Semantics, Applications, and Implementation of Program Generation. Proceedings, 2001. X, 219 pages. 2001.

Vol. 2197: O. Balet, G. Subsol, P. Torguet (Eds.), Virtual Storytelling. Proceedings, 2001. XI, 213 pages. 2001.

Vol. 2198: N. Zhong, Y. Yao, J. Liu, S. Ohsuga (Eds.), Web Intelligence: Research and Development. Proceedings, 2001. XVI, 615 pages. 2001. (Subseries LNAI).

Vol. 2199: J. Crespo, V. Maojo, F. Martin (Eds.), Medical Data Analysis. Proceedings, 2001. X, 311 pages. 2001.

Vol. 2200: G.I. Davida, Y. Frankel (Eds.), Information Security. Proceedings, 2001. XIII, 554 pages. 2001.

Vol. 2201: G.D. Abowd, B. Brumitt, S. Shafer (Eds.), Ubicomp 2001: Ubiquitous Computing. Proceedings, 2001. XIII, 372 pages. 2001.

Vol. 2202: A. Restivo, S. Ronchi Della Rocca, L. Roversi (Eds.), Theoretical Computer Science. Proceedings, 2001. XI, 440 pages. 2001.

Vol. 2203: A. Omicini, P. Petta, R. Tolksdorf (Eds.), Engineering Societies in the Agents World II. Proceedings, 2001. XI, 195 pages. 2001. (Subseries LNAI).

Vol. 2204: A. Brandstädt, V.B. Le (Eds.), Graph-Theoretic Concepts in Computer Science. Proceedings, 2001. X, 329 pages. 2001.

Vol. 2205: D.R. Montello (Ed.), Spatial Information Theory. Proceedings, 2001. XIV, 503 pages. 2001.

Vol. 2206: B. Reusch (Ed.), Computational Intelligence. Proceedings, 2001. XVII, 1003 pages. 2001.

Vol. 2207: I.W. Marshall, S. Nettles, N. Wakamiya (Eds.), Active Networks. Proceedings, 2001. IX, 165 pages. 2001.

Vol. 2208: W.J. Niessen, M.A. Viergever (Eds.), Medical Image Computing and Computer-Assisted Intervention – MICCAI 2001. Proceedings, 2001. XXXV, 1446 pages. 2001.

Vol. 2209: W. Jonker (Ed.), Databases in Telecommunications II. Proceedings, 2001. VII, 179 pages. 2001.

Vol. 2210: Y. Liu, K. Tanaka, M. Iwata, T. Higuchi, M. Yasunaga (Eds.), Evolvable Systems: From Biology to Hardware. Proceedings, 2001. XI, 341 pages. 2001.

Vol. 2211: T.A. Henzinger, C.M. Kirsch (Eds.), Embedded Software. Proceedings, 2001. IX, 504 pages. 2001.

Vol. 2212: W. Lee, L. Mé, A. Wespi (Eds.), Recent Advances in Intrusion Detection. Proceedings, 2001. X, 205 pages. 2001.

Vol. 2213: M.J. van Sinderen, L.J.M. Nieuwenhuis (Eds.), Protocols for Multimedia Systems. Proceedings, 2001. XII, 239 pages. 2001.

Vol. 2214: O. Boldt, H. Jürgensen (Eds.), Automata Implementation. Proceedings, 1999. VIII, 183 pages. 2001.

Vol. 2215: N. Kobayashi, B.C. Pierce (Eds.), Theoretical Aspects of Computer Software. Proceedings, 2001. XV, 561 pages. 2001.

Vol. 2216: E.S. Al-Shaer, G. Pacifici (Eds.), Management of Multimedia on the Internet. Proceedings, 2001. XIV, 373 pages. 2001.

Vol. 2217: T. Gomi (Ed.), Evolutionary Robotics. Proceedings, 2001. XI, 139 pages. 2001.

Vol. 2218: R. Guerraoui (Ed.), Middleware 2001. Proceedings, 2001. XIII, 395 pages. 2001.

Vol. 2219: S.T. Taft, R.A. Duff, R.L. Brukardt, E. Ploedereder (Eds.), Consolidated Ada Reference Manual. XXV, 560 pages. 2001.

Vol. 2220: C. Johnson (Ed.), Interactive Systems. Proceedings, 2001. XII, 219 pages. 2001.

Vol. 2221: D.G. Feitelson, L. Rudolph (Eds.), Job Scheduling Strategies for Parallel Processing. Proceedings, 2001. VII, 207 pages. 2001.

Vol. 2222: M.J. Wooldridge, G. Weiß, P. Ciancarini (Eds.) Agent-Oriented Software Engineering II. Proceedings, 2001. X, 319 pages. 2002.

Vol. 2223: P. Eades, T. Takaoka (Eds.), Algorithms and Computation. Proceedings, 2001. XIV, 780 pages. 2001.

Vol. 2224: H.S. Kunii, S. Jajodia, A. Sølvberg (Eds.), Conceptual Modeling – ER 2001. Proceedings, 2001. XIX, 614 pages. 2001.

Vol. 2225: N. Abe, R. Khardon, T. Zeugmann (Eds.), Algorithmic Learning Theory. Proceedings, 2001. XI, 379 pages. 2001. (Subseries LNAI).

Vol. 2226: K.P. Jantke, A. Shinohara (Eds.), Discovery Science. Proceedings, 2001. XII, 494 pages. 2001. (Subseries LNAI).

Vol. 2227: S. Boztaş, I.E. Shparlinski (Eds.), Applied Algebra, Algebraic Algorithms and Error-Correcting Codes. Proceedings, 2001. XII, 398 pages. 2001.

Vol. 2228: B. Monien, V.K. Prasanna, S. Vajapeyam (Eds.), High Performance Computing – HiPC 2001. Proceedings, 2001. XVIII, 438 pages. 2001.

Vol. 2229: S. Qing, T. Okamoto, J. Zhou (Eds.), Information and Communications Security. Proceedings, 2001. XIV, 504 pages. 2001.

Vol. 2230: T. Katila, I.E. Magnin, P. Clarysse, J. Montagnat, J. Nenonen (Eds.), Functional Imaging and Modeling of the Heart. Proceedings, 2001. XI, 158 pages. 2001.

Vol. 2231: A. Pasetti, Software Frameworks and Embedded Control Systems. XIV, 293 pages. 2002.

Vol. 2232: L. Fiege, G. Mühl, U. Wilhelm (Eds.), Electronic Commerce. Proceedings, 2001. X, 233 pages. 2001.

Vol. 2233: J. Crowcroft, M. Hofmann (Eds.), Networked Group Communication. Proceedings, 2001. X, 205 pages. 2001.

Vol. 2234: L. Pacholski, P. Ružička (Eds.), SOFSEM 2001: Theory and Practice of Informatics. Proceedings, 2001. XI, 347 pages. 2001.

Vol. 2235: C.S. Calude, G. Păun, G. Rozenberg, A. Salomaa (Eds.), Multiset Processing. VIII, 359 pages. 2001.

Vol. 2236: K. Drira, A. Martelli, T. Villemur (Eds.), Cooperative Environments for Distributed Systems Engineering. IX, 281 pages. 2001.

Vol. 2237: P. Codognet (Ed.), Logic Programming. Proceedings, 2001. XI, 365 pages. 2001.

Vol. 2239: T. Walsh (Ed.), Principles and Practice of Constraint Programming – CP 2001. Proceedings, 2001. XIV, 788 pages. 2001.

Vol. 2240: G.P. Picco (Ed.), Mobile Agents. Proceedings, 2001. XIII, 277 pages. 2001.

Vol. 2241: M. Jünger, D. Naddef (Eds.), Computational Combinatorial Optimization. IX, 305 pages. 2001.

Vol. 2242: C.A. Lee (Ed.), Grid Computing – GRID 2001. Proceedings, 2001. XII, 185 pages. 2001.

Vol. 2243: G. Bertrand, A. Imiya, R. Klette (Eds.), Digital and Image Geometry. VII, 455 pages. 2001.

Vol. 2244: D. Bjørner, M. Broy, A.V. Zamulin (Eds.), Perspectives of System Informatics. Proceedings, 2001. XIII, 548 pages. 2001.

Vol. 2245: R. Hariharan, M. Mukund, V. Vinay (Eds.), FST TCS 2001: Foundations of Software Technology and Theoretical Computer Science. Proceedings, 2001. XI, 347 pages. 2001.

Vol. 2246: R. Falcone, M. Singh, Y.-H. Tan (Eds.), Trust in Cyber-societies. VIII, 195 pages. 2001. (Subseries LNAI).

Vol. 2247: C. P. Rangan, C. Ding (Eds.), Progress in Cryptology – INDOCRYPT 2001. Proceedings, 2001. XIII, 351 pages. 2001.

Vol. 2248: C. Boyd (Ed.), Advances in Cryptology – ASIACRYPT 2001. Proceedings, 2001. XI, 603 pages. 2001.

Vol. 2249: K. Nagi, Transactional Agents. XVI, 205 pages. 2001.

Vol. 2250: R. Nieuwenhuis, A. Voronkov (Eds.), Logic for Programming, Artificial Intelligence, and Reasoning. Proceedings, 2001. XV, 738 pages. 2001. (Subseries LNAI).

Vol. 2251: Y.Y. Tang, V. Wickerhauser, P.C. Yuen, C.Li (Eds.), Wavelet Analysis and Its Applications. Proceedings, 2001. XIII, 450 pages. 2001.

Vol. 2252: J. Liu, P.C. Yuen, C. Li, J. Ng, T. Ishida (Eds.), Active Media Technology. Proceedings, 2001. XII, 402 pages. 2001.

Vol. 2253: T. Terano, T. Nishida, A. Namatame, S. Tsumoto, Y. Ohsawa, T. Washio (Eds.), New Frontiers in Artificial Intelligence. Proceedings, 2001. XXVII, 553 pages. 2001. (Subseries LNAI).

Vol. 2254: M.R. Little, L. Nigay (Eds.), Engineering for Human-Computer Interaction. Proceedings, 2001. XI, 359 pages. 2001.

Vol. 2255: J. Dean, A. Gravel (Eds.), COTS-Based Software Systems. Proceedings, 2002. XIV, 257 pages. 2002.

Vol. 2256: M. Stumptner, D. Corbett, M. Brooks (Eds.), AI 2001: Advances in Artificial Intelligence. Proceedings, 2001. XII, 666 pages. 2001. (Subseries LNAI).

Vol. 2257: S. Krishnamurthi, C.R. Ramakrishnan (Eds.), Practical Aspects of Declarative Languages. Proceedings, 2002. VIII, 351 pages. 2002.

Vol. 2258: P. Brazdil, A. Jorge (Eds.), Progress in Artificial Intelligence. Proceedings, 2001. XII, 418 pages. 2001. (Subseries LNAI).

Vol. 2259: S. Vaudenay, A.M. Youssef (Eds.), Selected Areas in Cryptography. Proceedings, 2001. XI, 359 pages. 2001.

Vol. 2260: B. Honary (Ed.), Cryptography and Coding. Proceedings, 2001. IX, 416 pages. 2001.

Vol. 2262: P. Müller, Modular Specification and Verification of Object-Oriented Programs. XIV, 292 pages. 2002.

Vol. 2263: T. Clark, J. Warmer (Eds.), Object Modeling with the OCL. VIII, 281 pages. 2002.

Vol. 2264: K. Steinhöfel (Ed.), Stochastic Algorithms: Foundations and Applications. Proceedings, 2001. VIII, 203 pages. 2001.

Vol. 2265: P. Mutzel, M. Jünger, S. Leipert (Eds.), Graph Drawing. Proceedings, 2001. XV, 524 pages. 2002.

Vol. 2266: S. Reich, M.T. Tzagarakis, P.M.E. De Bra (Eds.), Hypermedia: Openness, Structural Awareness, and Adaptivity. Proceedings, 2001. X, 335 pages. 2002.

Vol. 2267: M. Cerioli, G. Reggio (Eds.), Recent Trends in Algebraic Development Techniques. Proceedings, 2001. X, 345 pages. 2001.

Vol. 2271: B. Preneel (Ed.), Topics in Cryptology – CT-RSA 2002. Proceedings, 2002. X, 311 pages. 2002.

Vol. 2272: D. Bert, J.P. Bowen, M.C. Henson, K. Robinson (Eds.), ZB 2002: Formal Specification and Development in Z and B. Proceedings, 2002. XII, 535 pages. 2002.

Vol. 2273: A.R. Coden, E.W. Brown, S. Srinivasan (Eds.), Information Retrieval Techniques for Speech Applications. XI, 109 pages. 2002.

Vol. 2274: D. Naccache, P. Paillier (Eds.), Public Key Cryptography. Proceedings, 2002. XI, 385 pages. 2002.

Vol. 2275: N.R. Pal, M. Sugeno (Eds.), Advances in Soft Computing – AFSS 2002. Proceedings, 2002. XVI, 536 pages. 2002. (Subseries LNAI).

Vol. 2276: A. Gelbukh (Ed.), Computational Linguistics and Intelligent Text Processing. Proceedings, 2002. XIII, 444 pages. 2002.

Vol. 2277: P. Callaghan, Z. Luo, J. McKinna, R. Pollack (Eds.), Types for Proofs and Programs. Proceedings, 2000. VIII, 243 pages. 2002.

Vol. 2284: T. Eiter, K.-D. Schewe (Eds.), Foundations of Information and Knowledge Systems. Proceedings, 2002. X, 289 pages. 2002.

Vol. 2285: H. Alt, A. Ferreira (Eds.), STACS 2002. Proceedings, 2002. XIV, 660 pages. 2002.

Vol. 2300: W. Brauer, H. Ehrig, J. Karhumäki, A. Salomaa (Eds.), Formal and Natural Computing. XXXVI, 431 pages. 2002.